AF361678

Spinal Cord Injury and Transcutaneous Spinal Cord Stimulation

Spinal Cord Injury and Transcutaneous Spinal Cord Stimulation

Editors

Ursula S. Hofstoetter
Karen Minassian

MDPI • Basel • Beijing • Wuhan • Barcelona • Belgrade • Manchester • Tokyo • Cluj • Tianjin

Editors
Ursula S. Hofstoetter   Karen Minassian
Medical University of Vienna   Medical University of Vienna
Austria   Austria

Editorial Office
MDPI
St. Alban-Anlage 66
4052 Basel, Switzerland

This is a reprint of articles from the Special Issue published online in the open access journal *Journal of Clinical Medicine* (ISSN 2077-0383) (available at: https://www.mdpi.com/journal/jcm/special_issues/Spinal_Cord_Injury).

For citation purposes, cite each article independently as indicated on the article page online and as indicated below:

LastName, A.A.; LastName, B.B.; LastName, C.C. Article Title. *Journal Name* **Year**, *Volume Number*, Page Range.

ISBN 978-3-0365-4755-8 (Hbk)
ISBN 978-3-0365-4756-5 (PDF)

Contents

About the Editors

Ursula S. Hofstoetter

Ursula S. Hofstoetter is a professor at the Center for Medical Physics and Biomedical Engineering of the Medical University of Vienna. Her scientific approach is inspired by her background in technical mathematics and her longstanding collaborations with leading neurorehabilitation centers and clinical research units across the globe. Her work focuses on human spinal sensorimotor circuits, their functional alterations after upper motoneuron injury or disease, and their neuromodulation via electrical spinal cord stimulation. Her research has contributed to unraveling the directly recruited neural structures and the transsynaptically activated spinal circuits. Ursula S. Hofstoetter has co-developed the method of transcutaneous spinal cord stimulation in Vienna. She is internationally recognized for her work with transcutaneous spinal cord stimulation to augment residual locomotor function and to control spinal spasticity in individuals with spinal cord injury and multiple sclerosis.

Karen Minassian

Karen Minassian is a professor at the Center for Medical Physics and Biomedical Engineering of the Medical University of Vienna. He is a pioneer of the recent advances in electrical spinal cord stimulation in individuals with spinal cord injury. His main interest is in the understanding of the neural control of movement with a focus on human spinal cord locomotor circuits. With a background in physics and mathematics, he has used computer simulations as well as neurophysiological methods to identify the neuronal structures electrically activated by spinal cord stimulation and to unravel some of the mechanisms underlying the generation of movement in otherwise paralyzed legs. His research has paved the way for the recent high-profile studies on epidural electrical stimulation. In parallel, Karen Minassian developed transcutaneous spinal cord stimulation, a non-invasive method that can be used as a neuromodulation tool as well as for human neurophysiological studies in Vienna.

Journal of *Clinical Medicine*

Editorial

Transcutaneous Spinal Cord Stimulation: Advances in an Emerging Non-Invasive Strategy for Neuromodulation

Ursula S. Hofstoetter * and Karen Minassian *

Center for Medical Physics and Biomedical Engineering, Medical University of Vienna, 1090 Vienna, Austria
* Correspondence: ursula.hofstoetter@meduniwien.ac.at (U.S.H.); karen.minassian@meduniwien.ac.at (K.M.)

Citation: Hofstoetter, U.S.; Minassian, K. Transcutaneous Spinal Cord Stimulation: Advances in an Emerging Non-Invasive Strategy for Neuromodulation. *J. Clin. Med.* **2022**, *11*, 3836. https://doi.org/10.3390/jcm11133836

Received: 29 June 2022
Accepted: 30 June 2022
Published: 1 July 2022

Publisher's Note: MDPI stays neutral with regard to jurisdictional claims in published maps and institutional affiliations.

Recent studies of epidural electrical spinal cord stimulation have shown the enabling and, in some cases, the recovery of motor functions thought to be irreversibly lost due to severe spinal cord injury. These findings have marked the dawn of a new era in neurorehabilitation, in which unprecedented levels of improvements have become attainable even in the chronic stage of a lesion. With the development of transcutaneous spinal cord stimulation, a clinically accessible technique has complemented the current landscape of state-of-the-art neuromodulative therapeutic options. This method activates a subpopulation of the same neural structures of the spinal cord as engaged by epidural stimulation. As a non-invasive intervention, it holds the great potential to accelerate the wider application of electrical neuromodulation in the treatment of central nervous system diseases and injuries. Yet its firm establishment and lasting acceptance in clinical practice will not only hinge on the demonstration of safety and efficacy, but also on the delineation of the underlying physiological mechanisms. This will require an advance in our understanding of its immediate effects on neuronal circuits both in the intact and injured spinal cord. In parallel, there is a need to investigate clinical outcomes induced by repeated-administration regimens that go beyond the mere alleviation of symptoms. Importantly, such longer-lasting beneficial effects are indicative of structural and physiological plastic adaptations at various levels of the central nervous system. The present Special Issue is an ensemble of 16 contributions by 92 peers in the field. The articles shed light on the conceptual framework of the interplay between transcutaneous spinal cord stimulation and (altered) central nervous system function, seek to advance its usability, and explore untapped areas of application in neurorehabilitation following spinal cord injury.

A key question of studies focusing on improving lower-extremity motor function by spinal cord stimulation is to gain a better understanding about the integration of (residual) descending voluntary drive, externally applied electrical inputs, and the activity of sensorimotor circuits residing within the lumbosacral spinal cord. An important step in this direction is provided by Malloy and colleagues by introducing a clinically relevant rat model in which they adapted the procedure of transcutaneous spinal cord stimulation to activate neural targets and evoke short-latency spinal reflexes similar to those in humans [1]. In spinalized rats, their stimulation setup could be safely and stably applied over several weeks and with measureable modulatory effects on spinal reflex gain. The question of lumbar spinal sensorimotor integration is further addressed by Steele and colleagues [2]. In neurologically intact individuals, they used transcutaneous spinal cord stimulation as a non-invasive tool to assess spinal activation profiles during phases of preparation or execution of voluntary lower-limb motor tasks and demonstrated characteristic spatiotemporal patterns of increased or decreased spinal excitability. These findings are fundamental to the characterization of alterations in spinal-network function caused by injury or disease of the central nervous system. Such alterations are demonstrated by Calvert and co-workers in individuals with spinal cord injury [3]. During attempted voluntary movements of the lower limbs, spinal reflexes evoked either by epidural or transcutaneous spinal cord stimulation were significantly inhibited across muscles, irrespective of their functional role.

Megía-García and colleagues studied the effects of single 10-min sessions of transcutaneous spinal cord stimulation applied at 30 Hz versus sham stimulation on lower-limb motor evoked potentials induced by transcranial magnetic stimulation in neurologically intact individuals [4]. They found increased amplitudes of motor evoked potentials in quadriceps during and after transcutaneous spinal cord stimulation, but not during or after sham stimulation. No effects on motor evoked potentials in tibialis anterior were observed. Their findings present an important interim step towards a better understanding of the spatial segmental effects of stimulation with associated clinical implications.

A comparatively new development in transcutaneous spinal cord stimulation necessitating basic mechanistic studies is its application over the cervical spinal cord with the aim to improve arm and hand function. Wecht and colleagues [5] investigated in individuals with and without chronic cervical spinal cord injury whether transcutaneous spinal cord stimulation paired with stimulation at other levels of the nervous system would enhance synaptic transmission in spinal circuits serving hand function. They found that subthreshold spinal stimulation magnified hand muscle responses to motor cortex stimulation, but not H reflexes and F waves induced by median nerve stimulation. Appropriately timed cortical and transcutaneous spinal cord stimulation may hence facilitate convergent sensorimotor transmission in the cervical spinal cord. The attainable neuromodulative outcomes of transcutaneous cervical spinal cord stimulation further depend on the applied stimulation intensity as well as on the level of voluntary participation in hand training, as shown by Kumru and co-workers in neurologically intact individuals [6]. Stimulation at 90% motor threshold led to higher maximum muscle grip strength, F-wave persistency, and maximum F wave to maximum M wave amplitude ratios, respectively, when compared to stimulation at 80 and 110% motor threshold. Stimulation at 90% motor threshold combined with training at 100% maximal volitional contraction in hand grip strength induced better hand function than with training at 50% maximal volitional contraction. The effects of sub-motor threshold transcutaneous cervical spinal cord stimulation were further examined by Sasaki and colleagues [7]. They studied whether stimulation applied for ten minutes in able-bodied individuals would alter corticospinal and spinal reflex activity at rest, yet found no modulation of motor evoked potentials or posterior root-muscle reflexes during or after the intervention. Likewise, McGeady and colleagues, who applied cervical stimulation for 10 min at the individually maximum tolerable intensity, found no consistent alterations in cortical oscillatory dynamics across their cohort of neurologically intact participants [8]. However, a weak inhibitory effect at cortical level was observed in those individuals who received the stimulation at the highest intensity levels.

While transcutaneous spinal cord stimulation is a clinically accessible method, a few aspects need to be considered to avoid potential pitfalls that could negatively impact its efficacy as a therapeutic intervention. This is partially because the stimulation conditions are sensitive to biophysical changes in the volume conductor in-between the surface electrodes that determine the electric field acting on the targeted neural structures. Binder and colleagues addressed this issue by studying the effect of extended, neutral, and flexed spine alignments in different body positions on the elicitation of H reflexes and posterior root-muscle reflexes in neurologically intact individuals [9]. They showed that, with neutral or extended spine alignments, the target neural structures of transcutaneous spinal cord stimulation in the posterior roots were most reliably activated and recommend body positions that allow easy maintenance of such alignment for therapeutic applications. Technological developments to further ease the use of transcutaneous spinal cord stimulation in clinical environments are a necessary prerequisite for its wide acceptance and lasting integration into rehabilitation practice. Salchow-Hömmen and colleagues tackled this question by introducing a novel algorithm that allows to automatically calibrate the stimulation setup and determine required stimulation intensities for antispasticity applications for each individual treated, all within a few minutes only [10]. The spasticity-alleviating effect of transcutaneous spinal cord stimulation as a viable non-pharmacological approach was further investigated by Sandler and colleagues [11]. In a randomized crossover trial including

32 individuals with motor-incomplete spinal cord injury, they tested the effects of single sessions of either transcutaneous spinal cord stimulation or whole-body vibration. Both methods reduced quadriceps spasticity for time periods beyond the intervention. Estes and colleagues combined transcutaneous spinal cord stimulation and locomotor training in individuals with sub-acute motor-incomplete spinal cord injury to enhance walking function and alleviate spasticity and compared the results to a paradigm employing sham stimulation and locomotor training [12]. In the verum group, walking outcomes were significantly improved after the intervention period of two weeks. No alterations in spasticity were seen in either group, which was partially attributed to the variability in spasticity encountered in the study participants. The influence of transcutaneous spinal cord stimulation on voluntary movement and locomotion in chronic motor-incomplete spinal cord injury were further addressed by Meyer and colleagues by studying the immediate effects of stimulation applied at different frequencies [13]. They found an increased maximum dorsiflexion angle and range of movement during rhythmic ankle movements of the more affected lower limb with 30 Hz stimulation compared to baseline, but not with 15 or 50 Hz stimulation. Electrophysiological assessments further showed a significant reduction of pathological components of polysynaptic spinal reflexes during stimulation at 30 Hz. The effects on walking were variable, with improvements seen in the subject with the highest walking scores as well as in a subgroup of the participants with the lowest locomotor function. Al'joboori and colleagues studied the outcomes of an 8-week sit-to-stand training paradigm with and without transcutaneous spinal cord stimulation in a small cohort of individuals with motor-complete or incomplete spinal cord injury [14]. While unassisted standing was not achieved in any participant, motor scores were improved, and volitional lower-limb movements were partially recovered in three individuals in whom the training had been complemented by stimulation. No such changes were observed in the control group. The importance of rehabilitation paradigms combining training and electrical stimulation is also highlighted by the work of Kumru and co-workers, who targeted the cervical spinal cord to enhance hand function [15]. Single sessions of transcutaneous cervical spinal cord stimulation applied during hand training in neurologically intact individuals retained hand grip force and increased spinal and corticospinal excitability for at least an hour. Stimulation alone increased spinal but not corticospinal excitability and had no effect on hand grip force, and training alone reduced both hand grip force and corticospinal excitability.

The Special Issue is rounded off by a review contributed by Barss and colleagues [16], in which they trace the development of transcutaneous spinal cord stimulation as a neuromodulation intervention after spinal cord injury. They elaborate on the efficacy of the stimulation applied to distinct levels of the spinal cord to induce multi-segmental effects and on how multi-site stimulation may facilitate spinal reflex and corticospinal network activity. The review provides an overview of current potentials and limitations of transcutaneous spinal cord stimulation directed to both the cervical and the lumbar spinal cord.

Conflicts of Interest: The authors declare no conflict of interest.

References

1. Malloy, D.C.; Knikou, M.; Côté, M.-P. Adapting Human-Based Transcutaneous Spinal Cord Stimulation to Develop a Clinically Relevant Animal Model. *J. Clin. Med.* **2022**, *11*, 2023. [CrossRef] [PubMed]
2. Steele, A.G.; Atkinson, D.A.; Varghese, B.; Oh, J.; Markley, R.L.; Sayenko, D.G. Characterization of Spinal Sensorimotor Network Using Transcutaneous Spinal Stimulation during Voluntary Movement Preparation and Performance. *J. Clin. Med.* **2021**, *10*, 5958. [CrossRef] [PubMed]
3. Calvert, J.S.; Gill, M.L.; Linde, M.B.; Veith, D.D.; Thoreson, A.R.; Lopez, C.; Lee, K.H.; Gerasimenko, Y.P.; Edgerton, V.R.; Lavrov, I.A.; et al. Voluntary Modulation of Evoked Responses Generated by Epidural and Transcutaneous Spinal Stimulation in Humans with Spinal Cord Injury. *J. Clin. Med.* **2021**, *10*, 4898. [CrossRef] [PubMed]
4. Megía-García, Á.; Serrano-Muñoz, D.; Taylor, J.; Avendaño-Coy, J.; Comino-Suárez, N.; Gómez-Soriano, J. Transcutaneous Spinal Cord Stimulation Enhances Quadriceps Motor Evoked Potential in Healthy Participants: A Double-Blind Randomized Controlled Study. *J. Clin. Med.* **2020**, *9*, 3275. [CrossRef] [PubMed]

5.   Wecht, J.; Savage, W.; Famodimu, G.; Mendez, G.; Levine, J.; Maher, M.; Weir, J.; Wecht, J.; Carmel, J.; Wu, Y.; et al. Posteroanterior Cervical Transcutaneous Spinal Cord Stimulation: Interactions with Cortical and Peripheral Nerve Stimulation. *J. Clin. Med.* **2021**, *10*, 5304. [CrossRef] [PubMed]

6.   Kumru, H.; Rodríguez-Cañón, M.; Edgerton, V.R.; García, L.; Flores, Á.; Soriano, I.; Opisso, E.; Gerasimenko, Y.; Navarro, X.; García-Alías, G.; et al. Transcutaneous Electrical Neuromodulation of the Cervical Spinal Cord Depends Both on the Stimulation Intensity and the Degree of Voluntary Activity for Training. A Pilot Study. *J. Clin. Med.* **2021**, *10*, 3278. [CrossRef] [PubMed]

7.   Sasaki, A.; de Freitas, R.M.; Sayenko, D.G.; Masugi, Y.; Nomura, T.; Nakazawa, K.; Milosevic, M. Low-Intensity and Short-Duration Continuous Cervical Transcutaneous Spinal Cord Stimulation Intervention Does Not Prime the Corticospinal and Spinal Reflex Pathways in Able-Bodied Subjects. *J. Clin. Med.* **2021**, *10*, 3633. [CrossRef] [PubMed]

8.   McGeady, C.; Alam, M.; Zheng, Y.-P.; Vučković, A. Effect of Cervical Transcutaneous Spinal Cord Stimulation on Sensorimotor Cortical Activity during Upper-Limb Movements in Healthy Individuals. *J. Clin. Med.* **2022**, *11*, 1043. [CrossRef] [PubMed]

9.   Binder, V.E.; Hofstoetter, U.S.; Rienmüller, A.; Száva, Z.; Krenn, M.J.; Minassian, K.; Danner, S.M. Influence of Spine Curvature on the Efficacy of Transcutaneous Lumbar Spinal Cord Stimulation. *J. Clin. Med.* **2021**, *10*, 5543. [CrossRef] [PubMed]

10.   Salchow-Hömmen, C.; Schauer, T.; Müller, P.; Kühn, A.; Hofstoetter, U.; Wenger, N. Algorithms for Automated Calibration of Transcutaneous Spinal Cord Stimulation to Facilitate Clinical Applications. *J. Clin. Med.* **2021**, *10*, 5464. [CrossRef] [PubMed]

11.   Sandler, E.B.; Condon, K.; Field-Fote, E.C. Efficacy of Transcutaneous Spinal Stimulation versus Whole Body Vibration for Spasticity Reduction in Persons with Spinal Cord Injury. *J. Clin. Med.* **2021**, *10*, 3267. [CrossRef] [PubMed]

12.   Estes, S.; Zarkou, A.; Hope, J.M.; Suri, C.; Field-Fote, E.C. Combined Transcutaneous Spinal Stimulation and Locomotor Training to Improve Walking Function and Reduce Spasticity in Subacute Spinal Cord Injury: A Randomized Study of Clinical Feasibility and Efficacy. *J. Clin. Med.* **2021**, *10*, 1167. [CrossRef] [PubMed]

13.   Meyer, C.; Hofstoetter, U.S.; Hubli, M.; Hassani, R.H.; Rinaldo, C.; Curt, A.; Bolliger, M. Immediate Effects of Transcutaneous Spinal Cord Stimulation on Motor Function in Chronic, Sensorimotor Incomplete Spinal Cord Injury. *J. Clin. Med.* **2020**, *9*, 3541. [CrossRef] [PubMed]

14.   Al'Joboori, Y.; Massey, S.J.; Knight, S.L.; Donaldson, N.D.N.; Duffell, L.D. The Effects of Adding Transcutaneous Spinal Cord Stimulation (tSCS) to Sit-To-Stand Training in People with Spinal Cord Injury: A Pilot Study. *J. Clin. Med.* **2020**, *9*, 2765. [CrossRef] [PubMed]

15.   Kumru, H.; Flores, Á.; Rodríguez-Cañón, M.; Edgerton, V.R.; García, L.; Benito-Penalva, J.; Navarro, X.; Gerasimenko, Y.; García-Alías, G.; Vidal, J. Cervical Electrical Neuromodulation Effectively Enhances Hand Motor Output in Healthy Subjects by Engaging a Use-Dependent Intervention. *J. Clin. Med.* **2021**, *10*, 195. [CrossRef] [PubMed]

16.   Barss, T.S.; Parhizi, B.; Porter, J.; Mushahwar, V.K. Neural Substrates of Transcutaneous Spinal Cord Stimulation: Neuromodulation across Multiple Segments of the Spinal Cord. *J. Clin. Med.* **2022**, *11*, 639. [CrossRef] [PubMed]

Journal of
Clinical Medicine

MDPI

Article

Adapting Human-Based Transcutaneous Spinal Cord Stimulation to Develop a Clinically Relevant Animal Model

Dillon C. Malloy [1], Maria Knikou [2] and Marie-Pascale Côté [1,*]

[1] Marion Murray Spinal Cord Research Center, Department of Neurobiology and Anatomy, Drexel University, Philadelphia, PA 19129, USA; dcm332@drexel.edu

[2] Klab4Recovery Research Laboratory, Department of Physical Therapy, College of Staten Island, New York, NY 10314, USA; maria.knikou@csi.cuny.edu

* Correspondence: mc849@drexel.edu; Tel.: +1-215-991-8598

Abstract: Transcutaneous spinal cord stimulation (tSCS) as a neuromodulatory strategy has received great attention as a method to promote functional recovery after spinal cord injury (SCI). However, due to the noninvasive nature of tSCS, investigations have primarily focused on human applications. This leaves a critical need for the development of a suitable animal model to further our understanding of this therapeutic intervention in terms of functional and neuroanatomical plasticity and to optimize stimulation protocols. The objective of this study is to establish a new animal model of thoracolumbar tSCS that (1) can accurately recapitulate studies in healthy humans and (2) can receive a repeated and stable tSCS treatment after SCI with minimal restraint, while the electrode remains consistently positioned. We show that our model displays bilateral evoked potentials in multisegmental leg muscles characteristically comparable to humans. Our data also suggest that tSCS mainly activates dorsal root structures like in humans, thereby accounting for the different electrode-to-body-size ratio between the two species. Finally, a repeated tSCS treatment protocol in the awake rat after a complete spinal cord transection is feasible, tolerable, and safe, even with minimal body restraint. Additionally, repeated tSCS was capable of modulating motor output after SCI, providing an avenue to further investigate stimulation-based neuroplasticity and optimize treatment.

Keywords: spinal cord injury; transcutaneous spinal cord stimulation; neuromodulation; electrical stimulation; evoked potentials; lumbar spinal cord

Citation: Malloy, D.C.; Knikou, M.; Côté, M.-P. Adapting Human-Based Transcutaneous Spinal Cord Stimulation to Develop a Clinically Relevant Animal Model. *J. Clin. Med.* **2022**, *11*, 2023. https://doi.org/10.3390/jcm11072023

Academic Editors: Ursula S. Hofstoetter and Karen Minassian

Received: 1 March 2022
Accepted: 31 March 2022
Published: 5 April 2022

Publisher's Note: MDPI stays neutral with regard to jurisdictional claims in published maps and institutional affiliations.

1. Introduction

Sensory and motor deficits following spinal cord injury (SCI) persist with limited functional recovery regardless of extensive efforts to optimize current therapeutic interventions. Neuromodulation strategies, in particular spinal cord stimulation, have become an increasingly popular and promising approach that can be used alone or in conjunction with well-established treatments such as locomotor training to promote functional recovery after SCI [1,2]. While animal models of epidural stimulation are abundant, transcutaneous spinal cord stimulation (tSCS), because of its noninvasive nature, has mostly been directly investigated in able-bodied and SCI individuals [3–7].

This has resulted in the emergence of valuable information about the feasibility and safety of this method, its access to key spinal circuitry, the benefits for patients with SCI, and its potential use in a clinical setting. Amongst the functional benefits of tSCS in SCI individuals are improvements in functional motor output [8–16], reduced hyperreflexia and spasticity [15,17–19], and improvements in volitional motor control [9,11–14]. Although critical to our understanding for therapeutic applications of tSCS, evidence-based human-only approaches are hindered by the limited number of study participants, heterogeneity of injuries, and complexity of clearly identifying neuroplastic changes in humans. The lack of knowledge on the specific neurophysiological mechanisms contributing to motor recovery with tSCS and their anatomical and molecular correlates prevents the optimization of tSCS protocols and its transition to a broader clinical population. Therefore, there is a critical need to validate a suitable preclinical animal model to further our understanding of tSCS and potentiate its use to promote functional plasticity after SCI.

In animal models, transcutaneous direct current stimulation over the thoracolumbar region has mostly been utilized [20–22], which is critically different from the alternated constant current and associated neuronal mechanisms of plasticity [23,24]. In the case of direct current stimulation over the thoracolumbar region, patients report discomfort and irritable sensations such as burning and tingling of the skin, and physical adverse effects include persistent skin irritations and lesions resembling burns in the area under the electrode [25,26]. In none of our applications of constant alternated current with 1 ms at 0.2 Hz did we observe blisters or burning sensation. Furthermore, the stable delivery of repeated tSCS in rodents over time has been performed at the cervical level [20–22,27], which has very different physiological and anatomical features, and more importantly, does not address plasticity and functional recovery in the lumbar circuitry. To our knowledge, a single study performed repeated tSCS at the thoracolumbar level in rodents, and animals were heavily restrained to ensure electrode stability [20]. The current lack of breadth in animal models for tSCS limits the progress and understanding of this intervention required to support its potential implementation into the SCI community.

In order for tSCS to optimally activate lumbar spinal neuronal networks and loco-motor centers in humans, a clinically relevant animal model is in great need to allow a thorough and systematic investigation of the most effective stimulation parameters as well as identification of the mechanisms at play. The purpose of this study is to establish an effective translational approach by (1) developing a rodent model in which responses to tSCS display similar electrophysiological features as in healthy humans, and (2) determine the feasibility of delivering repeated tSCS at the thoracolumbar level in awake rats with minimal restraint and distress.

Here, we show that we were able to successfully scale down human-based tSCS to develop an animal model that activates similar neural structures and displays similar transspinal evoked potentials (TEPs) in leg muscles following tSCS as in humans. We further show the feasibility and stability of delivering tSCS over time in awake rodents, which ultimately increased the motor output of ankle extensor and flexor muscles. The validation of this model will allow us to more precisely investigate the neuroplastic changes and mechanisms of action with tSCS that otherwise cannot be learned from human stud-ies alone.

2. Materials and Methods

All animal procedures were performed in accordance with the National Institutes of Health (NIH) guidelines for the care and use of laboratory animals, and experimental pro-tocols were approved by Drexel University College of Medicine Institutional Animal Care and Use Committee. Twenty-one adult female Sprague Dawley rats (240–260 g; Charles River Laboratories, Wilmington, MA, USA) were used for all experimental procedures. Animals were housed 2–3 per cage with ad libitum food and water under 12-h light/dark cycle in temperature-controlled facilities accredited by the Association for Assessment and Accreditation of Laboratory Animal Care (AAALAC). All animals were given a one-week acclimation period upon arrival before any procedures were performed.

2.1. Intact Animals—Experiments

2.1.1. Transcutaneous Electrode Fabrication and Stimulation Set-Up

We adapted human-based stimulation electrodes and set-up to our rat model. Re-usable, self-adhering hydrogel electrodes (Uni-Patch StarBurst Square, Balego, St. Paul, MN, USA) were cut down to 4 cm × 1 cm rectangles. Animals were briefly anesthetized with gaseous isoflurane in oxygen (1–4%), shaved over the back and abdomen, wiped clean with alcohol prep pads (70% isopropyl), and fitted with transcutaneous electrodes

One electrode (cathode) was placed over the T10-L2 thoracolumbar spine equally between paravertebral sides and no lower than the tips of the T13 floating ribs. Two of the same electrodes, connected to function as a single electrode (anode), were placed bilaterally over the abdomen. To ensure adherence and constant placement, the electrodes were covered with 3M Tegaderm transparent film (3M, St. Paul, MN, USA), and the body of the animal was lightly swaddled with self-adhesive athletic wrap (3M, St. Paul, MN, USA). This set-up was used for both the repeated tSCS treatment in SCI rats and for the terminal experiment in intact and SCI rats.

2.1.2. Stimulation and Recordings

Rats ($n = 9$) were anesthetized with gaseous isoflurane in oxygen (1–4%) and fitted with transcutaneous electrodes as described above. Electromyographic (EMG) needle electrodes (Neuroline Subdermal, Ambu A/S, Ballerup, DK, USA) were placed in the left tibialis anterior (L-TA), left medal gastrocnemius (L-MG), and right tibialis anterior (R-TA) muscles, bipolar wire electrodes (Cooner Wire, Chatsworth, CA, USA) were placed in the left plantar muscles of the foot (L-Pl) on the plantar surface of the left hind paw, and ground electrodes were inserted into the skin of the arm. Rats were held at approximately 1.5% isoflurane in oxygen for the duration of the experiment.

TEPs were evoked by single monophasic 1 ms pulses through the transcutaneous cathodal electrode and recorded in L-Pl, L-MG, L-TA, and R-TA muscles. The stimulation was delivered using the DS7A constant current isolated stimulator (Digitimer Ltd., Hertfordshire, UK), which was triggered by customized scripts using the Signal software version 6.0 through a 1401 analog-to-digital data acquisition board system (CED; Cambridge Electronics Design Ltd., Cambridge, UK). EMG recordings were amplified ($\times 1000$; model 1700, A–M Systems, Sequin, WA, USA) and bandpass filtered (1 Hz–10 kHz). Signals were digitized (10 kHz) and fed to Signal software. TEPs were first recorded in response to a range of increasing stimulus intensities to construct a recruitment curve and determine motor threshold, response latency, and maximum response amplitudes for each muscle. In addition, stimulation trains (30 pulses) were delivered at 100 Hz at an intensity of 1.4 times TEP threshold (1.4 T). After completing terminal experiments, rats were sacrificed with an overdose of Euthasol (150 mg/kg, i.p.).

2.1.3. Data Analysis

The recruitment curve was plotted by expressing the peak-to-peak amplitude as a function of the stimulus intensity. Peak-to-peak amplitude was measured from the maximum negative peak to the maximum positive peak within the duration of the response regardless of the number of peaks. To determine the input–output relationship, a sigmoid function (Systat SigmaPlot 14.0, Inpixon, Palo Alto, CA, USA) was fitted to individual TEP recruitment curves (i.e., each muscle of each animal) to predict maximal amplitude, slope, and stimulation intensity required to reach 50% of maximal amplitude. These parameters were used to normalize individual sigmoid functions to the predicted threshold and maximal amplitude of the curve. Group averages were calculated from the individual normalized values and used to establish a group sigmoid function.

Response latency (onset of the responses) and duration of the responses were measured at maximal amplitude as determined by the recruitment curves for each individual muscle. Latency was determined by measuring the time between stimulus and response onset, while duration was determined by the time EMG activity varied from baseline.

TEP amplitude during high-frequency stimulation trains was measured by calculating the peak-to-peak amplitude of TEPs for each of the 30 pulses and was represented as a percentage of the amplitude of the first response.

2.1.4. Statistical Analysis

Significant differences between muscles were determined using a one-way analysis of variance (ANOVA) followed by the Holm–Sidak post hoc multiple comparison test unless stated otherwise. Kruskal–Wallis one-way ANOVA followed by Dunn's post hoc multiple comparison test was used if normality or equal variance tests failed. For TEP amplitude in response to a high-frequency stimulation train, a repeated measures one-way ANOVA was utilized. Statistical analyses were performed using GraphPad Prism software version 7.04 (GraphPad Software, San Diego, CA, USA). For all tests, significance was determined when $p < 0.05$, and values are reported as the group average $\pm$ standard error of the mean.

2.2. Spinal Cord-Injured Animals—Procedures and Experiments

2.2.1. Surgical Procedures and Postoperative Care

Rats ($n = 12$) underwent a complete spinal cord transection at the thoracic level (T10) under aseptic conditions. Gaseous isoflurane in oxygen (1–4%) was used as an anesthetic prior to and throughout the duration of the surgery. A T10–T11 laminectomy was performed, the dura was carefully split open, and the spinal cord was cut with small scissors. The completeness of the spinal transection was confirmed by examining the ventral floor of the spinal canal. Gel foam was placed into the cavity between rostral and caudal portions of the spinal cord to achieve hemostasis. Back muscles and skin incision were sutured accordingly using appropriately sized materials. Rats were singly housed for 3 days following surgical procedures before being paired for the remainder of the study. Animals received one dose of slow-release buprenorphine (0.05 mg/kg, s.c.) as an analgesic prior to surgery end and received saline (5 mL, s.c.) and Baytril (100 mg/kg, s.c.) postoperatively for 7 days to prevent dehydration and infection. Bladders were manually expressed at least twice a day for the duration of the study.

2.2.2. Repeated Transcutaneous Stimulation in SCI Animals

SCI rats were randomized into one of two treatment groups: repeated transcutaneous stimulation (SCI + tSCS group, $n = 6$) or no stimulation (SCI, $n = 6$). Starting 5 days post-injury, animals from the SCI + tSCS group were fitted with transcutaneous electrodes as described above (Section 2.1.1) and secured in a modified, custom-built apparatus to allow them to lie prone with hindlimbs hanging below at rest. The motor threshold (T) was evaluated visually and with light manual touch and was determined as the lowest intensity eliciting a twitch of the ankle. Stimulations were evoked using a DS3 constant current isolated stimulator (Digitimer Ltd., Hertfordshire, UK) and a customized script written in Signal (CED). The stimulation protocol consisted of single, monophasic pulses of 1 ms in duration delivered at 0.2 Hz. Stimulation intensity alternated in bouts of 3 min between suprathreshold (1.2 T) and subthreshold (0.8 T) for a total duration of 18 min per session. Untreated SCI animals were similarly secured to the apparatus for an equal amount of time but were not stimulated. Sessions were repeated 3 times a week for 4 weeks before the terminal experiment.

2.2.3. Terminal Experiments and Analysis

TEPs were elicited and recorded similar to intact animals in L-MG, L-TA, and R-TA muscles. A one-way ANOVA followed by Holm–Sidak post hoc test was run to compare latency and duration across muscles and between SCI and SCI + tSCS groups. A two-way repeated measure ANOVA followed by Holm–Sidak post hoc test was utilized to assess the effect of tSCS on TEP amplitude at increasing stimulation intensities expressed as xMT. Significance was determined when $p < 0.05$, and values are reported as the group average $\pm$ standard error of the mean.

2.3. Human Experiments

All procedures were performed in accordance with NIH guidelines for human research, and experimental protocol was approved by the CUNY-wide Institutional Review Board committee. Eleven healthy persons (5 males, 6 females; 24 to 34 age range) participated in the study.

2.3.1. Stimulation and Recordings

We adopted similar experimental procedures as we have previously employed in humans [6]. Briefly, a self-adhering hydrogel electrode (UniPatch EP84169, 10.2 cm × 5.1 cm, Wabash, MN, USA) was placed over the T10-L2 vertebrae, and two electrodes (anode) connected to function as one were placed either over the iliac crests or abdominals based on self-reported comfort. Stimulation and recordings were performed with subjects supine, legs at midline, and hips/knees at 30° of flexion.

2.3.2. Data Analysis and Statistics

In healthy humans, the recruitment curves of EMG potentials evoked by tSCS were constructed from below threshold intensities to maximal intensities that allowed for establishing motor thresholds, latencies, and maximum responses as well as the slope of the curve confined to occur at TEPs equivalent to 50% of the maximal responses. Stimulation was a single 1 ms pulse at 0.2 Hz delivered by a constant current stimulator (DS7A, Digitimer Ltd., Welwyn Garden City, UK), triggered by Spike 2 scripts through a 36-channel Power 1401 plus analog-to-digital data acquisition interface running Spike 2 (CED Ltd., Cambridge, UK). Single differential bipolar electrodes (Motion Lab Systems Inc., Baton Rouge, LA, USA) were used to record responses from ankle and thigh muscles. EMG recordings were amplified (×1000) and bandpass filtered (1 Hz–10 kHz).

A sigmoid function (Systat SigmaPlot 11, Inpixon, Plato Alto, CA, USA) was fitted to the TEPs measured as the area under the full wave rectified waveform and plotted as a function of the stimulation intensity. This was performed separately for responses recorded for each muscle and subject. Through the sigmoid function we established the m function of the slope and predicted stimuli corresponding to 50% of the maximal amplitude. These values were used to establish the slope and stimulation intensity corresponding to threshold [16]. For each muscle, the stimulation intensities and TEPs were normalized to the predicted threshold intensity and maximal amplitude, respectively. Averages of normalized TEPs for all 11 subjects were calculated in steps of 0.05 from 0.6 up to 2.5 times the stimulation threshold. Latency of TEPs was established with the cumulative sum (CUSUM) statistical method [28,29]. Each TEP was rectified and averaged, and CUSUM was applied to detect change in the series of datum points while taking into consideration 60 ms of pre-stimulation background EMG activity ± 2 standard deviation from the mean reference level [28]. The first point that the EMG signal was above the standard deviation of the EMG signal was taken as latency. Latency values are reported as the group average ± standard error of the mean.

3. Results

3.1. Adapting Electrode Configurations

We adapted human-based stimulation electrode configurations to our rat model. Reusable, self-adhering hydrogel electrodes were similarly positioned (Figure 1) and covered with Tegaderm transparent film to ensure adherence and constant placement. Electrodes were cut down from 10.2 cm × 5.1 cm for humans to 4 cm × 1 cm rectangles to fit the rat spinal cord size. The cathode was placed over the thoracolumbar spine equally between paravertebral sides over the T10-L2 vertebrae. The anode, composed of two similar electrodes, was connected to function as a single electrode and placed bilaterally over the abdomen. The animal was then lightly swaddled with self-adhesive athletic wrap.

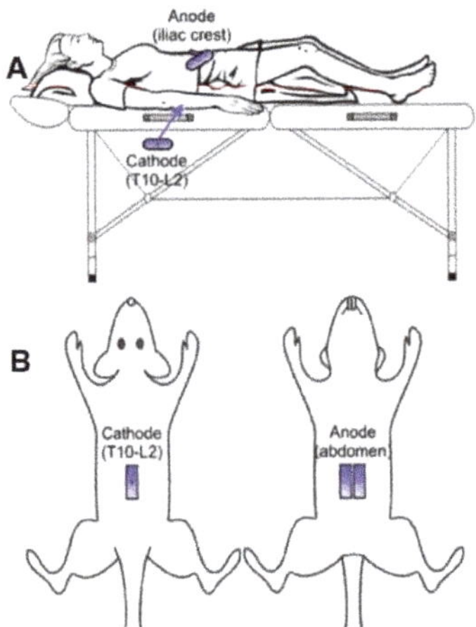

Figure 1. Transcutaneous electrode configuration. The cathode is placed over the T10-L2 vertebrae and the anodes over the iliac crests or abdomen in human (**A**) and rats (**B**).

3.2. TEPs Are Characteristically Similar across Muscles and between Models

To confirm the validity of our animal model, we tested the ability to produce stable and reproducible transspinal evoked potentials (TEPs) in leg muscles of intact animals following a stimulation delivered through the transcutaneous electrode [14,30–33]. As we have extensively described in humans [6], a single monophasic 1 ms duration pulse evoked responses bilaterally in rat hindlimb muscles, including the ankle flexor TA, the ankle extensor MG, and the Pl muscle (Figure 2A). Healthy humans display responses with similar characteristics (Figure 2B). The evoked responses were similar bilaterally, as depicted from the TEP recorded in the left and right TA, with similar shape and amplitude, suggesting accurate placement of the electrode at midline to ensure equal activation of the left and right motor pools. The latency of the responses recorded from a variety of hindlimb muscles ranged from 2.2 to 5.2 ms in rats and 6.8 to 22.5 ms in humans (Figure 2, see also Section 3.4). This range of latencies is consistent with spinally-induced responses. More importantly, the plantar muscle of the foot response displayed a longer latency (Figure 2A, arrow) compared to MG and TA ankle muscles in rat (grey area). This is consistent with the more caudal location of the Pl motor pool. Similarly, rectus femoris (RF) displayed a shorter latency (Figure 2B, arrow) than MG and TA in humans (grey area). Differences in latencies between different muscles and between rats and humans can be accounted for by the difference in size, conduction velocity, and location of the motor pool.

Additionally, TEPs oftentimes displayed an increasing number of peaks with increasing stimulation intensity. An example is depicted in Figure 3. The TEPs elicited in the TA muscle were initially biphasic, with two peaks at low stimulation intensity. As the stimulation intensity increased, the number of peaks increased from 3 to 5. This suggests the ability of tSCS to activate a larger proportion of the spinal circuitry, likely including the recruitment of spinal interneurons in both rat and human models. It is worth noting that not all subjects (animals or humans) displayed this feature. While a more systematic investigation is necessary, this suggest that the excitability of motoneurons (subliminal fringe) and interneurons differs between subjects following tSCS.

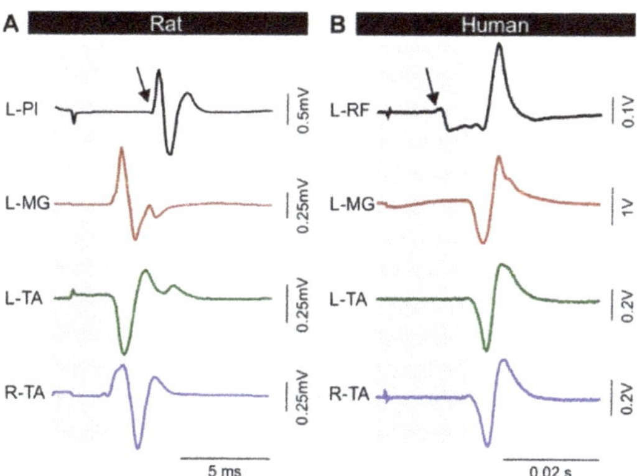

Figure 2. Transspinal evoked potential (TEP). Typical TEPs recorded from leg muscles in rat (**A**) and human (**B**). The response latency for ankle muscles (grey area) is different from muscle at other joints (arrow). L, left; R, right; Pl, plantar muscle; MG, medial gastrocnemius; TA, tibialis anterior; RF, rectus femoris.

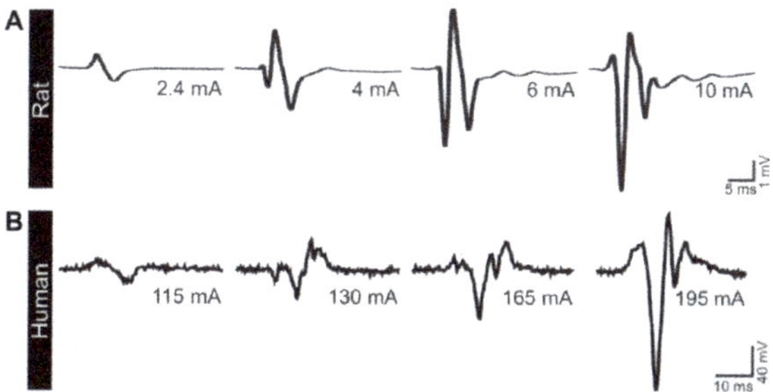

Figure 3. tSCS can activate complex spinal circuitry at higher intensity. As the stimulation intensity is augmented, tSCS-evoked TEPs recorded in TA display an increased number of peaks in rats (**A**) and humans (**B**).

3.3. Excitability of Motor Pools in Response to tSCS

We then evaluated the excitability of multisegmental motor pools innervating the legs in response to tSCS. As expected, the amplitude of TEPs increased with augmenting stimulation intensity (Figure 4A) to eventually reach a plateau at higher stimulation intensities. The recruitment curve followed a sigmoid function for all muscles recorded (Figure 4B), similar to the well-established sigmoid shape of the recruitment curves for the soleus H-reflex and M-wave, cortically-induced MEPs, and TEPs, which we have previously reported in healthy and SCI human subjects [16,34,35]. This suggests the ability of tSCS to activate motoneurons in rat hindlimbs according to the typical recruitment order of motoneurons in humans.

Overall, the recruitment of motoneurons in response to tSCS in rats (Figure 5A) is similar to that observed in humans (Figure 5B) with function of the slope m, slope, and predicted stimulation intensity corresponding to threshold, 50% of maximal, and at maximal amplitudes for each TEP (Table 1). Note the excellent sigmoid function between amplitudes and intensities, as depicted by the R^2, and the similar parameters of TEPs recorded across different muscles.

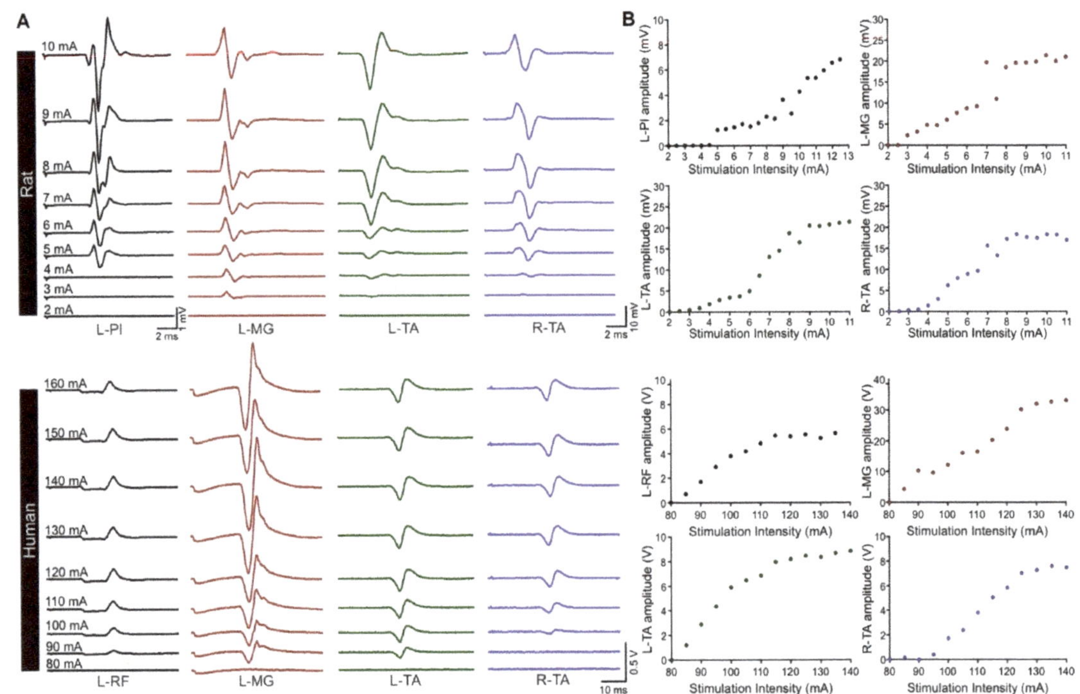

Figure 4. Representative examples of TEP recruitment curves. (**A**) TEPs were recorded from leg muscles at increasing stimulation intensity in a rat (**top**) and a human (**bottom**). (**B**) The amplitude of TEPs in all muscles increases with augmenting stimulaton intensity. L, left; R, right; Pl, plantar muscle; MG, medial gastrocnemius; TA, tibialis anterior; RF, rectus femoris.

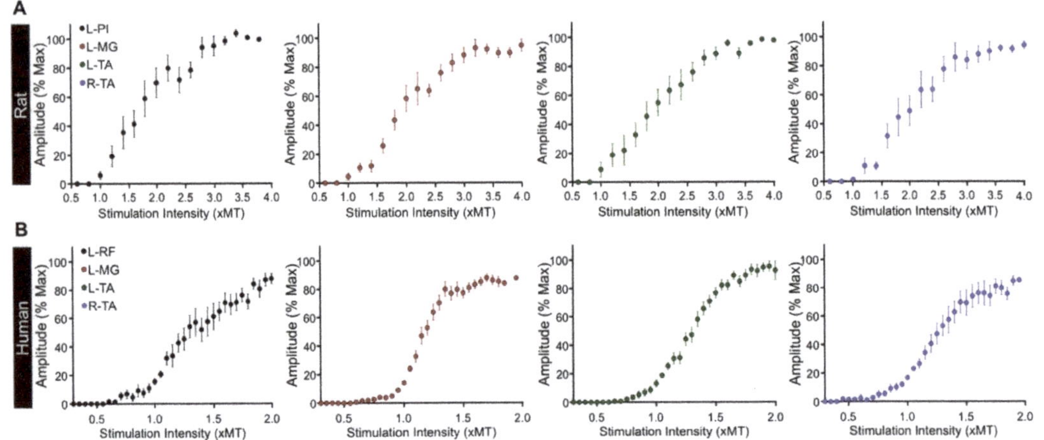

Figure 5. Recruitment curves of transspinal evoked potentials (TEPs). Group average of the input–output relationship in rat (**A**) and human (**B**) leg muscles. L, left; R, right; Pl, plantar muscle; MG, medial gastrocnemius; TA, tibialis anterior; RF, rectus femoris.

Table 1. Parameters of sigmoid function for the TEP recruitment curves in intact animals and healthy humans.

		R^2	Max	m	S50	Slope	S-Threshold	S-Max
Rats	L-Pl	0.958 ± 0.013	2.29 ± 1.03	1.08 ± 0.21	6.03 ± 0.92	2.36 ± 0.46	4.21 ± 0.55	8.94 ± 1.27
	L-MG	0.977 ± 0.003	14.62 ± 2.12	1.36 ± 0.32	5.17 ± 0.39	2.00 ± 0.31	3.17 ± 0.24	7.18 ± 0.66
	L-TA	0.984 ± 0.004	12.76 ± 2.85	1.42 ± 0.21	4.39 ± 0.48	1.59 ± 0.16	2.80 ± 0.41	5.98 ± 0.59
	R-TA	0.986 ± 0.003	14.94 ± 2.63	2.13 ± 0.60	4.35 ± 0.60	1.45 ± 0.31	2.91 ± 0.38	5.80 ± 0.87
Humans	L-RF	0.923 ± 0.018	7.90 ± 1.94	0.08 ± 0.03	170.84 ± 14.07	52.37 ± 14.82	118.47 ± 10.88	223.21 ± 36.64
	L-MG	0.947 ± 0.013	20.17 ± 2.46	0.08 ± 0.02	175.95 ± 14.07	31.56 ± 4.76	144.39 ± 11.56	207.51 ± 17.53
	L-TA	0.953 ± 0.005	7.16 ± 0.83	0.05 ± 0.01	170.63 ± 13.08	53.66 ± 8.82	116.97 ± 9.27	224.28 ± 20.29
	R-TA	0.950 ± 0.010	9.68 ± 1.52	0.05 ± 0.01	177.73 ± 16.29	58.01 ± 9.98	119.71 ± 11.12	235.74 ± 24.62

Average $\pm$ SEM predicted values from the sigmoid fit with stimulation intensities plotted against TEP amplitude. Max is the maximal amplitude of the TEP; m is the slope parameter of the sigmoid function; S50 is the stimulus intensity required to elicit a TEP equivalent to 50% of the maximal amplitude (mA); slope is the slope of the sigmoid relationship confined to occur at S50; S-threshold and S-max are predicted stimulation intensities (mA) corresponding to threshold and maximal amplitudes, respectively. L, left; R, right; Pl, plantar muscle; MG, medial gastrocnemius; TA, tibialis anterior; RF, rectus femoris.

3.4. Latency and Duration of TEPs

TEP latency (Figure 6A) was not significantly different between ipsilateral ankle flexor (L-TA, 2.74 ± 0.10 ms) and extensor muscles (L-MG, 2.70 ± 0.09 ms) or between bilateral ankle muscles (L-TA/MG and R-TA, 2.79 ± 0.14 ms; $p > 0.05$, one-way ANOVA). However, TEP onset latency for the L-Pl muscle was significantly delayed (4.78 ± 0.11 ms) compared to ankle muscles ($F_{3,32} = 86.7$, $p < 0.0001$, one-way ANOVA). This is in agreement with the more distal location of the Pl muscle and caudal location of the motor pool as compared to more proximal muscles such as the TA and MG. Similarly, TEP onset latency in humans was clearly shorter for proximal thigh muscles (RF, 9.41 ± 1.49 ms) as compared to distal ankle muscles (TA, 16.49 ± 2.12 ms and MG, 16.37 ± 2.11 ms; Figure 6A). One-way ANOVA showed that the latency was significantly different across muscles ($F_{4,50} = 32.607$, $p < 0.01$). Holm–Sidak multiple comparisons showed that the latency of L-RF was significantly different from ankle muscles ($p < 0.001$), while the latency of ankle flexors and extensors was similar ($p > 0.05$).

The response duration (Figure 6B) at maximal TEP amplitude was not significantly different between ipsilateral or bilateral ankle muscles (L-TA, 7.96 ± 0.48 ms; L-MG, 7.09 ± 0.37 ms; R-TA, 7.36 ± 0.34 ms) ($p > 0.05$, Kruskal–Wallis one-way ANOVA). The response duration of L-Pl TEP (5.54 ± 0.27 ms) was, however, significantly less than ankle muscles (H (3) = 16.08, $p < 0.05$, Kruskal–Wallis one-way ANOVA). This was predictable, as this is a small muscle which has a limited number of muscle fibers within the plantar surface of the hind paw. It is noteworthy that similar results were obtained when latency and duration was measured at oui 30% of the TEP maximal response (not shown). In humans, TEP duration measured at maximal stimulation intensities (Figure 6B) was different between L-RF and L-MG as well as between R-TA and L-MG and between L-TA and L-MG (H (3) = 19.71, $p < 0.05$, Kruskal–Wallis one-way ANOVA on ranks), suggesting that more interneurons are recruited for responses recorded from ankle and hip flexors following cathodal thoracolumbar tSCS in healthy humans.

3.5. Rat tSCS Activates Primary Afferents

To confirm that tSCS in rats is mostly mediated by primary afferents as reported in humans [4,36–40], we measured the latency and amplitude of TEPs induced by high-frequency stimulation trains (30 p, 100 Hz, 1.4 T). The onset latency of TEPs was similar across all 30 pulses (Kruskal–Wallis one-way ANOVA, H (29) = 16.572, $p = 0.968$) with an average of (5.19 ± 0.23 ms), and no different from TEPs evoked by single pulse (Figure 6A,

4.78 ± 0.11). As expected from the repetitive stimulation of primary afferents in rats [41,42] and humans [43], TEP amplitude decreased with the second response at 71.05 ± 15.04% of the first response, and all subsequent responses continually showed significant depression (Figure 7; $F_{3,18}$ = 18.42, $p < 0.0001$, one-way ANOVA) as compared to the first response with an average amplitude of 13.58 ± 5.49%. This suggests that tSCS in our rat model is also, at least partly, mediated by the activation of primary afferents.

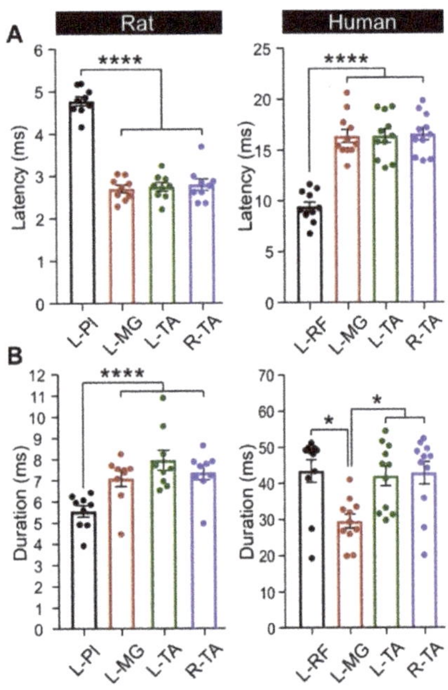

Figure 6. TEP latency and duration at maximal output. (**A**) L-Pl response latency was significantly longer in rats as compared to L-MG and L- and R-TA. In humans, L-RF latency was shorter as compared to ankle mucles. (**B**) In rats, the response duration was significanty shorter in L-Pl as compared to ankle muscles. In humans, MG response duration was significantly shorter than all other muscles. L, left; R, right; Pl, plantar muscle; MG, medial gastrocnemius; TA, tibialis anterior; RF, rectus femoris. * $p < 0.05$; **** $p < 0.0001$.

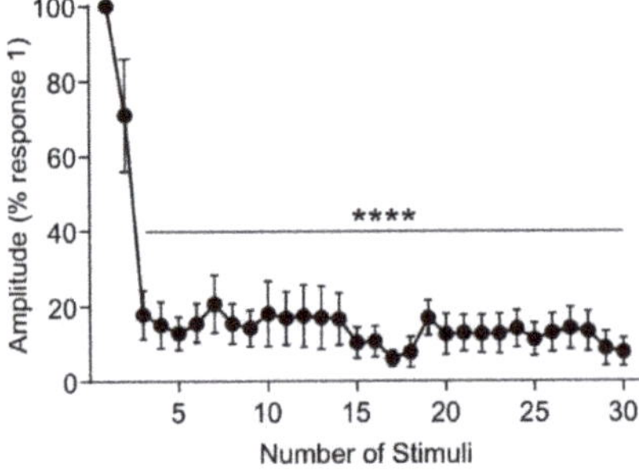

Figure 7. High-frequency pulse train with tSCS. TEP amplitude was recorded in response to a 30-pulse train of high frequency (100 Hz) tSCS at suprathreshold (1.4 T, motor threshold) intensity. TEP amplitude was significantly decreased compared to pulse number 1 from pulses 3 through 30. **** $p < 0.0001$.

3.6. Repeated tSCS in the Awake Rat Is Feasible, Tolerable, and Safe

After validating the similarity between tSCS responses in neurologically intact rats and humans, we tested the possibility to deliver tSCS as a treatment in awake SCI animals. We were able to test the feasibility, tolerability, and safety of tSCS in awake animals by providing chronic, repeated treatment 3 days per week for 4 weeks. By securing rats in a custom-built apparatus in a physiologically normal resting state with hindlimbs hanging below at rest (Figure 8), we were able to stimulate animals without movements or changes in body position that would jeopardize the placement of the transcutaneous electrodes. This was confirmed by measuring the stimulation intensity required to reach motor threshold over time, where intensity was consistent throughout and not significantly different from the first session (not shown). This suggests that the placement of the transcutaneous electrode was consistent and stable over time with repeated placement. The stimulation protocol was well-tolerated for the 18 min duration of each treatment session. In addition, animals did not show any visual signs of pain or discomfort during stimulation bouts whether the stimulation intensity was above or below motor threshold, including but not limited to vocalization, wincing, orbital tightening, or ear folding [44]. Therefore, repeated tSCS in the awake rat proved to be feasible to perform and reproduce in a chronic setting, tolerable throughout each repeated treatment session over time, and safe with minimal risk of tissue damage or discomfort similar to humans.

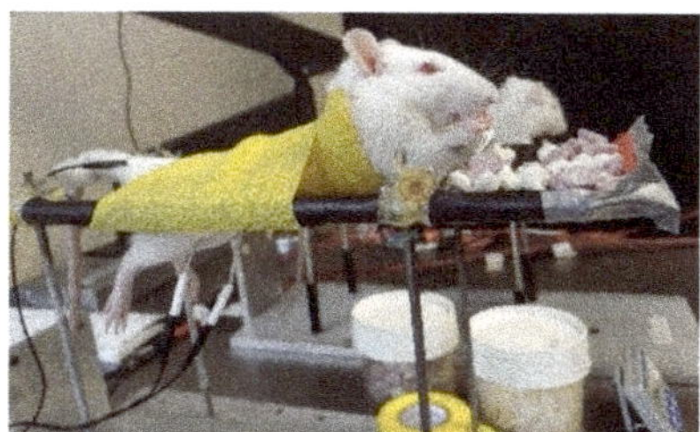

Figure 8. Transcutaneous spinal cord stimulation in the awake rodent.

3.7. Repeated tSCS Increases Motor Output in Ankle Muscles of the Spinalized Rat

We have recently shown that repeated tSCS increases motor output based on TEP recruitment curves in people with motor complete or incomplete SCI [16]. To investigate the potential of repeated tSCS to produce similar results in spinalized animals, we built TEP recruitment curves for the MG and TA muscles, as constructed in intact animals (Figures 4 and 5). MG and TA TEPs were similar in shape, latency, and duration whether the animals had received tSCS or not ($p > 0.05$, one-way ANOVA, not shown). Response latency and duration for SCI animals were 2.66 ± 0.23 ms and 5.82 ± 0.61 ms (MG) and 2.75 ± 0.14 ms and 5.99 ± 0.32 ms (TA), respectively, and were 2.51 ± 0.15 ms and 5.23 ± 0.35 ms (MG) and 2.72 ± 0.15 ms and 5.51 ± 0.20 ms (TA), respectively, for SCI + tSCS animals. A two-way repeated measure ANOVA revealed an interaction between groups (SCI vs. SCI + tSCS) and stimulation intensities (from 0.8 to 2.6 in 0.2 xMT increments) for both MG ($F_{10,81} = 5.5190$, $p < 0.001$) and TA ($F_{10,109} = 2.165$, $p = 0.025$). A Holm–Sidak post hoc test identified that repeated tSCS significantly increased TEP amplitude in the MG at intensities ranging from 1.2 to 2 T (Figure 9A) and in the TA from 1.4 to 1.8 T (Figure 9B) as compared to untreated SCI animals. These results further support that the actions of tSCS include increased motor output after SCI. This also indicates that repeated tSCS in SCI rats can recapitulate human-based treatment paradigms and produce similar findings as in SCI humans after tSCS [16,17].

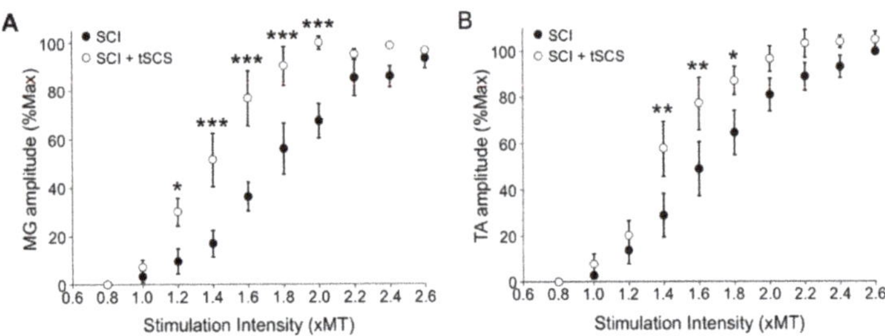

Figure 9. Repeated tSCS increases motor output after SCI. The recruitment curves of TEPs recorded in MG (**A**) and TA (**B**) show that the amplitude is larger in SCI + tSCS than SCI. * $p < 0.05$, ** $p < 0.01$, *** $p < 0.001$.

4. Discussion

The development and application of clinically relevant animal models for noninvasive tSCS are paramount to understand this therapeutic intervention at the neuroanatomical level. However, it is necessary that these animal models have similarities to human-based tSCS for adequate comparison and extrapolation of results between the two species.

Appropriately scaling down tSCS to rats had certain challenges. The selectivity of tSCS is primarily determined by spatial conditions, especially the design and the placement of stimulating electrodes. Spatial selectivity is also strongly limited by the distance between the electrode and target neuron, the diverse distribution of tissue conductivity, and the resulting distribution of the electrical field [5,45–47]. Importantly, nonhomogeneous electrical conduction properties, such as bony structures of the spine, also differ between rats and humans and could have a significant influence on the generated electrical fields. Considering this and the different electrode-to-body-size ratio in rats as compared to humans, it was critical to ensure that the same neural structures were stimulated with tSCS in rats and that TEPs displayed similar neurophysiological characteristics.

While the neuronal circuits and pathways activated by tSCS are not yet fully understood, it is believed to activate similar neuronal structures as epidural stimulation [40]. It has been suggested to mainly excite primary afferents in the dorsal roots leading to transsynaptic excitation of motoneurons and spinal interneuronal networks over multiple spinal segments close to and far from the stimulation site [4,36–39]. However, the presence of orthodromic and antidromic volleys traveling across the mixed peripheral nerve needs to be determined given the summation of soleus H-reflex and soleus TEP action potentials on surface EMGs as well as the depression of soleus H-reflex or soleus TEP based on the relative timing between the two stimuli [48]. As expected from the stimulation of primary afferents, the amplitude of TEPs evoked by a high-frequency train of tSCS was significantly depressed in intact rats. This response contrasts with direct motoneuronal stimulation, which would either cause facilitation or lack of depression [49,50]. Finally, if the conduction of the electrical current was extended through soft tissues located outside the spinal networks, delayed EMG responses would have been expected but were not observed.

The similar recruitment curves in rat hindlimb muscles and healthy human leg muscles (Figure 5) following tSCS support an orderly recruitment of motoneurons. The net system gain may thus be estimated from the slope of the sigmoid function, which, in turn, can potentially reflect activation of motoneurons residing within the subliminal fringe. Transcutaneous spinal cord stimulation generated TEPs in bilateral hindlimb muscles at different joints with biphasic, triphasic, and polyphasic waveforms at increasing stimulation intensities that are characteristics of responses evoked in human leg muscles [5,6,36,48]. The increasing number of phases at increasing intensities suggest the activation of spinal interneurons, as supported from the polysynaptic effects on soleus H-reflex when it is evoked at its maximal amplitude [48]. We also observed a preferential activation of proximal and distal motor pools based on their location along the lumbar spinal cord in both rats and humans, with distal limb muscles having longer latencies compared to those more proximal to the stimulation site. This is comparable between the two species when corrected for distance traveled and conduction velocity differences [6,51–55]. Lastly, we should note that more research is needed to determine whether the maximal TEP reflects depolarization of the whole motoneuron pool, as is the case for the maximal M-wave [56].

Since the development of this model is ultimately meant to be used as a treatment after neurological injuries, we tested a protocol in awake SCI animals using the same stimulation configuration as initially used in intact animals during terminal experiments. Throughout the stimulation session, the animals were secured to a custom-built support harness and displayed no discomfort whether the stimulation intensity applied was above or below TEP threshold. We have more than 10 years of experience using this type of restraint in SCI rats for bicycling up to 1 h per day with no issue [57–59]. This is significant progress in delivering a clinically relevant treatment that does not necessitate anesthesia or immobilization of the animal [20,60].

We further investigated the effect of repeated tSCS over 4 weeks using alternated subthreshold and suprathreshold 0.2 Hz stimulation over the lumbar enlargement of rats with a complete SCI. This protocol was prioritized to match that used in humans by Knikou in both able-bodied and SCI individuals [16,17] and confirm that the effects of repeated tSCS can be reproduced in a rodent model of complete transection. Repeated tSCS increased net motor output, as assessed by TEP recruitment curves, supporting a similar effect in rats and SCI individuals [16]. Now that we have validated the feasibility of repeated tSCS stimulation in a rodent model and its relevance to humans in terms of spinal excitability, it will allow us to move forward and utilize this treatment intervention of tSCS in animals with a contusion injury. Our next objective is to use tSCS in anatomically incomplete spinal rats to further bolster the clinical relevance of this model across both intervention and injury.

Together, we have provided compelling evidence that our rat model shares common neurophysiological characteristics with similar physiological properties following tSCS to those observed in human subjects. We have also demonstrated that repeated stimulation can be performed in the awake SCI animal and that this treatment increased motor output in both flexor and extensor muscles of the hindlimbs. This strongly supports that our animal model can be effectively utilized to further our understanding of neuroplasticity induced by tSCS after SCI and how it affects functional recovery. This will be instrumental to optimize this therapeutic intervention and accelerate its transition to the clinic.

Author Contributions: Conceptualization, M.-P.C. and M.K.; methodology, M.-P.C. and M.K.; software, M.-P.C. and M.K.; validation, D.C.M., M.-P.C. and M.K.; formal analysis, D.C.M., M.-P.C., and M.K.; investigation, M.-P.C. and M.K.; resources, M.-P.C. and M.K.; data curation, D.C.M., M.-P.C. and M.K.; writing—original draft preparation, D.C.M., M.-P.C. and M.K.; writing—review and editing, D.C.M., M.-P.C. and M.K.; project administration, M.-P.C. and M.K.; funding acquisition, M.-P.C. and M.K. All authors have read and agreed to the published version of the manuscript.

Funding: This work was supported by grants from the National Institute of Neurological Disorders (RO1 NS119475) and the Craig H. Neilsen Foundation (647897) awarded to M.-P.C. and performed at Drexel University. This research was also funded by the National Institute of Child Health and Human Development (RO1 HD100544) and the New York State Department of Health (NYSDOH), Spinal Cord Injury Research Program (SCIRP), grant number 647897 awarded to M.K. and performed at the City University of New York. Publication made possible in part by support from the Drexel University Libraries Open Access Publishing Fund.

Institutional Review Board Statement: The study was conducted according to the guidelines of the Declaration of Helsinki and approved by the Institutional Review Board (IRB) of the City University of New York (CUNY-wide IRB committee protocol number 2019-0806 (approved 19 December 2019) and protocol number 2022-0003 (approved 5 January 2022). The study was conducted according to the guidelines of the Declaration of Helsinki and approved by the IACUC of Drexel University (protocol number 20666 (approved 13 December 2017).

J. Clin. Med. **2022**, 11, 2023

Informed Consent Statement: Informed consent was obtained from all subjects involved in the study.

Data Availability Statement: The data presented in this study are available on request from the corresponding author.

Acknowledgments: We would like to thank Kyle Yeakle for technical assistance.

Conflicts of Interest: The authors declare no conflict of interest. The funders had no role in the design of the study; in the collection, analyses, or interpretation of data; in the writing of the manuscript, or in the decision to publish the results.

References

1. Hofstoetter, U.S.; Knikou, M.; Guertin, P.A.; Minassian, K. Probing the Human Spinal Locomotor Circuits by Phasic Step-Induced Feedback and by Tonic Electrical and Pharmacological Neuromodulation. *Curr. Pharm. Des.* **2017**, *23*, 1805–1820. [CrossRef] [PubMed]
2. Smith, A.C.; Knikou, M. A Review on Locomotor Training after Spinal Cord Injury: Reorganization of Spinal Neuronal Circuits and Recovery of Motor Function. *Neural Plast.* **2016**, *2016*, 1216258. [CrossRef] [PubMed]
3. Einhorn, J.; Li, A.; Hazan, R.; Knikou, M. Cervicothoracic Multisegmental Transpinal Evoked Potentials in Humans. *PLoS ONE* **2013**, *8*, e76940. [CrossRef] [PubMed]
4. Minassian, K.; Persy, I.; Rattay, F.; Dimitrijevic, M.R.; Ms, C.H.; Kern, H. Posterior root–muscle reflexes elicited by transcutaneous stimulation of the human lumbosacral cord. *Muscle Nerve* **2006**, *35*, 327–336. [CrossRef]
5. Krenn, M.; Toth, A.; Danner, S.M.; Hofstoetter, U.S.; Minassian, K.; Mayr, W. Selectivity of transcutaneous stimulation of lumbar posterior roots at different spinal levels in humans. *Biomed. Technol.* **2013**, *58* (Suppl. 1). [CrossRef]
6. Knikou, M. Neurophysiological characterization of transpinal evoked potentials in human leg muscles. *Bioelectromagnetics* **2013**, *34*, 630–640. [CrossRef]
7. Gerasimenko, Y.; Gorodnichev, R.; Machueva, E.; Pivovarova, E.; Semyenov, D.; Savochin, A.; Roy, R.R.; Edgerton, V.R. Novel and Direct Access to the Human Locomotor Spinal Circuitry. *J. Neurosci.* **2010**, *30*, 3700–3708. [CrossRef]
8. Gerasimenko, Y.; Gorodnichev, R.; Moshonkina, T.; Sayenko, D.; Gad, P.; Edgerton, V.R. Transcutaneous electrical spinal-cord stimulation in humans. *Ann. Phys. Rehabil. Med.* **2015**, *58*, 225–231. [CrossRef]
9. Gerasimenko, Y.P.; Lu, D.C.; Modaber, M.; Zdunowski, S.; Gad, P.; Sayenko, D.G.; Morikawa, E.; Haakana, P.; Ferguson, A.; Roy, R.R.; et al. Noninvasive Reactivation of Motor Descending Control after Paralysis. *J. Neurotrauma* **2015**, *32*, 1968–1980. [CrossRef]
10. Gerasimenko, Y.; Gorodnichev, R.; Puhov, A.; Moshonkina, T.; Savochin, A.; Selionov, V.; Roy, R.R.; Lu, D.C.; Edgerton, V.R. Initiation and modulation of locomotor circuitry output with multisite transcutaneous electrical stimulation of the spinal cord in noninjured humans. *J. Neurophysiol.* **2015**, *113*, 834–842. [CrossRef]
11. Gad, P.; Gerasimenko, Y.; Zdunowski, S.; Turner, A.; Sayenko, D.; Lu, D.C.; Edgerton, V.R. Weight Bearing Over-ground Stepping in an Exoskeleton with Non-invasive Spinal Cord Neuromodulation after Motor Complete Paraplegia. *Front. Neurosci.* **2017**, *11*, 333. [CrossRef] [PubMed]
12. Sayenko, D.G.; Rath, M.; Ferguson, A.; Burdick, J.W.; Havton, L.A.; Edgerton, V.R.; Gerasimenko, Y.P. Self-Assisted Standing Enabled by Non-Invasive Spinal Stimulation after Spinal Cord Injury. *J. Neurotrauma* **2019**, *36*, 1435–1450. [CrossRef] [PubMed]
13. Gad, P.; Lee, S.; Terrafranca, N.; Zhong, H.; Turner, A.; Gerasimenko, Y.; Edgerton, V.R. Non-Invasive Activation of Cervical Spinal Networks after Severe Paralysis. *J. Neurotrauma* **2018**, *35*, 2145–2158. [CrossRef] [PubMed]
14. Inanici, F.; Samejima, S.; Gad, P.; Edgerton, V.R.; Hofstetter, C.P.; Moritz, C.T. Transcutaneous Electrical Spinal Stimulation Promotes Long-Term Recovery of Upper Extremity Function in Chronic Tetraplegia. *IEEE Trans. Neural Syst. Rehabil. Eng.* **2018**, *26*, 1272–1278. [CrossRef]
15. Minassian, K.; Hofstoetter, U.S.; Danner, S.M.; Mayr, W.; Bruce, J.A.; McKay, W.B.; Tansey, K.E. Spinal Rhythm Generation by Step-Induced Feedback and Transcutaneous Posterior Root Stimulation in Complete Spinal Cord–Injured Individuals. *Neurorehabilit. Neural Repair* **2015**, *30*, 233–243. [CrossRef]
16. Murray, L.M.; Knikou, M. Transspinal stimulation increases motoneuron output of multiple segments in human spinal cord injury. *PLoS ONE* **2019**, *14*, e0213696. [CrossRef]
17. Knikou, M.; Murray, L.M. Repeated transspinal stimulation decreases soleus H-reflex excitability and restores spinal inhibition in human spinal cord injury. *PLoS ONE* **2019**, *14*, e0223135. [CrossRef]
18. Hofstoetter, U.S.; Freundl, B.; Danner, S.M.; Krenn, M.J.; Mayr, W.; Binder, H.; Minassian, K. Transcutaneous Spinal Cord Stimulation Induces Temporary Attenuation of Spasticity in Individuals with Spinal Cord Injury. *J. Neurotrauma* **2020**, *37*, 481–493. [CrossRef]
19. Hofstoetter, U.S.; McKay, W.B.; Tansey, K.E.; Mayr, W.; Kern, H.; Minassian, K. Modification of spasticity by transcutaneous spinal cord stimulation in individuals with incomplete spinal cord injury. *J. Spinal Cord Med.* **2013**, *37*, 202–211. [CrossRef]
20. Mekhael, W.; Begum, S.; Samaddar, S.; Hassan, M.; Toruno, P.; Ahmed, M.; Gorin, A.; Maisano, M.; Ayad, M.; Ahmed, Z. Repeated anodal trans-spinal direct current stimulation results in long-term reduction of spasticity in mice with spinal cord injury. *J. Physiol.* **2019**, *597*, 2201–2223. [CrossRef]

21. Zareen, N.; Shinozaki, M.; Ryan, D.; Alexander, H.; Amer, A.; Truong, D.; Khadka, N.; Sarkar, A.; Naeem, S.; Bikson, M.; et al. Motor cortex and spinal cord neuromodulation promote corticospinal tract axonal outgrowth and motor recovery after cervical contusion spinal cord injury. *Exp. Neurol.* **2017**, *297*, 179–189. [CrossRef] [PubMed]

22. Song, W.; Truong, D.Q.; Bikson, M.; Martin, J.H. Transspinal direct current stimulation immediately modifies motor cortex sensorimotor maps. *J. Neurophysiol.* **2015**, *113*, 2801–2811. [CrossRef] [PubMed]

23. Nitsche, M.A.; Cohen, L.G.; Wassermann, E.M.; Priori, A.; Lang, N.; Antal, A.; Paulus, W.; Hummel, F.; Boggio, P.S.; Fregni, F.; et al. Transcranial direct current stimulation: State of the art 2008. *Brain Stimul.* **2008**, *1*, 206–223. [CrossRef] [PubMed]

24. Priori, A.; Ciocca, M.; Parazzini, M.; Vergari, M.; Ferrucci, R. Transcranial cerebellar direct current stimulation and transcutaneous spinal cord direct current stimulation as innovative tools for neuroscientists. *J. Physiol.* **2014**, *592*, 3345–3369. [CrossRef]

25. Matsumoto, H.; Ugawa, Y. Adverse events of tDCS and tACS: A review. *Clin. Neurophysiol. Pract.* **2017**, *2*, 19–25. [CrossRef]

26. Murray, L.M.; Tahayori, B.; Knikou, M. Transspinal Direct Current Stimulation Produces Persistent Plasticity in Human Motor Pathways. *Sci. Rep.* **2018**, *8*, 717. [CrossRef]

27. Sharif, H.; Alexander, H.; Azam, A.; Martin, J.H. Dual motor cortex and spinal cord neuromodulation improves rehabilitation efficacy and restores skilled locomotor function in a rat cervical contusion injury model. *Exp. Neurol.* **2021**, *341*, 113715. [CrossRef]

28. Ellaway, P. Cumulative sum technique and its application to the analysis of peristimulus time histograms. *Electroencephalogr. Clin. Neurophysiol.* **1978**, *45*, 302–304. [CrossRef]

29. Brinkworth, R.; Türker, K.S. A method for quantifying reflex responses from intra-muscular and surface electromyogram. *J. Neurosci. Methods* **2003**, *122*, 179–193. [CrossRef]

30. Mishra, A.M.; Pal, A.; Gupta, D.; Carmel, J.B. Paired motor cortex and cervical epidural electrical stimulation timed to converge in the spinal cord promotes lasting increases in motor responses. *J. Physiol.* **2017**, *595*, 6953–6968. [CrossRef]

31. Kumru, H.; Rodríguez-Cañón, M.; Edgerton, V.; García, L.; Flores, A.; Soriano, I.; Opisso, E.; Gerasimenko, Y.; Navarro, X.; García-Alías, G.; et al. Transcutaneous Electrical Neuromodulation of the Cervical Spinal Cord Depends Both on the Stimulation Intensity and the Degree of Voluntary Activity for Training. A Pilot Study. *J. Clin. Med.* **2021**, *10*, 3278. [CrossRef] [PubMed]

32. Alam, M.; Garcia-Alias, G.; Jin, B.; Keyes, J.; Zhong, H.; Roy, R.R.; Gerasimenko, Y.; Lu, D.C.; Edgerton, V.R. Electrical neuromodulation of the cervical spinal cord facilitates forelimb skilled function recovery in spinal cord injured rats. *Exp. Neurol.* **2017**, *291*, 141–150. [CrossRef]

33. Fadeev, F.; Eremeev, A.; Bashirov, F.; Shevchenko, R.; Izmailov, A.; Markosyan, V.; Sokolov, M.; Kalistratova, J.; Khalitova, A.; Garifulin, R.; et al. Combined Supra- and Sub-Lesional Epidural Electrical Stimulation for Restoration of the Motor Functions after Spinal Cord Injury in Mini Pigs. *Brain Sci.* **2020**, *10*, 744. [CrossRef] [PubMed]

34. Knikou, M. Transpinal and Transcortical Stimulation Alter Corticospinal Excitability and Increase Spinal Output. *PLoS ONE* **2014**, *9*, e102313. [CrossRef] [PubMed]

35. Knikou, M.; Dixon, L.; Santora, D.; Ibrahim, M.M. Transspinal constant-current long-lasting stimulation: A new method to induce cortical and corticospinal plasticity. *J. Neurophysiol.* **2015**, *114*, 1486–1499. [CrossRef] [PubMed]

36. Maruyama, Y.; Shimoji, K.; Shimizu, H.; Kuribayashi, H.; Fujioka, H. Human spinal cord potentials evoked by different sources of stimulation and conduction velocities along the cord. *J. Neurophysiol.* **1982**, *48*, 1098–1107. [CrossRef] [PubMed]

37. De Noordhout, A.M.; Rothwell, J.; Thompson, P.D.; Day, B.L.; Marsden, C.D. Percutaneous electrical stimulation of lumbosacral roots in man. *J. Neurol. Neurosurg. Psychiatry* **1988**, *51*, 174–181. [CrossRef]

38. Danner, S.M.; Krenn, M.; Hofstoetter, U.S.; Toth, A.; Mayr, W.; Minassian, K. Body Position Influences Which Neural Structures Are Recruited by Lumbar Transcutaneous Spinal Cord Stimulation. *PLoS ONE* **2016**, *11*, e0147479. [CrossRef]

39. Minassian, K.; Hofstoetter, U.; Rattay, F. Transcutaneous lumbar posterior root stimulation for motor control studies and modification of motor activity after spinal cord injury. In *Restorative Neurology of Spinal Cord Injury*; Dimitrijevic, M.R., Byron, A., Vrbova, G., McKay, W.B., Eds.; Oxford University Press: New York, NY, USA, 2011; pp. 226–255.

40. Hofstoetter, U.S.; Freundl, B.; Binder, H.; Minassian, K. Common neural structures activated by epidural and transcutaneous lumbar spinal cord stimulation: Elicitation of posterior root-muscle reflexes. *PLoS ONE* **2018**, *13*, e0192013. [CrossRef]

41. Lloyd, D.P.C. Monosynaptic reflex response of individual motoneurons as a function of frequency. *J. Gen. Physiol.* **1957**, *40*, 435–450. [CrossRef]

42. Eccles, J.C.; Rall, W. Effects induced in a monosynaptic reflex path by its activation. *J. Neurophysiol.* **1951**, *14*, 353–376. [CrossRef]

43. Rothwell, J.; Day, B.L.; Berardelli, A.; Marsden, C.D. Habituation and conditioning of the human long latency stretch reflex. *Exp. Brain Res.* **1986**, *63*, 197–204. [CrossRef] [PubMed]

44. Sotocinal, S.G.; Sorge, R.E.; Zaloum, A.; Tuttle, A.H.; Martin, L.; Wieskopf, J.S.; Mapplebeck, J.C.; Wei, P.; Zhan, S.; Zhang, S.; et al. The Rat Grimace Scale: A Partially Automated Method for Quantifying Pain in the Laboratory Rat via Facial Expressions. *Mol. Pain* **2011**, *7*, 55. [CrossRef] [PubMed]

45. Sayenko, D.G.; Atkinson, D.A.; Dy, C.J.; Gurley, K.M.; Smith, V.L.; Angeli, C.; Harkema, S.J.; Edgerton, V.R.; Gerasimenko, Y.P. Spinal segment-specific transcutaneous stimulation differentially shapes activation pattern among motor pools in humans. *J. Appl. Physiol.* **2015**, *118*, 1364–1374. [CrossRef]

46. Roy, F.D.; Gibson, G.; Stein, R.B. Effect of percutaneous stimulation at different spinal levels on the activation of sensory and motor roots. *Exp. Brain Res.* **2012**, *223*, 281–289. [CrossRef] [PubMed]

47. Krenn, M.; Hofstoetter, U.S.; Danner, S.M.; Minassian, K.; Mayr, W. Multi-Electrode Array for Transcutaneous Lumbar Posterior Root Stimulation. *Artif. Organs* **2015**, *39*, 834–840. [CrossRef] [PubMed]

48. Knikou, M.; Murray, L. Neural interactions between transspinal evoked potentials and muscle spindle afferents in humans. *J. Electromyogr. Kinesiol.* **2018**, *43*, 174–183. [CrossRef]

49. Sharpe, A.N.; Jackson, A. Upper-limb muscle responses to epidural, subdural and intraspinal stimulation of the cervical spinal cord. *J. Neural Eng.* **2014**, *11*, 016005. [CrossRef]

50. Jackson, A.; Baker, S.N.; Fetz, E.E. Tests for presynaptic modulation of corticospinal terminals from peripheral afferents and pyramidal tract in the macaque. *J. Physiol.* **2006**, *573*, 107–120. [CrossRef]

51. Thomas, P.K.; Sears, T.A.; Gilliatt, R.W. The range of conduction velocity in normal motor nerve fibres to the small muscles of the hand and foot. *J. Neurol. Neurosurg. Psychiatry* **1959**, *22*, 175–181. [CrossRef]

52. Yap, C.B.; Hirota, T. Sciatic nerve motor conduction velocity study. *J. Neurol. Neurosurg. Psychiatry* **1967**, *30*, 233–239. [CrossRef] [PubMed]

53. Buchthal, F.; Rosenfalck, A. Evoked action potentials and conduction velocity in human sensory nerves. *Brain Res.* **1966**, *3*, 1–122. [CrossRef]

54. Huseyinoglu, N.; Ozaydin, I.; Huseyinoglu, U.; Yayla, S.; Aksoy, O. Minimally Invasive Motor Nerve Conduction Study of the Rat Sciatic and Tail Nerves. *Kafkas Üniversitesi Veteriner Fakültesi Dergisi* **2013**, *19*, 943–948. [CrossRef]

55. Hofstoetter, U.S.; Danner, S.M.; Minassian, K. Paraspinal magnetic and transcutaneous electrical stimulation. In *Encyclopedia of Computational Neuroscience*; Jaeger, D., Jung, R., Eds.; Springer: New York, NY, USA, 2014.

56. Knikou, M. The H-reflex as a probe: Pathways and pitfalls. *J. Neurosci. Methods* **2008**, *171*, 1–12. [CrossRef] [PubMed]

57. Cote, M.-P.; Azzam, G.A.; Lemay, M.A.; Zhukareva, V.; Houlé, J.D. Activity-Dependent Increase in Neurotrophic Factors Is Associated with an Enhanced Modulation of Spinal Reflexes after Spinal Cord Injury. *J. Neurotrauma* **2011**, *28*, 299–309. [CrossRef] [PubMed]

58. Cote, M.-P.; Gandhi, S.; Zambrotta, M.; Houlé, J.D. Exercise Modulates Chloride Homeostasis after Spinal Cord Injury. *J. Neurosci.* **2014**, *34*, 8976–8987. [CrossRef]

59. Beverungen, H.; Klaszky, S.C.; Klaszky, M.; Côté, M.-P. Rehabilitation Decreases Spasticity by Restoring Chloride Homeostasis through the Brain-Derived Neurotrophic Factor–KCC2 Pathway after Spinal Cord Injury. *J. Neurotrauma* **2020**, *37*, 846–859. [CrossRef]

60. Ahmed, Z. Trans-spinal direct current stimulation modifies spinal cord excitability through synaptic and axonal mechanisms. *Physiol. Rep.* **2014**, *2*, e12157. [CrossRef]

Journal of
Clinical Medicine

Article

Characterization of Spinal Sensorimotor Network Using Transcutaneous Spinal Stimulation during Voluntary Movement Preparation and Performance

Alexander G. Steele [1,2], Darryn A. Atkinson [1,3], Blesson Varghese [1], Jeonghoon Oh [1], Rachel L. Markley [1] and Dimitry G. Sayenko [1,*]

[1] Center for Neuroregeneration, Department of Neurosurgery, Houston Methodist Research Institute, 6550 Fannin Street, Houston, TX 77030, USA; agsteele@houstonmethodist.org (A.G.S.); datkinson@usa.edu (D.A.A.); bvarghese6@houstonmethodist.org (B.V.); joh@houstonmethodist.org (J.O.); rmarkley@houstonmethodist.org (R.L.M.)

[2] Department of Electrical and Computer Engineering, University of Houston, E413 Engineering Bldg 2, 4726 Calhoun Road, Houston, TX 77204, USA

[3] College of Rehabilitative Sciences, University of St. Augustine for Health Sciences, 5401 La Crosse Avenue, Austin, TX 78739, USA

* Correspondence: dgsayenko@houstonmethodist.org; Tel.: +1-713-363-9910

Abstract: Transcutaneous electrical spinal stimulation (TSS) can be used to selectively activate motor pools based on their anatomical arrangements in the lumbosacral enlargement. These spatial patterns of spinal motor activation may have important clinical implications, especially when there is a need to target specific muscle groups. However, our understanding of the net effects and interplay between the motor pools projecting to agonist and antagonist muscles during the preparation and performance of voluntary movements is still limited. The present study was designed to systematically investigate and differentiate the multi-segmental convergence of supraspinal inputs on the lumbosacral neural network before and during the execution of voluntary leg movements in neurologically intact participants. During the experiments, participants (N = 13) performed isometric (1) knee flexion and (2) extension, as well as (3) plantarflexion and (4) dorsiflexion. TSS consisting of a pair pulse with 50 ms interstimulus interval was delivered over the T12-L1 vertebrae during the muscle contractions, as well as within 50 to 250 ms following the auditory or tactile stimuli, to characterize the temporal profiles of net spinal motor output during movement preparation. Facilitation of evoked motor potentials in the ipsilateral agonists and contralateral antagonists emerged as early as 50 ms following the cue and increased prior to movement onset. These results suggest that the descending drive modulates the activity of the inter-neuronal circuitry within spinal sensorimotor networks in specific, functionally relevant spatiotemporal patterns, which has a direct implication for the characterization of the state of those networks in individuals with neurological conditions.

Keywords: spinal cord; spinal stimulation; corticospinal tract; functional connectivity; movement; sensorimotor networks; task dependence

Citation: Steele, A.G.; Atkinson, D.A.; Varghese, B.; Oh, J.; Markley, R.L.; Sayenko, D.G. Characterization of Spinal Sensorimotor Network Using Transcutaneous Spinal Stimulation during Voluntary Movement Preparation and Performance. *J. Clin. Med.* **2021**, *10*, 5958. https://doi.org/10.3390/jcm10245958

Academic Editors: Ursula S. Hofstoetter and Karen Minassian

Received: 9 November 2021
Accepted: 15 December 2021
Published: 18 December 2021

Publisher's Note: MDPI stays neutral with regard to jurisdictional claims in published maps and institutional affiliations.

1. Introduction

In individuals with a spinal cord injury (SCI), many sub-functional neural connections between the brain and spinal cord can remain intact across the injury, despite a clinical diagnosis of "complete" loss of sensorimotor function [1–4]. These connections are not robust enough to drive clinically detectable function; however, they are capable of influencing the excitability of spinal sensorimotor networks below the lesion [5–8]. These spinal networks can be activated by electrical spinal neuromodulation [9–12]. Notably, when combined with rehabilitation, epidural (ESS) or transcutaneous (TSS) spinal stimulation can promote functional recovery and increase overall well-being [13–17]. Computational [18,19] and electrophysiological [20–22] studies have demonstrated that the stimulus pulses recruit spinal circuits [21,23] that are involved in the regulation of rhythm- and pattern-generating neural networks [24–26]. It has been proposed that spinal stimulation modulates the physiological states of the spinal cord below the lesion, enabling descending and sensory information processing to generate coordinated and robust motor outputs, even after chronic "complete" (most likely "incomplete") paralysis [14,27–29]. The complex interactions between stimulus pulses, descending commands originating above the SCI (and passing through the lesion site), and ongoing afferent inputs, which produce functional movement, have yet to be characterized. Furthermore, our understanding of physiological states of the intact spinal cord in humans is still limited; quantitative characterization of the net effects of descending commands and interplay between the agonist and antagonist motor pools within the lumbosacral spinal segments during voluntary movements is needed. In patients, electrophysiological evaluation of supraspinal-spinal and spinal inter-neuronal networks may be useful in characterizing the extent of residual function, as well as in identifying the neuroplastic changes associated with improved motor performance after neurological injuries and disorders [30,31].

Electrically induced soleus or quadricep H-reflexes have been used in earlier studies to document changes in spinal motoneuronal excitability associated with processes in the spinal sensorimotor network during voluntary movement preparation and execution in healthy [32–36] and spastic [34] individuals. The H-reflex was shown to be facilitated about 80 to 400 ms before the onset of a movement in the contracting muscles, due to a decrease in presynaptic inhibition to Ia fibers terminating onto discharging motoneurons [32–34]. Interestingly, the H-reflex magnitude in a given muscle can also be modulated by activity in other muscles [34]. For instance, the soleus H-reflex increased prior to and during ipsilateral plantarflexion, and it decreased prior to and during ipsilateral dorsiflexion. Similar, although less pronounced, changes were observed prior to and during contractions of the contralateral muscles. In later experiments, Hultborn et al. [35] proposed that the presynaptic inhibition of Ia afferent terminals on motoneurons of contracting muscles was decreased, permitting Ia activity to contribute to the excitation of voluntarily activated motoneurons; whereas, the presynaptic inhibition of motoneurons of muscles not involved in the contraction was increased. It was concluded that the control of presynaptic inhibition of Ia fibers prior to the movements is supraspinal in origin and is organized to aid in achieving selectivity of muscle activation. At the same time, Hultborn et al. [35] pointed out that during movement, the Ia-Ib discharge occurring from the contracting muscle(s) also activates presynaptic interneurons, and it therefore provides a peripheral source of modulation of presynaptic inhibition of the motoneurons of the muscles not involved in the contraction.

The complex interaction of spinal inter-neuronal circuitry with descending drive in the preparatory stage, and during the execution of voluntary movements (or movement attempts), is particularly intriguing in the context of neuromuscular rehabilitation after paralysis. As the first step, we sought to investigate the multi-segmental convergence of the descending drive on the lumbosacral neural networks in neurologically intact participants using double-pulse TSS. Spinally evoked motor potentials are reminiscent of the monosynaptic H-reflex [37,38], but owing to the convergence of the sensory fibers at the lumbosacral enlargement, stimulation delivered over the posterior roots entering the spinal cord can generate evoked potentials in multiple proximal and distal leg muscles, bilaterally and simultaneously. We hypothesized that descending input prior to and during voluntary movements would modulate activity of the inter-neuronal circuitry within spinal sensorimotor networks in specific, functionally relevant spatiotemporal patterns. To assess this hypothesis, we examined the effects of knee flexion and extension, as well as plantarflexion and dorsiflexion, on the amplitude of spinally evoked motor potentials in multiple leg muscles.

2. Methods

2.1. Participants

Thirteen neurologically intact adults (6 females, 7 males; height: 170.5 ± 9.6 cm, weight: 69.0 ± 8.7 kg, age: 26.5 ± 5.4 years old) were recruited to participate in this study. Specifically, the healthy volunteers between 21 and 70 years old, who declared to have no medical history or current diagnostic of, or therapy for, neurological or orthopedic disorders were invited to participate in the study. The information about the study was posted on the laboratory website and on the Texas Medical Center campus. Written informed consent to the experimental procedures, which were approved by the Houston Methodist Research Institute institutional review board (Study ID: Pro00019704), was obtained from each participant.

2.2. Experimental Setup

TSS was delivered to the skin over the lumbosacral spinal enlargement using a constant current stimulator DS8R (Digitimer Ltd., Hertfordshire, UK). Stimulation was administered using conductive self-adhesive electrodes (PALS, Axelgaard Manufacturing Co. Ltd., Fallbrook, CA, USA). The cathode (diameter: 5 cm) was placed at midline between the T12 and L1 spinous processes, and two oval anodes (size: 7.5 cm $\times$ 13 cm) were placed on the abdomen, symmetrical to the sagittal plane.

To confirm the sensory route (i.e., via the posterior roots) of the delivered spinal stimulation and the transsynaptic transmission of the stimuli on the motor pools of interest, we used double-pulse TSS [39–41]. Double-pulse TSS was delivered using two 1 ms biphasic square wave pulses with an inter-stimulus interval of 50 ms. Stimulation began at 30 mA and increased in increments of 5 mA until motor responses were observed in the lower limb muscles. The location of the electrode was adjusted in a rostro-caudal direction as needed to obtain responses in proximal and distal limb muscles with the minimum difference in motor threshold stimulation intensity.

The auditory conditioning cue was delivered using a square 250 ms tone burst at 3000 Hz with an intensity of 70 dB generated via the PowerLab data acquisition system (ADInstruments, New South Wales, Australia). For the tactile condition cue, electrical stimulation was delivered to the skin over the metatarsal region on the plantar surface of the left foot using a constant current stimulator DS8R (Digitimer Ltd., UK). Stimulation was administered using conductive self-adhesive electrodes (PALS, Axelgaard Manufacturing Co. Ltd., USA). Two electrodes (diameter: 3.2 cm) were placed over the fibular sesamoid (cathode) and the center of the medial longitudinal arch (anode). Stimulation was delivered using a single monophasic 500 µs square wave pulse. Tactile stimulation began at 2 mA and was increased in increments of 1 mA until the participant reported a distinct sensation when stimulation was applied. Stimulation did not produce visible contractions of plantar muscles.

Trigno Avanti wireless surface electromyography (EMG) electrodes (Delsys Inc., Natick, MA, USA; common-mode rejection ratio < 80 dB; size: 27 mm $\times$ 37 mm $\times$ 13 mm; input impedance: >1015 $\Omega/0.2$ pF) were placed at eight different sites: longitudinally over the left and right vastus lateralis (VL), medial hamstrings (MH), tibialis anterior (TA), and the lateral portion of the soleus (SOL) muscles, as seen in Figure 1A. EMG data were amplified using a Trigno Avanti amplifier (Delsys Inc, Natick, MA.; gain: 909; bandwidth: 20 to 450 Hz) and recorded at a sampling frequency of 2000 Hz using a PowerLab data acquisition system (ADInstruments, New South Wales, Australia).

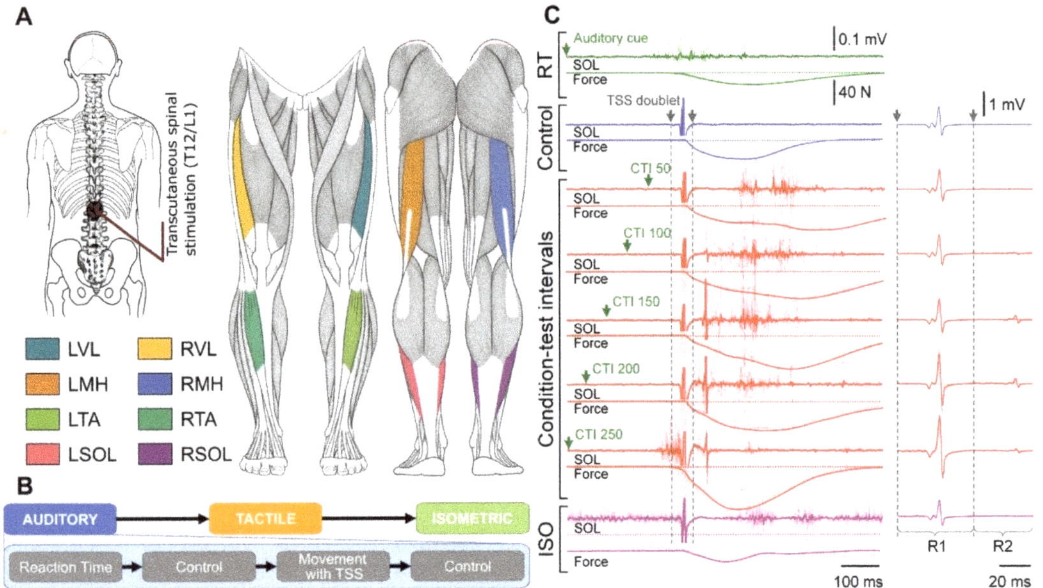

Figure 1. Experimental setup: transcutaneous spinal stimulation (**A**) was applied midline between the T12 and L1 spinous processes. EMG responses were recorded bilaterally from the vastus lateralis (VL), medial hamstring (MH), tibialis anterior (TA), and soleus (SOL) muscles. (**B**) Three experimental conditions were used during each movement task: an auditory cue, a tactile cue, and isometric contraction while TSS doublets were applied. (**C**) All CTIs are shown in red for the left SOL during the plantarflexion task for the auditory condition. Green designates reaction time (RT), blue designates control, and purple designates isometric contraction (ISO). The left and right panels use different time and amplitude scales to better visualize the waveform for voluntary SOL activation during plantarflexion (left panel) and R1 and R2 responses (right panel). Arrows indicate each pulse during the TSS doublets. The dotted horizontal line marks the initial force level prior to movement (left panel). Individual traces are shown with thin lines, and the average response is designated with a bold line.

2.3. Experimental Procedure

The tests were performed in the supine position. The participants' legs were placed in the Exolab apparatus (Antex Lab LLC, Moscow, Russia) supporting the hip, knee, and ankle joints at 155, 90, and 90 degrees, respectively. Force output during isometric left knee extension, left knee flexion, left dorsiflexion, and left plantarflexion was measured using calibrated, two-sided precision load cells with a capacity of 50 kg (Futek, Irvine, CA, USA). Measurements were recorded at a sampling frequency of 2000 Hz using a PowerLab data acquisition system (ADInstruments, New South Wales, Australia). Initially, double-pulse TSS were applied at a subthreshold intensity, then increased in 5 mA steps with three pulses delivered at each intensity and continued until the magnitude of spinally evoked motor potentials in all muscles were observed to plateau or be at maximum tolerated intensity. This allowed for the generation of recruitment curves for each muscle, which were used to identify stimulation intensities and supra-threshold values of responses for conditioning experiments [30,31].

Figure 1B outlines the following experimental conditions. First, ten repetitions of fast isometric contractions on the auditory cue "as quickly as possible" were performed to calculate the baseline reaction time. Participants were asked to perform a fast isometric contraction of the muscle being tested and produce the 5 to 10% maximum voluntary contraction force. Specifically, the participants were instructed to perform isolated contractions of the agonist muscles, while trying not to engage synergists and antagonists. For instance, during knee extension, the participants were asked to contract only the quadriceps muscles while maintaining the knee flexors and calf muscles relaxed. Next, double-pulse TSS were applied ten times at 5 s intervals, in a relaxed condition (baseline control). After that, auditory cues were given every 5 to 7 s with pseudorandom spacing between them to reduce the possibility of anticipation of the following cue. The participants were asked to perform the predetermined movement on the cue as quickly as possible, then return to rest. TSS was applied at a condition-test interval (CTI) of 50, 100, 150, 200, and 250 ms after the auditory cue was given. CTIs were randomized to reduce carry-over effects. Each movement task was repeated for a total of ten times per CTI. A final ten double-pulse TSS were done at 5 s intervals with no movement being performed (post-conditioned control). Finally, the participants were instructed to maintain the isometric contraction of the muscle being tested, producing the 5 to 10% maximum voluntary contraction force, while double-pulse TSS was applied 2 s after the start of isometric contraction. After 2 s of the isometric contraction, a cue was given for the participant to relax the muscles. The isometric contraction test was performed ten times per movement condition with a minimum of 5 s of rest between repetitions. Example muscle responses from the left SOL obtained at different CTIs are shown in Figure 1C.

The experiment was then repeated using the tactile cue, except for the isometric contraction portion of the protocol.

2.4. Statistical Analysis

Statistical analysis was performed using the open-source software RStudio version 1.4.1717 (RStudio PBC, Boston, MA, USA). The minimum sample size of the study was calculated before the initiation of the experiment using the R package PowerUpR and the estimated effect size from previous studies. The minimum sample size was found to be n_{min} = 10. The calculated value for each CTI was an average of the 10 responses for that condition, except in the case of outliers (>2.5 standard deviations from the mean), which were excluded from the analysis (<0.01% of the data). Data from each CTI were compared to the average of 20 responses from the corresponding muscle when no movement was performed, where 10 responses were recorded prior to the start of the movement, and 10 responses were recorded after the completion of that movement. For each comparison, the normality of the difference values was tested with the Shapiro–Wilk test. If values were normally distributed ($p > 0.05$), a two-tailed paired sample Student's t-test was used for analysis. If normality was violated, a Wilcoxon signed-rank test was performed. Significance set at $p < 0.05$ and a Bonferroni correction was applied to account for multiple comparisons. Second responses (R2) were removed if the first response (R1) produced a peak-to-peak value below 0.02 mV or where R2 > R1, which would suggest contamination from volitional muscle activation (<0.05% of the data). The R2/R1 ratio is given as the ratio between R2 and R1 for that pair of responses. Probability maps of the modulation of agonist, synergist, and antagonist motor pools during selected movements were constructed using the number of instances where a response for a given CTI and muscle was higher than any of the control responses and was divided by the total number of comparisons.

3. Results

3.1. Reaction Time

The average reaction time across movements during the auditory cue was determined to be 219.1 ± 41.8 ms, while the average reaction time across movements during the tactile cue was 200.9 ± 52.0 ms. The overall average reaction time for both cues across movements was 210.2 ± 47.7 ms. There were no significant differences ($p < 0.05$) in reaction time between movements or cue.

3.2. Auditory Condition

Figure 2 exemplifies evoked response changes for R1 and R2 for each muscle and movement task during the 200 ms CTI auditory condition for a representative participant. Changes in R1 can be seen in the agonist for each movement task when compared across tasks. Some movement tasks caused visible changes in contralateral responses, such as knee extension, which results in facilitation of the contralateral knee flexors (RMH), when compared to other movement tasks for that muscle.

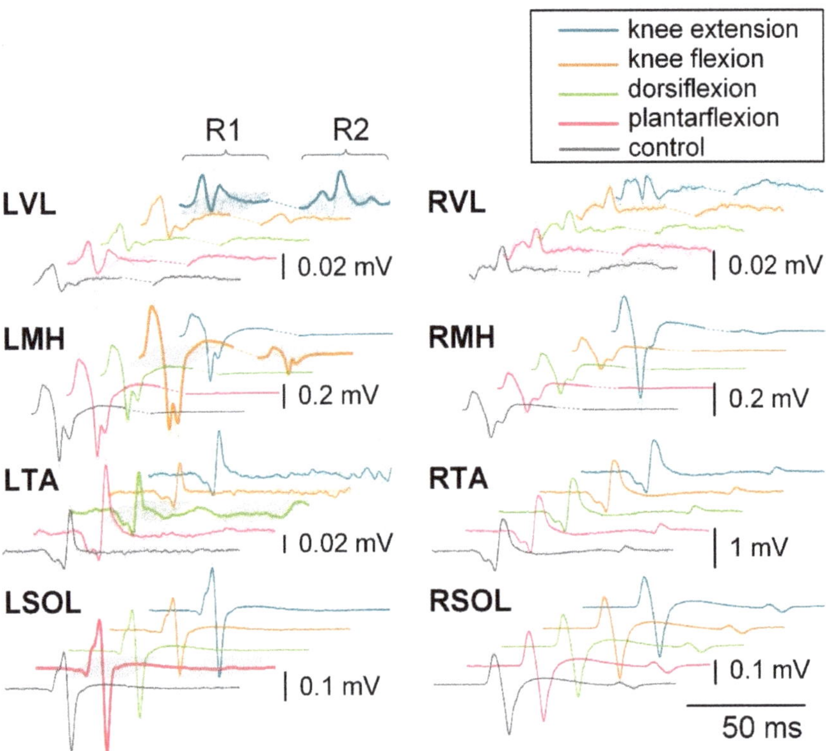

Figure 2. R1 and R2 responses for each muscle during knee extension (blue), knee flexion (orange), dorsiflexion (green), plantarflexion (pink), and control (gray) during the 200 ms CTI auditory condition for a representative participant. Individual responses are traces represented with thin dotted lines, while the average is shown with a thin solid line. The bold line signifies the agonist for that movement. Different amplitude scales are used for each muscle to better visualize the waveforms.

Figure 3 presents the patterns of evoked response modulation during each movement preparation following auditory cues. Only muscle responses that were significantly different from control responses ($p < 0.05$) are shown (see Supplemental Data S1 for all muscle responses). Muscles are listed along the bottom of each plot and the CTI tested are across the top. Data were normalized to the control condition (stimulation with no movement), and 100% of the average control response is equal to the distance between tick marks. The width of the connection represents the amount of facilitation/inhibition for a given muscle and CTI.

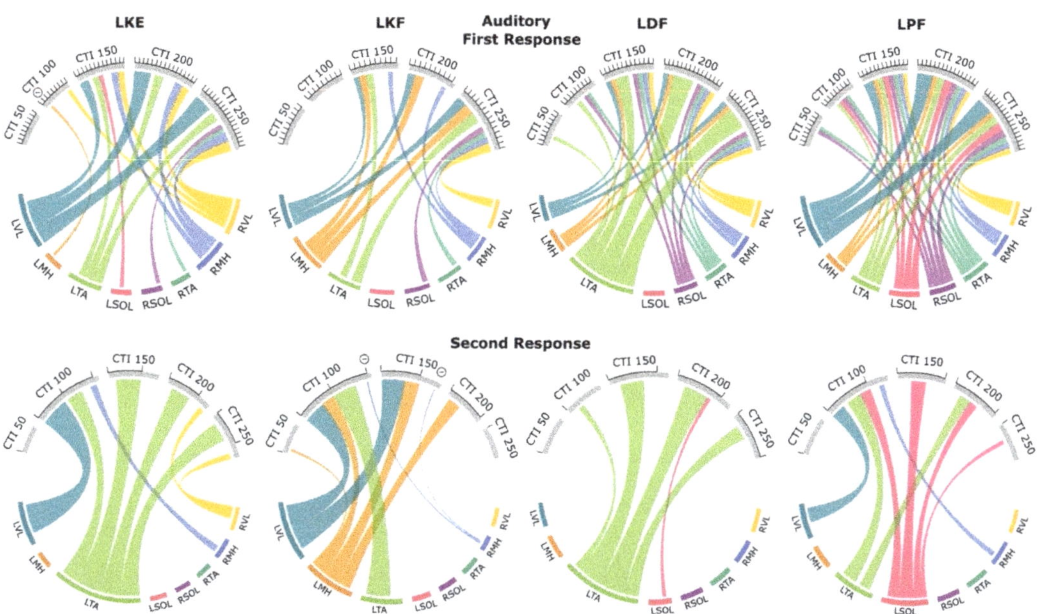

Figure 3. Excitability and inhibitory interactions during left knee extension (LKE), left knee flexion (LKF), left dorsiflexion (LDF) and left plantarflexion (LPF) movements using auditory cues: averaged significant ($p < 0.05$) facilitations or inhibitions during first (**top**) or second (**bottom**) responses during auditory cues. Each first response is scaled to the control for that movement condition, where the average control response is 100% and corresponds to the distance between tick marks. Second responses are given as the ratio between the second and first response for that pair of responses. Individual muscles are grouped along the bottom of each plot. Thickness of connections denotes the amount of inhibition or facilitation found where significant inhibition is designated with a circled minus sign.

During knee extension movement using the auditory cue, we found R1 facilitation of all muscles for at least one CTI. The knee flexion movement produced R1 facilitation (top) of the agonist (LMH), the antagonist (LVL), LTA, RVL, RMH, RTA, and RSOL for at least one CTI. Dorsiflexion facilitated R1 responses for at least one CTI for all muscles apart from the antagonist (LSOL). The plantarflexion movement caused R1 facilitation of all muscles for at least one CTI. Facilitation of R2/R1 ratio (bottom) was found at the 100 ms CTI for most movements, apart from knee flexion, which first significance was found at the 50 ms CTI. Inhibition of the R1 response of the antagonist (LMH) was found during the knee extension movement at the 50 ms CTI. Inhibition of the R2/R1 ratio was found during knee flexion for the contralateral knee flexors (RMH) during the 100 ms and 150 ms CTIs. In general, the modulation of the evoked responses increased with increasing CTI duration up to the 250 ms CTI, for which double-pulse TSS was often delivered immediately after movement initiation (Figure 1C).

3.2.1. First Responses (R1s)

Prior to knee extension, the largest facilitation of R1s occurred in the agonist (LVL). Facilitation of the response in LVL averaged 258% of control at 150 ms CTI. Facilitation increased during both 200 and 250 ms CTIs by 437% and 468%, respectively. R1s in contralateral knee flexors and extensors were also facilitated; responses in the RVL were 127% of control responses during the 100 ms CTI condition, with a maximum increase of 273% during the 250 ms CTI condition. In the RMH, a similar pattern of facilitation was observed, which first reached statistical significance at the150 ms CTI with 187%, compared to the control condition, and 246% during the 250 ms CTI. The antagonist (LMH) was the only muscle to show inhibition in the R1s, with an amplitude decrease of 85% at the 100 ms CTI.

Similarly, prior to knee flexion, the R1s in the agonist (LMH) increased in amplitude even after movement initiation at the 250 ms CTI (234%). In this case, facilitation of the antagonist (LVL) was 214% after movement initiation at the 250 ms CTI, and R1 in the contralateral knee extensor (RVL) was facilitated up to 192% at the 250 ms CTI.

Prior to dorsiflexion, facilitation of the R1s occurred in the agonist (LTA) at all CTIs with 121% facilitation at the 50 ms CTI, and it increased to 141%, 350%, 554%, and 556% during the 100, 150, 200, and 250 ms CTIs, respectively. No modulation of the R1 in the antagonist (LSOL) was observed. The first increase in LVL occurred at the 150 ms CTI with 137% and increased to 147% at the 250 ms CTI. The second largest evoked facilitation for the movement was measured from the RVL, which was 168% of the control response at the 250 ms CTI. Facilitation of the contralateral plantar flexor (RSOL) was similar, with 133% at the 250 ms CTI.

Prior to plantarflexion, R1s in the agonist (LSOL) were first significantly facilitated at the 100 ms CTI, with 110% of the control, and increased to 153%, 203%, and 215% for the 150, 200, and 250 ms CTIs, respectively. The largest facilitation was observed in the LVL for the condition, with significant facilitation of responses beginning at the 150 ms CTI (270%) up to the 250 ms CTI (437%).

3.2.2. Second Responses (R2s)

The R2/R1 ratio increased prior to knee extension for the agonist (LVL) during the 100 ms CTI, with the R2/R1 ratio of 33.8% compared to the control value of 17.0%. The R2/R1 ratio in the contralateral knee flexor (RMH) increased to 24.2% for the same CTI. The R2/R1 ratio in the contralateral knee extensor (RVL) increased during the 200 ms CTI (31.1%) and 250 ms CTI (43.4%) when compared to the control ratio (21.3%). Prior to knee flexion, the R2/R1 ratio in the agonist (LMH) increased starting at the 100 ms CTI (43.2%) and was 30.5%, 22.2%, and 45.1% at the 150, 200, and 250 ms CTIs, respectively, while the control ratio was 2.6%. Prior to dorsiflexion, the R2/R1 ratio in the agonist (LTA) increased starting at the 100 ms CTI (15.2%) and remained significantly higher with a value of 55.2%, 57.0%, and 33.9% at the 150, 200, and 250 ms CTIs, while the control ratio was 8.9%. Prior to plantarflexion, the R2/R1 ratio in the agonist (LSOL) increased up to 20.6%, 22.2%, 11.5%, and 5.2% at the 100, 150, 200, and 250 ms CTIs, while the control ratio was 3.3%. The R2/R1 ratio in the antagonist (LTA) was also higher at the 100 and 200 ms CTIs (24.7% and 18.9%, respectively), while the control ratio was 11.6%.

3.3. Tactile Condition

Figure 4 exemplifies evoked response changes found in R1 and R2 for each muscle and movement task during the 200 ms CTI tactile condition for a representative participant. Facilitation of R1, and in some cases R2, can be seen for each movement when compared across movement tasks for individual muscles. For example, knee flexion caused facilitation of R2 for the agonist (LMH), while other movements had no visible R2 response.

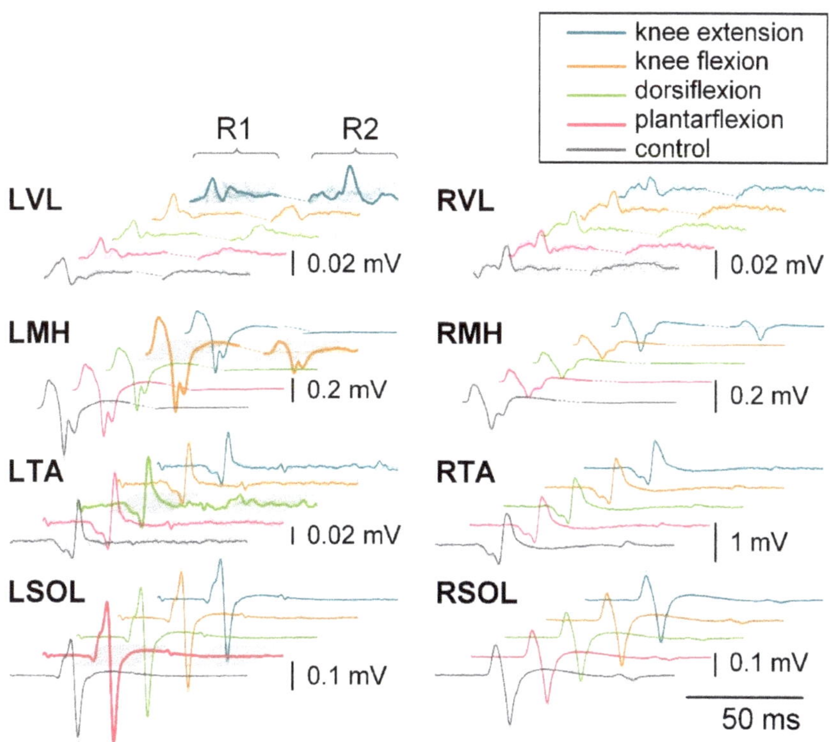

Figure 4. R1 and R2 responses for each muscle during knee extension (blue), knee flexion (orange), dorsiflexion (green), plantarflexion (pink), and control (gray) during the 200 ms CTI tactile condition for a representative participant. Individual responses are traces that are represented with a thin dotted line, while the average is shown with a thin solid line. The bold line signifies the agonist for that movement. Different scales are used for each muscle to better visualize the waveforms.

Figure 5 shows the patterns of evoked response modulation during preparation for each movement following tactile cues. Only muscle responses that were significantly different from control ($p < 0.05$) are shown (see Supplemental Data S2 for all muscle responses). During the tactile cue knee extension movement, we found R1 facilitation of the agonist (LVL), LTA, RVL, RMH, RTA, and RSOL for at least one CTI. The knee flexion movement produced R1 facilitation of the agonist (LMH), the antagonist (LVL), RVL, RMH, RTA, and RSOL for at least one CTI. Dorsiflexion during the tactile cue facilitated R1 responses for at least one CTI for all muscles. The plantarflexion movement also caused R1 facilitation of all muscles for at least one CTI. Facilitation of R2/R1 ratio was found at the 100 ms CTI for most movements, apart from knee flexion. No inhibition was found for the R1s or the R2/R1 ratio. In general, the modulation of the evoked responses increased with increasing CTI duration up to the 250 ms CTI, for which double-pulse TSS was often delivered immediately after movement initiation.

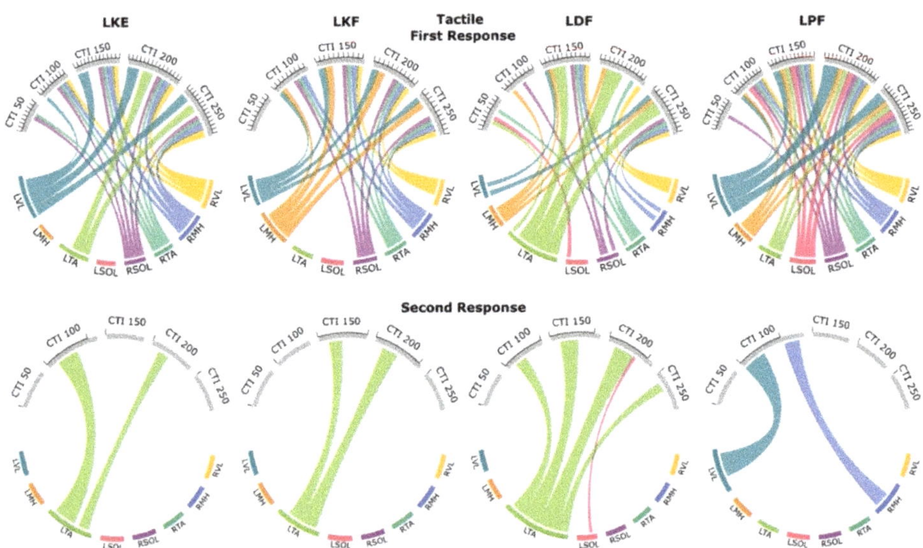

Figure 5. Excitability and inhibitory interactions during left knee extension (LKE), left knee flexion (LKF), left dorsiflexion (LDF) and left plantarflexion (LPF) movements using tactile cues: averaged significant (*p* < 0.05) facilitation or inhibition during first (**top**) or second (**bottom**) responses during tactile cues. Each first response is scaled to the control condition where control is 100% and corresponds to the distance between tick marks. Second responses are given as the ratio between second and first response for that pair of responses. Individual muscles are grouped along the bottom of each plot. Thickness of connections denotes the amount of inhibition or facilitation found where significant inhibition is designated with a circled minus sign.

3.3.1. First Responses (R1s)

Prior to the knee extension, R1s were facilitated for the agonist (LVL) as early as the 100 ms CTI, with a 111% increase when compared to control. The facilitation of the LVL response peaked at the 200 ms CTI, with a 335% increase, and after movement initiation was 265% for the 250 ms CTI. The maximum evoked facilitation for LTA was 313% at the 200 ms CTI and 257% for the 250 ms CTI. The earliest significant facilitation for contralateral knee flexor (RMH) was at the 100 ms CTI, which increased to 123% of control, then peaked at the 150 ms CTI with 190%, and remained consistent at 187% and 186% for both the 200 and 250 ms CTI. The largest facilitation for the contralateral side was for the RVL, which reached an evoked maximum of 206% increase at the 250 ms CTI.

The R1s prior to knee flexion followed a similar pattern. The agonist (LMH) response was facilitated at the 100 ms CTI and a 135% response increase. Maximum evoked facilitation for the antagonist (LVL) was at the 150 ms CTI (197%) and remained consistently high at the 200 and 250 ms CTIs, with 195% and 173% of control value, respectively. The similar facilitation was found for the LTA with an evoked maximum of 172% at the 200 ms CTI and, after movement initiation, was 141% at the 250 ms CTI. The greatest magnitude of facilitation for the contralateral side was the contralateral knee flexor (RMH) with an evoked maximum of 140% during the 150 ms CTI, while similar levels of facilitation were found for the contralateral knee extensor (RVL) with an evoked maximum of 137% at the 200 ms CTI.

Prior to dorsiflexion, the greatest magnitude of facilitation was evoked from the agonist (LTA) at the 200 ms CTI, which was 439% of control, and after movement initiation was 329% at the 250 ms CTI. On the contralateral side, the largest facilitation found was RMH at 133% during the 150 ms CTI, and 125% and 130% at the 200 and 250 ms CTIs, respectively.

Similar levels of facilitation were measured in the RVL, with an evoked maximum of 131% at the 200 ms CTI, and after movement initiation was 127% at the 250 ms CTI.

Prior to plantarflexion, the agonist (LSOL) increased to 174% at the 200 ms CTI and, after movement initiation, was 163% at the 250 ms CTI. The largest ipsilateral activation occurred in the LVL, with an evoked maximum recorded facilitation at the 200 ms CTI, which was a 439% increase in amplitude, and after movement initiation was 313% for the 250 ms CTI. The largest contralateral response was observed in the RMH with an increase of 169% at the 200 ms CTI and 162% at the 250 ms CTI. The greatest magnitude of facilitation was measured in the RSOL (147%) and RTA (138%) at the 200 ms CTI.

3.3.2. Second Responses (R2s)

The R2/R1 ratio in the LTA increased prior to knee extension at the 100 ms CTI (51.4%) and the 200 ms CTI (25.9%), whereas the control ratio was 7.6%. During preparation for knee flexion, the R2/R1 ratio increased in LTA only at the 150 ms CTI (30.1%) and the 200 ms CTI (37.4%), while the control ratio was 9.0%. Preparation for dorsiflexion caused changes in the R2/R1 ratio in the agonist (LTA) at the CTIs of 100, 150, 200, and after movement onset at 250 ms, with values of 31.8%, 46.0%, 51.4%, and 24.1%, while the control ratio was 9.8%. The R2/R1 ratio in the antagonist (LSOL) at the 200 ms CTI reached 10.3%, while the control ratio for the muscle was 3.2%. Prior to plantarflexion, the R2/R1 ratio in the LVL and RMH was increased at the 100 ms and 150 ms CTIs to 31.0% and 34.7%, while the control ratios were 19.4% and 4.6%, respectively.

3.4. Isometric Contraction Response

Figure 6 exemplifies evoked response changes found in R1 and R2 for each muscle and movement task during the isometric contraction condition for a representative participant. While significant R1 facilitation was observed, the magnitude was diminished when compared to the auditory and tactile conditions.

Figure 7 shows the patterns of evoked response modulation during isometric contraction tasks for each movement. Only muscle responses that were significantly different from control responses ($p < 0.05$) are presented (see Supplemental Data S3 for all muscle responses). During isometric knee extension, we found R1 facilitation of the agonist (LVL) with inhibition of the LMH, LSOL, RVL, RMH, RTA, and RSOL. Isometric knee flexion produced facilitation of R1s in all muscles except for the RVL and caused inhibition in the LTA and LSOL. Isometric dorsiflexion facilitated R1 responses in all muscles, apart from the LMH, and caused inhibition of the LVL and LSOL. Isometric plantarflexion caused R1 facilitation in the agonist (LSOL) and antagonist (LTA) along with RVL, RTA, and RSOL. Plantarflexion did result in inhibition in any muscle tested. Facilitation of the R2/R1 ratio was found for most movements and at least one muscle apart from plantarflexion. Inhibition of the R2/R1 ratio was found during knee flexion for the RVL, dorsiflexion for the RVL and RSOL, and plantarflexion for RVL, RTA, and RSOL.

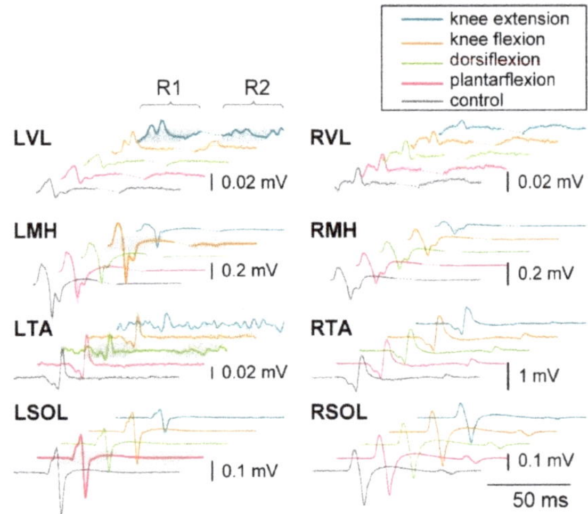

Figure 6. R1 and R2 responses for each muscle during knee extension (blue), knee flexion (orange), dorsiflexion (green), plantarflexion (pink), and control (gray) during the isometric contraction condition for a representative participant. Individual responses are traces that are represented with a thin dotted line, while the average is shown with a thin solid line. The bold line signifies the agonist for that movement. Different scales are used for each muscle to better visualize the waveforms.

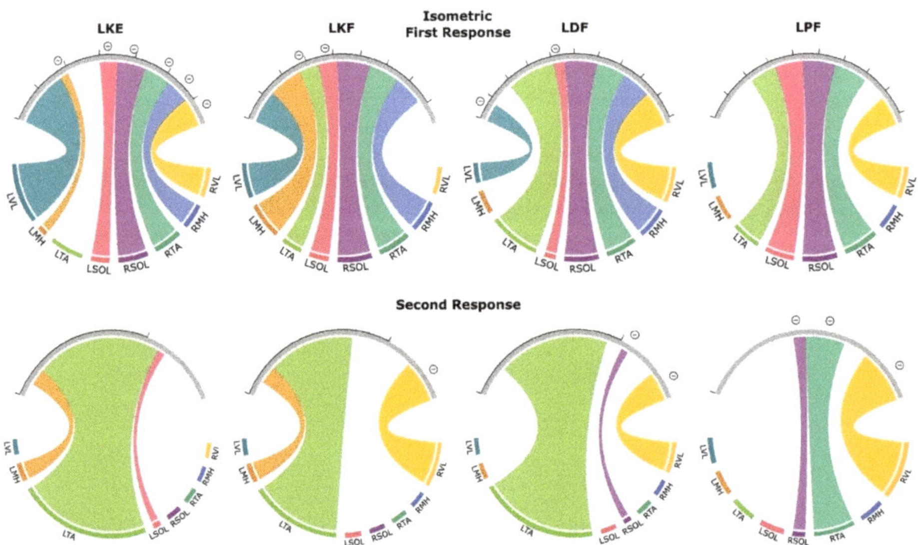

Figure 7. Excitability and inhibitory interactions during left knee extension (LKE), left knee flexion (LKF), left dorsiflexion (LDF) and left plantarflexion (LPF) isometric contraction: averaged significant ($p < 0.05$) facilitations or inhibitions during first (**top**) or second (**bottom**) responses during tactile cues. Each first response is scaled to the control condition where control is 100% and corresponds to the distance between tick marks. Second responses are given as the ratio between second and first response for that pair of responses. Individual muscles are grouped along the bottom of each plot. Thickness of connections denotes the amount of inhibition or facilitation found where significant inhibition is designated with a circled minus sign.

3.4.1. First Responses (R1s)

Isometric knee extension caused the facilitation of R1s in only the agonist muscle (LVL) (169% of control). This was lower ($p < 0.05$) when compared to the 200 ms CTI in the auditory condition (437% of control). The movement caused the inhibition of the antagonist (LMH) (28% of control) and the LSOL (85% of control). All muscles on the contralateral side were inhibited. The contralateral knee flexor (RMH) was most inhibited, averaging 71% of the control. The contralateral knee extensor (RVL) had an R1 of 84% of control, while RTA and RSOL averaged 83% and 85%, respectively.

During isometric knee flexion, the agonist (LMH) was facilitated (132% of control), which was reduced ($p < 0.05$) when compared to the 200 ms CTI in the auditory condition, where facilitation reached 218% of control. Two muscles were inhibited: the LTA (78%) and the LSOL (74%).

Isometric dorsiflexion caused facilitation in the agonist (LTA) (221% of control) but was lower ($p < 0.01$) than during the 200 ms CTI in the auditory condition (554% of control). The movement also caused inhibition in the antagonist (LSOL) (49% of control).

Isometric plantarflexion caused facilitation in the agonist (LSOL) (117% of control), which was reduced ($p < 0.01$) compared to the 200 ms CTI auditory condition (203% of control) (an example of this can be seen in Figure 1C). The movement also caused facilitation in the agonist (LTA) (118% of control).

3.4.2. Second Responses (R2s)

Isometric knee extension caused facilitation in the antagonist (LMH) with an average R2/R1 ratio of 27.8%, while the control ratio was 13.4%. During knee flexion, the agonist (LMH) facilitated with an average R2/R1 ratio of 21%, while the average control ratio was 10.3%. The contralateral antagonist (RVL) was inhibited, with a R2/R1 ratio decreasing to 32.1%, while the average control ratio was 59.5%. Isometric dorsiflexion caused facilitation in the agonist (LTA), which had an average R2/R1 ratio of 35.9%, while the control ratio was 28.6%. The R2/R1 ratio in the contralateral plantar flexor (RSOL) was decreased to 3.2% when compared to the control ratio which was 18.2%. During isometric plantarflexion the R2/R1 ratio in the contralateral plantar flexor (RSOL) decreased to 3.7%, while the control ratio was 27.4%. The contralateral dorsiflexor (RTA) was also inhibited with a R2/R1 ratio of 11.8%, while the ratio of control responses was 38.7%.

3.5. Probability Maps of Modulated Responses

To demonstrate temporal changes of the excitability of lumbosacral motor pools projecting to ipsilateral and contralateral knee and ankle flexors and extensors prior to and during the movements for the auditory, tactile, and isometric conditions, we created the probability maps of modulated R1s (Figure 8). Facilitation is shown using the muscle agonist color, and inhibition is shown using the muscle antagonist color.

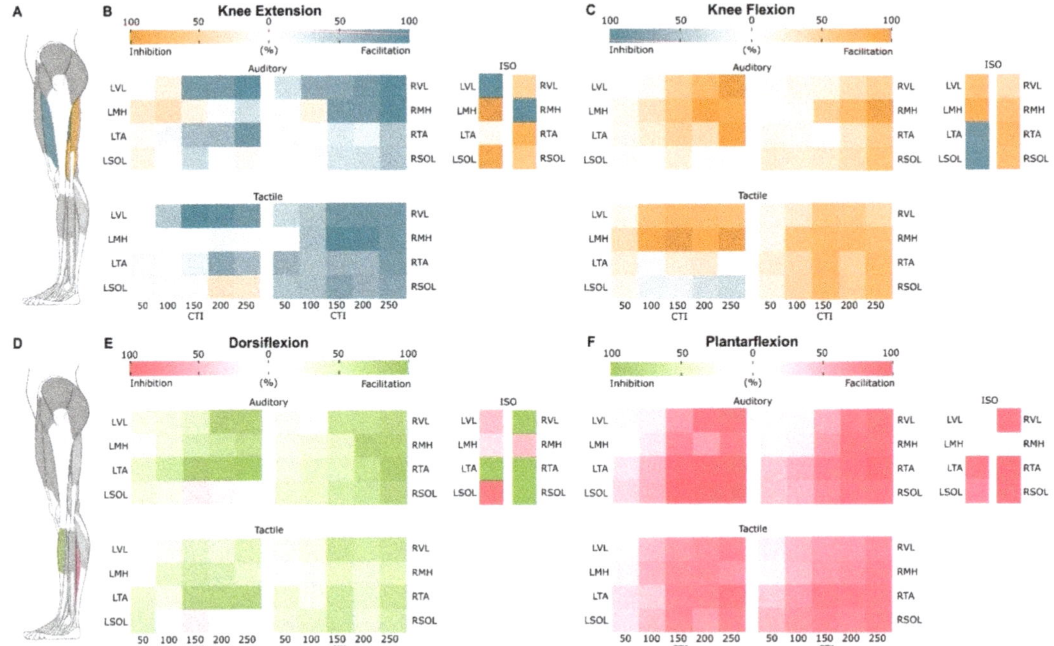

Figure 8. Motoneuronal excitability map: temporal changes in excitability during (**B**) left knee extension (LKE), (**C**) left knee flexion (LKF), (**E**) left dorsiflexion (LDF), and (**F**) left plantarflexion during auditory and tactile conditions, as well as isometric contraction. Each row represents one muscle, and each column represents one CTI. Facilitation and inhibition are denoted using the muscle agonist and antagonist colors, respectively (**A,D**). Note that a high probability of inhibition in antagonists during knee extension and dorsiflexion, as well as a high probability of co-facilitation in antagonists during plantarflexion and knee flexion, occurred across conditions.

3.5.1. Knee Extension

During the auditory condition, the probability of R1 facilitation increased for the agonist (LVL) as CTI increased until it reached 100% facilitation at the 250 ms CTI. The antagonist (LMH) had a higher chance of inhibition up to the 150 ms CTI, but at the 250 ms CTI there was a higher (21%) probability of facilitation. The contralateral knee extensor (RVL) at the 250 ms CTI had 100% probability of facilitation. The contralateral knee flexor (RMH) had a higher probability of facilitation with a likelihood of 100% at the 250 ms CTI. Only at the 100 ms CTI was there a higher probability (13%) of inhibition.

During the tactile condition, the probability of facilitation peaked at the 150 ms CTI for the agonist (LVL) with a value of 98% and 87% at the 250 ms CTI. The antagonist (LMH) had a higher probability of facilitation, but it was less than 10% for all CTIs. The probability for facilitation of the contralateral knee extensor (RVL) also peaked at the 150 ms CTI with an 88% chance of facilitation. The maximum likelihood for facilitation of the contralateral knee flexor (RMH) was 96% occurring at the 150 ms CTI.

The isometric contraction condition had the largest differences between the agonist and antagonist muscles. The agonist had (LVL) a 100% chance of facilitation, while the antagonist (LMH) had a 100% chance of inhibition. The contralateral knee extensor (RVL) had a higher probability (27%) of inhibition, and there was a 100% chance of facilitation of the contralateral knee flexor (RMH).

3.5.2. Knee Flexion

During the auditory condition, the probability of facilitation increased for the agonist (LMH) as the CTI increased with a maximum of 95% at the 200 ms CTI and was 88% at the 250 ms CTI. Facilitation increased for the antagonist (LVL) with a maximum of 97% at the 250 ms CTI. The contralateral knee flexor (RMH) had a higher probability of facilitation as CTI increased to a maximum at the 250 ms CTI with a 98% chance of facilitation. The contralateral knee extensor (RVL) had a maximum probability of facilitation of 65% at the 250 ms CTI.

During the tactile condition, the probability of facilitation of the agonist (LMH) reached a maximum of 88% at the 150 ms CTI. The antagonist (LVL) had a maximum of 57% probability for facilitation at the 150 ms CTI. The contralateral knee extensor (RVL) was found to have a higher probability (48%) of facilitation at the 150 ms CTI. The contralateral knee flexor (RMH) had a peak facilitation probability of 55% at the 150 ms CTI.

The isometric condition had a higher probability of facilitation in both the agonist (LMH) and the antagonist (LVL), with a likelihood of 48% and 65%, respectively. The contralateral knee flexor (RMH) and knee extensor (RVL) had a 53% and 32% probability for facilitation, respectively.

3.5.3. Dorsiflexion

During the auditory condition, the agonist (LTA) had an increased probability of facilitation with a maximum of 96% at the 250 ms CTI. The antagonist (LSOL) had a more dynamic likelihood profile, with a slight probability of facilitation at the 50 and 100 ms CTIs (12% and 8%, respectively), but the 150 ms CTI had a 34% probability of inhibition, which dropped to 18% and 3% for the 200 and 250 ms CTIs, respectively. The contralateral dorsiflexor (RTA) reached a maximum of 76% probability of facilitation at the 250 ms CTI. The contralateral plantar flexor (RSOL) had an increased probability of facilitation as CTI increased, with a maximum of 60% chance of facilitation at the 250 ms CTI.

The tactile condition resulted in a higher probability of facilitation of the agonist (LTA), with a maximum of 73% at the 250 ms CTI. The antagonist (LSOL) probability was similar to what was found during the auditory condition, where there was a 31% chance of facilitation at the 50 ms CTI and 36% chance of inhibition at the 150 ms CTI. Both the contralateral dorsiflexor (RTA) and plantar flexor (RSOL) reached a maximum of 31% and 33% probability for facilitation at the 150 ms CTI, respectively.

During isometric contraction condition, the agonist (LTA) had a 99% chance of facilitation while the antagonist (LSOL) had a 100% chance of inhibition. The contralateral dorsiflexor (RTA) had a 99% chance of facilitation, and the contralateral plantar flexor (RSOL) had an 89% chance of facilitation.

3.5.4. Plantarflexion

During the auditory condition, the probability of facilitation for the agonist (LSOL) increased as the CTI increased, with a maximum of 100% at the 200 and 250 ms CTIs. The antagonist (LTA) had a higher probability of facilitation as the CTI increased, with a maximum probability of 100% at the 250 ms CTI. The contralateral plantar flexor (RSOL) had a similar probability of response but had a maximum probability of facilitation at the 250 ms CTI of 86%, while the contralateral dorsiflexor (RTA) had a similar likelihood with a maximum of 87% probability of facilitation at the 250 ms CTI.

The tactile condition resulted in a maximum (74%) facilitation probability at the 150 ms CTI for the agonist (LSOL). The antagonist (LTA) had a higher probability of facilitation as the CTI increased to a maximum of 86% probability of facilitation at the 250 ms CTI. The contralateral plantar flexor (RSOL) had a higher probability of facilitation as the CTI increased, with a maximum of 78% at the 250 ms CTI. The contralateral dorsiflexor (RTA) also had a higher probability of facilitation as the CTI increased with a maximum of 77% probability for facilitation at the 250 ms CTI.

During isometric contraction, the probability of R1 facilitation for the agonist (LSOL) was 64%, while the antagonist (LTA) had a facilitation probability of 71%. The contralateral plantar flexor (RSOL) and dorsiflexor (RTA) were found to have a probability for facilitation of 98% and 100%, respectively.

4. Discussion

To better understand the descending modulation of spinal excitability prior to and during voluntary movements, we investigated the multi-segmental convergence of supraspinal inputs on the lumbosacral neural network. We demonstrated distinct changes in excitability of agonist and antagonist motor pools preceding voluntary movements, as well as during isometric contraction of the lower limb muscles, in neurologically intact individuals. Prior to the movements, the modulation of spinally evoked motor potentials was the most prominent in the primary agonists and antagonists, and this was also observed in other muscles. Across all muscles, the pattern of modulation of evoked responses was similar following either the auditory or tactile cues. During isometric contractions, the evoked responses were also modulated, but they were less pronounced and more agonist-antagonist specific. The second evoked responses (R2), which were delivered 50 ms following the first response (R1) and are inhibited at rest, were facilitated during the tested conditions and were the most prominent in the synergists. These results demonstrate generalized effects of the descending drive on lumbosacral motoneuronal and inter-neuronal circuitries.

4.1. Modulation of Spinally Evoked Motor Potentials Prior to the Movements

Our results demonstrate the net effects of the descending inputs on the excitability of ipsilateral and contralateral motor pools prior to movement, starting as early as 50 ms following an auditory or tactile cue. The corticospinal system is thought to be responsible for the regulation of excitability of agonist and antagonist muscles during initiation and maintenance of the voluntary motor tasks [42]. Prior studies have demonstrated the role of Ia inputs to motoneurons during movements, and they have shown that the presynaptic inhibition of Ia terminals can substantially reduce the size of monosynaptic Ia excitatory postsynaptic potentials in motoneurons [43]. These processes can also be modulated from motor centers in the brain [44]. In later research, Hultborn et al. (1987) confirmed the supraspinal origin of inhibition of the interneurons mediating presynaptic inhibition to Ia fibers terminating onto voluntarily discharging motoneurons. It has been also proposed that there is an inverse descending facilitation of the interneurons mediating presynaptic inhibition on Ia fibers to motoneurons innervating muscles that remain relaxed during a selective contraction [35]. The modulation of presynaptic inhibition seen in these experiments implies that there is a tonic presynaptic inhibition of Ia terminals at rest, controlled from supraspinal motor centers, and these descending influences are removed or affected following spinal lesion [39,45]. Thus, it seems likely that the facilitation we observed prior to the movements was the result of the descending downregulation of inhibitory interneurons (disinhibition), rather than the direct excitation of specific motor pools. This arrangement allows corticospinal projections to the lumbosacral spinal cord to function as a substrate for the coordinated and automatic control of the lower limbs with incorporation of ongoing afferent input, rather than directly relaying the cortical commands on specific motor pools during each individual movement (e.g., steps). The distinct physiological and functional role of the corticospinal projections in the modulation of the lumbar sensorimotor network was demonstrated in rodents [46]. Specifically, it has been proposed that, at lumbar level, the main role of the corticospinal tract is the modulation of sensory inputs, which is a critical component of the selective regulation of sensory feedback used to ensure well-coordinated movements of the hindlimbs. In contrast, in the forelimbs, the corticospinal tract can directly induce motor contractions, independently of its supraspinal collaterals, through spinal targets located at the cervical level. Similar distinction in the function of the corticospinal system projecting to the cervical and lumbar spinal cord has also been proposed in primates [47]. Although the organization of the corticospinal tract undoubtedly differs between species, our findings support the previous works and likely provide the basis to characterize the function and role of descending regulation of spinal sensorimotor networks.

4.2. Modulation of Spinally Evoked Motor Potentials during Voluntary Contractions

Previous studies have reported the inhibition of spinally evoked motor potentials during passive muscle stretching and the facilitation of the responses during voluntary upper limb muscle contractions [48]. Modulation of evoked potentials in the lower limb muscles during voluntary contractions has also been studied [37,49,50], and, overall, it demonstrates the facilitation of agonists and the inhibition of antagonists. At the same time, the interplay between the ipsilateral and contralateral agonists and antagonists remains unclear. For instance, Minassian et al. (2007a) demonstrated the inhibition of evoked potentials in TA, SOL, and MH during the dorsiflexion and facilitation of ipsilateral SOL and VL during plantarflexion; on the contralateral side, only VL was facilitated during both plantarflexion and dorsiflexion, and the responses from the rest of the muscles were unchanged compared to control. Hofstoetter et al. (2008) demonstrated the facilitation of evoked response responses in ipsilateral TA and VL and inhibition in SOL and MH during dorsiflexion; whereas changes during plantarflexion were not significant in any muscle (note, however, that the sample size was small in this study with $n = 3$). Saito et al. (2021) found inhibition in MH during dorsiflexion and facilitation in SOL during knee extension. In our experiments, during isometric knee extension and dorsiflexion, we observed facilitation in the agonists (i.e., in VL and TA, respectively) and inhibition of the antagonists (i.e., in MH and SOL, respectively); whereas, during isometric knee flexion and plantarflexion, both the agonist and antagonist were facilitated. On the contralateral side, isometric knee flexion, plantarflexion, and dorsiflexion resulted in the facilitation of the evoked responses in the majority of muscles, whereas isometric knee extension caused inhibition of the responses in all recorded contralateral muscles (Figure 7). The inconsistency between the previous and present studies can be attributed to the functional flexibility of the spinal circuitry, required to produce movements involving many different potential combinations of muscle contractions, as well as spinal mechanisms triggered by specific contracting muscles [50], including heteronymous recurrent inhibition originated from the agonists and affecting the circuitry of synergists [51], inter-segmental inhibitory force feedback [52,53], and presynaptic mechanisms [35]. Differences in the motor tasks, fatigue, lower limb joints' angle position, number of joints involved in the movement, and posture of the study participants in the above studies could contribute to the inconsistent findings. These considerations diminish the value of using maintained muscle contractions during spinal stimulation as the experimental paradigm to quantify the function and activity of supraspinal connections and spinal inter-neuronal network. In addition, motor tasks requiring the research subject to maintain individual muscle contractions may not be feasible for clinical populations with severe neurological impairments, such as SCI.

4.3. Probability of Modulated Responses in Agonists and Antagonists

Our results demonstrate temporal changes in the excitability of motoneurons prior to movement onset, where the probability of facilitation of the muscle agonist peaks at 150 to 200 ms following the auditory or tactile cue, which is about 100 to 50 ms prior to the onset of contraction. We found that a high probability of inhibition in antagonists occurred prior to and during dorsiflexion and knee extension, while a high probability of co-facilitation in antagonists occurred prior to and during plantarflexion and knee flexion (Figure 8). The difference in modulation of excitability in antagonists during different movements can be explained by the functional role of the muscles executing the movement. For instance, during standing, the anteriorly located dorsiflexors and knee extensors counteract posterior shifts of the body center of mass and provide fast and accurate postural adjustments, thus requiring more dynamic and "fine-tuned" coordination between the agonists and antagonists. Conversely, the posteriorly located plantar flexors and knee flexors continuously resist the gravity-induced torque pulling the body forward, and lower leg joint stability can "gain" from co-contracted antagonists [54–56]. Similarly, during stepping, co-facilitation of the ipsilateral agonist-antagonist couple during plantarflexion, occurring in the stance phase of the gait cycle, would increase ankle stiffness thus improving the joint stability. It is worth noting that the manifest difference in the recruitment rate of evoked responses of the anteriorly and posteriorly located leg muscles was also observed previously when examining the maximal slope of their recruitment curves [22], indicating different tonic level of presynaptic inhibition of Ia terminals and distinct neurophysiological properties of the circuitry in the leg muscles.

4.4. Modulation of Second Response Prior to and during the Movements

We sought to investigate post-activation depression in lumbosacral motor pools prior to and during movement initiation as another measurement of the functional state of corticospinal and spinal inter-neuronal networks in an intact spinal cord. We suggested that, in addition to being modulated as a function of the motor state, the dynamic of the R2 recovery in different muscles would differ between conditions and movements, given that a spinal stimulus results in bilateral, multi-segmental activation of lumbosacral roots. A reduction of the evoked potential amplitude, in a previously activated sensory-motor synapse attributed to post-activation- or homosynaptic-depression, is caused by a combination of presynaptic [57] and postsynaptic [58] inhibition mechanisms. Studies using the H-reflex demonstrated that repeated stimulation inhibits spinal motoneurons, likely because of their afterhyperpolarization [59], recurrent inhibition [60], and changes in the Ia afferent terminal leading to a transient reduction in neurotransmitter release [61–63]. Similar to the H-reflex, spinally evoked motor potentials exhibit post-activation depression in that the second pulse delivered within the next 20–2000 ms elicits a smaller response [39–41]. Post-activation depression can be reduced when the agonist muscle is voluntarily activated [40,64,65]. The mechanisms of this reduction are not well understood and are attributed to an increase in the neurotransmitter release due to increased frequency of impulses fired by muscle spindles in the contracting muscle [61].

We demonstrated, for the first time, the dynamics of the second response (R2) modulation in different leg muscles prior to and during voluntary movements. The most prominent changes in reduction of post-activation depression in our experiments were observed prior to the movement, when the R2/R1 ratio increased up to 51.4% at the CTIs between 100 and 200 ms in quiescent agonist muscles. These findings are coherent with previous observations that the ongoing descending drive decreases the amount of presynaptic inhibition of Ia fibers acting on spinal motoneurons [66,67].

4.5. Clinical Implications

Traditionally, rehabilitation clinicians have utilized subjective neuromotor assessments in a limited set of muscles for the determination of spared motor function, which is suggested to reflect the function of descending tracts and confer rehabilitative prognosis after SCI [68–71]. Among the descending tracts, the corticospinal system is a compelling target that is not only "neurophysiologically testable" [7,27,28,42] but is also one of the major biomarkers of neuromuscular recovery after SCI [72,73]. A key limitation of the existing clinical assessment tools is the inability to quantify residual corticospinal connectivity, whose influence on spinal networks is insufficient to produce muscle activation within the constraints of the examination [4–6,31,74,75]. It is remarkable that in our study, the observed multiplex changes of spinal motoneuronal excitability were occurring prior to the actual movement. This not only reflects the level of function of corticospinal and spinal inter-neuronal networks, but it also allows clinicians and researchers to perform objective neuromotor assessment even in the absence of visible voluntary contractions.

Furthermore, our findings on the interaction of the descending drive to the lumbosacral sensorimotor network and spinal stimulation suggest that disinhibition within spinal networks, occurring at the onset of voluntary efforts, may be critical in the context of the administration of neuromodulatory approaches using repetitive spinal stimulation administered to regain motor function in clinical populations. Currently, the dominating hypothesis in the field is that "artificial" afferent inputs, induced by epidural or non-invasive electrical stimulation of the spinal cord, can elevate the functional state of excitability of spinal networks, re-enabling or augmenting voluntary movement functions [8,14,16,23,27]. Conversely, it has been demonstrated that the repetitive electrical stimuli applied to the spinal cord at rest enhances intrinsic inhibitory mechanisms within the sensorimotor networks, as revealed by reduced spasticity following ESS [76,77] or TSS [78,79]. Based on these observations, one can propose that the excitatory or inhibitory effects of spinal stimulation on neuromotor outcomes depend on the presence or absence of descending drive to the spinal circuitry. Our data demonstrate that, prior to and during voluntary isometric contractions, the excitability of motor pools projecting to agonists and antagonists, as well as to the contralateral muscles, is, in general, elevated. Thus, it can be proposed it is the descending drive (residual in case of incomplete SCI) that in fact elevates the state of spinal inter-neuronal network excitability, while electrical spinal stimuli target specific populations of inter- and/or motoneurons and serve to regain given functions. Indeed, it is known that ESS and TSS demonstrate different patterns of specificity in activating selected motor pools [15,22,80–82], likely due to the different local and wide field stimulation capacity. To the best of our knowledge, every study utilizing ESS during voluntary effort after motor complete SCI has been successful so far in regaining muscle-specific control below the lesion in the presence of stimulation [14,15,82–84], while the fine and selective voluntary activation of specific agonists (with minimum co-contraction of antagonists) below the spinal lesion has yet to be shown with TSS. These observations may challenge or at least refine the current view on the enhancing vs. inhibitory effects of spinal stimulation on spinal excitability, if applied alone and without voluntary effort or activity. Whether our current findings change the existing activity-based neuromodulation approaches in neuro-restoration is a question for further research in clinical populations; however, this underlines the significance of proper and likely individual selection of the stimulation location, intensity, frequency, and, most importantly, timing between voluntary effort and spinal stimuli, to target specific movements or functions.

Changes in the functional state of excitability of spinal neuronal networks due to neuromodulation have been proposed as the primary mechanism of motor recovery following SCI [8,14,23,27,28]. The term "functional state" of spinal neuronal excitability can be reflective of the resting membrane potential of motoneurons, interneurons, as well as their responsiveness to afferent inputs and descending drive. Although all the above factors most likely play a role in neurophysiological and functional effects during neuromodulation, their quantitative characterization is required. Without that, attributing the effects to "changes in functional state" remains vague and difficult to reproduce in large sample size trials. We propose that the spatiotemporal characterization of excitability in response to descending commands provides the way to characterize the functional state of spinal sensorimotor networks. Investigation of the relationship between the degree of remaining supraspinal connections with spinal networks and the ability to voluntarily control muscles of the lower limbs in individuals with SCI will be the next step to quantify the residual and regained function throughout the course of neurorehabilitation. Finally, the demonstrated feasibility of evaluating spinally evoked motor potentials' modulation during voluntary movement attempts following the auditory and tactile cues offers opportunities to probe specific sensorimotor pathways in individuals with various neurological disorders and injuries, including SCI, stroke, multiple sclerosis, hereditary spastic paraplegia, and various forms of peripheral neuropathy.

5. Conclusions

We demonstrated the modulation of the excitatory state of spinal sensorimotor networks in response to descending commands during both the preparatory stage and execution of voluntary movements. Observed changes in spinal motoneuronal excitability during our experiments were likely triggered by presynaptic and postsynaptic mechanisms. Movement-specific modulation of motoneuronal disinhibition within spinal networks was prominent and consistent prior to the actual movements. Disinhibition of ipsilateral and contralateral agonists and antagonists indicate that the corticospinal projections to the lumbosacral spinal cord serve an essential, yet indirect, component of the selective tuning of motor output based on sensory feedback to ensure coordinated and patterned movement. During isometric muscle contractions, the afferent discharge occurring from the contracting muscles influences the regulation of inter-neuronal networks, and further modulates the agonist-antagonist activation. The spatiotemporal characterization of multi-segmental motoneuronal excitability in response to descending commands provides the way to characterize the functional state of spinal sensorimotor networks. The excitation of spinal sensorimotor networks caused by the descending drive can be further manipulated by spinal stimulation targeted to specific motor pools to enable a desired motor output.

Supplementary Materials: The following are available online at https://www.mdpi.com/article/10.3390/jcm10245958/s1, Data S1: Modulation of evoked responses during the auditory condition, Data S2: Modulation of evoked responses during the tactile condition, Data S3: Modulation of evoked responses during the isometric condition.

Author Contributions: Conceptualization, D.G.S.; Data curation, A.G.S., D.A.A., B.V., J.O., R.L.M. and D.G.S.; Formal analysis, A.G.S., B.V. and D.G.S.; Funding acquisition, D.G.S.; Investigation, D.G.S.; Methodology, D.A.A. and D.G.S.; Project administration, R.L.M.; Supervision, D.G.S.; Validation, D.A.A. and D.G.S.; Visualization, A.G.S., D.A.A. and D.G.S.; Writing—original draft, A.G.S. and D.G.S.; Writing—review & editing, A.G.S., D.A.A. and D.G.S. All authors have read and agreed to the published version of the manuscript.

Funding: Sources of funding for the work reported here include the National Institutes of Health grant 1 R01 NS119587-01A1, Craig H. Neilsen Research Grant (733278), and Wings for Life Foundation (227). In addition, this work was in part supported by philanthropic funding from Paula and Rusty Walter and the Walter Oil & Gas Corporation. The funders were not involved in the design of the study, the collection, analysis, and interpretation of the experimental data, the writing of this article, or the decision to submit this article for publication.

Institutional Review Board Statement: The study was conducted according to the guidelines of the Declaration of Helsinki, and approved by the Institutional Review Board of Houston Methodist Research Institute (Study ID: Pro00019704).

Informed Consent Statement: Written informed consent to the experimental procedures was obtained from all participants involved in the study.

Data Availability Statement: The data presented in this study are available on request from the corresponding author.

Conflicts of Interest: The authors declare no conflict of interest.

References

1. Dobkin, B.H.; Havton, L.A. Basic advances and new avenues in therapy of spinal cord injury. *Annu. Rev. Med.* **2004**, *55*, 255–282. [CrossRef] [PubMed]
2. Edgerton, V.R.; Roy, R.R. Physiology of motor deficits and the potential of motor recovery after a spinal cord injury. In *The Physiology of Exercise in Spinal Cord Injury*; Taylor, J.A., Ed.; Springer: Boston, MA, USA, 2016; pp. 13–35.
3. Kakulas, B.A. Pathology of spinal injuries. *Cent. Nerv. Syst. Trauma* **1984**, *1*, 117–129. [CrossRef] [PubMed]
4. Dimitrijevic, M.R. Residual motor functions in spinal cord injury. *Adv. Neurol.* **1988**, *47*, 138–155. [PubMed]
5. Sherwood, A.M.; Dimitrijevic, M.R.; McKay, W.B. Evidence of subclinical brain influence in clinically complete spinal cord injury: Discomplete sci. *J. Neurol. Sci.* **1992**, *110*, 90–98. [CrossRef]
6. McKay, W.B.; Lim, H.K.; Priebe, M.M.; Stokic, D.S.; Sherwood, A.M. Clinical neurophysiological assessment of residual motor control in post-spinal cord injury paralysis. *Neurorehabilit. Neural Repair* **2004**, *18*, 144–153. [CrossRef]
7. Mayr, W.; Krenn, M.; Dimitrijevic, M.R. Motor control of human spinal cord disconnected from the brain and under external movement. *Adv. Exp. Med. Biol.* **2016**, *957*, 159–171. [PubMed]
8. Taccola, G.; Sayenko, D.; Gad, P.; Gerasimenko, Y.; Edgerton, V.R. And yet it moves: Recovery of volitional control after spinal cord injury. *Prog. Neurobiol.* **2018**, *160*, 64–81. [CrossRef]

9. Edgerton, V.R.; Courtine, G.; Gerasimenko, Y.P.; Lavrov, I.; Ichiyama, R.M.; Fong, A.J.; Cai, L.L.; Otoshi, C.K.; Tillakaratne, N.J.; Burdick, J.W.; et al. Training locomotor networks. *Brain Res. Rev.* **2008**, *57*, 241–254. [CrossRef]
10. Fong, A.J.; Roy, R.R.; Ichiyama, R.M.; Lavrov, I.; Courtine, G.; Gerasimenko, Y.; Tai, Y.C.; Burdick, J.; Edgerton, V.R. Recovery of control of posture and locomotion after a spinal cord injury: Solutions staring us in the face. *Prog. Brain Res.* **2009**, *175*, 393–418.
11. Courtine, G.; Gerasimenko, Y.; van den Brand, R.; Yew, A.; Musienko, P.; Zhong, H.; Song, B.; Ao, Y.; Ichiyama, R.M.; Lavrov, I.; et al. Transformation of nonfunctional spinal circuits into functional states after the loss of brain input. *Nat. Neurosci.* **2009**, *12*, 1333–1342. [CrossRef] [PubMed]
12. Rossignol, S.; Frigon, A. Recovery of locomotion after spinal cord injury: Some facts and mechanisms. *Annu. Rev. Neurosci.* **2011**, *34*, 413–440. [CrossRef]
13. Angeli, C.A.; Boakye, M.; Morton, R.A.; Vogt, J.; Benton, K.; Chen, Y.; Ferreira, C.K.; Harkema, S.J. Recovery of over-ground walking after chronic motor complete spinal cord injury. *N. Engl. J. Med.* **2018**, *379*, 1244–1250. [CrossRef]
14. Gill, M.L.; Grahn, P.J.; Calvert, J.S.; Linde, M.B.; Lavrov, I.A.; Strommen, J.A.; Beck, L.A.; Sayenko, D.G.; Van Straaten, M.G.; Drubach, D.I.; et al. Neuromodulation of lumbosacral spinal networks enables independent stepping after complete paraplegia. *Nat. Med.* **2018**, *24*, 1677–1682. [CrossRef]
15. Wagner, F.B.; Mignardot, J.B.; Le Goff-Mignardot, C.G.; Demesmaeker, R.; Komi, S.; Capogrosso, M.; Rowald, A.; Seanez, I.; Caban, M.; Pirondini, E.; et al. Targeted neurotechnology restores walking in humans with spinal cord injury. *Nature* **2018**, *563*, 65–71. [CrossRef] [PubMed]
16. Gerasimenko, Y.P.; Lu, D.C.; Modaber, M.; Zdunowski, S.; Gad, P.; Sayenko, D.G.; Morikawa, E.; Haakana, P.; Ferguson, A.R.; Roy, R.R.; et al. Noninvasive reactivation of motor descending control after paralysis. *J. Neurotrauma* **2015**, *32*, 1968–1980. [CrossRef] [PubMed]
17. Sayenko, D.G.; Rath, M.; Ferguson, A.R.; Burdick, J.W.; Havton, L.A.; Edgerton, V.R.; Gerasimenko, Y.P. Self-assisted standing enabled by non-invasive spinal stimulation after spinal cord injury. *J. Neurotrauma* **2019**, *36*, 1435–1450. [CrossRef] [PubMed]
18. Rattay, F.; Minassian, K.; Dimitrijevic, M.R. Epidural electrical stimulation of posterior structures of the human lumbosacral cord: 2. Quantitative analysis by computer modeling. *Spinal Cord* **2000**, *38*, 473–489. [CrossRef]
19. Ladenbauer, J.; Minassian, K.; Hofstoetter, U.S.; Dimitrijevic, M.R.; Rattay, F. Stimulation of the human lumbar spinal cord with implanted and surface electrodes: A computer simulation study. *IEEE Trans. Neural Syst. Rehabil. Eng.* **2010**, *18*, 637–645. [CrossRef]
20. Minassian, K.; Jilge, B.; Rattay, F.; Pinter, M.M.; Binder, H.; Gerstenbrand, F.; Dimitrijevic, M.R. Stepping-like movements in humans with complete spinal cord injury induced by epidural stimulation of the lumbar cord: Electromyographic study of compound muscle action potentials. *Spinal Cord* **2004**, *42*, 401–416. [CrossRef]
21. Murg, M.; Binder, H.; Dimitrijevic, M.R. Epidural electric stimulation of posterior structures of the human lumbar spinal cord: 1. Muscle twitches—A functional method to define the site of stimulation. *Spinal Cord* **2000**, *38*, 394–402. [CrossRef] [PubMed]
22. Sayenko, D.G.; Atkinson, D.A.; Dy, C.J.; Gurley, K.M.; Smith, V.L.; Angeli, C.; Harkema, S.J.; Edgerton, V.R.; Gerasimenko, Y.P. Spinal segment-specific transcutaneous stimulation differentially shapes activation pattern among motor pools in humans. *J. Appl. Physiol.* **2015**, *118*, 1364–1374. [CrossRef]
23. Minassian, K.; McKay, W.B.; Binder, H.; Hofstoetter, U.S. Targeting lumbar spinal neural circuitry by epidural stimulation to restore motor function after spinal cord injury. *Neurotherapeutics* **2016**, *13*, 284–294. [CrossRef]
24. Minassian, K.; Persy, I.; Rattay, F.; Pinter, M.M.; Kern, H.; Dimitrijevic, M.R. Human lumbar cord circuitries can be activated by extrinsic tonic input to generate locomotor-like activity. *Hum. Mov. Sci.* **2007**, *26*, 275–295. [CrossRef] [PubMed]
25. Danner, S.M.; Hofstoetter, U.S.; Freundl, B.; Binder, H.; Mayr, W.; Rattay, F.; Minassian, K. Human spinal locomotor control is based on flexibly organized burst generators. *Brain* **2015**, *138*, 577–588. [CrossRef] [PubMed]
26. Hofstoetter, U.S.; Danner, S.M.; Freundl, B.; Binder, H.; Mayr, W.; Rattay, F.; Minassian, K. Periodic modulation of repetitively elicited monosynaptic reflexes of the human lumbosacral spinal cord. *J. Neurophysiol.* **2015**, *114*, 400–410. [CrossRef] [PubMed]
27. Harkema, S.; Gerasimenko, Y.; Hodes, J.; Burdick, J.; Angeli, C.; Chen, Y.; Ferreira, C.; Willhite, A.; Rejc, E.; Grossman, R.G.; et al. Effect of epidural stimulation of the lumbosacral spinal cord on voluntary movement, standing, and assisted stepping after motor complete paraplegia: A case study. *Lancet* **2011**, *377*, 1938–1947. [CrossRef]
28. Dimitrijevic, M.; Kakulas, B.; Vrbova, G. *Recent Achievements in Restorative Neurology: Progressive Neuromuscular Diseases*; Karger: New York, NY, USA, 1986.
29. Gerasimenko, Y.; Sayenko, D.; Gad, P.; Liu, C.T.; Tillakaratne, N.J.K.; Roy, R.R.; Kozlovskaya, I.; Edgerton, V.R. Feed-forwardness of spinal networks in posture and locomotion. *Neuroscientist* **2017**, *23*, 441–453. [CrossRef] [PubMed]
30. Sayenko, D.G.; Atkinson, D.A.; Mink, A.M.; Gurley, K.M.; Edgerton, V.R.; Harkema, S.J.; Gerasimenko, Y.P. Vestibulospinal and corticospinal modulation of lumbosacral network excitability in human subjects. *Front. Physiol.* **2018**, *9*, 1746. [CrossRef]
31. Atkinson, D.A.; Sayenko, D.G.; D'Amico, J.M.; Mink, A.; Lorenz, D.J.; Gerasimenko, Y.P.; Harkema, S. Interlimb conditioning of lumbosacral spinally evoked motor responses after spinal cord injury. *Clin. Neurophysiol.* **2020**, *131*, 1519–1532. [CrossRef]
32. Eichenberger, A.; Rüegg, D. Relation between the specific h reflex facilitation preceding a voluntary movement and movement parameters in man. *J. Physiol.* **1984**, *347*, 545–559. [CrossRef]
33. Gottlieb, G.L.; Agarwal, G.C.; Stark, L. Interactions between voluntary and postural mechanisms of human motor system. *J. Neurophysiol.* **1970**, *33*, 365–381. [CrossRef] [PubMed]

34. Pierrot-Deseilligny, E.; Lacert, P. Amplitude and variability of monosynaptic reflexes prior to various voluntary movements in normal and spastic man. In *Human Reflexes, Pathophysiology of Motor Systems, Methodology of Human Reflexes*; Karger Publishers: London, UK, 1973; Volume 3, pp. 538–549.

35. Hultborn, H.; Meunier, S.; Pierrot-Deseilligny, E.; Shindo, M. Changes in presynaptic inhibition of ia fibres at the onset of voluntary contraction in man. *J. Physiol.* **1987**, *389*, 757–772. [CrossRef]

36. Ivanchenko, Y.Z. Anticipatory changes in the soleus h reflex in humans related to the movements of the contralateral foot in upright stance. *Neurophysiology* **2014**, *46*, 134–138. [CrossRef]

37. Minassian, K.; Persy, I.; Rattay, F.; Dimitrijevic, M.R.; Hofer, C.; Kern, H. Posterior root-muscle reflexes elicited by transcutaneous stimulation of the human lumbosacral cord. *Muscle Nerve* **2007**, *35*, 327–336. [CrossRef] [PubMed]

38. Courtine, G.; Harkema, S.J.; Dy, C.J.; Gerasimenko, Y.P.; Dyhre-Poulsen, P. Modulation of multisegmental monosynaptic responses in a variety of leg muscles during walking and running in humans. *J. Physiol.* **2007**, *582*, 1125–1139. [CrossRef]

39. Hofstoetter, U.S.; Freundl, B.; Binder, H.; Minassian, K. Recovery cycles of posterior root-muscle reflexes evoked by transcutaneous spinal cord stimulation and of the h reflex in individuals with intact and injured spinal cord. *PLoS ONE* **2019**, *14*, e0227057. [CrossRef]

40. Andrews, J.C.; Stein, R.B.; Roy, F.D. Post-activation depression in the human soleus muscle using peripheral nerve and transcutaneous spinal stimulation. *Neurosci. Lett.* **2015**, *589*, 144–149. [CrossRef] [PubMed]

41. Roy, F.D.; Gibson, G.; Stein, R.B. Effect of percutaneous stimulation at different spinal levels on the activation of sensory and motor roots. *Exp. Brain Res.* **2012**, *223*, 281–289. [CrossRef] [PubMed]

42. McKay, W.B.; Lee, D.C.; Lim, H.K.; Holmes, S.A.; Sherwood, A.M. Neurophysiological examination of the corticospinal system and voluntary motor control in motor-incomplete human spinal cord injury. *Exp. Brain Res.* **2005**, *163*, 379–387. [CrossRef]

43. Eccles, J.C.; Eccles, R.M.; Magni, F. Central inhibitory action attributable to presynaptic depolarization produced by muscle afferent volleys. *J. Physiol.* **1961**, *159*, 147–166. [CrossRef] [PubMed]

44. Baldissera, F.; Hultborn, H.; Illert, M. Integration in spinal neuronal systems. *Compr. Physiol.* **2011**, *2*, 509–595.

45. Calancie, B.; Broton, J.G.; Klose, K.J.; Traad, M.; Difini, J.; Ayyar, D.R. Evidence that alterations in presynaptic inhibition contribute to segmental hypo- and hyperexcitability after spinal cord injury in man. *Electroencephalogr. Clin. Neurophysiol.* **1993**, *89*, 177–186. [CrossRef]

46. Moreno-Lopez, Y.; Bichara, C.; Delbecq, G.; Isope, P.; Cordero-Erausquin, M. The corticospinal tract primarily modulates sensory inputs in the mouse lumbar cord. *Elife* **2021**, *10*, e65304. [CrossRef] [PubMed]

47. Lemon, R. Recent advances in our understanding of the primate corticospinal system. *F1000Research* **2019**, *8*, 274–278. [CrossRef] [PubMed]

48. Milosevic, M.; Masugi, Y.; Sasaki, A.; Sayenko, D.G.; Nakazawa, K. On the reflex mechanisms of cervical transcutaneous spinal cord stimulation in human subjects. *J. Neurophysiol.* **2019**, *121*, 1672–1679. [CrossRef] [PubMed]

49. Hofstoetter, U.S.; Minassian, K.; Hofer, C.; Mayr, W.; Rattay, F.; Dimitrijevic, M.R. Modification of reflex responses to lumbar posterior root stimulation by motor tasks in healthy subjects. *Artif. Organs* **2008**, *32*, 644–648. [CrossRef] [PubMed]

50. Saito, A.; Nakagawa, K.; Masugi, Y.; Nakazawa, K. Intra-limb modulations of posterior root-muscle reflexes evoked from the lower-limb muscles during isometric voluntary contractions. *Exp. Brain Res.* **2021**, *239*, 3035–3043. [CrossRef]

51. Iles, J.; Pardoe, J. Changes in transmission in the pathway of heteronymous spinal recurrent inhibition from soleus to quadriceps motor neurons during movement in man. *Brain* **1999**, *122*, 1757–1764. [CrossRef] [PubMed]

52. Meunier, S.; Pierrot-Deseilligny, E.; Simonetta, M. Pattern of monosynaptic heteronymous ia connections in the human lower limb. *Exp. Brain Res.* **1993**, *96*, 534–544. [CrossRef] [PubMed]

53. Wilmink, R.J.; Nichols, T.R. Distribution of heterogenic reflexes among the quadriceps and triceps surae muscles of the cat hind limb. *J. Neurophysiol.* **2003**, *90*, 2310–2324. [CrossRef] [PubMed]

54. Koceja, D.M.; Trimble, M.H.; Earles, D.R. Inhibition of the soleus h-reflex in standing man. *Brain Res.* **1993**, *629*, 155–158. [CrossRef]

55. Masani, K.; Sayenko, D.G.; Vette, A.H. What triggers the continuous muscle activity during upright standing? *Gait Posture* **2013**, *37*, 72–77. [CrossRef] [PubMed]

56. Winter, D.A. *Biomechanics and Motor Control of Human Movement*, 4th ed.; Wiley: Hoboken, NJ, USA, 2009.

57. Crone, C.; Nielsen, J. Methodological implications of the post activation depression of the soleus h-reflex in man. *Exp. Brain Res.* **1989**, *78*, 28–32. [CrossRef] [PubMed]

58. Poon, D.E.; Roy, F.D.; Gorassini, M.A.; Stein, R.B. Interaction of paired cortical and peripheral nerve stimulation on human motor neurons. *Exp. Brain Res.* **2008**, *188*, 13–21. [CrossRef] [PubMed]

59. Matthews, P. Relationship of firing intervals of human motor units to the trajectory of post-spike after-hyperpolarization and synaptic noise. *J. Physiol.* **1996**, *492*, 597–628. [CrossRef] [PubMed]

60. Windhorst, U. On the role of recurrent inhibitory feedback in motor control. *Prog. Neurobiol.* **1996**, *49*, 517–587. [CrossRef]

61. Stein, R.B.; Thompson, A.K. Muscle reflexes in motion: How, what, and why? *Exerc. Sport Sci. Rev.* **2006**, *34*, 145–153. [CrossRef]

62. Hultborn, H.; Illert, M.; Nielsen, J.; Paul, A.; Ballegaard, M.; Wiese, H. On the mechanism of the post-activation depression of the h-reflex in human subjects. *Exp. Brain Res.* **1996**, *108*, 450–462. [CrossRef] [PubMed]

63. Katz, R.; Morin, C.; Pierrot-Deseilligny, E.; Hibino, R. Conditioning of h reflex by a preceding subthreshold tendon reflex stimulus. *J. Neurol. Neurosurg. Psychiatry* **1977**, *40*, 575–580. [CrossRef]

64. Burke, D.; Adams, R.W.; Skuse, N.F. The effects of voluntary contraction on the h reflex of human limb muscles. *Brain* **1989**, *112 Pt 2*, 417–433. [CrossRef]

65. Hultborn, H.; Nielsen, J.B. Modulation of transmitter release from ia afferents by their preceding activity—A "post-activation depression". In *Presynaptic Inhibition and Neural Control*; Oxford University Press: Oxford, UK, 1998; pp. 178–191.

66. Iles, J.F. Evidence for cutaneous and corticospinal modulation of presynaptic inhibition of ia afferents from the human lower limb. *J. Physiol.* **1996**, *491 Pt 1*, 197–207. [CrossRef]

67. Andrews, J.C.; Stein, R.B.; Roy, F.D. Reduced postactivation depression of soleus h reflex and root evoked potential after transcranial magnetic stimulation. *J. Neurophysiol.* **2015**, *114*, 485–492. [CrossRef] [PubMed]

68. Kirshblum, S.C.; Waring, W.; Biering-Sorensen, F.; Burns, S.P.; Johansen, M.; Schmidt-Read, M.; Donovan, W.; Graves, D.; Jha, A.; Jones, L.; et al. Reference for the 2011 revision of the international standards for neurological classification of spinal cord injury. *J. Spinal Cord Med.* **2011**, *34*, 547–554. [CrossRef] [PubMed]

69. Van Middendorp, J.J.; Hosman, A.J.; Donders, A.R.; Pouw, M.H.; Ditunno, J.F., Jr.; Curt, A.; Geurts, A.C.; Van de Meent, H. A clinical prediction rule for ambulation outcomes after traumatic spinal cord injury: A longitudinal cohort study. *Lancet* **2011**, *377*, 1004–1010. [CrossRef]

70. Zorner, B.; Blanckenhorn, W.U.; Dietz, V.; Curt, A. Clinical algorithm for improved prediction of ambulation and patient stratification after incomplete spinal cord injury. *J. Neurotrauma* **2010**, *27*, 241–252. [CrossRef] [PubMed]

71. Scivoletto, G.; Di Donna, V. Prediction of walking recovery after spinal cord injury. *Brain Res. Bull.* **2009**, *78*, 43–51. [CrossRef]

72. Freund, P.; Seif, M.; Weiskopf, N.; Friston, K.; Fehlings, M.G.; Thompson, A.J.; Curt, A. Mri in traumatic spinal cord injury: From clinical assessment to neuroimaging biomarkers. *Lancet Neurol.* **2019**, *18*, 1123–1135. [CrossRef]

73. Jo, H.J.; Perez, M.A. Corticospinal-motor neuronal plasticity promotes exercise-mediated recovery in humans with spinal cord injury. *Brain* **2020**, *143*, 1368–1382. [CrossRef]

74. Cioni, B.; Dimitrijevic, M.R.; McKay, W.B.; Sherwood, A.M. Voluntary supraspinal suppression of spinal reflex activity in paralyzed muscles of spinal cord injury patients. *Exp. Neurol.* **1986**, *93*, 574–583. [CrossRef]

75. Dimitrijevic, M.; McKay, W.; Sarjanovic, I.; Sherwood, A.; Svirtlit, L.; Vrbova, G. Co-activation of ipsi-and contralateral muscle groups during contraction of ankle dorsiflexors. *J. Neurol. Sci.* **1992**, *109*, 49–55. [CrossRef]

76. Barolat, G.; Myklebust, J.B.; Wenninger, W. Enhancement of voluntary motor function following spinal cord stimulation–case study. *Appl. Neurophysiol.* **1986**, *49*, 307–314. [CrossRef] [PubMed]

77. Pinter, M.; Gerstenbrand, F.; Dimitrijevic, M. Epidural electrical stimulation of posterior structures of the human lumbosacral cord: 3. Control of spasticity. *Spinal Cord* **2000**, *38*, 524–531. [CrossRef] [PubMed]

78. Hofstoetter, U.S.; McKay, W.B.; Tansey, K.E.; Mayr, W.; Kern, H.; Minassian, K. Modification of spasticity by transcutaneous spinal cord stimulation in individuals with incomplete spinal cord injury. *J. Spinal Cord Med.* **2014**, *37*, 202–211. [CrossRef] [PubMed]

79. Hofstoetter, U.S.; Freundl, B.; Danner, S.M.; Krenn, M.J.; Mayr, W.; Binder, H.; Minassian, K. Transcutaneous spinal cord stimulation induces temporary attenuation of spasticity in individuals with spinal cord injury. *J. Neurotrauma* **2020**, *37*, 481–493. [CrossRef]

80. Sayenko, D.G.; Angeli, C.; Harkema, S.J.; Edgerton, V.R.; Gerasimenko, Y.P. Neuromodulation of evoked muscle potentials induced by epidural spinal-cord stimulation in paralyzed individuals. *J. Neurophysiol.* **2014**, *111*, 1088–1099. [CrossRef] [PubMed]

81. Calvert, J.S.; Manson, G.A.; Grahn, P.J.; Sayenko, D.G. Preferential activation of spinal sensorimotor networks via lateralized transcutaneous spinal stimulation in neurologically intact humans. *J. Neurophysiol.* **2019**, *122*, 2111–2118. [CrossRef]

82. Grahn, P.J.; Lavrov, I.A.; Sayenko, D.G.; Van Straaten, M.G.; Gill, M.L.; Strommen, J.A.; Calvert, J.S.; Drubach, D.I.; Beck, L.A.; Linde, M.B.; et al. Enabling task-specific volitional motor functions via spinal cord neuromodulation in a human with paraplegia. *Mayo Clin. Proc.* **2017**, *92*, 544–554. [CrossRef] [PubMed]

83. Angeli, C.A.; Edgerton, V.R.; Gerasimenko, Y.P.; Harkema, S.J. Altering spinal cord excitability enables voluntary movements after chronic complete paralysis in humans. *Brain* **2014**, *137*, 1394–1409. [CrossRef]

84. Darrow, D.; Balser, D.; Netoff, T.I.; Krassioukov, A.; Phillips, A.; Parr, A.; Samadani, U. Epidural spinal cord stimulation facilitates immediate restoration of dormant motor and autonomic supraspinal pathways after chronic neurologically complete spinal cord injury. *J. Neurotrauma* **2019**, *36*, 2325–2336. [CrossRef] [PubMed]

Journal of
Clinical Medicine

Article

Voluntary Modulation of Evoked Responses Generated by Epidural and Transcutaneous Spinal Stimulation in Humans with Spinal Cord Injury

Jonathan S. Calvert [1], Megan L. Gill [2], Margaux B. Linde [2], Daniel D. Veith [2], Andrew R. Thoreson [2], Cesar Lopez [2], Kendall H. Lee [2,3,4], Yury P. Gerasimenko [5,6], Victor R. Edgerton [7,8,9,10,11,12], Igor A. Lavrov [3], Kristin D. Zhao [2,4], Peter J. Grahn [2,3] and Dimitry G. Sayenko [13,*]

[1] Mayo Clinic Graduate School of Biomedical Sciences, Mayo Clinic, Rochester, MN 55905, USA; jonathan_calvert@brown.edu
[2] Department of Physical Medicine and Rehabilitation, Mayo Clinic, Rochester, MN 55905, USA; gill.megan@mayo.edu (M.L.G.); linde.margaux@mayo.edu (M.B.L.); veith.daniel@mayo.edu (D.D.V.); thoreson.andrew@mayo.edu (A.R.T.); lopez.cesar@mayo.edu (C.L.); lee.kendall@mayo.edu (K.H.L.); zhao.kristin@mayo.edu (K.D.Z.); grahn.peter@mayo.edu (P.J.G.)
[3] Department of Neurologic Surgery, Mayo Clinic, Rochester, MN 55905, USA; lavrov.igor@mayo.edu
[4] Department of Physiology and Biomedical Engineering, Rochester, MN 55905, USA
[5] Pavlov Institute of Physiology of Russian Academy of Sciences, 199034 St. Petersburg, Russia; yury.gerasimenko@louisville.edu
[6] Department of Physiology and Biophysics, University of Louisville, Louisville, KY 40292, USA
[7] Department of Integrative Biology and Physiology, University of California Los Angeles, Los Angeles, CA 90095, USA; vre@ucla.edu
[8] Department of Neurobiology, University of California Los Angeles, Los Angeles, CA 90095, USA
[9] Department of Neurosurgery, University of California Los Angeles, Los Angeles, CA 90095, USA
[10] Brain Research Institute, University of California Los Angeles, Los Angeles, CA 90095, USA
[11] Institut Guttmann, Hospital de Neurorehabilitació, Institut Universitari Adscrit a la Universitat Autònoma de Barcelona, 08916 Badalona, Spain
[12] Centre for Neuroscience and Regenerative Medicine, Faculty of Science, University of Technology Sydney, Ultimo 2007, Australia
[13] Center for Neuroregeneration, Department of Neurosurgery, Houston Methodist Research Institute, Houston, TX 77030, USA
* Correspondence: dgsayenko@houstonmethodist.org; Tel.: +1-713-363-7949

Citation: Calvert, J.S.; Gill, M.L.; Linde, M.B.; Veith, D.D.; Thoreson, A.R.; Lopez, C.; Lee, K.H.; Gerasimenko, Y.P.; Edgerton, V.R.; Lavrov, I.A.; et al. Voluntary Modulation of Evoked Responses Generated by Epidural and Transcutaneous Spinal Stimulation in Humans with Spinal Cord Injury. *J. Clin. Med.* **2021**, *10*, 4898. https://doi.org/10.3390/jcm10214898

Academic Editors: Ursula S. Hofstoetter and Karen Minassian

Received: 12 September 2021
Accepted: 20 October 2021
Published: 24 October 2021

Publisher's Note: MDPI stays neutral with regard to jurisdictional claims in published maps and institutional affiliations.

Abstract: Transcutaneous (TSS) and epidural spinal stimulation (ESS) are electrophysiological techniques that have been used to investigate the interactions between exogenous electrical stimuli and spinal sensorimotor networks that integrate descending motor signals with afferent inputs from the periphery during motor tasks such as standing and stepping. Recently, pilot-phase clinical trials using ESS and TSS have demonstrated restoration of motor functions that were previously lost due to spinal cord injury (SCI). However, the spinal network interactions that occur in response to TSS or ESS pulses with spared descending connections across the site of SCI have yet to be characterized. Therefore, we examined the effects of delivering TSS or ESS pulses to the lumbosacral spinal cord in nine individuals with chronic SCI. During low-frequency stimulation, participants were instructed to relax or attempt maximum voluntary contraction to perform full leg flexion while supine. We observed similar lower-extremity neuromusculature activation during TSS and ESS when performed in the same participants while instructed to relax. Interestingly, when participants were instructed to attempt lower-extremity muscle contractions, both TSS- and ESS-evoked motor responses were significantly inhibited across all muscles. Participants with clinically complete SCI tested with ESS and participants with clinically incomplete SCI tested with TSS demonstrated greater ability to modulate evoked responses than participants with motor complete SCI tested with TSS, although this was not statistically significant due to a low number of subjects in each subgroup. These results suggest that descending commands combined with spinal stimulation may increase activity of inhibitory interneuronal circuitry within spinal sensorimotor networks in individuals with SCI, which may be relevant in the context of regaining functional motor outcomes.

Keywords: spinal cord injury; electrically evoked spinal motor potentials; spinal cord stimulation; neuromodulation

1. Introduction

Transcutaneous (TSS) and epidural spinal stimulation (ESS) are electrical neuromodulation approaches that have previously been used to modulate spinal sensorimotor networks in humans [1,2]. Both TSS and ESS have been shown to enable motor functions previously thought to be permanently lost in individuals with paraplegia due to spinal cord injury (SCI), such as voluntary movement of previously paralyzed limbs [3–8], standing [9–12], and stepping [13–15]. TSS and ESS are both hypothesized to increase the level of excitability below the injury level, allowing previously silent, intact neural tissue that remains following injury to access sensorimotor networks responsible for function below the injury [16,17]. TSS and ESS have been shown to recruit common neural structures in electrophysiological [18] and computational modeling studies [19,20]. However, the ability of individuals with SCI to modulate epidural and transcutaneous spinally evoked motor potentials has not been investigated in detail (Figure 1A).

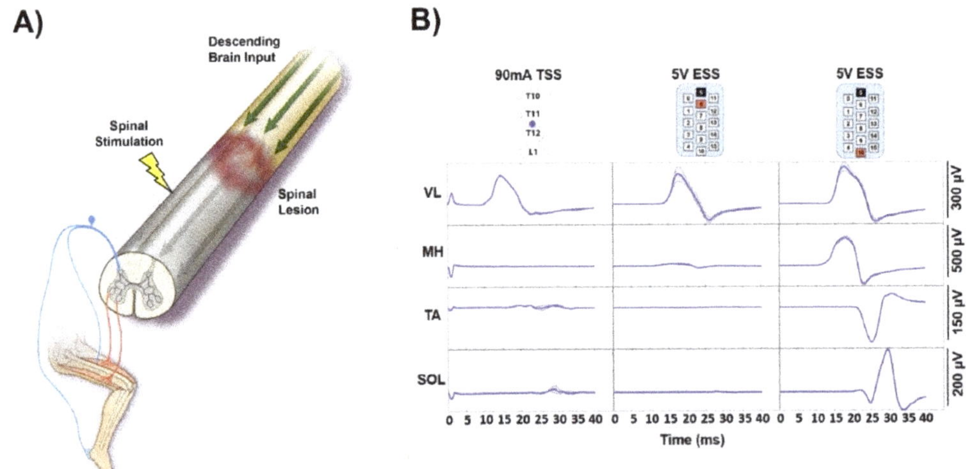

Figure 1. TSS- and ESS-Evoked Responses While Relaxed. (**A**). A diagram depicting inputs and outputs to the spinal cord during spinal stimulation. Descending brain input (green arrows) is interrupted by the spinal cord lesion. Spinal stimulation (yellow lightning bolt) is hypothesized to function by activating the dorsal roots carrying afferent proprioceptive information to the spinal cord. Afferent proprioceptive inputs (blue) enter the spinal cord and efferent motor outputs (red) exit the spinal cord and returns to the muscle. This figure is adapted with permission from a previous publication [6]. (**B**). While a study participant was instructed to relax while lying supine, stimulation was delivered to the same region of the spinal cord via transcutaneous spinal stimulation (TSS) and epidural spinal stimulation (ESS) using a focal and wide field. The dark line represents the average of at least three stimuli, and the shaded region indicates the ± standard deviation. VL—vastus lateralis, MH—medial hamstrings, TA—tibialis anterior, SOL—soleus, µV—microvolt, V—Volt, and mA—milliamp.

Previous reports of ESS and TSS have investigated spinally evoked responses via electromyography (EMG) of upper [21,22] and lower-extremity [4,23,24] musculature to characterize the effect of electrode location, different stimulation parameters, and body position on the motor thresholds and gain properties of sensorimotor networks. In these studies, stimulation was applied at low frequency ranges (0.2–2 Hz) in order to evaluate sensorimotor output while minimizing the effects of post-activation depression from frequent stimulation [25]. Previous reports indicate that some study participants, clinically diagnosed as having a motor complete SCI, were able to show signs of a non-specific, generalized increase in EMG activity below their injury level when asked to perform a full body muscle contraction by maximally flexing the muscles rostral to the SCI [7,13,26]. This has brought renewed focus to discomplete injuries, where study participants demonstrate motor activity via EMG in specific reinforcement tasks, despite being clinically classified in the ASIA (American Spinal Injury Association) Impairment Scale (AIS) as having a motor complete SCI [26,27]. Study participants without a SCI have demonstrated increased, as well as decreased, amplitude of TSS-evoked responses in some muscles during voluntary tasks [21,28,29]. However, the effect of voluntary control in individuals with SCI over TSS- or ESS-evoked responses has yet to be examined.

Here, we investigated the effect of voluntary control on TSS- and ESS-evoked responses in individuals with SCI at a range of injury severities. Participants were tested in two different conditions while supine: relaxed and while attempting maximal voluntary flexion of the lower extremities. During these tasks, spinally evoked motor potentials were recorded via EMG from the lower extremities. As previous work has demonstrated that individuals with SCI can increase the amplitude of EMG recordings taken from below the SCI, we hypothesized that voluntary attempts would increase spinally evoked response amplitude when compared to the relaxed condition.

2. Methods

2.1. Description of Participants

The experimental procedures described herein were approved by the respective University of California, Los Angeles (UCLA) and Mayo Clinic institutional review boards, and study participants provided written, informed consent to the experimental procedures. Data from two independent investigations were retrospectively analyzed via collaborative efforts from investigators at both institutions. Experiments were conducted in nine participants (seven at UCLA, two at Mayo Clinic) with chronic SCI (see Table 1 for full demographics). Study participants sustained an SCI at least two years prior to study enrollment. Two study participants were part of a study at the Mayo Clinic whose functional motor responses have previously been reported [4,7,13,30,31]. These publications focused on motor outputs during functional tasks such as voluntary control of lower-extremity muscles, stepping, standing, and sitting [7,13,30,31], as well as intraoperative recordings [4]. All data and analyses from these participants in this report were recorded at low (0.2–2 Hz) non-functional stimulation frequencies while the subjects were supine. All data contained within this manuscript have not previously been published. Briefly, these study participants performed six months of task-specific training, including body weight supported treadmill and over ground training without stimulation. At the initiation and conclusion of these six months, TSS was applied at the T10-L1 spinal vertebral levels to assess the sensorimotor connectivity of the lower-extremity musculature and spinally evoked motor responses prior to implantation of the epidural stimulator. Following these six months, participants were implanted with an epidural stimulator (Specify 5-6-5, Medtronic, Fridley, MN, USA) [4] and performed 12 months of multi-modal rehabilitation which paired task-specific rehabilitation with ESS [7]. The other seven participants were part of a study on the effects of TSS on trunk stability and self-assisted standing at the University of California, Los Angeles [12,32]. However, all data and analysis in this report are unpublished, and the study participants did not receive spinal stimulation prior to study enrollment.

Table 1. Study Participant Demographics.

Subject ID	Sex	Age	Injury Level	Time Since Injury	AIS Score	Stimulation Modality
N01	Male	26	T6	3 years	A	ESS, TSS
N02	Male	36	T3	6 years	A	ESS, TSS
N03	Male	22	C5	5 years	B	TSS
N04	Male	26	T2	8 years	A	TSS
N05	Female	32	C5	13 years	C	TSS
N06	Male	23	T2	4 years	A	TSS
N07	Male	25	T4	7 years	A	TSS
N08	Male	26	C4	7 years	C	TSS
N09	Male	28	T4	2 years	C	TSS

This table depicts the demographics of the study participants including their study ID, sex, age, injury level, time since injury, AIS (American Spinal Injury Association Impairment Score), and stimulation modality. ESS—epidural spinal stimulation; TSS—transcutaneous spinal stimulation.

2.2. Data Acquisition

Surface electromyogram (EMG) signals were recorded using bipolar self-adhesive electrodes placed longitudinally over the muscle belly of the vastus lateralis (VL), medial hamstrings (MH), tibialis anterior (TA), and soleus (SOL) muscles of each leg. Signals were differentially amplified and digitized at a sampling rate of 4000 samples per second (PowerLab, ADInstruments, Dunedin, New Zealand) and stored electronically (LabChart, ADInstruments, Dunedin, New Zealand). EMG data were analyzed offline using custom code written in MATLAB (Version R2020a, The Mathworks Inc., Natick, MA, USA) following application of a notch filter at 60 Hz and a 2nd order bandpass filter between 10 and 1000 Hz. All EMG recordings were synchronized to each pulse of TSS or ESS via stimulus artifact recorded from an electrode placed on the surface of the thoracolumbar spine.

Study participants were instructed to perform two experimental tasks with and without spinal stimulation: (1) to stay relaxed while lying supine to establish a control condition, and (2) to put forth maximum effort in attempting a single leg flexion maneuver including hip flexion, knee flexion, and ankle dorsiflexion simultaneously. A subset of subjects was also asked to perform joint-specific movements (e.g., plantarflexion, dorsiflexion) in the presence of stimulation. Each task was performed for at least three trials in each leg by each participant. During voluntary tasks, stimulation was delivered at a global motor threshold, which was defined as the stimulation amplitude where the peak-to-peak amplitude of all recorded muscles exceeded 20 µV responses.

2.3. Stimulation Procedures

Transcutaneous spinal stimulation was delivered either using a DS7A Biphasic Constant Current Stimulator (Digitimer, Hertfordshire, UK) or a custom-built, three channel constant-current stimulator. Stimulation was administered via self-adhesive electrodes (PALS, Axelgaard Manufacturing Co., Ltd., Fallbrook, CA, USA) with a diameter of 3.2 cm placed on the skin at the spinal midline between spinous processes from the T11 to L2 vertebrae to act as cathodes. Two 5 cm × 10 cm self-adhesive electrodes (PALS, Axelgaard Manufacturing Co., Ltd., Fallbrook, CA, USA) were placed symmetrically on the skin longitudinally over the abdomen for use as anodes. During TSS, stimuli were delivered as monophasic rectangular pulses with a 1 ms pulse width. Stimuli were delivered at 0–150 mA at stimulation frequencies between 0.2 and 2 Hz. A minimum of three stimuli were delivered during each trial.

Epidural spinal stimulation (ESS) was delivered using an implantable spinal cord stimulator (Specify 5-6-5, Medtronic, Fridley, MN, USA) placed between the T11-L1 vertebral bodies connected to an implanted pulse generator (RestoreSensor Sure-Scan MRI, Medtronic, Fridley, MN, USA). During ESS, stimuli were delivered as biphasic charge-balanced rectangular pulses with a 0.21 ms pulse width at a frequency of 0.2–2 Hz. Each electrode could be configured as a cathode, anode, or off. The electrode configurations were defined empirically based on the motor outputs of each subject, and were used to target specific rostral-caudal locations of the spinal cord that would enable either specific motor activation of proximal or distal lower-extremity musculature, or non-specific activation of multiple muscles of the lower extremity. ESS-evoked motor response recordings were captured during multiple ESS configurations and stimulation parameters with wide or local current distributions at the rostral and caudal ends of the electrode array (0–10 V).

2.4. Data Processing and Statistics

Mean and standard deviation values were calculated from at least 3 consecutive stimuli. Magnitudes of the spinally evoked potentials were calculated by measuring the area under the curve by applying a trapezoidal numerical integration to rectified EMG signals from 5 to 45 ms after the stimulus to capture the entire evoked response and prevent stimulation artifact contaminating the EMG signal. The evoked responses during voluntary contraction were normalized to the response in each muscle during the relaxed condition to account for individual differences during EMG collection in each participant. Statistically significant differences across the entire population of subjects were determined using the Wilcoxon signed-rank test for all EMG data ($p < 0.05$) using the signrank function in MATLAB, as the data were not normally distributed. The data used for the statistical tests were calculated by taking the average normalized area under the curve value of the first three evoked responses for each of the 9 subjects within the study population. After the average value was obtained for each participant, these data were entered into the signrank function to calculate the p-values for each recorded muscle. The paired, two-sided Wilcoxon signed-rank test was chosen over the Wilcoxon rank-sum test, as the data were from matched samples. However, comparisons across population subgroups did not have a large enough sample size to confirm statistical significance. Raw and processed datasets are available from the corresponding author upon request.

3. Results

3.1. Epidural and Transcutaneous Spinal Stimulation in the Same Participants

When stimulation was delivered at similar intensities at different electrode configurations, TSS applied at the T11/T12 intervertebral location and ESS applied at a focal, rostral portion of the electrode array (−5/+6) resulted in distinct evoked responses in the VL with relatively little activation in the other recorded muscles (MH, TA, SOL) (Figure 1B). When ESS was set with a wide field configuration (−5/+10) at the same stimulation intensity, all recorded muscles (VL, MH, TA, SOL) were activated.

3.2. Effect of Voluntary Effort on Spinally Evoked Responses

As shown in a representative ESS study participant and a representative TSS study participant, stimulation at motor threshold resulted in evoked responses in the leg muscles while the participants were relaxed (Figure 2A). However, when the participants were instructed to perform a full leg flexion, lower-extremity muscle responses were decreased compared to the relaxed condition. The data were normalized to compare across all participants, and the average area under the curve of the first three evoked responses was calculated for each of the nine study participants. When compared across the entire study population, the average area under the curve of the evoked responses was significantly lower across all recorded EMG muscles during the voluntary attempts to perform the leg flexion compared to the relaxed condition (mean ± standard error, p-value; VL: 0.6801 ± 0.1110, $p = 0.0117$; MH: 0.7084 ± 0.1157, $p = 0.0391$; TA: 0.6208 ± 0.1327, $p = 0.0391$; SOL: 0.4545 ± 0.1048, $p = 0.0039$) (Figure 2B). Furthermore, a representative subject who was asked to perform joint-specific movements (i.e., plantarflexion and dorsiflexion) demonstrated inhibition of the evoked potentials across muscles on both sides of the body during both plantarflexion and dorsiflexion (Figure 3).

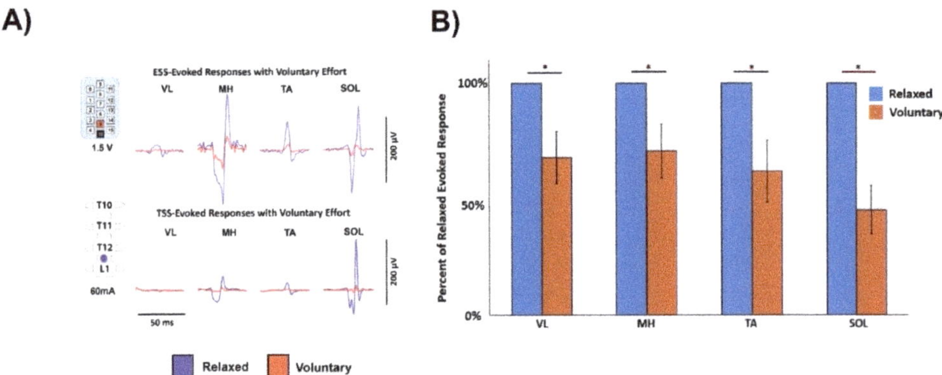

Figure 2. Inhibition of Evoked Response Amplitude During Voluntary Flexion. (**A**). Data from a representative study participant using ESS and a representative study participant using TSS while relaxed and while attempting maximum voluntary flexion of the lower extremities, which results in a decreased evoked response. Stimulation is delivered at the beginning of each trace. Blue indicates the relaxed condition and red indicates the voluntary flexion condition. (**B**). Grouped data from all participants within this study indicating significant decreases across all four recorded muscles when the voluntary flexion condition is compared to the relaxed condition. Data are normalized to the maximum EMG response in each muscle in each participant to compare across participants. Error bars represent the mean ± standard error. VL—vastus lateralis, MH—medial hamstrings, TA—tibialis anterior, SOL—soleus, µV—microvolt, V—Volt, mA—milliamp, and *—<0.05.

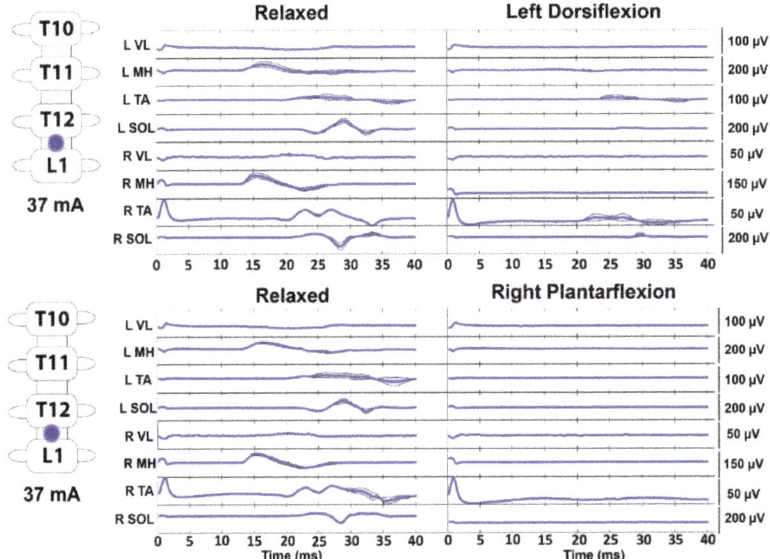

Figure 3. Joint-Specific Movements Decrease Motor-Evoked Responses. During T12/L1 stimulation in a representative study participant, motor-evoked responses were decreased across both the left and right lower extremities during attempts to voluntarily flex the ankle. Stimulation is delivered at the beginning of each trace. The dark line represents the average of at least three stimuli, and the shaded region indicates the ± standard deviation. VL—vastus lateralis, MH—medial hamstrings, TA—tibialis anterior, SOL—soleus, µV—microvolt, V—Volt, and mA—milliamp.

3.3. *Effect of Stimulation Modality and Injury Severity on Voluntary Modulation of Evoked Responses*

To examine if stimulation modality and injury severity had an effect on the ability to modulate the evoked responses, study participants were stratified into three groups: ESS with participants diagnosed with an AIS-A SCI, TSS with AIS-A SCI, and TSS with AIS-B/C SCI. When the evoked responses were averaged across the entire voluntary contraction, both participants with AIS-A tested with ESS decreased the amplitude of their evoked responses when instructed to perform a full leg flexion (Figure 4). All participants tested with TSS were exposed to stimulation with the cathode positioned between the T12-L1 vertebral bodies. Both ESS participants used a symmetric 9+/10− configuration. In all three AIS-A participants tested with TSS, the amplitude of the evoked responses in at least 3 out of 4 of the recorded muscles did not fall outside of the standard deviation of the normalized relaxed value. However, all four AIS-B/C participants tested with TSS demonstrated a reduction in the evoked responses amplitude compared to the normalized relaxed value in at least 3 out of 4 of the recorded muscles. However, statistical comparisons across subgroups could not be made due to the low number of study participants in each subgroup.

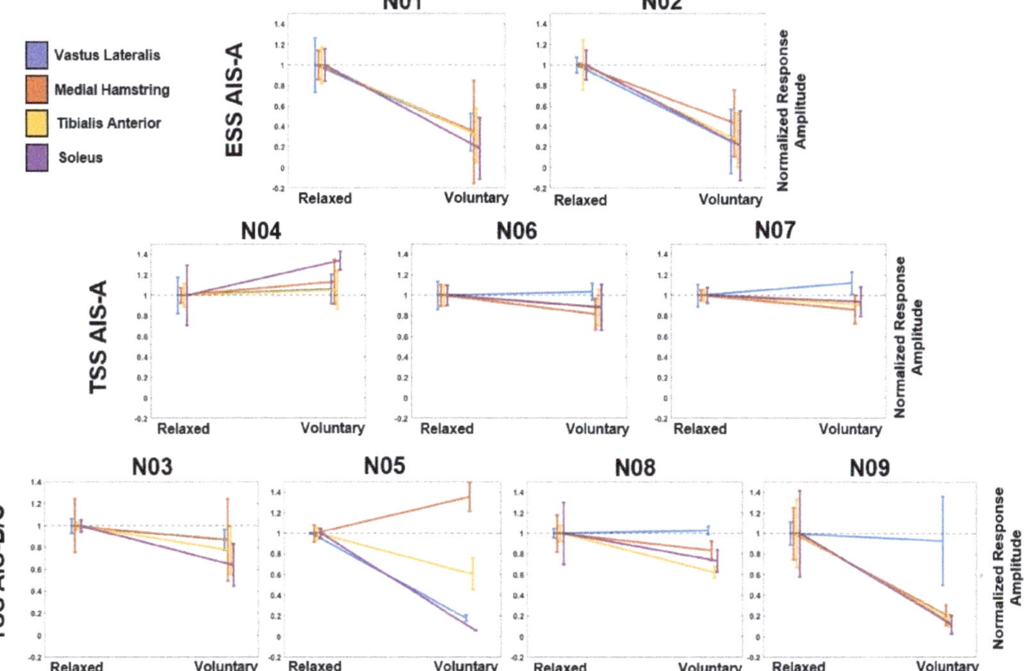

Figure 4. Evoked Response Modulation by ASIA Impairment Score. The first row indicates the two participants with clinically complete SCI tested with ESS. The second row indicates the three participants with clinically complete SCI tested with TSS. The third row indicates the four participants with clinically incomplete SCI tested with TSS. Data on the left of each plot refer to the average evoked response during the relaxed condition, and the data on the right refers to the average evoked response during the voluntary flexion condition. Data are normalized to the maximum EMG response in each muscle in each participant to compare across participants. The black dashed line indicates the average response of each muscle during the relaxed condition. Error bars represent the mean ± standard deviation. AIS—American Spinal Injury Association Impairment Score, VL—vastus lateralis (blue), MH—medial hamstrings (orange), TA—tibialis anterior (yellow), and SOL—soleus (purple).

4. Discussion

ESS and TSS have demonstrated improvements across a wide range of functions in individuals with SCI [3,4,6,9,11–15,32,33]. However, the complex interactions between stimulus pulses, the descending commands originating above the SCI and passing through the lesion site, and afferent inputs during movements to produce the functional spinal sensorimotor network outputs remain poorly understood. Here, we demonstrate the inhibition of evoked responses from ESS and TSS during voluntary attempts of individuals with severe SCI to move paralyzed limbs while lying supine.

In study participants who were stimulated with both TSS and ESS, similar evoked muscle responses were observed when the subjects were instructed to relax (Figure 1). ESS and TSS have previously been shown to activate common neural structures in electrophysiological studies [18]. Furthermore, ESS and TSS have both been shown to preferentially activate rostral-caudal and medio-lateral spinal motor pools [34–37], and both modalities are proposed to function, in part, through activation of dorsal roots entering the spinal cord [19,20,38]. However, it remains unknown what degree of specificity in activation of particular motor pools is necessary to achieve a given level of functional restoration of movement. It can be reasoned that either a specific or a broad activation pattern may be useful in engaging sensorimotor circuitry necessary for different functional tasks. Further studies are needed to demonstrate functional differences between TSS and ESS within the same individuals to effectively evaluate the advantages and disadvantages between these two modalities which may aid in choosing which strategy best fits a given individual's injury profile and goals. Based on the currently published data, the option to choose a modality will likely result in the most desirable patient-specific outcome.

Interestingly, when study participants were asked to voluntarily contract their lower limbs while stimulation was being delivered above motor threshold, the responses were inhibited (Figure 2). Furthermore, during joint-specific contractions, subjects inhibited all the recorded muscles bilaterally (Figure 3). Previous results using TSS in individuals without an SCI have indicated inhibition of responses during passive muscle stretching and muscle-tendon vibration, and facilitation of responses during voluntary muscle contraction [21,39]. Additionally, in previous TSS studies in individuals without an SCI, agonist lower-extremity muscle EMG responses were increased and antagonistic muscle responses were decreased while attempting voluntary movement [24,28]. Within our cohort of study participants with a severe SCI, it is possible that post SCI reorganization in sensorimotor mapping has altered electrophysiological outputs resulting in simultaneous activation and reciprocal inhibition of agonist and antagonistic muscles during voluntary attempts at leg flexion and joint-specific movement [40]. Interestingly, individuals with chronic SCI typically exhibit increased excitability as evidenced by spasticity and hyperreflexia following the period of areflexia and spinal shock immediately following injury [41]. Therefore, current treatments to address spasticity include pharmacological agents that are used to reduce the excitability of the spinal cord, such as baclofen [42]. Physical treatments such as stretching, range of motion exercises, and voluntary contraction in individuals with incomplete SCI have shown improvements in spasticity, likely from enhanced activation of spinal inhibitory pathways [43]. Therefore, the present data align with the concept of increased inhibitory responses during physical tasks as well as data using TSS to attenuate spasticity in individuals with SCI, which was hypothesized to work through pre-synaptic and/or post-synaptic pathways [33]. It is noteworthy that previous results have shown bilateral facilitation of evoked responses during TSS when paired with transcranial magnetic stimulation (TMS) or galvanic vestibular stimulation (GVS), which activate the corticospinal and vestibulospinal tracts, respectively [44–47]. However, the present data suggest that stimulation of spinal cord circuitry combined with ongoing voluntary commands through remaining neural pathways crossing the lesion can inhibit spinally evoked motor responses.

Furthermore, when study participants were stratified according to the stimulation modality that was used and their injury severity as measured by their AIS classification, different patterns of evoked potential modulation emerged. AIS-A participants were able to inhibit responses across all measured muscles in ESS; however, AIS-A participants tested with TSS did not demonstrate similar results. Interestingly, participants who were classified as clinically incomplete (AIS-B/C) could inhibit the responses in at least 3 out of 4 recorded muscles (Figure 4). However, these results could not be shown to be statistically significant due to the low number of subjects in each subgroup. Previous studies have indicated that study participants with motor complete or incomplete injuries could regain voluntary motor function while using ESS [3]. Additionally, previous studies have indicated that healthy individuals [28,48] and individuals with SCI [49] could modulate TSS-evoked responses during functional tasks. However, in this study, we analyze the effect of voluntary effort on evoked response amplitude in participants with both clinically complete and incomplete SCI. These results suggest that individuals with less severe injury may be able to exert greater modulation on evoked responses recorded at motor threshold in the lower extremity. However, these findings are in a small cohort of participants and further work needs to be done to understand how remaining spinal cord fiber composition may affect lower-extremity function when paired with neuromodulation therapies. Recent mechanistic studies have suggested that the recovery of function following SCI can be attributed to propriospinal [50,51] and reorganization of cortico-reticulo-spinal tracts [52]. Additionally, motor-evoked responses and muscles activated can be modulated based on the timing that the pulse is delivered within a movement in humans and animals with SCI, which may contribute to the findings presented here as the subjects remained in the supine position continuously attempting flexion across multiple joints [49,53]. Therefore, future work should focus on the role of effort at different stages from preparation to execution of the movement and identifying the contributions of different spinal tracts to the recovery of function within the SCI population.

SCI is a heterogeneous population and results may differ depending on location and severity of injury, time since injury, and age of participant, therefore, further studies into the voluntary modulation of TSS- and ESS-evoked responses across clinical diagnoses are warranted. All of our experiments used low-frequency (0.2–2 Hz) stimulation in order to evaluate the effects of stimulation and voluntary effort without post-activation depression due to frequent stimulation. However, recent studies demonstrating return of function with spinal stimulation in individuals with severe paralysis have been at higher frequencies [3,7,13,14], and the motor output during stimulation of the spinal networks at higher or lower frequency can be dramatically different, quantitatively and qualitatively [16]. It is plausible that during higher frequencies (e.g., above 25 Hz) of spinal stimulation, the excitation predominates inhibition [54], which results in voluntary movements in the presence of spinal stimulation. Additionally, recent results using TSS have indicated that repeated exposure to stimulation may increase motoneuron output [55]. All participants within this study were not trained to perform the task, and therefore may exhibit different results when part of a long-term study. Additionally, the stimulation paradigm used within TSS for this study was composed of monophasic pulses, whereas ESS was delivered using biphasic pulses. Furthermore, the global motor threshold for this study was intentionally set at a low value of 20 µV; to observe supra-motor threshold responses of motor pools projecting to different muscles, which due to their multi-segmental origin are expected to have different thresholds. This low threshold value may have affected the ability of the subjects to modulate the responses, and further work should be performed to elucidate the effect of voluntary effort on spinally evoked responses at a range of different stimulation intensities. Lastly, the results we report were generated while study participants were positioned supine; however, body positioning influences recruitment of neural structures during spinal stimulation, and future work should evaluate the effect of voluntary intent during different body positions and tasks [23,56].

5. Conclusions

In the present study, we found that individuals with severe SCI could modulate EMG outputs in their lower extremity, below their level of injury in the presence of spinal stimulation. During stimulation, both TSS and ESS pulses could elicit responses in lower-extremity musculature. Importantly, with low-frequency stimulation at motor threshold, both epidural and transcutaneous spinally evoked motor responses were inhibited, when participants voluntarily attempted to activate their lower-extremity muscles. However, study participants with clinically complete SCI using ESS and participants with clinically incomplete SCI using TSS demonstrated greater ability to modulate evoked responses than participants with clinically complete SCI using TSS. These results suggest the interaction of supraspinal and spinal mechanisms even in individuals with severe SCI.

Author Contributions: Conceptualization, J.S.C., P.J.G. and D.G.S.; methodology, J.S.C., M.L.G., M.B.L., D.D.V., A.R.T., C.L., I.A.L., P.J.G. and D.G.S.; validation, J.S.C., M.L.G., P.J.G. and D.G.S.; formal analysis, J.S.C., M.L.G., C.L., P.J.G. and D.G.S.; investigation, J.S.C., Y.P.G., V.R.E., I.A.L., K.D.Z., P.J.G. and D.G.S.; resources, K.H.L., K.D.Z. and D.G.S.; data curation, J.S.C., M.L.G., M.B.L., D.D.V., A.R.T., C.L., P.J.G. and D.G.S.; writing—original draft preparation, J.S.C. and D.G.S.; writing—review and editing, J.S.C., M.L.G., M.B.L., D.D.V., A.R.T., C.L., K.H.L., Y.P.G., V.R.E., I.A.L., K.D.Z., P.J.G. and D.G.S.; visualization, J.S.C., P.J.G. and D.G.S.; supervision, K.H.L., Y.P.G., V.R.E., I.A.L., K.D.Z., P.J.G. and D.G.S.; project administration, D.D.V., A.R.T., K.D.Z., P.J.G. and D.G.S.; funding acquisition, K.H.L., Y.P.G., V.R.E., K.D.Z., P.J.G. and D.G.S. All authors have read and agreed to the published version of the manuscript.

Funding: Sources of funding for the work reported here include the National Institutes of Health grant 1 R01 NS102920-01A1, The Grainger Foundation, Regenerative Medicine Minnesota, Jack Jablonski Bel13ve in Miracles Foundation, Mayo Clinic Graduate School of Biomedical Sciences, Mayo Clinic Center for Regenerative Medicine, Mayo Clinic Rehabilitation Medicine Research Center, Mayo Clinic Transform the Practice, Minnesota Office of Higher Education's Spinal Cord Injury and Traumatic Brain Injury Research Grant, Craig H. Neilsen Foundation, Dana and Albert R. Broccoli Charitable Foundation, Christopher and Dana Reeve Foundation, Walkabout Foundation, and Wings for Life Foundation.

Institutional Review Board Statement: The experimental procedures described herein were approved by the respective University of California, Los Angeles (UCLA) and Mayo Clinic institutional review boards.

Informed Consent Statement: Study participants provided written, informed consent to the experimental procedures.

Data Availability Statement: The data presented in this study are available on request from the corresponding author. The data are not publicly available due to potential patient privacy risks.

Acknowledgments: We thank the study participants for volunteering for this study. We also thank the Edgerton Neuromuscular Research Laboratory at UCLA, the Houston Methodist Neuromodulation & Recovery Laboratory, and the Mayo Clinic Assistive and Restorative Technology Laboratory research support staff for their contributions to study design, data collection and general study support.

Conflicts of Interest: Y.P.G., researcher on the study team, holds shareholder interest in NeuroRecovery Technologies and Cosyma. He holds certain inventorship rights on intellectual property licensed by the regents of the University of California to NeuroRecovery Technologies and its subsidiaries. V.R.E., researcher on the study team holds shareholder interest in SpineX and NeuroRecovery Technologies and holds certain inventorship rights on intellectual property licensed by The Regents of the University of California to NeuroRecovery Technologies and its subsidiaries.

References

1. Cho, N.; Squair, J.W.; Bloch, J.; Courtine, G. Neurorestorative interventions involving bioelectronic implants after spinal cord injury. *Bioelectron. Med.* **2019**, *5*, 10. [CrossRef] [PubMed]
2. Megía García, A.; Serrano-Muñoz, D.; Taylor, J.; Avendaño-Coy, J.; Gómez-Soriano, J. Transcutaneous Spinal Cord Stimulation and Motor Rehabilitation in Spinal Cord Injury: A Systematic Review. *Neurorehabil. Neural Repair* **2020**, *34*, 3–12. [CrossRef] [PubMed]
3. Angeli, C.A.; Edgerton, V.R.; Gerasimenko, Y.P.; Harkema, S.J. Altering spinal cord excitability enables voluntary movements after chronic complete paralysis in humans. *Brain* **2014**, *137*, 1394–1409. [CrossRef]

4. Calvert, J.S.; Grahn, P.J.; Strommen, J.A.; Lavrov, I.A.; Beck, L.A.; Gill, M.L.; Linde, M.B.; Brown, D.A.; Van Straaten, M.G.; Veith, D.D.; et al. Electrophysiological guidance of epidural electrode array implantation over the human lumbosacral spinal cord to enable motor function after chronic paralysis. *J. Neurotrauma* **2019**, *36*, 1451–1460. [CrossRef]
5. Darrow, D.; Balser, D.; Netoff, T.I.; Krassioukov, A.; Phillips, A.; Parr, A.; Samadani, U. Epidural Spinal Cord Stimulation Facilitates Immediate Restoration of Dormant Motor and Autonomic Supraspinal Pathways after Chronic Neurologically Complete Spinal Cord Injury. *J. Neurotrauma* **2019**, *36*, 2325–2336. [CrossRef] [PubMed]
6. Gerasimenko, Y.P.; Lu, D.C.; Modaber, M.; Zdunowski, S.; Gad, P.; Sayenko, D.G.; Morikawa, E.; Haakana, P.; Ferguson, A.R.; Roy, R.R.; et al. Noninvasive Reactivation of Motor Descending Control after Paralysis. *J. Neurotrauma* **2015**, *32*, 1968–1980. [CrossRef] [PubMed]
7. Grahn, P.J.; Lavrov, I.A.; Sayenko, D.G.; Van Straaten, M.G.; Gill, M.L.; Strommen, J.A.; Calvert, J.S.; Drubach, D.I.; Beck, L.A.; Linde, M.B.; et al. Enabling Task-Specific Volitional Motor Functions via Spinal Cord Neuromodulation in a Human With Paraplegia. *Mayo Clin. Proc.* **2017**, *92*, 544–554. [CrossRef] [PubMed]
8. Harkema, S.; Gerasimenko, Y.; Hodes, J.; Burdick, J.; Angeli, C.; Chen, Y.; Ferreira, C.; Willhite, A.; Rejc, E.; Grossman, R.G.; et al. Effect of epidural stimulation of the lumbosacral spinal cord on voluntary movement, standing, and assisted stepping after motor complete paraplegia: A case study. *Lancet* **2011**, *377*, 1938–1947. [CrossRef]
9. Rejc, E.; Angeli, C.; Harkema, S. Effects of Lumbosacral Spinal Cord Epidural Stimulation for Standing after Chronic Complete Paralysis in Humans. *PLoS ONE* **2015**, *10*, e0133998. [CrossRef]
10. Rejc, E.; Angeli, C.A.; Bryant, N.; Harkema, S.J. Effects of Stand and Step Training with Epidural Stimulation on Motor Function for Standing in Chronic Complete Paraplegics. *J. Neurotrauma* **2017**, *34*, 1787–1802. [CrossRef] [PubMed]
11. Rejc, E.; Angeli, C.A.; Atkinson, D.; Harkema, S.J. Motor recovery after activity-based training with spinal cord epidural stimulation in a chronic motor complete paraplegic. *Sci. Rep.* **2017**, *7*, 13476. [CrossRef] [PubMed]
12. Sayenko, D.; Rath, M.; Ferguson, A.R.; Burdick, J.; Havton, L.; Edgerton, V.R.; Gerasimenko, Y. Self-assisted standing enabled by non-invasive spinal stimulation after spinal cord injury. *J. Neurotrauma* **2018**, *36*, 1435–1450. [CrossRef] [PubMed]
13. Gill, M.L.; Grahn, P.J.; Calvert, J.S.; Linde, M.B.; Lavrov, I.A.; Strommen, J.A.; Beck, L.A.; Sayenko, D.G.; Van Straaten, M.G.; Drubach, D.I.; et al. Neuromodulation of lumbosacral spinal networks enables independent stepping after complete paraplegia. *Nat. Med.* **2018**, *24*, 1677–1682. [CrossRef] [PubMed]
14. Wagner, F.B.; Mignardot, J.; Le Goff-Mignardot, C.G.; Demesmaeker, R.; Komi, S.; Capogrosso, M.; Rowald, A.; Seáñez, I.; Caban, M.; Pirondini, E.; et al. Targeted neurotechnology restores walking in humans with spinal cord injury. *Nature* **2018**, *563*, 65–71. [CrossRef] [PubMed]
15. Angeli, C.A.; Boakye, M.; Morton, R.A.; Vogt, J.; Benton, K.; Chen, Y.; Ferreira, C.K.; Harkema, S.J. Recovery of Over-Ground Walking after Chronic Motor Complete Spinal Cord Injury. *N. Engl. J. Med.* **2018**, *379*, 1244–1250. [CrossRef] [PubMed]
16. Calvert, J.S.; Grahn, P.J.; Zhao, K.D.; Lee, K.H. Emergence of Epidural Electrical Stimulation to Facilitate Sensorimotor Network Functionality After Spinal Cord Injury. *Neuromodulation* **2019**, *22*, 244–252. [CrossRef]
17. Taccola, G.; Sayenko, D.; Gad, P.; Gerasimenko, Y.; Edgerton, V.R. And yet it moves: Recovery of volitional control after spinal cord injury. *Prog. Neurobiol.* **2018**, *160*, 64–81. [CrossRef] [PubMed]
18. Hofstoetter, U.S.; Freundl, B.; Binder, H.; Minassian, K. Common neural structures activated by epidural and transcutaneous lumbar spinal cord stimulation: Elicitation of posterior root-muscle reflexes. *PLoS ONE* **2018**, *13*, e0192013. [CrossRef] [PubMed]
19. Capogrosso, M.; Wenger, N.; Raspopovic, S.; Musienko, P.; Beauparlant, J.; Bassi Luciani, L.; Courtine, G.; Micera, S. A Computational Model for Epidural Electrical Stimulation of Spinal Sensorimotor Circuits. *J. Neurosci.* **2013**, *33*, 19326–19340. [CrossRef]
20. Ladenbauer, J.; Minassian, K.; Hofstoetter, U.S.; Dimitrijevic, M.R.; Rattay, F. Stimulation of the human lumbar spinal cord with implanted and surface electrodes: A computer simulation study. *IEEE Trans. Neural Syst. Rehabil. Eng.* **2010**, *18*, 637–645. [CrossRef]
21. Milosevic, M.; Masugi, Y.; Sasaki, A.; Sayenko, D.G.; Nakazawa, K. On the reflex mechanisms of cervical transcutaneous spinal cord stimulation in human subjects. *J. Neurophysiol.* **2019**, *121*, 1672–1679. [CrossRef]
22. Wu, Y.K.; Levine, J.M.; Wecht, J.R.; Maher, M.T.; LiMonta, J.M.; Saeed, S.; Santiago, T.M.; Bailey, E.; Kastuar, S.; Guber, K.S.; et al. Posteroanterior cervical transcutaneous spinal stimulation targets ventral and dorsal nerve roots. *Clin. Neurophysiol.* **2020**, *131*, 451–460. [CrossRef] [PubMed]
23. Danner, S.M.; Krenn, M.; Hofstoetter, U.S.; Toth, A.; Mayr, W.; Minassian, K. Body position influences which neural structures are recruited by lumbar transcutaneous spinal cord stimulation. *PLoS ONE* **2016**, *11*, e0147479. [CrossRef]
24. Minassian, K.; Persy, I.; Rattay, F.; Pinter, M.M.; Kern, H.; Dimitrijevic, M.R. Human lumbar cord circuitries can be activated by extrinsic tonic input to generate locomotor-like activity. *Hum. Mov. Sci.* **2007**, *26*, 275–295. [CrossRef] [PubMed]
25. Hofstoetter, U.S.; Freundl, B.; Binder, H.; Minassian, K. Recovery cycles of posterior root-muscle reflexes evoked by transcutaneous spinal cord stimulation and of the H reflex in individuals with intact and injured spinal cord. *PLoS ONE* **2019**, *14*, e0227057. [CrossRef] [PubMed]
26. Dimitrijevic, M.R.; Dimitrijevic, M.M.; Faganel, J.; Sherwood, A.M. Suprasegmentally induced motor unit activity in paralyzed muscles of patients with established spinal cord injury. *Ann. Neurol.* **1984**, *16*, 216–221. [CrossRef]
27. Moss, C.W.; Kilgore, K.L.; Peckham, P.H. A novel command signal for motor neuroprosthetic control. *Neurorehabil. Neural Repair* **2011**, *25*, 847–854. [CrossRef] [PubMed]

28. Hofstoetter, U.S.; Minassian, K.; Hofer, C.; Mayr, W.; Rattay, F.; Dimitrijevic, M.R. Modification of reflex responses to lumbar posterior root stimulation by motor tasks in healthy subjects. *Artif. Organs* **2008**, *32*, 644–648. [CrossRef] [PubMed]
29. Minassian, K.; Persy, I.; Rattay, F.; Dimitrijevic, M.R.; Hofer, C.; Kern, H. Posterior root-muscle preflexes elicited by transcutaneous stimulation of the human lumbosacral cord. *Muscle Nerve* **2007**, *35*, 327–336. [CrossRef] [PubMed]
30. Gill, M.L.; Linde, M.B.; Hale, R.F.; Lopez, C.; Fautsch, K.J.; Calvert, J.S.; Veith, D.D.; Beck, L.A.; Garlanger, K.L.; Sayenko, D.G.; et al. Alterations of Spinal Epidural Stimulation-Enabled Stepping by Descending Intentional Motor Commands and Proprioceptive Inputs in Humans With Spinal Cord Injury. *Front. Syst. Neurosci.* **2020**, *14*, 590231. [CrossRef]
31. Gill, M.; Linde, M.; Fautsch, K.; Hale, R.; Lopez, C.; Veith, D.; Calvert, J.; Beck, L.; Garlanger, K.; Edgerton, R.; et al. Epidural Electrical Stimulation of the Lumbosacral Spinal Cord Improves Trunk Stability During Seated Reaching in Two Humans With Severe Thoracic Spinal Cord Injury. *Front. Syst. Neurosci.* **2020**, *14*, 79. [CrossRef]
32. Rath, M.; Vette, A.H.; Ramasubramaniam, S.; Li, K.; Burdick, J.; Edgerton, V.R.; Gerasimenko, Y.P.; Sayenko, D.G. Trunk Stability Enabled by Noninvasive Spinal Electrical Stimulation after Spinal Cord Injury. *J. Neurotrauma* **2018**, *35*, 2540–2553. [CrossRef] [PubMed]
33. Hofstoetter, U.S.; Freundl, B.; Danner, S.M.; Krenn, M.J.; Mayr, W.; Binder, H.; Minassian, K. Transcutaneous Spinal Cord Stimulation Induces Temporary Attenuation of Spasticity in Individuals with Spinal Cord Injury. *J. Neurotrauma* **2020**, *37*, 481–493. [CrossRef] [PubMed]
34. Calvert, J.S.; Manson, G.A.; Grahn, P.J.; Sayenko, D.G. Preferential activation of spinal sensorimotor networks via lateralized transcutaneous spinal stimulation in neurologically intact humans. *J. Neurophysiol.* **2019**, *122*, 2111–2118. [CrossRef] [PubMed]
35. Krenn, M.; Toth, A.; Danner, S.M.; Hofstoetter, U.S.; Minassian, K.; Mayr, W. Selectivity of transcutaneous stimulation of lumbar posterior roots at different spinal levels in humans. *Biomed. Tech.* **2013**, *58* (Suppl. 1), 2–3. [CrossRef]
36. Sayenko, D.G.; Angeli, C.; Harkema, S.J.; Edgerton, V.R.; Gerasimenko, Y.P. Neuromodulation of evoked muscle potentials induced by epidural spinal-cord stimulation in paralyzed individuals. *J. Neurophysiol.* **2014**, *111*, 1088–1099. [CrossRef] [PubMed]
37. Sayenko, D.G.; Atkinson, D.A.; Dy, C.J.; Gurley, K.M.; Smith, V.L.; Angeli, C.; Harkema, S.J.; Edgerton, V.R.; Gerasimenko, Y.P. Spinal segment-specific transcutaneous stimulation differentially shapes activation pattern among motor pools in humans. *J. Appl. Physiol.* **2015**, *118*, 1364–1374. [CrossRef]
38. Cuellar, C.A.; Mendez, A.A.; Islam, R.; Calvert, J.S.; Grahn, P.J.; Knudsen, B.; Pham, T.; Lee, K.H.; Lavrov, I.A. The Role of Functional Neuroanatomy of the Lumbar Spinal Cord in Effect of Epidural Stimulation. *Front. Neuroanat.* **2017**, *11*, 82. [CrossRef]
39. Kato, T.; Sasaki, A.; Yokoyama, H.; Milosevic, M.; Nakazawa, K. Effects of neuromuscular electrical stimulation and voluntary commands on the spinal reflex excitability of remote limb muscles. *Exp. Brain Res.* **2019**, *237*, 3195–3205. [CrossRef]
40. Crone, C.; Johnsen, L.L.; Biering-Sørensen, F.; Nielsen, J.B. Appearance of reciprocal facilitation of ankle extensors from ankle flexors in patients with stroke or spinal cord injury. *Brain* **2003**, *126*, 495–507. [CrossRef]
41. Dietz, V. Behavior of spinal neurons deprived of supraspinal input. *Nat. Rev. Neurol.* **2010**, *6*, 167–174. [CrossRef]
42. Dario, A.; Tomei, G. A benefit-risk assessment of baclofen in severe spinal spasticity. *Drug Saf.* **2004**, *27*, 799–818. [CrossRef]
43. D'Amico, J.M.; Condliffe, E.G.; Martins, K.J.B.; Bennett, D.J.; Gorassini, M.A. Recovery of neuronal and network excitability after spinal cord injury and implications for spasticity. *Front. Integr. Neurosci.* **2014**, *8*, 36. [CrossRef]
44. Andrews, J.C.; Stein, R.B.; Roy, F.D. Reduced postactivation depression of soleus H reflex and root evoked potential after transcranial magnetic stimulation. *J. Neurophysiol.* **2015**, *114*, 485–492. [CrossRef] [PubMed]
45. Knikou, M. Transpinal and transcortical stimulation alter corticospinal excitability and increase spinal output. *PLoS ONE* **2014**, *9*, e102313. [CrossRef] [PubMed]
46. Roy, F.D.; Bosgra, D.; Stein, R.B. Interaction of transcutaneous spinal stimulation and transcranial magnetic stimulation in human leg muscles. *Exp. Brain Res.* **2014**, *232*, 1717–1728. [CrossRef] [PubMed]
47. Sayenko, D.; Atkinson, D.; Mink, A.; Gurley, K.; Edgerton, R.; Harkema, S.J.; Gerasimenko, Y. Vestibulospinal and corticospinal modulation of lumbosacral network excitability in human subjects. *Front. Physiol.* **2018**, *9*, 1746. [CrossRef]
48. Courtine, G.; Harkema, S.J.; Dy, C.J.; Gerasimenko, Y.P.; Dyhre-Poulsen, P. Modulation of multisegmental monosynaptic responses in a variety of leg muscles during walking and running in humans. *J. Physiol.* **2007**, *582*, 1125–1139. [CrossRef]
49. Dy, C.J.; Gerasimenko, Y.P.; Edgerton, V.R.; Dyhre-Poulsen, P.; Courtine, G.; Harkema, S.J. Phase-Dependent Modulation of Percutaneously Elicited Multisegmental Muscle Responses After Spinal Cord Injury. *J. Neurophysiol.* **2010**, *103*, 2808–2820. [CrossRef]
50. Courtine, G.; Song, B.; Roy, R.R.; Zhong, H.; Herrmann, J.E.; Ao, Y.; Qi, J.; Edgerton, V.R.; Sofroniew, M.V. Recovery of supraspinal control of stepping via indirect propriospinal relay connections after spinal cord injury. *Nat. Med.* **2008**, *14*, 69–74. [CrossRef]
51. Formento, E.; Minassian, K.; Wagner, F.; Mignardot, J.B.; Le Goff-Mignardot, C.G.; Rowald, A.; Bloch, J.; Micera, S.; Capogrosso, M.; Courtine, G. Electrical spinal cord stimulation must preserve proprioception to enable locomotion in humans with spinal cord injury. *Nat. Neurosci.* **2018**, *21*, 1728–1741. [CrossRef]
52. Asboth, L.; Friedli, L.; Beauparlant, J.; Martinez-Gonzalez, C.; Anil, S.; Rey, E.; Baud, L.; Pidpruzhnykova, G.; Anderson, M.A.; Shkorbatova, P.; et al. Cortico-reticulo-spinal circuit reorganization enables functional recovery after severe spinal cord contusion. *Nat. Neurosci.* **2018**, *21*, 576–588. [CrossRef] [PubMed]
53. Lavrov, I.; Dy, C.J.; Fong, A.J.; Gerasimenko, Y.; Courtine, G.; Zhong, H.; Roy, R.R.; Edgerton, V.R. Epidural Stimulation Induced Modulation of Spinal Locomotor Networks in Adult Spinal Rats. *J. Neurosci.* **2008**, *28*, 6022–6029. [CrossRef]

54. Hofstoetter, U.S.; Danner, S.M.; Freundl, B.; Binder, H.; Mayr, W.; Rattay, F.; Minassian, K. Periodic modulation of repetitively elicited monosynaptic reflexes of the human lumbosacral spinal cord. *J. Neurophysiol.* **2015**, *114*, 400–410. [CrossRef] [PubMed]
55. Murray, L.M.; Knikou, M. Transspinal stimulation increases motoneuron output of multiple segments in human spinal cord injury. *PLoS ONE* **2019**, *14*, e0213696. [CrossRef]
56. Militskova, A.; Mukhametova, E.; Fatykhova, E.; Sharifullin, S.; Cuellar, C.A.; Calvert, J.S.; Grahn, P.J.; Baltina, T.; Lavrov, I. Supraspinal and Afferent Signaling Facilitate Spinal Sensorimotor Network Excitability After Discomplete Spinal Cord Injury: A Case Report. *Front. Neurosci.* **2020**, *14*, 552. [CrossRef] [PubMed]

Journal of
Clinical Medicine

MDPI

Article

Transcutaneous Spinal Cord Stimulation Enhances Quadriceps Motor Evoked Potential in Healthy Participants: A Double-Blind Randomized Controlled Study

Álvaro Megía-García [1,2], Diego Serrano-Muñoz [2,*], Julian Taylor [3,4], Juan Avendaño-Coy [2], Natalia Comino-Suárez [2,5] and Julio Gómez-Soriano [2]

[1] Biomechanical and Technical Aids Unit, National Hospital for Paraplegia, SESCAM, 45071 Toledo, Spain; amegiag@externas.sescam.jccm.es

[2] Toledo Physiotherapy Research Group (GIFTO), Faculty of Physiotherapy and Nursing, Castilla La Mancha University, 45071 Toledo, Spain; juan.avendano@uclm.es (J.A.-C.); natalia.comino@cajal.csic.es (N.C.-S.); julio.soriano@uclm.es (J.G.-S.)

[3] Sensorimotor Function Group, National Hospital for Paraplegia, SESCAM, 45071 Toledo, Spain; juliantaylorgreen2@gmail.com

[4] Harris Manchester College, University of Oxford, Oxford OX1 3TD, UK

[5] Neural Rehabilitation Group, Cajal Institute, Spanish National Research Council (CSIC), 28002 Madrid, Spain

* Correspondence: diego.serrrano@uclm.es; Tel.: +34-925268800 (ext. 5830)

Received: 7 September 2020; Accepted: 12 October 2020; Published: 13 October 2020

Abstract: Transcutaneous electrical spinal cord stimulation (tSCS) is a non-invasive technique for neuromodulation and has therapeutic potential for motor rehabilitation following spinal cord injury. The main aim of the present study is to quantify the effect of a single session of tSCS on lower limb motor evoked potentials (MEPs) in healthy participants. A double-blind, sham-controlled, randomized, crossover, clinical trial was carried out in 15 participants. Two 10-min sessions of tSCS (active-tSCS and sham-tSCS) were applied at the T11-T12 vertebral level. Quadriceps (Q) and tibialis anterior (TA) muscle MEPs were recorded at baseline, during and after tSCS. Q and TA isometric maximal voluntary contraction was also recorded. A significant increase of the Q-MEP amplitude was observed during active-tSCS (1.96 ± 0.3 mV) when compared from baseline (1.40 ± 0.2 mV; $p = 0.01$) and when compared to sham-tSCS at the same time-point (1.13 ± 0.3 mV; $p = 0.03$). No significant modulation was identified for TA-MEP amplitude or for Q and TA isometric maximal voluntary isometric strength. In conclusion, tSCS applied over the T11-T12 vertebral level increased Q-MEP but not TA-MEP compared to sham stimulation. The specific neuromodulatory effect of tSCS on Q-MEP may reflect optimal excitation of this motor response at the interneuronal or motoneuronal level.

Keywords: transcutaneous spinal cord stimulation; evoked potentials motor; neuromodulation; motor activity

1. Introduction

Transcutaneous electrical spinal cord stimulation (tSCS) is a non-invasive technique designed to generate neuromodulation of the central nervous system at the spinal cord level. During the last ten years, studies have reported changes in excitability of neural networks organized at the spinal cord level by transcutaneous electrical stimulation applied via surface electrodes placed over the middle of back (T10-T11 spinal level) [1,2]. At the spinal level, tSCS modulates the excitability of sensorimotor circuits which are also affected by epidural stimulation [3]. Both techniques can evoke posterior root muscle (PRM) reflex responses measured at the target muscle [3]. The low PRM reflex threshold response to stimulation, suggests that tSCS activates large-to-medium diameter afferent fibers (Ia, Ib, II) within the sensory dorsal roots [1,3]. The similarity between PRM and Hoffmann reflex responses also suggests that tSCS activates proprioceptive Ia afferent fibers [4,5].

Several studies have also revealed that tSCS mediates several therapeutic effects in subjects with neurological impairments [6]. Thus, tonic tSCS, applied as a biphasic rectangular current at a frequency of 30–50 Hz, has shown a therapeutic potential to enhance voluntary motor activity [7,8], trunk stability [9], standing [10], gait function [7,11,12] and as a method to reduce spasticity [13,14]. However, the clinical effectiveness of this technique is still undetermined, and the stimulation parameters need to be optimized [6].

The results of a recently published review by our group, Megía-García et al. [6], suggest that the optimal site for spinal stimulation to enhance voluntary motor control of the lower extremities is over the T11-T12 vertebrae, which corresponds to the L1-L2 spinal cord segments. The idea of the existence of the central pattern generation (CPG) at this level [15], together with a multisegmental muscle activity in response to stimulation applied at this level [16], suggests that activation of local or propriospinal motor control mechanisms potentiate descending corticospinal acting on spinal motoneurons, thereby mediating the neuromodulatory effect of tSCS leading to improve motor function following spinal cord injury [15]. However, studies performed on healthy subjects have shown that the optimal stimulation site differs according to which target muscle group is activated [17–19]. Hence, greater activation of the anterior rectus femoris and vastus lateralis has been observed with T10-T11 (L1-L2 spinal cord segments) stimulation when compared to the hamstring, triceps surae, and tibialis anterior muscles which are activated optimally with T12-L1 stimulation (L5-S1 spinal cord segments) [19,20]. In line with these observations, the phenomenon of spatial summation has also been reported, with greater motor responses recorded from the target muscle with simultaneous electrical stimulation of contiguous spinal levels [21].

Although tSCS has been shown to be a safe technique with therapeutic potential for motor function rehabilitation following neurological diseases [6], it is necessary to understand how the tSCS technique can be optimized in the clinical practice. The validation of neurophysiological measurements which reflect changes in both cortical and spinal excitability, such as motor evoked potentials (MEPs) evoked by transcranial magnetic stimulation (TMS), and closer attention to appropriate sham-protocols are important first steps for developing future tSCS studies related to better neuromodulation with optimized stimulation parameters and electrode locations.

Until now, tSCS-mediated neuromodulation of the motor response has been measured indirectly through electromyogram recordings of ongoing voluntary muscle activity in combination with dynamometry, manual muscle testing, kinematics and functional assessments [7,8,22,23]. However, these measurement tools have a low sensitivity to single session tSCS-induced neuromodulation and cannot detect small changes in neuromuscular activity. Neurophysiological methods aimed at measuring changes in spinal excitability, such as the PRM reflex and the Hoffman reflex, have been used to assess neuromodulation after tSCS [24–27]. However, the extent to which tSCS can lead to a change in cortical and spinal excitability has only been studied in the upper limb [27]. Therefore, the measurement of MEPs evoked by TMS has been used in this study as the standard and sensitive neurophysiological method to assess changes in excitability of the corticospinal pathway in response to tSCS.

The main aim of the present study is to quantify the effect of a single session of active T11-T12 tSCS on lower limb MEP amplitude measured from the quadriceps and tibialis anterior muscles of healthy participants. We hypothesized that a single session of tSCS would increase corticospinal excitability with a preferential effect on those motoneurones closer to the level of spinal stimulation. As a secondary aim the efficacy of blinding both the sham and active tSCS was carefully analyzed using specialized metrics appropriate to randomized control trials.

2. Experimental Section

The present study was approved by the local Toledo Ethical Committee (Ref. No. 158; 2/11/2017) and the clinical trial was registered in the ClinicalTrials.gov Protocol Registration System (NCT04241406). Participants were informed about the protocol and signed the informed consent approved by the local ethics committee.

2.1. Design

A double-blinded, randomized, controlled, crossover clinical trial was designed. All participants (n = 15) received two randomized sessions of tSCS stimulation (active-tSCS and sham-tSCS). Both, participants and the assessor (AMG) were blinded, and the order for each recruited participant to receive either the sham or active tSCS intervention was randomized, with the consecutive order concealed in individual sealed envelopes. Randomization was performed using the web software www.randomizer.org. Intervention sessions were separated with a minimal 48 h washout period, which was longer than that used in similar studies (Aarskog et al., 2007; Buonocore and Camuzzini, 2007; Chen and Johnson, 2009; Dean et al., 2006).

In order to check the reliability of the neurophysiological outcomes, two measurements of quadriceps (Q) and tibialis anterior (TA) MEPs were recorded at baseline (PRE 1 and PRE 2) with a rest period of 2 min between them. Both Q and TA MEPs were recorded again during the intervention at 4 min from the onset of the tSCS intervention (DURING) and following the stimulus at 2 min (POST 1) and 4 min (POST 2). Postintervention measurements (POST 1 and POST 2) were recorded to determine the duration of the neuromodulatory effect following tSCS. Two measurements of isometric maximal voluntary contraction of the Q and TA muscles were recorded as secondary outcomes at PRE 1 and POST 2.

2.2. Procedures and Intervention

Fifteen healthy participants ≥ 18 years old, without injury to the central or peripheral nervous system, were recruited via non-probabilistic convenience sampling. The exclusion criteria for recruitment included musculoskeletal pathology of the lower limbs, metal or electronic implants, medications that influenced neural excitability (antiepileptic, antipsychotics, or antidepressants), allergy to the electrode material, epilepsy, and pregnancy.

Participants adopted a relaxed supine position without voluntary muscle contraction. Three self-adhesive surface electrodes (9 × 5 cm, ValuTrode, Axelgaard Manufacturing Co, Fallbrook, CA, USA) were used for tSCS. The anode was placed on the midline skin surface between the T11 and T12 spinous processes. Two interconnected cathodes were placed symmetrically on the abdomen, at both sides of the umbilicus. A rectangular foam (12 × 7 × 2 cm) was used to increase the pressure on the anode by strapping both the sponge and electrode around the torso, to improve skin contact. The electrical stimulator used for all interventions was an electrotherapeutical device (Enraf Nonius, Myomed 932, Rotterdam, The Netherlands).

2.2.1. Transcutaneous Spinal Cord Stimulation (Active-tSCS)

A symmetrical biphasic 30 Hz current of 1 ms pulse-width was applied at the threshold stimulation intensity required to evoke the PRM reflex. To calculate the PRM threshold the same device, electrode location, and waveform of current were used. The stimulus intensity was gradually increased in steps of 1 mA every 1 s until a minimally visible contraction was visually detected in the Q muscle. If subjects could not tolerate the stimulus intensity required to elicit the threshold PRM reflex, the stimulus was reduced to the intensity which was comfortable. The therapist informed the participant that the intensity could be set below the sensory threshold, with the possibility that the participant may or may not feel the stimulation current tSCS was applied for a total of 10 min with the subjects relaxed in the supine position.

2.2.2. Sham-tSCS

Sham stimulation consisted of the same stimulation parameters used for active tSCS except for the stimulus intensity, which was set at the sensory threshold, maintained for 30 s and then slowly decreased to zero, where the stimulus intensity was set for the remaining 10 min intervention. This method has been previously validated as an appropriate sham stimulus for controlled studies (Deyo et al., 1990; Petrie and Hazleman, 1985) [28,29]. The participants received the same instructions that given to the active tSCS session.

2.3. Blinding Assessment

Both, the intervention, and evaluation process of the protocol were carried out by two independent researchers (NCS and AMG, respectively).

The success of participant and assessor blinding were evaluated through a questionnaire [30] after each intervention (active-tSCS and sham-tSCS), by assessing whether they had received the sham or active tSCS with five questions: (1) "Strongly believe the applied intervention is new treatment"; (2) "Somewhat believe the applied intervention is new treatment"; (3) "Somewhat believe the applied intervention is a placebo", (4) "Strongly believe the applied intervention is a placebo", or (5) "Do not Know".

2.4. Primary and Secondary Outcome Measures

2.4.1. Motor Evoked Potential Recordings

Motor evoked potentials were elicited by transcranial magnetic stimulation (TMS, Magstim Rapid 2, Magstim Company Ltd., UK) using a double-cone coil. The optimal stimulation site (hot-spot, area where TMS elicited the largest MEP) was identified for both the Q and TA muscles individually with reference to the standard CZ point (approximately 1 cm lateral and 2 cm posterior to CZ). The coil was then fixed, and the rim of the coil was marked with a pen on the scalp, so that the stimulation site was maintained constant. MEP threshold was defined as the minimal TMS intensity required to evoke a motor response (>0.1 mV peak-to-peak amplitude) during slight tonic contraction of the target muscle (approximately 20% of the isometric MVC) [28,29]. Test MEPs during the protocol were recorded during 20% MVC as an average of in response to 10 single-pulse stimuli applied at 120% of the MEP threshold. A maximal voluntary contraction was performed individually for Q and TA to record 100% EMG activity for each muscle, so that the 20% EMG activity could be estimated and used as a visual feedback to instruct the subjects to maintain that level of muscle contraction. Once the subjects had learnt to contract either the Q or TA muscles at 20% maximal voluntary activation, TMS-evoked MEPs were performed. EMG was recorded using bipolar silver chloride electrodes (×1000 amplification) filtered with a built-in 20–450 Hz bandpass filter (Signal Conditioning Electrodes v2.3, Delsys Inc., USA). Electrodes were placed over the rectus femoris of the belly of Q muscle and the proximal third region TA following the SENIAM recommendations. Average MEP peak-to-peak amplitude and latency were analyzed as the primary outcome measure of tSCS.

2.4.2. Isometric Maximal Voluntary Contraction Strength

Isometric MVC strength was measured using the hand-held dynamometer Micro Fet 2TM (Hoggan Scientific, LLC, Utah, USA,), a method with intra and inter-rater reliability has been reported [31]. Assessment of isometric MVC strength was performed with participants in either the sitting or supine position to evaluate Q and TA muscles,, respectively. Participants were instructed to hold the side of the table for stabilization and contract the test muscle as hard and as fast as they could against the dynamometer and to maintain the contraction until the assessor requested them to "stop". Each test measured MVC strength during three to five seconds. Three test trials, with a 1 min interval resting period, were performed for each muscle group. Average peak force was used for statistical analysis.

2.5. Statistical Analysis

The sample size was calculated based on Q-MEP amplitude as the main variable. A mean difference of 0.5 mV (approx. 30%) between groups was expected, with a standard deviation of 0.065 was considered from a previous preliminary unpublished report, with a type I error (α) of 0.05 and a power of 80%. A sample size of 15 subjects was calculated.

Statistical analysis was performed using the commercial software package SPSS v22. Assessment of the relative reliability of baseline MEPs for PRE1-PRE2 was achieved with the intraclass correlation coefficient (ICC) that indicates the error in measurements as a proportion of the total variance in scores. An ICC of over 0.90 was defined as a high; from 0.80 to 0.90 as moderate; and below 0.80 as low reliability [32,33]. Moreover, the homogeneity between sessions for MEPs and PRM baseline threshold was analyzed with a t-student test. For statistical analysis of MEPs, the average of PRE1 and PRE 2 was considered as the unique baseline (PRE) measure. A two-way repeated-measures ANOVA, with "time" as one factor (PRE, DURING, POST1, POST2) and the "intervention" factor (tSCS and Sham) was performed to compare differences in MEP amplitude latency. Similarly, a two-way repeated-measures ANOVA was carried out for dynamometer data collected at PRE1 and POST2. A paired *t*-test comparison with a Bonferroni correction for multiple comparisons was used to highlight specific differences between time and interventions. To analyze the blinding outcome variable, James' Blinding Index (BI) [34] and Bang's BI [30] were obtained using Stata v15.0 (Stata Corp, TX, USA). James' BI is used to infer the overall blinding success in RCTs. However, Bang's BI is used to characterize and evaluate the blinding situation in each trial arm independently. James' BI ranges from 0 to 1 (0 representing total lack of blinding, 1 representing complete blinding, and 0.5 representing completely random blinding). To interpret the results, this study considered a lack of blinding if the upper bound of the confidence interval (CI) was below 0.5. Bang's BI can be directly interpreted as the proportion of the unblinding in each arm [30]. It ranges between −1 and 1, with 0 as a null value indicating the most desirable situation representing random complete blinding. Therefore, when one-sided CI did not cover the 0 value, the study was regarded as lacking blinding.

3. Results

Fifteen participants were recruited and completed the study. Nine were female (60%) and 6 males, with a mean age of 25.2 years old (SD 3.8), weight of 60.9 kg (SD 12.62), height of 1.67 m (SD 0.11), and a mean body mass index of 21.6 (SD 2.47). None of the subjects withdrew from the study (Figure 1. CONSORT flow diagram). No sex significant differences were found when comparing the proportion of the subjects ($p > 0.05$) nor when comparing the effect of tSCS during the stimulation ($p > 0.1$) and after the stimulation ($p > 0.2$) for both Q-MEP and TA-MEP outcomes.

With regard to baseline neurophysiological parameters, mean MEP thresholds for Q (t = 0.00, $p = 1.00$) and TA (t = −0.187, $p = 0.85$), and PRM reflex threshold (t = −0.58, $p = 0.56$) were similar before application of sham or active tSCS (Table 1). The averaged tSCS intensity during the active-tSCS session was 27.6 mA (8.9), that was 27.5% less than the PRM reflex threshold (Table 1). Basal peak-to-peak amplitude MEP measured at PRE1 and PRE2 were highly reliable for both active and sham interventions. During the active tSCS session, a 0.91 ICC was registered for Q-MEP amplitude and a 0.97 ICC for TA-MEP amplitude. During the sham intervention, a 0.89 ICC was registered for Q-MEP amplitude and a 0.97 ICC for TA-MEP amplitude were calculated. Regarding Q-MEP latency, a 0.97 ICC was calculated during the active tSCS session and an ICC of 0.97 during sham intervention. The latency of TA-MEPs showed an ICC of 0.95 during active tSCS intervention and 0.92 during the sham session.

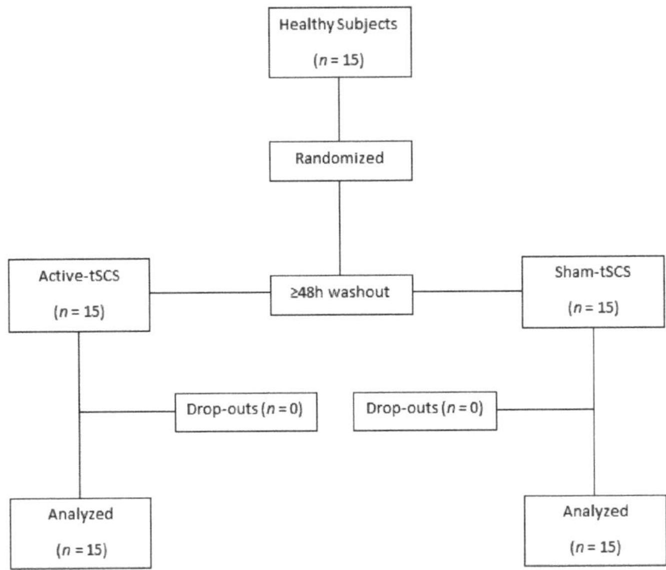

Figure 1. Consolidated Standards of Reporting Trials (CONSORT) flow diagram.

Table 1. Neurophysiological assessment at baseline time for each session.

	Active-tSCS Session				Sham-tSCS Session		
Subject	MEP-t Q (%)	MEP-t TA (%)	PRM-t (mA)	Intensity tSCS (mA)	MEP-t Q (%)	MEP-t TA (%)	PRM-t (mA)
01	55	50	45	45	55	55	41
02	31	31	40	25	32	32	40
03	42	35	29	29	35	35	29
04	45	40	33	33	55	44	32
05	50	50	33	33	55	55	33
06	42	42	35	26	40	40	30
07	32	32	50	37	30	30	40
08	32	32	40	21.5	45	35	43
09	47	42	35	14	37	35	29
10	35	35	35	27	33	38	35
11	42	37	41	39	38	33	45
12	35	35	40	17.5	34	34	40
13	38	35	38	22	38	35	38
14	35	35	36	13	35	35	26
15	54	40	44	32	53	42	46
Mean	41	38.1	38.3	27.6	41	38.5	36.5
(SD)	(7.9)	(5.9)	(5.2)	(8.9)	(9.1)	(7.6)	(6.3)

PRM = posterior root-muscle reflex; Q = quadriceps; SD = Standard deviation; −t = threshold; TA = tibialis anterior; tSCS = transcutaneous spinal cord stimulation.

3.1. Effect of tSCS on Peak-To-Peak MEP Amplitude

The effect of active or sham tSCS on MEPs recorded from subject (#7) (Figure 2) is illustrated on the Q and TA muscle (Figure 2), recorded at PRE 2, during the intervention (DURING), and after the intervention finished (POST 1). Mean peak-to-peak amplitude and the change in MEP amplitude DURING and POST tSCS are represented in Figure 3A–D, respectively. In the quantitative analysis, the two-way ANOVA showed no significant differences in Q-MEP amplitude for either the "intervention" factor (F = 1.178, p = 0.28) or the "time" factor (F = 1.176, p = 0.18). However, a significant "intervention-time" interaction was found for Q-MEP amplitude (F = 3.88, p = 0.02). A post-hoc pairwise comparison of Q-MEP amplitude revealed a significant increase during active-tSCS (1.96 mV; SD: 0.3) when compared to the averaged PRE amplitude (1.40 mV; SD: 0.2; $p = 0.01$). Furthermore, a higher amplitude of Q-MEPs during active-tSCS (1.96 mV; SD: 0.3) was observed when compared to sham-tSCS at the same time-point (1.13 mV; SD: 0.3; $p = 0.03$; Figure 3A). Regarding TA-MEP amplitude, although a greater amplitude during the active tSCS intervention (Figure 3B) was identified (3.06 mV; SD: 2.0) with respect to the PRE baseline amplitude (2.67 mV; SD: 0.4) and higher than the equivalent MEP amplitude during sham tSCS (2.45 mV, SD: 1.1), no statistical significant differences were observed for the factors analyzed ("intervention" F = 0.268, $p = 0.609$; nor "time" TA F = 0.339, $p = 0.797$; nor "intervention-time" interaction F = 1.137, $p = 0.333$).

When the specific effect of each intervention was calculated subtracting the baseline activity in each time-point (Figure 3C,D), the pairwise comparison between groups with the Bonferroni correction revealed a mean difference of 0.73 mV (SD: 0.24; $p = 0.05$) during the stimulation for the Q-MEP, without significant differences for the POST-1 (0.001 mV; SD: 0.19; $p = 0.94$) and POST-2 (0.25 mV; SD: 0.16; $p = 0.13$) time-points (Figure 3C). For the specific effect of the intervention over TA-MEP, a non-significant difference between groups of 0.91 mV (SD: 0.62; $p = 0.15$) was evidenced during the stimulation. Neither significant differences were found for the POST-1 (0.69 mV; SD: 0.62; $p = 0.21$) and POST-2 (0.33 mV; SD: 0.53; $p = 0.53$) time-points (Figure 3D).

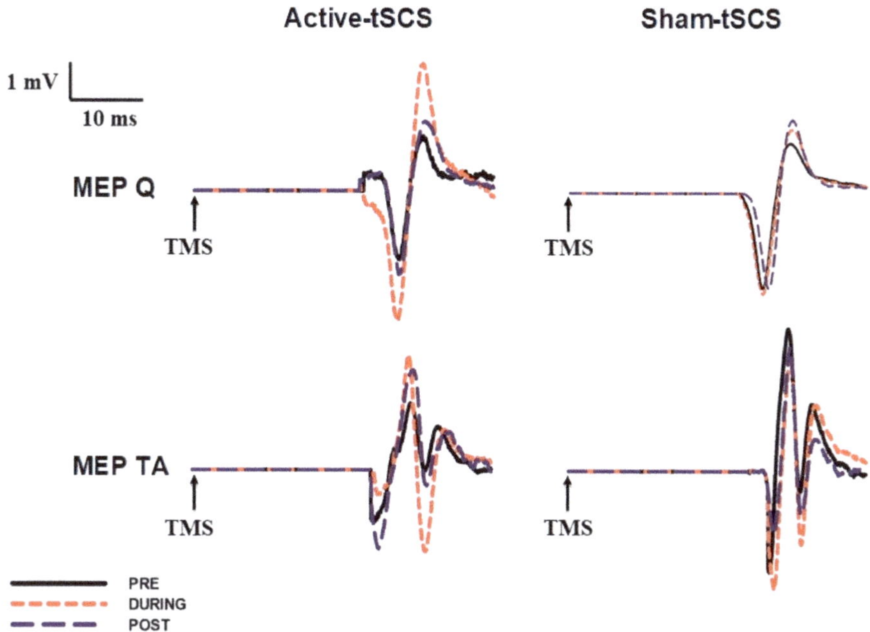

Figure 2. Representative motor-evoked potentials from participant #7 recorded from the Quadriceps and Tibialis Anterior muscles PRE, DURING and POST active or sham tSCS.

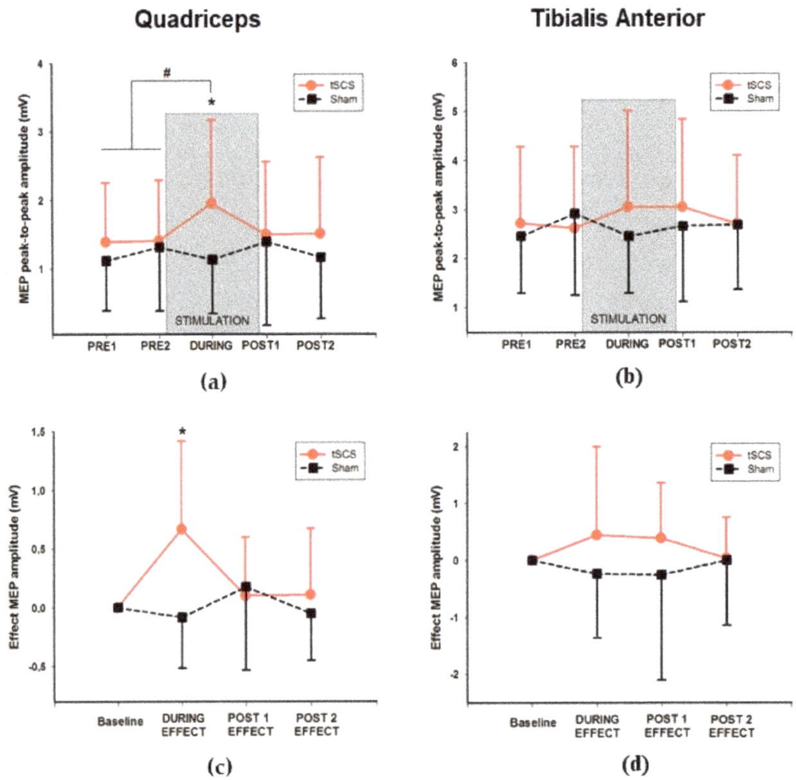

Figure 3. MEP peak-to-peak amplitude: (**a**) MEP of quadriceps muscle at each evaluation time; (**b**) MEP of tibialis anterior muscle at each evaluation time; (**c**) change obtained in MEP of quadriceps during and after tSCS respect to baseline; (**d**) change obtained in MEP of tibialis anterior during and after tSCS respect to baseline. * $p \leq 0.05$ when compared to sham-tSCS; # $p \leq 0.05$ when compared from baseline.

3.2. Effect on MEP Latency

Table 2 reveals that the application of tSCS intervention had no effect on MEP latencies for the Q and TA muscles. The two-way ANOVA revealed no significant differences neither for the "time" factor (Q muscle: F = 1.952, $p = 0.13$; TA muscle: F = 0.89, $p = 0.45$), "intervention" factor (Q muscle: F = 0.549, $p = 0.46$; TA muscle: F = 2.621, $p = 0.11$), nor "intervention-time" interaction (Q muscle F = 0.937, $p = 0.42$; TA muscle: F = 0.254, $p = 0.86$).

Table 2. Latencies MEPs following active-tSCS or sham-stimulation. Mean and standard deviation (SD).

		PRE 1	**PRE 2**	**DURING**	**POST 1**	**POST 2**
MEP Q (ms)	Active-tSCS	21.74 (1.98)	21.71 (2.07)	21.77 (2.11)	21.81 (2.01)	21.72 (2.04)
	Sham-tSCS	22.49 (2.78)	22.53 (3.1)	22.70 (2.77)	22.35 (3.04)	22.48 (3.07)
MEP TA (ms)	Active-tSCS	28.83 (1.81)	29.03 (1.7)	28.71 (1.75)	28.77 (1.79)	28.72 (1.67)
	Sham-tSCS	29.54 (2.66)	29.97 (1.97)	29.92 (2.27)	29.72 (2.00)	29.80 (1.86)

3.3. Effect on Isometric Maximal Voluntary Contraction Strength

Table 3 shows the effect of tSCS on mean peak isometric MVC force recorded from the Q and TA muscles before and after the active and sham interventions. No significant differences were observed in MVC strength when pre-post values were compared for active-tSCS and sham-tSCS (Q: F = 0.583, p = 0.45; TA: F = 0.268, p = 0.60). In the same way, no significant differences were observed for the "time" factor (Q: F= 0.017; p = 0.89; TA: F = 0.339; p = 0.797) nor the "intervention-time" interaction (Q: F = 0.981, p = 0.33; TA: F = 0.808, p = 0.38).

Table 3. Isometric maximal voluntary contraction strength following tSCS or Sham. Mean (SD).

	Active-tSCS			Sham-tSCS		
	PRE	**POST**	**% of Change**	**PRE**	**POST**	**% of Change**
Strength Q (Kgs)	76.18 (16.62)	77.91 (18.38)	3.10 (13.23) p = 0.428	82.84 (17.84)	81.30 (21.50)	−2.32 (11.47) p = 0.562
Strength TA (Kgs)	73.46 (11.96)	73.00 (14.69)	−1.09 (6.43) p = 0.699	82.84 (17.84)	81.30 (21.50)	1.55 (6.11) p = 0.392

Kgs = kilograms; POST = postintervention; PRE = preintervention; Q = quadriceps; SD = standard deviation, TA = tibialis anterior.

3.4. Adverse Effects

No moderate or severe adverse events were reported in any of the participants. Thus, the stimulation was determined as a safe procedure. However, 66.6% (n = 10) of participants could not tolerate the optimal intensity previously determined (PRM threshold), considering the intervention as "very uncomfortable". Two participants reported paraesthesia in lower limb during stimulation. Moreover, two patients reported DOMS (delayed onset muscle soreness) in quadriceps muscle evaluated 24 h after stimulation which could be generated by the voluntary isometric muscle contractions performed during the strength evaluation.

3.5. Blinding and the Assessment of Its Success

The guesses of the participants and assessor are shown in Table 4 in a 2 × 5 table format (Table 4). Furthermore, Table 5 shows James' BI and Bang's BI values obtained for the study subjects and assessor. According to the interpretation of data established by James et al. [34] and Bang et al. [30], when blinding was analyzed globally by James's BI a lack of blinding was observed in participants and assessor. However, when each treatment arm (Active tSCS and Sham tSCS) was analyzed independently by Bang's BI a successful blinding was observed for the participants and assessor in the sham group and the active group, respectively.

Table 4. Assessment of the blinding. Absolutes values and percentages of answers selected by participants and assessor.

	Participants' Guess, n (%)					
Assignment	**Strongly Active tSCS**	**Somewhat Active tSCS**	**Strongly Sham tSCS**	**Somewhat Sham tSCS**	**Do not Know**	**Total**
Active tSCS	9 (30.0)	4 (13.3)	0 (0)	0 (0)	2 (6.7)	15 (50.0)
Sham tSCS	4 (13.3)	4 (13.3)	2 (6.7)	4 (13.3)	1 (3.3)	15 (50.0)
Total	13 (43.3)	8 (26.7)	2 (6.7)	4 (13.3)	3 (10.0)	30 (100.0)
	Assessor's Guess, n (%)					
Assignment	**Strongly Active tSCS**	**Somewhat Active tSCS**	**Strongly Sham tSCS**	**Somewhat Sham tSCS**	**Do not Know**	**Total**
Active tSCS	3 (10.0)	6 (20.0)	1 (3.3)	3 (10.0)	2 (6.7)	15 (50.0)
Sham tSCS	0 (0.0)	0 (0.0)	4 (13.3)	9 (30.0)	2 (6.7)	15 (50.0)
Total	3 (10.0)	6 (20.0)	5 (16.7)	12 (40.0)	4 (13.3)	30 (100.0)

Table 5. Statistical analysis of blinding assessment.

Participants Results				
Methods	**Index**	***p*-Value**	**95% Confidence Interval**	**Conclusion**
James	0.36	0.031	0.24 to 0.48	Unblinded
Bang-Active/2 × 5	0.73	<0.001	0.58 to 0.89	Unblinded
Bang-Placebo/2 × 5	−0.07	0.63	−0.40 to 0.27	Blinded
Assessor Results				
Methods	**Index**	***p*-Value**	**95% Confidence Interval**	**Conclusion**
James	0.27	0.001	0.14 to 0.40	Unblinded
Bang-Active/2 × 5	0.17	0.18	−0.13 to 0.46	Blinded
Bang-Placebo/2 × 5	0.73	<0.001	0.58 to 0.89	Unblinded

4. Discussion

The present study shows that a single session of 10 min of tSCS increases the peak-to-peak MEP amplitude of Quadriceps muscle, but without a change in muscle strength. Alpha-motoneurons located more distally below the level of tSCS, namely those innervating the TA muscle, did not show any change in MEP amplitude. This is the first study that describes an increase in corticospinal-evoked lower limb motor-evoked potentials after a single session of active-tSCS when compared to a sham tSCS. There is an urgent need to optimize the motor neuromodulatory effect of tSCS in subjects with SCI using neurophysiological measures that detect changes in both cortical and spinal excitability.

A number of studies in healthy individuals have observed that tSCS located over the T10-T11 vertebral level (L1-L2 spinal level) evokes multi-segmental responses from lower limb motoneurons [17,19–21]. However, both the excitation threshold and the motor response magnitude differ according to the specific positioning of the spinal stimulus [5,21,35]. Our study observed a significant increase of MEP amplitude in the Q muscle, which is innervated by the L1-L2 spinal level that corresponded to the level of tSCS. In contrast, the TA MEP amplitude, innervated by the L4-S1 spinal level, was not modulated by tSCS. Computational studies of the effect of tSCS suggest that afferent fibers with the lowest threshold are activated by spinal stimulation focused at the dorsal root entry zone, which is accessible via the intervertebral foramen [1,36–38]. In the case of TA, the afferents that optimally activate this muscle are located more distally to the site of the stimulating electrode. Another potential explanation for no change in TA MEP amplitude with tSCS is that the applied current intensity was insufficient to stimulate the afferents required to facilitate descending drive to the TA muscle, because the intensity was set with reference to the Q PRM reflex threshold. Previous studies have reported a higher PRM reflex threshold for the TA muscle [17,19,20,35]. In this study it is possible that the tSCS current intensity was not optimal to increase corticospinal drive to TA motoneurons.

Motor evoked potentials from the Q and TA muscles were recorded at 20% of MVC so as to reduce the variability due to changes in corticospinal excitability [39]. Knee extensor MEPs are reliably recorded in the active Q [40] in response to single pulse TMS, which permits measurement of MEP amplitude as a standard measure of corticospinal excitability [41]. This is the first study to identify a selective effect of tSCS on Q MEPs compared to distal lower limb muscles such as the TA. Recently, an effect of non-invasive spinal stimulation applied at 30 Hz pulses, with a 5 kHz carrier frequency, has been identified on MEPs recorded from the upper limb muscles following TMS in cases of cervical spinal cord injury [27]. Previously low frequency stimulation of afferents has been shown to increase corticospinal excitability in the upper limb [42]. In the current study tSCS was applied at 30 Hz at the spinal lumbar level, which led to an increase in CST excitability during spinal stimulation. The TA MEP also revealed a non-significant increase in amplitude both during and immediately after spinal stimulation.

Non-invasive spinal stimulation may activate several afferent systems which are known to modulate descending excitation of spinal systems involved in motor control. Activation of the skin of the anterior aspect of the thigh activates both excitatory and inhibitory spinal interneurons-including transcortical pathways directed to biceps femoris (BF) and Q muscles [43]. Activation of cutaneous afferents within the sural nerve have been shown to facilitate TMS evoked TA muscle activity [44]. Cutaneous afferents are also known to activate transcortical pathways as suggested in studies of people with dorsal column injury or corticospinal tract abnormalities [45], although activation of subcortical pathways may also be possible [27,46]. Cutaneous volleys evoke muscle responses at a transcortical latency followed by a longer-latency excitatory response [47,48] and ongoing voluntary contraction of lower limb muscles can also potentiate this longer-latency excitatory cutaneomuscular response [48].

A role for group I proprioceptive afferents in tSCS-mediated Q muscle facilitation is also possible, especially as spatial facilitation of corticospinal evoked responses is increased with TMS and simultaneous afferent activation following common peroneal nerve activation, mediated at an interneuronal site [49]. Furthermore, during weak voluntary contraction of the Q muscle, common peroneal nerve [50] or femoral nerve [51] group I mediated facilitation of the Q H-reflex was increased. These studies suggest that the main role of descending facilitation of feedback inhibitory interneurons to lumbar propriospinal neurons during weak Quadriceps contraction is to focus corticospinal control of a limited number of motoneurons required for muscle activation.

The increase of the Q MEP amplitude was first detected 4 min after the onset of the 10 min stimulation, and supports previous studies that have observed an immediate increase of voluntary EMG activity, voluntary movement, and gait function in subjects with spinal cord injury as soon as the current is switched on [8,52]. In our study, the increase in peak-to-peak Q and TA MEP amplitude was not maintained up to 2 and 4 min after tSCS. In contrast in a recent study a 20-min single session of tSCS increased the amplitude of subcortical motor evoked responses in arm muscles of people with a high level of SCI for up to 75 min [27]. Moreover, other studies have identified prolonged post intervention effects of tSCS based on improvement of spasticity [13] or the subjective perception of voluntary movement [8]. Other studies have identified improved upper arm function at 3–6 months after 20–30 repeated sessions of tSCS measured with handgrip strength, upper limb EMG activity and an improvement in general functionality (Upper Extremity Motor scores and Action Research Arm Test) [23,53]. These studies applied tSCS for greater than 20 min, which in addition to our study, supports the hypothesis that repeated tSCS interventions of longer than 20 min is required to elicit immediate postintervention effects that lead to an effective therapeutic option.

To date, the vast majority of studies have fixed the intensity of tSCS by using the subjective perception of sensations made by participants or more directly by detecting muscle contraction in response to the stimulus [6]. Recently, Serrano-Munoz et al. [54] demonstrated the practical limitations of adjusting the intensity of stimulation current based on subjective perception and highlighted the importance of defining objective parameters to adjust current intensity. In this study, PRM reflex threshold was used as an objective measure to adjust stimulation intensity. However, only 44% of the participants could tolerate this objectively defined intensity when tSCS was applied at this intensity at 30 Hz (see Table 1). The other participants received a more tolerable stimulus intensity based on their subjective tolerance threshold. Further protocols are necessary to explore a wide range of tSCS frequencies and intensities to improve the tolerance threshold of subjects in order to optimize the effect of neuromodulation on motor function.

The study was performed with the subject in a relaxed supine position, a posture which has been shown to be associated with the most effective tSCS neuromodulatory effect [55]. Indeed, currently the most optimal translation of the tSCS technique to clinical practice is to adopt supine-position stimulation strategies in combination with other activity-based neurorehabilitation tasks. However future studies should examine position-dependent tSCS efficacy using the MEP technique with the subject performing tasks such as sitting or standing, which are more compatible with actual rehabilitation techniques.

Although MEPs evoked by TMS have been demonstrated to be a reliable tool to quantify the excitability of the corticospinal pathway [40,56], it is questionable if MEPs are sensitive enough to reflect functionally relevant changes evoked by the tSCS. In the present study, both TA and Q MEPs were stable at baseline, and were reproducible when measured at the two pre-evaluation test periods. Stable Q MEPs recorded before tSCS together with the absence of changes during the sham intervention support the significant neuromodulatory effect of active tSCS on Q muscle activity. Furthermore, this finding supports the use of MEPs in future studies designed to optimize tSCS parameters including the specific location of the electrodes.

Sham-controlled interventions have been applied previously to control the effect of peripheral electrical stimulation [28,29] and electrical stimulation of the CNS [57,58]. To our knowledge this is the first study of tSCS where the success of the blinding procedure has been formally assessed using the specialized blinding questionnaire [30], which was completed by both the assessor and the participants. The sham protocol was evaluated as a good method to blind participants who were to receive the sham tSCS, but not when they received active intervention, because 73% of participants correctly guess the assignment. This issue may have arisen due to fact that the previous experience of participants with electrical stimulation was not controlled appropriately, possibly because of the perception of the high stimulus intensity. Previous studies of transcranial direct current stimulation (tDCS) have shown that the application of high intensity electrical stimulation hinders the blinding of participants [59,60]. In contrast, the current study has shown that the blinding procedure was evaluated as good when used to mask the assessor using the active tSCS intervention, but not with sham stimulus application, because the assignment was correctly guessed by the assessor in 73% of participants. Although James's Blinding Index indicated that blinding was not generally effective as assessed by participants and the assessor, the upper 95% CI limit was 0.48 for participants and 0.40 for the assessor, close to the value of 0.5 which is defined as successful. Finally, it is necessary to highlight that crossover studies are inherently difficult to blind compared to more effective parallel design studies. Although further improvement in intervention blinding techniques are still required for the design of future randomized controlled trials, the current sham protocol based on applying only 30 s of active tSCS intensity appears to be a good option to blind both subjects and assessors to the intervention.

No serious side effects were observed after the single 10-min tSCS session in this study. In a recent systematic analysis of the effect of tSCS on motor function after spinal cord injury, this technique has been shown to be painless, well tolerated and safe [6]. However, in our study, eight participants reported discomfort around the electrodes when tSCS commenced. In most cases, this discomfort was related to a strong contraction of abdominal muscle generated by the electrical stimulation. For this reason, 10 participants could not tolerate the intensity of current previously set according to the PRM reflex threshold. Considering that the participants of this study received a single session of tSCS, it is possible that stimulus training sessions are necessary so that participants become accustomed to the tSCS and to improve tolerance. Thus, in a recent study which involved the self-administration of the tSCS applied over 6 week in subjects with spinal cord injury has been shown to be safe and well tolerated in a home-based setting [13]. Future studies should consider employing repeated sessions of tSCS where the intensity is increased incrementally to improve subject tolerance and therefore allowing higher current density and more effective neuromodulation.

5. Conclusions

A single 10 min session of active tSCS applied over the T11-T12 vertebral level increased the quadriceps muscle MEP but not for the TA muscle, when compared to sham stimulation. The neuromodulatory effect of the motor response was specific to the stimulation of the metameric level innervating the Q muscle. In this study we have shown that MEP amplitude is a reliable and sensitive technique to demonstrate changes in the excitability of the corticospinal tract after active tSCS. The used method for sham stimulation could be a feasible option for future tSCS interventions. More randomized sham-controlled clinical trials will be needed to optimize the efficacy of active tSCS to promote motor recovery in subjects with spinal cord injury.

Author Contributions: Conceptualization, J.G.-S., D.S.-M. and J.T.; methodology, Á.M.-G., J.G.-S., D.S.-M. and J.T.; formal analysis, Á.M.-G. and J.A.-C.; investigation, Á.M.-G. and N.C.-S.; writing—original draft preparation, Á.M.-G., D.S.-M. and J.G.-S.; writing—review and editing, Á.M.-G., D.S.-M., J.G.-S., J.T. and J.A.-C.; visualization, D.S.-M., N.C.-S. and J.A.-C.; supervision, J.G.-S. and J.T.; project administration, J.G.-S. and J.T.; funding acquisition, J.G.S. All authors have read and agreed to the published version of the manuscript.

Funding: This research was funded by the "Instituto de Salud Carlos III" and "Fondos FEDER" (NEUROTRAIN project, grant number PI17/00581). DSM has received funding by the European Regional Development Fund (2019/7375).

Acknowledgments: We are grateful to all volunteers for their altruism to participate in this study.

Conflicts of Interest: The authors declare no conflict of interest. The funders had no role in the design of the study; in the collection, analyses, or interpretation of data; in the writing of the manuscript, or in the decision to publish the results.

References

1. Danner, S.M.; Hofstoetter, U.S.; Ladenbauer, J.; Rattay, F.; Minassian, K. Can the human lumbar posterior columns be stimulated by transcutaneous spinal cord stimulation? A modeling study. *Artif. Organs* **2011**, *35*, 257–262. [CrossRef] [PubMed]

2. Gorodnichev, R.M.; Pivovarova, E.A.; Puhov, A.; Moiseev, S.A.; Savochin, A.A.; Moshonkina, T.R.; Chsherbakova, N.A.; Kilimnik, V.A.; Selionov, V.A.; Kozlovskaya, I.B. Transcutaneous electrical stimulation of the spinal cord: A noninvasive tool for the activation of stepping pattern generators in humans. *Hum. Physiol.* **2012**, *38*, 158–167. [CrossRef]

3. Hofstoetter, U.S.; Freundl, B.; Binder, H.; Minassian, K. Common neural structures activated by epidural and transcutaneous lumbar spinal cord stimulation: Elicitation of posterior root-muscle reflexes. *PLoS ONE* **2018**, *13*, e0192013. [CrossRef] [PubMed]

4. Minassian, K.; Persy, I.; Rattay, F.; Dimitrijevic, M.R.; Hofer, C.; Kern, H. Posterior root-muscle preflexes elicited by transcutaneous stimulation of the human lumbosacral cord. *Muscle Nerve* **2007**, *35*, 327–336. [CrossRef] [PubMed]

5. Kitano, K.; Koceja, D.M. Spinal reflex in human lower leg muscles evoked by transcutaneous spinal cord stimulation. *J. Neurosci. Methods* **2009**, *180*, 111–115. [CrossRef]

6. Megía García, A.; Serrano-Muñoz, D.; Taylor, J.; Avendaño-Coy, J.; Gómez-Soriano, J. Transcutaneous Spinal Cord Stimulation and Motor Rehabilitation in Spinal Cord Injury: A Systematic Review. *Neurorehabil. Neural Repair* **2020**, *34*, 3–12. [CrossRef]

7. Hofstoetter, U.S.; Hofer, C.; Kern, H.; Danner, S.M.; Mayr, W.; Dimitrijevic, M.R.; Minassian, K. Effects of transcutaneous spinal cord stimulation on voluntary locomotor activity in an incomplete spinal cord injured individual. *Biomed. Tech.* **2013**. [CrossRef]

8. Hofstoetter, U.S.; Krenn, M.; Danner, S.M.; Hofer, C.; Kern, H.; Mckay, W.B.; Mayr, W.; Minassian, K. Augmentation of Voluntary Locomotor Activity by Transcutaneous Spinal Cord Stimulation in Motor-Incomplete Spinal Cord-Injured Individuals. *Artif. Organs* **2015**, *39*, 176–186. [CrossRef]

9. Rath, M.; Vette, A.H.; Ramasubramaniam, S.; Li, K.; Burdick, J.; Edgerton, V.R.; Gerasimenko, Y.P.; Sayenko, D.G. Trunk stability enabled by noninvasive spinal electrical stimulation after spinal cord injury. *J. Neurotrauma* **2018**, *35*, 2540–2553. [CrossRef]

10. Sayenko, D.G.; Rath, M.; Ferguson, A.R.; Burdick, J.W.; Havton, L.A.; Edgerton, V.R.; Gerasimenko, Y.P. Self-assisted standing enabled by non-invasive spinal stimulation after spinal cord injury. *J. Neurotrauma* **2019**, *36*, 1435–1450. [CrossRef]

11. Minassian, K.; Hofstoetter, U.S.; Danner, S.M.; Mayr, W.; Bruce, J.A.; McKay, W.B.; Tansey, K.E. Spinal rhythm generation by step-induced feedback and transcutaneous posterior root stimulation in complete spinal cord-injured individuals. *Neurorehabil. Neural Repair* **2016**. [CrossRef] [PubMed]

12. Gad, P.; Gerasimenko, Y.; Zdunowski, S.; Turner, A.; Sayenko, D.; Lu, D.C.; Edgerton, V.R. Weight bearing over-ground stepping in an exoskeleton with non-invasive spinal cord neuromodulation after motor complete paraplegia. *Front. Neurosci.* **2017**, *11*, 333. [CrossRef] [PubMed]

13. Hofstoetter, U.S.; Freundl, B.; Danner, S.M.; Krenn, M.J.; Mayr, W.; Binder, H.; Minassian, K. Transcutaneous Spinal Cord Stimulation Induces Temporary Attenuation of Spasticity in Individuals with Spinal Cord Injury. *J. Neurotrauma* **2020**, *37*, 481–493. [CrossRef]

14. Hofstoetter, U.S.; McKay, W.B.; Tansey, K.E.; Mayr, W.; Kern, H.; Minassian, K. Modification of spasticity by transcutaneous spinal cord stimulation in individuals with incomplete spinal cord injury. *J. Spinal Cord Med.* **2014**, *37*, 202–211. [CrossRef]

15. MacKay-Lyons, M. Central pattern generation of locomotion: A review of the evidence. *Phys. Ther.* **2002**, *82*, 69–83. [CrossRef]

16. Sabbahi, M.A.; Sengul, Y.S. Thoracolumbar multisegmental motor responses in the upper and lower limbs in healthy subjects. *Spinal Cord* **2011**. [CrossRef] [PubMed]

17. Krenn, M.; Hofstoetter, U.S.; Danner, S.M.; Minassian, K.; Mayr, W. Multi-Electrode Array for Transcutaneous Lumbar Posterior Root Stimulation. *Artif. Organs* **2015**, *39*, 834–840. [CrossRef] [PubMed]

18. Bocci, T.; Caleo, M.; Tognazzi, S.; Francini, N.; Briscese, L.; Maffei, L.; Rossi, S.; Priori, A.; Sartucci, F. Evidence for metaplasticity in the human visual cortex. *J. Neural Transm.* **2014**, *121*, 221–231. [CrossRef]

19. Sayenko, D.G.; Atkinson, D.A.; Dy, C.J.; Gurley, K.M.; Smith, V.L.; Angeli, C.; Harkema, S.J.; Edgerton, V.R.; Gerasimenko, Y.P. Spinal segment-specific transcutaneous stimulation differentially shapes activation pattern among motor pools in humans. *J. Appl. Physiol.* **2015**. [CrossRef]

20. Krenn, M.; Toth, A.; Danner, S.M.; Hofstoetter, U.S.; Minassian, K.; Mayr, W. Selectivity of transcutaneous stimulation of lumbar posterior roots at different spinal levels in humans. *Biomed. Tech.* **2013**. [CrossRef]

21. Gerasimenko, Y.; Gorodnichev, R.; Puhov, A.; Moshonkina, T.; Savochin, A.; Selionov, V.; Roy, R.R.; Lu, D.C.; Edgerton, V.R. Initiation and modulation of locomotor circuitry output with multisite transcutaneous electrical stimulation of the spinal cord in noninjured humans. *J. Neurophysiol.* **2015**, *113*, 834–842. [CrossRef] [PubMed]

22. Gad, P.; Lee, S.; Terrafranca, N.; Zhong, H.; Turner, A.; Gerasimenko, Y.; Edgerton, V.R. Non-invasive activation of cervical spinal networks after severe paralysis. *J. Neurotrauma* **2018**, *35*, 2145–2158. [CrossRef] [PubMed]

23. Freyvert, Y.; Yong, N.A.; Morikawa, E.; Zdunowski, S.; Sarino, M.E.; Gerasimenko, Y.; Edgerton, V.R.; Lu, D.C. Engaging cervical spinal circuitry with non-invasive spinal stimulation and buspirone to restore hand function in chronic motor complete patients. *Sci. Rep.* **2018**, *8*, 1–10. [CrossRef] [PubMed]

24. Hofstoetter, U.S.; Freundl, B.; Binder, H.; Minassian, K. Recovery cycles of posterior root-muscle reflexes evoked by transcutaneous spinal cord stimulation and of the H reflex in individuals with intact and injured spinal cord. *PLoS ONE* **2019**, *14*, 1–20. [CrossRef] [PubMed]

25. Hofstoetter, U.Ṣ.; Minassian, K.; Hofer, C.; Mayr, W.; Rattay, F.; Dimitrijevic, M.R. Modification of reflex responses to lumbar posterior root stimulation by motor tasks in healthy subjects. *Artif. Organs* **2008**, *32*, 644–648. [CrossRef] [PubMed]

26. Andrews, J.C.; Stein, R.B.; Roy, F.D. Post-activation depression in the human soleus muscle using peripheral nerve and transcutaneous spinal stimulation. *Neurosci. Lett.* **2015**, *589*, 144–149. [CrossRef]

27. Benavides, F.D.; Jo, H.J.; Lundell, H.; Edgerton, V.R.; Gerasimenko, Y.; Perez, M.A. Cortical and Subcortical Effects of Transcutaneous Spinal Cord Stimulation in Humans with Tetraplegia. *J. Neurosci.* **2020**, *40*, 2633–2643. [CrossRef]

28. Petrie, J.; Hazleman, B. Credibility of placebo transcutaneous nerve stimulation and acupuncture. *Clin. Exp. Rheumatol.* **1985**, *3*, 151–153.

29. Deyo, R.A.; Walsh, N.E.; Schoenfeld, L.S.; Ramamurthy, S. Can trials of physical treatments be blinded? The example of transcutaneous electrical nerve stimulation for chronic pain. *Am. J. Phys. Med. Rehabil.* **1990**, *69*, 6–10. [CrossRef]

30. Bang, H.; Flaherty, S.P.; Kolahi, J.; Park, J. Blinding assessment in clinical trials: A review of statistical methods and a proposal of blinding assessment protocol. *Clin. Res. Regul. Aff.* **2010**, *27*, 42–51. [CrossRef]

31. Mentiplay, B.F.; Perraton, L.G.; Bower, K.J.; Adair, B.; Pua, Y.-H.; Williams, G.P.; McGaw, R.; Clark, R.A. Assessment of lower limb muscle strength and power using hand-held and fixed dynamometry: A reliability and validity study. *PLoS ONE* **2015**, *10*, e0140822. [CrossRef]

32. Hopkins, W.G. Sports Medicine 30: 1–15 July 2000. *Sport. Med.* **2000**, *30*, 1–15. [CrossRef]

33. Abián-Vicén, J.; Aparicio-Garciá, C.; Ruiz-Lázaro, P.; Simón-Martínez, C.; Bravo-Esteban, E.; Gómez-Soriano, J. Test-retest reliability and responsiveness of a comprehensive protocol for the assessment of muscle tone of the ankle plantar flexors in healthy subjects. *Isokinet. Exerc. Sci.* **2019**, *27*, 107–115. [CrossRef]

34. James, K.E.; Bloch, D.A.; Lee, K.K.; Kraemer, H.C.; Fuller, R.K. An index for assessing blindness in a multi-centre clinical trial: Disulfiram for alcohol cessation—A VA cooperative study. *Stat. Med.* **1996**, *15*, 1421–1434. [CrossRef]

35. Knikou, M. Neurophysiological characterization of transpinal evoked potentials in human leg muscles. *Bioelectromagnetics* **2013**. [CrossRef] [PubMed]

36. Hofstoetter, U.S.; Danner, S.M.; Minassian, K. Paraspinal Magnetic and Transcutaneous Electrical Stimulation. *Encycl. Comput. Neuroscience.* **2014**. [CrossRef]

37. Capogrosso, M.; Wenger, N.; Raspopovic, S.; Musienko, P.; Beauparlant, J.; Luciani, L.B.; Courtine, G.; Micera, S. A computational model for epidural electrical stimulation of spinal sensorimotor circuits. *J. Neurosci.* **2013**, *33*, 19326–19340. [CrossRef]

38. Ladenbauer, J.; Minassian, K.; Hofstoetter, U.S.; Dimitrijevic, M.R.; Rattay, F. Stimulation of the human lumbar spinal cord with implanted and surface electrodes: A computer simulation study. *IEEE Trans. Neural Syst. Rehabil. Eng.* **2010**, *18*, 637–645. [CrossRef]

39. Darling, W.G.; Wolf, S.L.; Butler, A.J. Variability of motor potentials evoked by transcranial magnetic stimulation depends on muscle activation. *Exp. Brain Res.* **2006**, *174*, 376–385. [CrossRef]

40. O'leary, T.J.; Morris, M.G.; Collett, J.; Howells, K. Reliability of single and paired-pulse transcranial magnetic stimulation in the vastus lateralis muscle. *Muscle Nerve* **2015**, *52*, 605–615. [CrossRef]

41. Burke, M.J.; Fried, P.J.; Pascual-Leone, A. Transcranial magnetic stimulation: Neurophysiological and clinical applications. In *Handbook of Clinical Neurology*; Elsevier: Amsterdam, The Netherlands, 2019; Volume 163, pp. 73–92. ISBN 0072-9752.

42. Kaelin-Lang, A.; Luft, A.R.; Sawaki, L.; Burstein, A.H.; Sohn, Y.H.; Cohen, L.G. Modulation of human corticomotor excitability by somatosensory input. *J. Physiol.* **2002**, *540*, 623–633. [CrossRef] [PubMed]

43. Hagbarth, K.E.; Finer, B.L. The plasticity of human withdrawal reflexes to noxious skin stimuli in lower limbs. In *Progress in Brain Research*; Elsevier: Amsterdam, The Netherlands, 1963; Volume 1, pp. 65–81. ISBN 0079-6123.

44. Nielsen, J.; Petersen, N.; Fedirchuk, B. Evidence suggesting a transcortical pathway from cutaneous foot afferents to tibialis anterior motoneurones in man. *J. Physiol.* **1997**, *501*, 473–484. [CrossRef] [PubMed]

45. Jenner, J.R.; Stephens, J.A. Cutaneous reflex responses and their central nervous pathways studied in man. *J. Physiol.* **1982**, *333*, 405–419. [CrossRef] [PubMed]

46. Shimamura, M.; Mori, S.; Yamauchi, T. Effects of spino-bulbo-spinal reflex volleys on extensor motoneurons of hindlimb in cats. *J. Neurophysiol.* **1967**, *30*, 319–332. [CrossRef] [PubMed]

47. Aniss, A.M.; Gandevia, S.C.; Burke, D. Reflex responses in active muscles elicited by stimulation of low-threshold afferents from the human foot. *J. Neurophysiol.* **1992**, *67*, 1375–1384. [CrossRef]

48. Gibbs, J.; Harrison, L.M.; Stephens, J.A. Cutaneomuscular reflexes recorded from the lower limb in man during different tasks. *J. Physiol.* **1995**, *487*, 237–242. [CrossRef]

49. Marchand-Pauvert, V.; Simonetta-Moreau, M.; Pierrot-Deseilligny, E. Cortical control of spinal pathways mediating group II excitation to human thigh motoneurones. *J. Physiol.* **1999**, *517*, 301–313. [CrossRef]

50. Forget, R.; Hultborn, H.; Meunier, S.; Pantieri, R.; Pierrot-Deseilligny, E. Facilitation of quadriceps motoneurones by group I afferents from pretibial flexors in man. 2. Changes occurring during voluntary contraction. *Exp. Brain Res.* **1989**, *78*, 21–27. [CrossRef]

51. Hultborn, H.; Meunier, S.; Pierrot-Deseilligny, E.; Shindo, M. Changes in polysynaptic Ia excitation to quadriceps motoneurones during voluntary contraction in man. *Exp. Brain Res.* **1986**, *63*, 436–438. [CrossRef]

52. Gerasimenko, Y.P.; Lu, D.C.; Modaber, M.; Zdunowski, S.; Gad, P.; Sayenko, D.G.; Morikawa, E.; Haakana, P.; Ferguson, A.R.; Roy, R.R.; et al. Noninvasive Reactivation of Motor Descending Control after Paralysis. *J. Neurotrauma* **2015**, *32*, 1968–1980. [CrossRef]

53. Inanici, F.; Samejima, S.; Gad, P.; Edgerton, V.R.; Hofstetter, C.P.; Moritz, C.T. Transcutaneous electrical spinal stimulation promotes long-term recovery of upper extremity function in chronic tetraplegia. *IEEE Trans. Neural Syst. Rehabil. Eng.* **2018**, *26*, 1272–1278. [CrossRef] [PubMed]

54. Serrano-Munoz, D.; Gómez-Soriano, J.; Bravo-Esteban, E.; Vazquez-Farinas, M.; Taylor, J.; Avendano-Coy, J. Intensity matters: Therapist-dependent dose of spinal transcutaneous electrical nerve stimulation. *PLoS ONE* **2017**, *12*, e0189734. [CrossRef] [PubMed]

55. Danner, S.M.; Krenn, M.; Hofstoetter, U.S.; Toth, A.; Mayr, W.; Minassian, K. Body position influences which neural structures are recruited by lumbar transcutaneous spinal cord stimulation. *PLoS ONE* **2016**. [CrossRef] [PubMed]

56. Groppa, S.; Oliviero, A.; Eisen, A.; Quartarone, A.; Cohen, L.G.; Mall, V.; Kaelin-Lang, A.; Mima, T.; Rossi, S.; Thickbroom, G.W. A practical guide to diagnostic transcranial magnetic stimulation: Report of an IFCN committee. *Clin. Neurophysiol.* **2012**, *123*, 858–882. [CrossRef]

57. Geiger, M.; Supiot, A.; Zory, R.; Aegerter, P.; Pradon, D.; Roche, N. The effect of transcranial direct current stimulation (tDCS) on locomotion and balance in patients with chronic stroke: Study protocol for a randomised controlled trial. *Trials* **2017**, *18*, 492. [CrossRef]

58. Straudi, S.; Buja, S.; Baroni, A.; Pavarelli, C.; Pranovi, G.; Fregni, F.; Basaglia, N. The effects of transcranial direct current stimulation (tDCS) combined with group exercise treatment in subjects with chronic low back pain: A pilot randomized control trial. *Clin. Rehabil.* **2018**, *32*, 1348–1356. [CrossRef]

59. O'connell, N.E.; Cossar, J.; Marston, L.; Wand, B.M.; Bunce, D.; Moseley, G.L.; De Souza, L.H. Rethinking clinical trials of transcranial direct current stimulation: Participant and assessor blinding is inadequate at intensities of 2mA. *PLoS ONE* **2012**, *7*, e47514. [CrossRef]

60. Palm, U.; Reisinger, E.; Keeser, D.; Kuo, M.-F.; Pogarell, O.; Leicht, G.; Mulert, C.; Nitsche, M.A.; Padberg, F. Evaluation of sham transcranial direct current stimulation for randomized, placebo-controlled clinical trials. *Brain Stimul.* **2013**, *6*, 690–695. [CrossRef]

Journal of
Clinical Medicine

MDPI

Article

Posteroanterior Cervical Transcutaneous Spinal Cord Stimulation: Interactions with Cortical and Peripheral Nerve Stimulation

Jaclyn R. Wecht [1,†], William M. Savage [1,†], Grace O. Famodimu [1], Gregory A. Mendez [1], Jonah M. Levine [1], Matthew T. Maher [1], Joseph P. Weir [2], Jill M. Wecht [1,3], Jason B. Carmel [4], Yu-Kuang Wu [1,3] and Noam Y. Harel [1,3,5,*]

[1]   James J. Peters VA Medical Center, Bronx, NY 10468, USA; jwecht@student.nymc.edu (J.R.W.); ws2384@columbia.edu (W.M.S.); grace.famodimu@va.gov (G.O.F.); gregory.mendez1@va.gov (G.A.M.); jonah.levine@va.gov (J.M.L.); matthew.maher@va.gov (M.T.M.); jill.wecht@va.gov (J.M.W.); yu-kuang.wu@va.gov (Y.-K.W.)
[2]   Department of Health, Sport & Exercise Sciences, University of Kansas, Lawrence, KS 66045, USA; joseph.weir@ku.edu
[3]   Department of Rehabilitation and Human Performance, Icahn School of Medicine at Mount Sinai, New York, NY 10029, USA
[4]   Department of Orthopedic Surgery, Columbia University, New York, NY 10032, USA; jbc28@columbia.edu
[5]   Department of Neurology, Icahn School of Medicine at Mount Sinai, New York, NY 10029, USA
*   Correspondence: noam.harel@mssm.edu
†   Contributed equally.

Citation: Wecht, J.R.; Savage, W.M.; Famodimu, G.O.; Mendez, G.A.; Levine, J.M.; Maher, M.T.; Weir, J.P.; Wecht, J.M.; Carmel, J.B.; Wu, Y.-K.; et al. Posteroanterior Cervical Transcutaneous Spinal Cord Stimulation: Interactions with Cortical and Peripheral Nerve Stimulation. *J. Clin. Med.* **2021**, 10, 5304. https://doi.org/10.3390/jcm10225304

Academic Editors: Ursula S. Hofstoetter and Karen Minassian

Received: 4 October 2021
Accepted: 9 November 2021
Published: 15 November 2021

Abstract: Transcutaneous spinal cord stimulation (TSCS) has demonstrated potential to beneficially modulate spinal cord motor and autonomic circuitry. We are interested in pairing cervical TSCS with other forms of nervous system stimulation to enhance synaptic plasticity in circuits serving hand function. We use a novel configuration for cervical TSCS in which the anode is placed anteriorly over ~C4–C5 and the cathode posteriorly over ~T2–T4. We measured the effects of single pulses of TSCS paired with single pulses of motor cortex or median nerve stimulation timed to arrive at the cervical spinal cord at varying intervals. In 13 participants with and 15 participants without chronic cervical spinal cord injury, we observed that subthreshold TSCS facilitates hand muscle responses to motor cortex stimulation, with a tendency toward greater facilitation when TSCS is timed to arrive at cervical synapses simultaneously or up to 10 milliseconds after cortical stimulus arrival. Single pulses of subthreshold TSCS had no effect on the amplitudes of median H-reflex responses or F-wave responses. These findings support a model in which TSCS paired with appropriately timed cortical stimulation has the potential to facilitate convergent transmission between descending motor circuits, segmental afferents, and spinal motor neurons serving the hand. Studies with larger numbers of participants and repetitively paired cortical and spinal stimulation are needed.

Keywords: spinal cord stimulation; cervical spinal cord injury; motor evoked potentials

1. Introduction

Both invasive and non-invasive forms of repetitive electrical spinal cord stimulation have shown great promise in amplifying supraspinal influence over the sublesional cord after spinal cord injury (SCI) [1–9]. Notably, a non-invasive approach to spinal cord stimulation carries significantly lower risk with greater potential for widespread implementation, especially at the cervical level.

Transcutaneous spinal cord stimulation (TSCS) paradigms generally involve cathodal stimulation over the cord, with anodes usually placed over the iliac crests or abdomen [5–7,9]. We and others have demonstrated that single-pulse cervical TSCS can be safely performed using a posteroanterior configuration with the cathode placed over the upper thoracic spinous processes and the anode placed over the anterior surface of the neck [10,11]. This posteroanterior TSCS configuration easily elicits muscle responses across multiple cervical myotomes through a mix of sensory afferent and motor efferent circuit activation. At low stimulus intensities, posteroanterior TSCS appears to activate predominantly sensory afferent circuits, whereas at higher stimulus intensities, motor efferents are directly activated [11].

Sensory afferent fiber activation likely mediates the beneficial effects of both epidural and transcutaneous spinal cord stimulation [9,12–15]. We are interested in exploring the potential for pairing TSCS-mediated afferent activation with appropriately timed descending motor signals to enhance synaptic potentiation through heterosynaptic summation. In this study, we measured upper extremity muscle responses to single-pulse transcranial magnetic stimulation (TMS) conditioned with subthreshold cervical TSCS at varying interstimulus intervals in individuals with chronic cervical SCI and able-bodied volunteers. To provide further insight into posteroanterior TSCS effects on motor neuron excitability and synaptic transmission within the cervical cord, we also measured interactions between TSCS and peripheral nerve stimulation at either supramaximal (F-wave responses) or submaximal (H-reflex responses) stimulus intensity. We hypothesized that low-intensity posteroanterior cervical TSCS would increase upper extremity muscle responses to motor cortical stimulation in a timing-dependent manner.

2. Materials and Methods

2.1. Design

This prospective human research study was approved by the Institutional Review Board of the James J. Peters VA Medical Center, Bronx, NY (MIRB# 01743). All applicable institutional and governmental regulations concerning the ethical participation of human volunteers were followed during the course of this research. This manuscript reports data from TSCS-conditioning experiments registered at clinicaltrials.gov (NCT03414424).

2.2. Participants

Individuals between ages 21 and 75 without neurological injury (able-bodied or AB) and those with chronic cervical SCI were eligible for participation. For participants with SCI, inclusion criteria included duration of injury greater than 12 months, level of injury between C2 and C8, and incomplete paresis of intrinsic muscles in either hand. All participants required detectable F-wave responses of left or right abductor pollicis brevis (APB) muscle to median nerve stimulation or first dorsal interosseous (FDI) muscle to ulnar nerve stimulation, and detectable motor evoked potentials (greater than 50 µV) in left or right APB or FDI muscle to TMS. Exclusion criteria included ventilator dependence, open tracheostomy site or other open lesions over the neck, shoulders, or arms, multiple sclerosis, stroke, amyotrophic lateral sclerosis, or other serious neurological disorder, hemorrhagic brain injury, seizures, medications that increase seizure risk, self-reported recurrent spontaneous bouts of symptomatic autonomic dysreflexia, significant coronary artery or cardiac conduction disease, bipolar disorder, active psychosis, pregnancy, or implanted electrical or ferromagnetic devices [16]. Participant ID numbers were assigned in order of study enrollment.

2.3. General Protocol

Sessions were performed at a consistent time of day in each participant, with attempts to maintain consistent timing of caffeine intake if applicable. Stimulation was delivered with participants in an upright seated position in an adjustable TMS chair (Magventure) or in an individual's personal wheelchair if preferred. Arms were flexed at roughly 90 degrees, with the hands pronated and relaxed on a pillow cushion placed in the participant's lap. For AB participants, stimulation was targeted toward the dominant arm. For those with SCI, stimulation was targeted toward the arm with lower motor thresholds and more consistent electrophysiological responses to central and peripheral stimulation. Blood pressure, heart rate, pulse oximetry, and symptoms, such as a headache, chest tightness, shortness of breath, and palpitations, were monitored and recorded every three minutes during TSCS, and no less than every 15 min during other portions of the protocol. Subjective symptoms related to TMS such as headache, confusion, hearing loss, etc., were assessed according to questions suggested by the International Federation of Clinical Neurophysiology [17]. To further evaluate potential autonomic effects of TSCS, one session incorporated continuous beat-to-beat hemodynamic monitoring, described below.

2.4. Electromyography (EMG)

EMG was recorded using surface sensors with $\times 300$ preamplification, 15–2000 Hz bandwidth, and internal grounding (Motion Lab Systems Z03-002). EMG was collected at a sample rate of 5000 Hz via digital acquisition board and customized LabVIEW software (National Instruments USB-6363). This manuscript reports data for the target arm APB and flexor carpi radialis (FCR). Note that two participants (#18 and #27) had unreliable APB responses, so the target arm first dorsal interosseous (FDI) was analyzed.

2.5. Transcutaneous Spinal Cord Stimulation (TSCS)

Stimulation was delivered using 5×10 cm electrodes (Natus 019-422200). The cathode electrode was placed longitudinally over the posterior midline with the cephalad edge ~4 cm caudal to the C7 spinous process, corresponding to the T2–T4 vertebral levels posteriorly. The anode electrode was placed horizontally over the anterior midline with the caudal edge ~2–3 cm superior to the sternal notch, corresponding to the C4–C5 levels anteriorly [11]. Two 5×10 cm electrodes over the distal clavicles were connected to a common ground.

Stimulation (2 ms biphasic pulses) was delivered using constant-current stimulators (Digitimer DS7A or DS8R). Resting motor threshold (RMT) was determined as the intensity (in mA) required to elicit a potential in the APB muscle of at least 50 μV in 5 out of 10 repetitions. All subsequent TSCS testing was performed at specified intensities normalized to each individual's APB RMT.

2.6. Transcranial Magnetic Stimulation (TMS)

A MagPro X100 system (Magventure) with 80 mm winged coil (D-B80) was used. The magnet was oriented at a 45-degree angle from the sagittal plane, centered over the hand motor cortex hotspot for maximal APB response. Coil and hotspot positioning were tracked using an optical-based neural navigation system (Brainsight 2.4, Rogue Research, Montreal, QC, Canada). RMT was determined as the percent of maximal stimulator output required to elicit an MEP of at least 50 μV in the resting APB muscle in 5 out of 10 repetitions. All subsequent TMS testing was completed at specified percentages of TMS intensity normalized to each individual's APB RMT.

2.7. Peripheral Nerve Stimulation (PNS)

Stimulation was delivered using constant-current stimulators (Digitimer DS7A or DS8R) via dual surface electrodes (Natus 019-429400) placed over the median nerve at the wrist (F-waves), or the median nerve at the elbow (H-reflex). For F-waves, monophasic 0.2 ms duration pulses were delivered at supramaximal intensity 25 times at 0.5 Hz to record both direct (M-wave) and late (F-wave) responses at the APB. The minimal F-wave latency was used to calculate the peripheral motor conduction time (PMCT) as $(\text{Latency}_M + \text{Latency}_F - 1) \div 2$ [18]. Central motor conduction time (CMCT) was calculated as MEP latency (at 120% of TMS RMT) minus PMCT. For H-reflexes, monophasic 1.0 ms duration pulses were delivered across a range of submaximal intensities at 0.2 Hz to determine the maximal and 50%-maximal H-reflex amplitudes at the FCR muscle. The H-reflex conduction time (HRCT) was calculated as $(\text{Latency}_H - \text{Latency}_M - 1) \div 2$.

F-wave persistence was calculated based on the percentage of total positive F-waves, defined as an F-wave with an amplitude above 20 μV. F-wave peak-to-peak amplitudes were normalized to the maximal compound motor action potential (CMAP) from that session [19].

2.8. Hemodynamic Data Collection

Seven AB volunteers and five participants with SCI underwent an extra experiment with continuous hemodynamic data collection. Prior to initiation of study procedures, participants were asked to empty their bladder and to loosen any tight-fitting clothing or belt. Although the participant rested quietly in the seated position, instrumentation was applied, which included: (1) a three-lead ECG and respiration monitor with electrodes placed at the right and left mid-axillary lines in the 5th intercostal space and at the right anterior axillary line (Model RESP 1 with EKG: UFI, Morro Bay, CA, USA), and (2) a finger BP monitor was placed on the index finger or middle finger of the non-dominant hand (AB) or the non-targeted hand (SCI) (Finapres Medical Systems, Amsterdam, Netherlands) for simultaneous assessment of continuous beat-to-beat BP and HR. After instrumentation, a 5-min baseline assessment of HR and BP was recorded in the seated position prior to stimulation. Beat-to-beat HR and BP were then recorded for 1 min prior to TSCS, during TSCS, and 1 min post TSCS. TSCS was delivered at three intensities relative to RMT in random order: one subthreshold (70%) and two suprathreshold (125% and 175%), at 0.1–0.2 Hz for 6–10 repetitions. There was a 1–2 min rest period between delivery of each TSCS intensity. Beat-to-beat BP and HR signals were sampled at 500 Hz using customized data acquisition programs written with LabVIEW software (version 2014 SP1, National Instruments, Austin, TX, USA). The raw BP and HR data files were stored for offline data analysis conducted using customized software programs written with LabVIEW graphical software.

2.9. TSCS-TMS Interactions

To test whether subthreshold TSCS could facilitate response to suprathreshold TMS (120% of motor threshold), single pulses of TMS were delivered either alone (control) or conditioned with single pulses of TSCS delivered across a range of intensities and interstimulus intervals. TSCS was delivered at 50%, 70%, or 90% RMT, timed to arrive at cervical synapses at intervals ranging from 25 ms prior to TMS arrival to 10 ms after TMS arrival (Table 1). Cervical cord arrival timing was calculated utilizing participant-specific CMCT and PMCT, with subthreshold TSCS conduction time set at 1.5–2 ms for all participants [11,20,21]. Unpaired TSCS pulses were delivered as further controls. Paired or unpaired pulses were delivered at 0.1 Hz in pseudorandom order. Each condition was repeated 8 times per session except for unconditioned TMS, which was repeated 16 times in session 1 and 10 times in session 2. To reduce participant burden (nearly 1000 paired pulse paradigms delivered across experiments), we prioritized more repetitions of unconditioned TMS, as this was the control against which all paired TSCS-TMS paradigms were compared. A subset of the paired TSCS-TMS paradigms was repeated on two separate days to accumulate more repetitions.

2.10. TSCS-PNS Interactions: F-Waves

PNS was delivered over the median nerve at the wrist of the target hand at supramaximal intensity, recording over the APB muscle (except over the ulnar nerve at the wrist, recording over the FDI muscle in two participants (#18 and #27)). PNS was delivered either alone (control) or conditioned with single pulses of TSCS delivered across a range of intensities and interstimulus intervals. TSCS was delivered at 50%, 70%, or 175% RMT. Subthreshold TSCS pulses were timed to arrive at cervical synapses at intervals ranging from 200 ms to 2 ms prior to PNS arrival, or simultaneously with PNS arrival at cervical motor neurons (Table 2). Suprathreshold TSCS pulses were delivered to either arrive at cervical motor neurons 10 ms prior to or simultaneously with PNS arrival to test for collisional interference or facilitation of F-wave responses, respectively [21]. Unpaired TSCS pulses were delivered as further controls. Paired or unpaired pulses were delivered at 0.1 Hz in pseudorandom order. This experiment was repeated on two separate days with partially overlapping conditions to confirm reliability. Each condition was repeated 25 times per session except for unconditioned TSCS, which was repeated 8 times in session 1 and 7 times in session 2.

Table 1. Scheme for conditioning experiments. Conditioning TSCS pulses were delivered at the indicated intensities (in % of TSCS resting motor threshold) at various synaptic delays relative to test pulses (TMS, F-wave, or H-reflex). Synaptic delay represents the time of TSCS pulse arrival at cervical motor neurons relative to test pulse arrival. Negative numbers indicate TSCS pulse arrival before test pulse arrival.

Synaptic Delay (ms)	TSCS Intensity (% RMT)		
	TMS	**F**	**H**
−200		50	
−100			50
−50		50	50
−25	50, 70, 90	50	50
−10	50, 70, 90	50, 70, 175	50
−5		50, 70	50
−2	50, 70, 90	50, 70	50
0	50, 70, 90	50, 175	50
2	50, 70, 90		
5	50, 70, 90		
10	50, 70, 90		

TSCS: Transcutaneous spinal cord stimulation; RMT: Resting motor threshold; TMS, transcranial magnetic stimulation; F, F-wave; H, H-reflex.

2.11. TSCS-PNS Interactions: H-Reflexes

PNS was delivered over the median nerve at the elbow of the target hand, recording at the FCR muscle. PNS intensity was calibrated to result in H-reflex amplitude ~50% of H_{max}. PNS was delivered either alone (control) or conditioned with single pulses of TSCS stimuli (50% RMT intensity) timed to arrive at cervical synapses at intervals ranging from 100 ms to 2 ms prior to PNS arrival, or simultaneously with PNS arrival (Table 2). Unpaired TSCS pulses were delivered as further controls. Paired or unpaired pulses were delivered at 0.1 Hz in pseudorandom order. Each condition was repeated 10 times during one session.

2.12. Data Analysis

Pulse responses were discarded if spontaneous muscular activity was observed within the 50 ms prior to stimulation, if the participant was noted to move during a stimulation, if background electrical noise was above 50 μV, or if the TMS coil was noted to be off-target during experimentation.

Peak-to-peak amplitudes were quantified for all responses to TSCS, TMS, and peripheral nerve stimulation. For conditioning paradigms, each participant's amplitudes of TSCS-conditioned responses (or unpaired TSCS responses) were normalized to that participant's average amplitudes for unconditioned TMS, F-wave, or H-reflex responses. Furthermore, due to the definition of RMT as the intensity at which half of stimuli evoke a response of >50 μV, some TSCS stimuli at 70% and 90% of threshold evoked detectable responses. If at a given TSCS intensity, the average response amplitude across multiple pulses per session was greater than 30% of the response to TMS, then all TSCS-conditioned responses to TMS at that intensity were discarded for that session.

2.13. Statistical Analysis

Data are reported as the mean ± standard error of the mean. Due to non-normal data distributions, non-parametric Mann-Whitney U Tests were used to compare differences between AB and SCI groups for response amplitudes to TMS alone, TSCS alone, and peripheral nerve stimulation (F or H) alone.

Table 2. Participant demographics. NT—non-traumatic; DOI—duration of injury (years); Level—neurological level of injury; Grade—SCI severity according to International Standards for the Neurological Classification of SCI.

SCI ID	Gender	Age	Trauma/NT	DOI (Years)	Level	Grade	Baclofen Use (Oral)
1	M	64	T	35	C4	D	No
3	M	54	T	13	C5	C	No
5	F	22	NT	1.5	C5	C	Yes
12	M	43	T	2	C4	D	Yes
15	M	56	T	20	C7	D	No
16	M	71	T	1.5	C3	D	Yes
17	M	54	T	3	C5	D	No
18	M	38	T	13	C3	C	No
19	F	62	T	4	C3	D	No
23	M	32	T	2	C5	C	No
25	M	26	T	3	C3	B	No
27	F	34	T	2	C3	A	Yes
28	M	63	T	4	C3	C	Yes
AB ID							
2	M	46					
6	M	22					
7	M	55					
8	M	58					
9	F	52					
10	M	47					
11	M	60					
13	F	22					
14	M	22					
20	M	24					
21	M	45					
22	M	26					
24	M	24					
26	M	51					
29	F	27					

For the TSCS conditioning paradigms, linear mixed modeling was performed using a maximum likelihood estimation approach. Fixed effects were: group (AB vs. SCI), TSCS intensity (50%, 70%, 90%, and/or 175%), and synaptic interval (7 levels ranging from −200 ms to +10 ms depending on paradigm). Participants and sessions were random effects. Main and interaction effects were subjected to analysis of variance (ANOVA) using Satterthwaite's method. Significance was set at an alpha level of 0.05.

Excel (Microsoft, Redmond, Washington, DC, USA), SPSS Version 28 (IBM, Armonk, New York, NY, USA), and R (https://www.R-project.org/, accessed on 4 October 2021) were used for all analyses.

Individual-level data are included as Supplementary Tables S2–S8.

3. Results

3.1. Participants

In total, 30 participants (15 AB, 15 SCI; 23 males, 7 females) enrolled, and 28 (15 AB, 13 SCI) passed screening (Table 2). The groups did not differ significantly for age, which ranged from 22 to 71 years old. Of the 13 SCI participants, 12 had traumatic SCI, one had idiopathic transverse myelitis. Data for TMS were excluded from two participants (23 and 27) who were found after screening to either have unacceptable electrical background EMG activity or unreliable responses in resting muscle during TMS. Thus, analysis of TMS results included 15 AB and 11 SCI participants, whereas analysis of TSCS and PNS results included 15 AB and 13 SCI participants.

3.2. TMS Responses

SCI participants showed significantly higher mean ± SEM RMT at the APB muscle (52.2% ± 4.2% maximum stimulator output) than AB participants (40.7% ± 1.7% maximum stimulator output) (p = 0.024, Mann-Whitney U test) (Figure 1; Table 3).

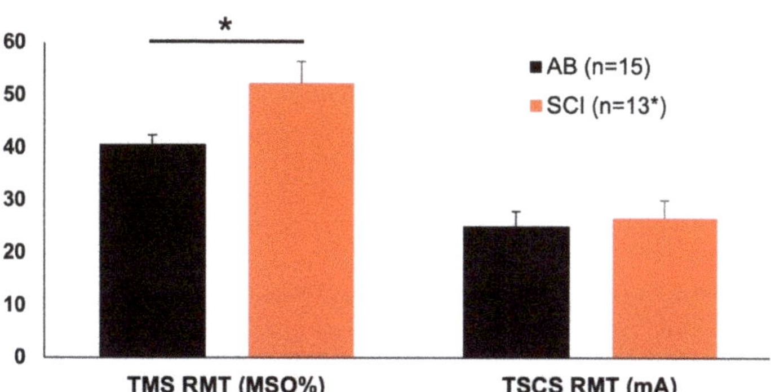

Figure 1. Resting thresholds are higher in SCI participants for TMS but not TSCS. Resting motor threshold (RMT) for the abductor pollicis brevis (in two SCI participants, the first dorsal interosseous). Note that two SCI participants had unobtainable transcranial magnetic stimulation (TMS) responses, whereas all participants responded to transcutaneous spinal cord stimulation (TSCS). Hence, the asterisk next to "13" in the legend. MSO%, percent of maximal stimulator output. mA, milliamperes. Mean and SEM shown. *, p = 0.024.

SCI participants showed significantly lower mean ± SEM amplitudes at 120% of RMT (0.183 mV ± 0.057 mV) than AB participants (0.549 mV ± 0.051 mV) (p < 0.001, Mann-Whitney U test).

3.3. TSCS Responses

As observed in our prior work, there was no difference in TSCS response thresholds between AB and SCI participants. Mean ± SEM TSCS RMT at the APB muscle was 25.1 mA ± 2.8 mA for AB participants and 26.5 ± 3.5 mA for SCI participants (non-significant, Mann-Whitney U test) (Figure 1; Table 3).

3.4. PNS (F-Wave) Responses

SCI participants showed larger amplitude, more persistent F-waves than AB participants. Relative to M_{max}, the amplitude of unconditioned F-waves was 0.019 ± 0.002 for AB participants and 0.093 ± 0.027 for SCI participants (p = 0.014, Mann-Whitney U test). Correspondingly, mean F-wave persistence tended to be lower for AB participants (68.1% ± 4.9%) than for SCI participants (77.0% ± 9.3%) (p = 0.069, Mann-Whitney U test) (Table 3).

3.5. PNS (H-Reflex) Responses

Only 7 AB and 6 SCI participants had reliable FCR H-reflex responses. Relative to M_{max}, maximal H-reflex amplitude was similar in AB (0.293 ± 0.063) and SCI participants (0.229 ± 0.082) (non-significant, Mann-Whitney U test) (Table 3).

Table 3. Responses to unconditioned cortical and spinal stimuli.

Participant	Group	TMS RMT	TMS120 Ampl	TSCS RMT
2	AB	32.0	0.848	27.0
6	AB	38.0	0.714	26.3
7	AB	35.0	0.696	24.3
8	AB	46.0	0.581	45.0
9	AB	35.5	0.289	26.2
10	AB	30.0	0.665	30.8
11	AB	43.5	0.383	45.0
13	AB	35.0	0.511	4.0
14	AB	49.5	0.413	22.3
20	AB	41.0	0.176	16.7
21	AB	40.5	0.466	10.2
22	AB	41.5	0.518	26.7
24	AB	46.0	0.801	21.7
26	AB	52.0	0.758	24.5
29	AB	44.5	0.415	25.7
AB	**Mean**	**40.7**	**0.549**	**25.1**
	SEM	1.7	0.051	2.8
1	SCI	53.0	0.182	15.0
3	SCI	53.5	0.093	33.3
5	SCI	30.5	0.261	5.8
12	SCI	62.0	0.058	27.8
15	SCI	34.0	0.713	38.0
16	SCI	72.0	0.065	38.7
17	SCI	51.0	0.182	42.5
18	SCI	36.5	0.074	23.0
19	SCI	61.0	0.186	4.8
23	SCI			40.1
25	SCI	71.0	0.058	33.7
27	SCI			18.7
28	SCI	50.0	0.142	23.2
SCI	**Mean**	**52.2**	**0.183**	**26.5**
	SEM	4.2	0.057	3.5

TMS: transcranial magnetic stimulation; RMT: Resting motor threshold; TSCS: Transcutaneous spinal cord stimulation. Mean values for each group listed in boldface.

3.6. TSCS-TMS Interactions

TMS-evoked APB muscle amplitudes were facilitated by subthreshold TSCS, with a significant main effect of TSCS intensity (F = 4.401, p = 0.013) and synapse delay (F = 2.520, p = 0.020) but no main effect of group (F = 0.268, p = 0.607). There was significant interaction between group and TSCS intensity (F = 4.205, p = 0.015) but not between group and synapse delay (F = 0.583, p = 0.744) or TSCS intensity and synapse delay (F = 1.174, p = 0.297). Though not statistically significant on pairwise comparisons, TSCS-mediated facilitation tended to be stronger when TSCS pulses arrived at cervical synapses simultaneously or up to 10 ms after TMS pulse arrival (Figure 2, Table 4). Increasing TSCS intensity tended to facilitate TMS responses more in SCI than in AB participants.

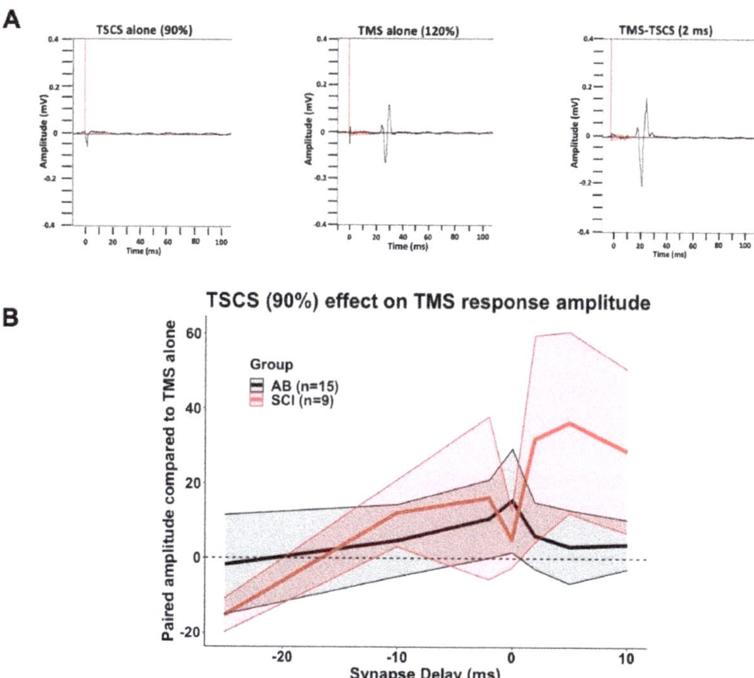

Figure 2. Subthreshold TSCS acutely facilitates TMS-evoked potentials. Suprathreshold (120%) TMS and subthreshold TSCS were given alone or in combination as depicted in Table 1. Response amplitudes were compared to the response to TMS alone (0). Synapse delay represents the time of TSCS pulse arrival at cervical motor neurons relative to test pulse arrival. Negative numbers indicate TSCS pulse arrival before test pulse arrival. (**A**). Representative waves from participant #29. (**B**). Effect of conditioning TSCS at 90% RMT. Black line (grey shading) indicates mean (SEM) for AB participants. Red line (pink shading) indicates mean (SEM) for SCI participants.

Table 4. TMS responses to conditioning TSCS pulses.

Group	TSCS Intensity (% RMT)	Synapse Delay	Compared to TMS Alone (%)	SEM
AB	50%	−25	0.2%	9.4
		−10	15.4%	16.7
		−2	0.6%	8.9
		0	−11.4%	8.9
		2	−0.5%	7.1
		5	8.1%	16.2
		10	8.7%	9.1
		n/a	−98.9%	0.6
	70%	−25	−2.7%	4.8
		−10	2.9%	11.0
		−2	0.7%	6.3
		0	4.4%	7.8
		2	12.6%	11.2

Table 4. *Cont.*

Group	TSCS Intensity (% RMT)	Synapse Delay	Compared to TMS Alone (%)	SEM
		5	9.4%	12.2
		10	20.4%	19.6
		n/a	−97.4%	1.1
	90%	−25	−1.8%	13.3
		−10	4.7%	9.6
		−2	10.6%	10.5
		0	15.5%	13.9
		2	6.1%	8.9
		5	3.1%	9.8
		10	3.7%	6.6
		n/a	−93.1%	1.9
	n/a	n/a	0.0%	0.0
SCI	50%	−25	2.5%	6.4
		−10	4.7%	8.7
		−2	−1.1%	6.4
		0	−4.5%	4.4
		2	13.3%	12.9
		5	−9.1%	6.1
		10	0.6%	4.7
		n/a	−94.5%	2.1
	70%	−25	−10.1%	6.6
		−10	1.5%	7.3
		−2	−2.9%	6.5
		0	5.8%	5.4
		2	7.6%	12.8
		5	9.6%	8.1
		10	17.8%	8.2
		n/a	−92.0%	3.1
	90%	−25	−15.3%	4.5
		−10	12.1%	9.0
		−2	16.2%	21.8
		0	5.1%	7.6
		2	32.0%	27.8
		5	36.4%	24.3
		10	28.8%	21.9
		n/a	−88.2%	3.1
	n/a	n/a	0.0%	0.0

TSCS: Transcutaneous spinal cord stimulation; RMT: Resting motor threshold; n/a: not applicable.

3.7. TSCS-PNS Interactions: F-Waves

Subthreshold TSCS did not substantially affect F-wave response amplitudes in either AB or SCI participants (Table 5). There was no main effect of TSCS intensity (F = 0.377,

$p = 0.540$) or group (F = 3.612, $p = 0.062$). There was a main effect of synapse delay (F = 2.148, $p = 0.048$). However, there were no interaction effects, and no significant effect of any individual synapse delay on pairwise comparisons. Subthreshold TSCS tended to have more effect on F-wave persistence than amplitude. There was a main effect of synapse delay (F = 2.543, $p = 0.021$) and an interaction effect between group and synapse delay (F = 2.684, $p = 0.015$) but no main effect of group (F = 0.898, $p = 0.347$) or intensity (F = 0.186, $p = 0.666$). There was no significant effect of any individual synapse delay on pairwise comparisons.

Suprathreshold (175% RMT) TSCS, previously shown to directly activate motor efferents [11], facilitated an F-wave response when TSCS arrived at cervical motor neurons simultaneously to retrograde F-wave arrival. Suprathreshold TSCS interfered with F-wave transmission when TSCS was timed to arrive at cervical motor neurons 10 ms prior to retrograde F-wave arrival (Figure 3). The difference in amplitude and in persistence between simultaneous and early (collisional) TSCS arrival was significant across both AB and SCI participants: main effect of synapse delay (F = 9.879, $p = 0.003$ for amplitude; F = 19.804, $p < 0.0005$ for persistence); no main effect of group or group × synapse delay interaction.

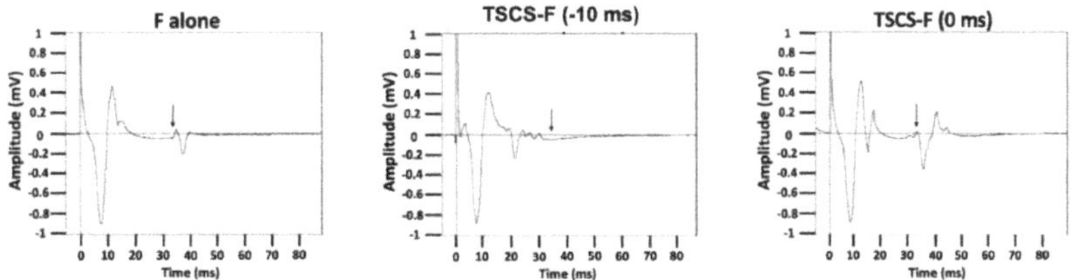

Figure 3. Suprathreshold TSCS collides with or facilitates F-wave responses depending on timing. Supramaximal median nerve stimulation was delivered to generate F-wave responses. Conditioning TSCS pulses at 175% of RMT were timed to arrive at cervical motor neurons either 10 ms prior to or simultaneously with retrograde median nerve pulse arrival.

3.8. TSCS-PNS Interactions: H-Reflexes

Subthreshold TSCS did not significantly affect H-reflex responses at any synaptic interval in either AB or SCI participants (Supplementary Table S1).

3.9. Safety and Hemodynamic Responses

No serious adverse events occurred during this study. The most common mild adverse events were transient headache and neck soreness (5 incidents each). These symptoms were more often related to TMS than to TSCS.

Seven AB and five SCI participants underwent hemodynamic monitoring while receiving TSCS at 70%, 125%, or 175% of RMT. Results of repeated measures ANOVA indicated no significant main or interaction effects for condition (intensity) or group (AB, SCI) on heart rate or systolic BP (Supplementary Figure S1). The largest change in BP (a decrease of, roughly, 18 mm Hg systolic) was noted in AB #6, a 22 year-old man with no neurological history. This participant moved his hands excessively during the experiment, which interfered with the beat-to-beat recording on the 3rd digit of his non-target hand (see Methods) and caused an artifactual drop in BP readings.

Table 5. F-wave responses to conditioning TSCS pulses.

Group	TSCS Intensity (% RMT)	Synapse Delay	Ampl Compared to F Alone (%)	SEM	Persistence Compared to F Alone (%)	SEM
AB	50%	−200	23.1%	11.56	8.9%	6.0
		−50	6.7%	8.6	5.7%	6.3
		−25	1.7%	8.3	−2.8%	6.6
		−10	7.2%	7.2	−4.3%	5.5
		−5	−3.8%	11.6	−5.8%	5.5
		−2	12.0%	10.8	0.3%	5.0
		0	0.8%	7.3	1.8%	6.0
		n/a	−92.8%	2.1	n/a	n/a
	70%	−10	4.3%	7.7	−2.6%	8.0
		−5	16.8%	12.4	9.6%	10.1
		−2	5.8%	9.9	0.2%	5.9
		n/a	−77.4%	6.5	n/a	n/a
	175%	−10	−18.5%	14.6	−27.0%	9.3
		0	328.5%	136.1	39.5%	18.6
		n/a	1229.1%	340.4	n/a	n/a
	n/a	n/a	0.0%	0.00	0.0%	0.00
SCI	50%	−200	9.5%	7.7	6.0%	2.9
		−50	−1.5%	11.6	11.2%	8.9
		−25	−11.0%	9.0	−4.5%	9.4
		−10	−11.5%	8.9	−7.2%	8.0
		−5	−1.9%	6.4	5.7%	3.1
		−2	17.2%	16.9	9.2%	7.2
		0	−9.9%	11.3	−9.9%	9.8
		n/a	−77.7%	9.5	n/a	n/a
	70%	−10	−17.6%	6.5	−5.8%	4.0
		−5	−11.2%	4.5	−5.2%	2.9
		−2	−6.7%	6.2	3.1%	4.9
		n/a	−81.8%	7.7	n/a	n/a
	175%	−10	−27.7%	22.1	−7.0%	37.0
		0	371.4%	189.1	75.4%	35.8
		n/a	1649.4%	604.7	n/a	n/a
	n/a	n/a	0.0%	0.00	0.0%	0.00

TSCS: Transcutaneous spinal cord stimulation; RMT: Resting motor threshold; n/a: not applicable.

4. Discussion

Repetitive stimulation of the brain, spinal cord, and/or peripheral nerves has long been known to affect ensuing muscle activity. The optimal combination of timing, intensity, frequency, waveforms, and participant characteristics have yet to be determined despite numerous studies. Far less study has been directed toward paired stimulation at two sites of the nervous system to magnify the muscular response to stimulation. Paired associative stimulation (PAS) times brain and peripheral nerve stimuli to converge in sensorimotor cortex [22–24]. Rather than cortical convergence, studies have also demonstrated that spinal convergence of segmental stimuli with descending cortical stimuli may have summative effects. In the lumbar spine, suprathreshold TSCS, presumably activating dorsal sensory afferent fibers, facilitated responses in the soleus and hamstring muscles but not tibialis anterior when spinal impulses temporally converged at spinal motor neurons with cortical impulses [25]. Convergent arrival at spinal motor neurons of suprathreshold afferent peripheral stimuli and descending stimuli from TMS leads to summation of responses at several hand [26] and leg muscles [27]. A related paradigm (spike timing-dependent plasticity) in which peripheral stimulation is delivered retrogradely through motor axons timed to converge in the spinal cord with descending cortical stimulation also facilitates hand [20,28,29] and leg [30,31] responses. Invasive experiments in rhesus monkeys further support the concept of facilitating spinal motor neuron responses through heterosynaptic summation partly through dorsal afferent spinal input [32,33].

To increase the applicability of spinal stimulation toward physical rehabilitation protocols, spinal stimulation needs to be delivered at subthreshold intensities, (1) so as not to interfere with ongoing movements, and (2) because noxious suprathreshold stimuli disrupt motor learning [34]. Multiple studies in rodents have shown that subthreshold spinal stimulation facilitates upper limb responses to motor cortex stimulation [15,35–37], though not with uniform results [38]. Furthermore, a recent study of spinal stimulation combined with volitional handgrip exercise in healthy humans showed that subthreshold stimulation achieved greater facilitation of spinal and corticospinal responses than suprathreshold stimulation [39]. To our knowledge, the current study is the first experiment in humans with and without SCI to measure the effect of paired suprathreshold cortical with subthreshold cervical spinal cord stimulation across a range of interstimulus intervals. This study intended to further our understanding of how cervical TSCS delivered in a novel posteroanterior configuration [10,11] interacts with other forms of peripheral and central nervous system stimulation, and to explore the potential for using subthreshold cervical TSCS to facilitate motor responses to brain stimulation.

Based on the studies cited above, we hypothesized that subthreshold TSCS can amplify hand muscle responses to motor cortex stimulation through heterosynaptic summation [24,25]. We expected single pulses of subthreshold TSCS to increase the amplitude of hand muscle responses to single pulses of suprathreshold motor cortex TMS when the pulses were timed to temporally converge at cervical motor synapses. Furthermore, we expected that single pulses of subthreshold TSCS would not affect lower motor neuron responses to F-wave stimulation (non-synaptic), and that single pulses of subthreshold TSCS would reduce the FCR response to H-reflex stimulation (synaptic). These findings would support a model in which TSCS facilitates convergent transmission between descending motor circuits, segmental afferents, and spinal motor neurons rather than directly affecting intrinsic spinal motor neuron excitability [25,40].

Our findings in this study did not fully confirm our hypotheses. Though TMS responses tended to be higher when afferent TSCS arrived at cervical cord synapses simultaneously to up to 10 milliseconds after TMS, these effects were somewhat variable and not statistically significant. For intervals up to ~2 ms between TMS and subsequent electrical signal arrival at the cervical cord, there is established evidence of associative facilitation [15,20,28,37]. For 5–10 ms intervals, we speculate that the arriving TSCS pulse interacts with the TMS volley of indirect waves in a way that facilitates the resulting MEP [40]. One major technical factor that affected the consistency of our results was the way in which RMT was defined. Analogous to the method for determining TMS RMT, we defined TSCS RMT as the intensity required to produce a response >50 μV in 5 out of 10 trials. We often see responses of 20–40 μV to stimuli at 90% and even 70% of RMT, the main intensities of interest used in this study. Thus, even though we discarded data when TSCS responses were greater than 30% of the TMS response, we cannot exclude a non-specific partially suprathreshold facilitation across a wider range of interstimulus intervals than intended [25].

In an unpaired stimulation, we, unsurprisingly, observed higher TMS thresholds and lower TMS-evoked muscle response amplitudes in SCI participants and confirmed our prior findings that TSCS thresholds are similar in AB and SCI participants [11]. We also observed higher F-wave amplitude and persistence in SCI participants than in AB participants. This suggests an increased state of spinal motor neuron excitability after SCI, as has been noted by others [41], especially in the context of lesions in the rostral cervical cord [42]—of the 13 participants with SCI in our study, 6 had SCI at C3, and 2 had SCI at C4. Conversely, we observed a non-significantly higher unpaired H-reflex amplitude in AB participants than in SCI participants, which was surprising given that H-reflexes are usually increased below the level of SCI.

Results of pairing subthreshold TSCS with F-wave responses were mixed. Overall, synapse delay between TSCS and F-wave arrival at the cervical cord appeared to reduce F-wave amplitude and persistence in SCI but not AB participants, though this did not reach significance on pairwise comparisons. We speculate this may have resulted from the higher baseline F-wave amplitude and persistence in SCI participants. The lack of a consistent effect of subthreshold TSCS on F-wave responses corresponds to findings that vibratory stimuli activating Ia afferents do not affect F-waves [43], and is consistent with a model that at low stimulus intensity, TSCS activates dorsal afferent fibers, modulating spinal motor neurons through indirect transsynaptic pathways [10,11,25,44].

Our results with suprathreshold TSCS shed more insight into routes of posteroanterior TSCS transmission. Our previous data had shown that as posteroanterior cervical TSCS intensity increased, post-activation depression of ensuing pulses decreased, suggesting more efferent motor fiber activation [11]. The finding in the current study that TSCS delivery at 175% of RMT 10 ms prior to peripheral F-wave stimulation reduced F-wave amplitude and persistence indicates collisional interference between the anterograde TSCS pulse and retrograde F-wave along motor axons [21]. Likewise, when the interstimulus interval was adjusted for convergent arrival of suprathreshold TSCS and retrograde F-wave at spinal motor neurons, the F-wave amplitude and persistence increased. Both of these findings further support the model that high-intensity posteroanterior cervical TSCS directly activates motor nerve roots.

H-reflex transmission is carried by Ia afferents that synapse with segmental spinal motor neurons. These synapses are susceptible to homosynaptic and post-activation depression, i.e., when H-reflex stimuli are delivered in intervals of roughly five seconds or less, responses are blunted after the first response [9,10,45]. Multiple groups have shown that at most intensities, especially subthreshold intensity, TSCS also activates large-diameter dorsal afferent fibers, and are alternatively termed posterior root reflexes [44]. In this study, TSCS pulses were delivered 200 ms or less prior to H-reflex pulses, well within the window of post-activation depression. Hence, since TSCS pulses presumably emulate H-reflex transmission, we thus expected conditioning subthreshold TSCS to decrease H-reflex amplitudes across all interstimulus intervals in both AB and SCI populations. A reduction in H-reflex amplitude would be clinically relevant in the SCI population, as more than 60% of people with SCI experience spasticity [46,47]. In fact, repetitive TSCS of the lumbar spine has been shown to reduce spasticity after SCI [48,49]. However, we did not see significant changes in H-reflex amplitudes when conditioned with TSCS in the current study. Only 7 AB and 6 SCI participants demonstrated easily distinguishable FCR H-reflexes in this study, a number prone to either Type I or Type II error and too small to make definitive conclusions. To reduce participant burden in terms of overall number of pulses delivered, the TSCS-H-reflex conditioning experiments in our study only applied TSCS at 50% of RMT, possibly at too low an intensity to achieve post-activation depression.

4.1. Limitations

Multiple limitations affected this study. The small sample size limited our ability to determine reliable associations between injury characteristics and results of different stimulation paradigms. Conditioning experiments involving TSCS paired with peripheral nerve stimulation predominantly applied TSCS at only 50% of RMT, perhaps too low of an intensity to mediate significant effects. Conversely, as discussed earlier, the traditional definition of motor threshold we used (a response of >50 µV in 5 out of 10 trials) resulted in 'subthreshold' TSCS pulses leading to action potentials in several participants, complicating our ability to focus on subthreshold TSCS conditioning. To address this, we discarded data from sessions in which a participant's response amplitude to unconditioned TSCS was greater than 30% of unconditioned TMS amplitude. To reduce participant burden while undergoing multiple different conditioning paradigms, many of the TSCS–TMS paradigms were only measured 8 times per participant, further increasing variability [50]. Likewise, we did not test interstimulus intervals in which TSCS arrived at cervical spinal synapses greater than 10 ms after TMS arrival-in retrospect, this prevented us from mapping a full curve of increased and decreased facilitation across interstimulus intervals in the TSCS-conditioned TMS experiments.

4.2. Further Information and Experiments Needed

Though we focused on the APB in this study and normalized all stimulus intensities to APB motor thresholds, we recorded other muscles in both upper extremities. This raw data have yet to be fully analyzed and will be disseminated separately. The effects of TSCS and TSCS-conditioned TMS on autonomic parameters, such as blood pressure and cerebral blood flow, need to be carefully measured. Future experiments incorporating subthreshold TSCS should probably use a different, less stringent definition for motor threshold than traditionally used in TMS experiments. For example, an amplitude cutoff of 20 µV rather than 50 µV, or a cutoff of 2 rather than 5 positive responses out of 10 threshold trials.

5. Conclusions

We measured the effects of single pulses of posteroanterior cervical TSCS paired with single pulses of motor cortex or median nerve stimulation timed to arrive at the cervical spinal cord at varying intervals. In 13 participants with and 15 participants without chronic cervical spinal cord injury, we observed that subthreshold TSCS facilitates hand muscle responses to motor cortex stimulation, with a tendency toward greater facilitation when TSCS is timed to arrive at cervical synapses simultaneously or up to 10 milliseconds after cortical stimulus arrival. When paired with median nerve stimulation, single pulses of subthreshold TSCS had no effect on the amplitude of H-reflex responses or F-wave responses. Though variability was high, the overall findings suggest that TSCS paired with appropriately timed cortical stimulation has the potential to facilitate convergent transmission between descending motor, segmental afferent, and spinal motor neurons serving the hand. Since TMS travels via similar descending motor pathways that mediate volitional movement, studying TSCS-conditioned TMS responses can produce findings with potential for direct clinical translation. To improve reliability, further studies with larger numbers of participants and repetitively paired cortical and cervical spinal stimulation are needed.

Supplementary Materials: The following are available online at https://www.mdpi.com/article/10.3390/jcm10225304/s1, Figure S1: Effects of TSCS on change in heart rate and blood pressure, Table S1: H-reflex responses to conditioning TSCS; Table S2: TSCS thresholds by individual and session; Table S3: TMS thresholds by individual and session; Table S4: F-wave amplitude and persistence by individual and session; Table S5: H-reflex amplitude by individual and session; Table S6: TSCS-TMS pairing by individual and session; Table S7: TSCS-F pairing by individual and session; Table S8: TSCS-H pairing by individual and session.

Author Contributions: Conceptualization, J.B.C. and N.Y.H.; methodology, J.M.W., J.B.C., Y.-K.W., N.Y.H.; software, J.P.W., Y.-K.W.; formal analysis, J.R.W., W.M.S., G.O.F., G.A.M., J.M.L., M.T.M., J.P.W., J.M.W., Y.-K.W., N.Y.H.; investigation, J.R.W., W.M.S., G.O.F., G.A.M., J.M.L., M.T.M., J.M.W., Y.-K.W., N.Y.H.; resources, N.Y.H.; data curation, Y.-K.W., N.Y.H.; writing—original draft preparation, J.R.W., W.M.S., J.M.L., M.T.M., J.M.W., N.Y.H.; writing—review and editing, J.R.W., W.M.S., J.M.L., M.T.M., J.P.W., J.M.W., J.B.C., N.Y.H.; supervision, J.M.W., Y.-K.W., N.Y.H.; funding acquisition, N.Y.H. All authors have read and agreed to the published version of the manuscript.

J. Clin. Med. **2021**, 10, 5304

Funding: This research was funded predominantly by Craig H. Neilsen Foundation, grant number 457648, and partially through National Institute of Neurological Disorders and Stroke, grant number 1R01NS124224.

Institutional Review Board Statement: The study was conducted according to the guidelines of the Declaration of Helsinki, and approved by the Institutional Review Board of James J. Peters Veterans Affairs Medical Center (MIRB# 01743, 5 December 2017).

Informed Consent Statement: Informed consent was obtained from all subjects involved in the study.

Data Availability Statement: The individual-level participant data presented in this study are available in the Supplementary Material and on Figshare at https://figshare.com/articles/dataset/dx_doi_org_10_6084_m9_figshare_6025748/6025748 (accessed on 4 October 2021).

Acknowledgments: We thank the research participants for their patience. We thank Kenlys Fajardo, Melissa Veale, and William Zerbarini for administrative support.

Conflicts of Interest: The authors declare no conflict of interest. The funders had no role in the design of the study aside from comments during the peer review process prior to study initiation. The funder had no role in the collection, analyses, or interpretation of data; in the writing of the manuscript, or in the decision to publish the results.

References

1. Harkema, S.; Gerasimenko, Y.; Hodes, J.; Burdick, J.; Angeli, C.; Chen, Y.; Ferreira, C.; Willhite, A.; Rejc, E.; Grossman, R.G.; et al. Effect of epidural stimulation of the lumbosacral spinal cord on voluntary movement, standing, and assisted stepping after motor complete paraplegia: A case study. *Lancet* **2011**, *377*, 1938–1947. [CrossRef]
2. Angeli, C.A.; Edgerton, V.R.; Gerasimenko, Y.P.; Harkema, S.J. Altering spinal cord excitability enables voluntary movements after chronic complete paralysis in humans. *Brain* **2014**, *137*, 1394–1409. [CrossRef]
3. Angeli, C.A.; Boakye, M.; Morton, R.A.; Vogt, J.; Benton, K.; Chen, Y.; Ferreira, C.K.; Harkema, S.J. Recovery of Over-Ground Walking after Chronic Motor Complete Spinal Cord Injury. *N. Engl. J. Med.* **2018**, *379*, 1244–1250. [CrossRef]
4. Gad, P.; Lee, S.; Terrafranca, N.; Zhong, H.; Turner, A.; Gerasimenko, Y.; Edgerton, V.R. Non-Invasive Activation of Cervical Spinal Networks after Severe Paralysis. *J. Neurotrauma* **2018**, *35*, 2145–2158. [CrossRef]
5. Inanici, F.; Samejima, S.; Gad, P.; Edgerton, V.R.; Hofstetter, C.P.; Moritz, C.T. Transcutaneous Electrical Spinal Stimulation Promotes Long-Term Recovery of Upper Extremity Function in Chronic Tetraplegia. *IEEE Trans. Neural Syst. Rehabil. Eng.* **2018**, *26*, 1272–1278. [CrossRef] [PubMed]
6. Inanici, F.; Brighton, L.N.; Samejima, S.; Hofstetter, C.P.; Moritz, C.T. Transcutaneous Spinal Cord Stimulation Restores Hand and Arm Function after Spinal Cord Injury. *IEEE Trans. Neural Syst. Rehabil. Eng.* **2021**, *29*, 310–319. [CrossRef] [PubMed]
7. Hofstoetter, U.S.; Krenn, M.; Danner, S.M.; Hofer, C.; Kern, H.; McKay, W.B.; Mayr, W.; Minassian, K. Augmentation of Voluntary Locomotor Activity by Transcutaneous Spinal Cord Stimulation in Motor-Incomplete Spinal Cord-Injured Individuals. *Artif. Organs* **2015**, *39*, E176–E186. [CrossRef]
8. Gerasimenko, Y.P.; Lu, D.C.; Modaber, M.; Zdunowski, S.; Gad, P.; Sayenko, D.G.; Morikawa, E.; Haakana, P.; Ferguson, A.; Roy, R.R.; et al. Noninvasive Reactivation of Motor Descending Control after Paralysis. *J. Neurotrauma* **2015**, *32*, 1968–1980. [CrossRef] [PubMed]
9. Minassian, K.; Hofstoetter, U.S.; Danner, S.M.; Mayr, W.; Bruce, J.A.; McKay, W.B.; Tansey, K.E. Spinal Rhythm Generation by Step-Induced Feedback and Transcutaneous Posterior Root Stimulation in Complete Spinal Cord–Injured Individuals. *Neurorehabilit. Neural Repair* **2016**, *30*, 233–243. [CrossRef] [PubMed]
10. Milosevic, M.; Masugi, Y.; Sasaki, A.; Sayenko, D.G.; Nakazawa, K. On the reflex mechanisms of cervical transcutaneous spinal cord stimulation in human subjects. *J. Neurophysiol.* **2019**, *121*, 1672–1679. [CrossRef]
11. Wu, Y.-K.; Levine, J.M.; Wecht, J.R.; Maher, M.T.; Limonta, J.M.; Saeed, S.; Santiago, T.M.; Bailey, E.; Kastuar, S.; Guber, K.S.; et al. Posteroanterior cervical transcutaneous spinal stimulation targets ventral and dorsal nerve roots. *Clin. Neurophysiol.* **2020**, *131*, 451–460. [CrossRef]
12. Formento, E.; Minassian, K.; Wagner, F.; Mignardot, J.B.; Le Goff-Mignardot, C.G.; Rowald, A.; Bloch, J.; Micera, S.; Capogrosso, M.; Courtine, G. Electrical spinal cord stimulation must preserve proprioception to enable locomotion in humans with spinal cord injury. *Nat. Neurosci.* **2018**, *21*, 1728–1741. [CrossRef] [PubMed]
13. Minassian, K.; Persy, I.; Rattay, F.; Pinter, M.; Kern, H.; Dimitrijevic, M. Human lumbar cord circuitries can be activated by extrinsic tonic input to generate locomotor-like activity. *Hum. Mov. Sci.* **2007**, *26*, 275–295. [CrossRef]
14. Gerasimenko, Y.; Gorodnichev, R.; Machueva, E.; Pivovarova, E.; Semyenov, D.; Savochin, A.; Roy, R.R.; Edgerton, V.R. Novel and Direct Access to the Human Locomotor Spinal Circuitry. *J. Neurosci.* **2010**, *30*, 3700–3708. [CrossRef] [PubMed]
15. Mishra, A.M.; Pal, A.; Gupta, D.; Carmel, J.B. Paired motor cortex and cervical epidural electrical stimulation timed to converge in the spinal cord promotes lasting increases in motor responses. *J. Physiol.* **2017**, *595*, 6953–6968. [CrossRef] [PubMed]

16. Rossi, S.; Hallett, M.; Rossini, P.M.; Pascual-Leone, A. Safety, ethical considerations, and application guidelines for the use of transcranial magnetic stimulation in clinical practice and research. *Clin. Neurophysiol.* **2009**, *120*, 2008–2039. [CrossRef] [PubMed]
17. Rossi, S.; Hallett, M.; Rossini, P.M.; Pascual-Leone, A. Screening questionnaire before TMS: An update. *Clin. Neurophysiol.* **2011**, *122*, 1686. [CrossRef]
18. Robinson, L.R.; Jantra, P.; MacLean, I.C. Central motor conduction times using transcranial stimulation and F Wave latencies. *Muscle Nerve* **1988**, *11*, 174–180. [CrossRef]
19. Bunday, K.L.; Tazoe, T.; Rothwell, J.C.; Perez, M.A. Subcortical Control of Precision Grip after Human Spinal Cord Injury. *J. Neurosci.* **2014**, *34*, 7341–7350. [CrossRef]
20. Bunday, K.L.; Perez, M.A. Motor Recovery after Spinal Cord Injury Enhanced by Strengthening Corticospinal Synaptic Transmission. *Curr. Biol.* **2012**, *22*, 2355–2361. [CrossRef]
21. Mills, K.; Murray, N. Electrical stimulation over the human vertebral column: Which neural elements are excited? *Electroencephalogr. Clin. Neurophysiol.* **1986**, *63*, 582–589. [CrossRef]
22. Stefan, K.; Kunesch, E.; Cohen, L.G.; Benecke, R.; Classen, J. Induction of plasticity in the human motor cortex by paired associative stimulation. *Brain* **2000**, *123*, 572–584. [CrossRef] [PubMed]
23. Dixon, L.; Ibrahim, M.M.; Santora, D.; Knikou, M. Paired associative transspinal and transcortical stimulation produces plasticity in human cortical and spinal neuronal circuits. *J. Neurophysiol.* **2016**, *116*, 904–916. [CrossRef]
24. Harel, N.Y.; Carmel, J.B. Paired Stimulation to Promote Lasting Augmentation of Corticospinal Circuits. *Neural Plast.* **2016**, *2016*, 7043767. [CrossRef] [PubMed]
25. Roy, F.D.; Bosgra, D.; Stein, R.B. Interaction of transcutaneous spinal stimulation and transcranial magnetic stimulation in human leg muscles. *Exp. Brain Res.* **2014**, *232*, 1717–1728. [CrossRef]
26. Ziemann, U.; Ilić, T.V.; Alle, H.; Meintzschel, F. Estimated magnitude and interactions of cortico-motoneuronal and Ia afferent input to spinal motoneurones of the human hand. *Neurosci. Lett.* **2004**, *364*, 48–52. [CrossRef]
27. Cortes, M.; Thickbroom, G.W.; Valls-Sole, J.; Pascual-Leone, A.; Edwards, D.J. Spinal associative stimulation: A non-invasive stimulation paradigm to modulate spinal excitability. *Clin. Neurophysiol.* **2011**, *122*, 2254–2259. [CrossRef] [PubMed]
28. Taylor, J.L.; Martin, P.G. Voluntary Motor Output Is Altered by Spike-Timing-Dependent Changes in the Human Corticospinal Pathway. *J. Neurosci.* **2009**, *29*, 11708–11716. [CrossRef]
29. D'amico, J.M.; Dongés, S.C.; Taylor, J.L. Paired corticospinal-motoneuronal stimulation increases maximal voluntary acti-vation of human adductor pollicis. *J. Neurophysiol.* **2018**, *119*, 369–376. [CrossRef]
30. Urbin, M.A.; Ozdemir, R.A.; Tazoe, T.; Perez, M.A. Spike-timing-dependent plasticity in lower-limb motoneurons after human spinal cord injury. *J. Neurophysiol.* **2017**, *118*, 2171–2180. [CrossRef]
31. Shulga, A.; Lioumis, P.; Kirveskari, E.; Savolainen, S.; Mäkelä, J.P.; Ylinen, A. The use of F-response in defining interstimulus intervals appropriate for LTP-like plasticity induction in lower limb spinal paired associative stimulation. *J. Neurosci. Methods* **2015**, *242*, 112–117. [CrossRef]
32. Sharpe, A.N.; Jackson, A. Upper-limb muscle responses to epidural, subdural and intraspinal stimulation of the cervical spinal cord. *J. Neural Eng.* **2014**, *11*, 016005. [CrossRef]
33. Jackson, A.; Zimmermann, J. Neural interfaces for the brain and spinal cord—Restoring motor function. *Nat. Rev. Neurol.* **2012**, *8*, 690–699. [CrossRef]
34. Mercier, C.; Roosink, M.; Bouffard, J.; Bouyer, L. Promoting Gait Recovery and Limiting Neuropathic Pain After Spinal Cord Injury. *Neurorehabilit. Neural Repair* **2017**, *31*, 315–322. [CrossRef] [PubMed]
35. Yang, Q.; Ramamurthy, A.; Lall, S.; Santos, J.; Ratnadurai-Giridharan, S.; Lopane, M.; Zareen, N.; Alexander, H.; Ryan, D.; Martin, J.H.; et al. Independent replication of motor cortex and cervical spinal cord electrical stimulation to promote forelimb motor function after spinal cord injury in rats. *Exp. Neurol.* **2019**, *320*, 112962. [CrossRef]
36. Zareen, N.; Shinozaki, M.; Ryan, D.; Alexander, H.; Amer, A.; Truong, D.; Khadka, N.; Sarkar, A.; Naeem, S.; Bikson, M.; et al. Motor cortex and spinal cord neuromodulation promote corticospinal tract axonal outgrowth and motor recovery after cervical contusion spinal cord injury. *Exp. Neurol.* **2017**, *297*, 179–189. [CrossRef]
37. Pal, A.; Park, H.; Ramamurthy, A.; Asan, S.; Bethea, T.; Johnkutty, M.; Carmel, J. Paired motor cortex and spinal cord epidural stimulation strengthens sensorimotor connections and improves forelimb function after cervical spinal cord injury in rats. *bioRxiv* **2020**. [CrossRef]
38. Ting, W.K.-C.; Huot-Lavoie, M.; Ethier, C. Paired Associative Stimulation Fails to Induce Plasticity in Freely Behaving Intact Rats. *eNeuro* **2020**, *7*. [CrossRef] [PubMed]
39. Kumru, H.; Rodríguez-Cañón, M.; Edgerton, V.; García, L.; Flores, Á.; Soriano, I.; Opisso, E.; Gerasimenko, Y.; Navarro, X.; García-Alías, G.; et al. Transcutaneous Electrical Neuromodulation of the Cervical Spinal Cord Depends Both on the Stimulation Intensity and the Degree of Voluntary Activity for Training. A Pilot Study. *J. Clin. Med.* **2021**, *10*, 3278. [CrossRef]
40. Thickbroom, G.W.; Byrnes, M.L.; Edwards, D.; Mastaglia, F.L. Repetitive paired-pulse TMS at I-wave periodicity markedly increases corticospinal excitability: A new technique for modulating synaptic plasticity. *Clin. Neurophysiol.* **2006**, *117*, 61–66. [CrossRef]
41. Vastano, R.; Perez, M.A. Changes in motoneuron excitability during voluntary muscle activity in humans with spinal cord injury. *J. Neurophysiol.* **2020**, *123*, 454–461. [CrossRef]

42. Thomas, C.K.; Häger, C.K.; Klein, C.S. Increases in human motoneuron excitability after cervical spinal cord injury depend on the level of injury. *J. Neurophysiol.* **2017**, *117*, 684–691. [CrossRef]
43. Espiritu, M.G.; Lin, C.S.-Y.; Burke, D. Motoneuron excitability and the F wave. *Muscle Nerve* **2003**, *27*, 720–727. [CrossRef]
44. Minassian, K.; Persy, I.; Rattay, F.; Dimitrijevic, M.R.; Hofer, C.; Kern, H. Posterior root–muscle reflexes elicited by transcutaneous stimulation of the human lumbosacral cord. *Muscle Nerve* **2007**, *35*, 327–336. [CrossRef] [PubMed]
45. Murray, L.M.; Knikou, M. Repeated cathodal transspinal pulse and direct current stimulation modulate cortical and corticospinal excitability differently in healthy humans. *Exp. Brain Res.* **2019**, *237*, 1841–1852. [CrossRef]
46. Holtz, K.A.; Lipson, R.; Noonan, V.K.; Kwon, B.K.; Mills, P.B. Prevalence and Effect of Problematic Spasticity after Traumatic Spinal Cord Injury. *Arch. Phys. Med. Rehabil.* **2017**, *98*, 1132–1138. [CrossRef] [PubMed]
47. Sköld, C.; Levi, R.; Seiger, Å. Spasticity after traumatic spinal cord injury: Nature, severity, and location. *Arch. Phys. Med. Rehabil.* **1999**, *80*, 1548–1557. [CrossRef]
48. Hofstoetter, U.S.; McKay, W.B.; Tansey, K.E.; Mayr, W.; Kern, H.; Minassian, K. Modification of spasticity by transcutaneous spinal cord stimulation in individuals with incomplete spinal cord injury. *J. Spinal Cord Med.* **2014**, *37*, 202–211. [CrossRef]
49. Hofstoetter, U.S.; Freundl, B.; Danner, S.M.; Krenn, M.J.; Mayr, W.; Binder, H.; Minassian, K. Transcutaneous Spinal Cord Stimulation Induces Temporary Attenuation of Spasticity in Individuals with Spinal Cord Injury. *J. Neurotrauma* **2020**, *37*, 481–493. [CrossRef]
50. Goldsworthy, M.; Hordacre, B.; Ridding, M. Minimum number of trials required for within- and between-session reliability of TMS measures of corticospinal excitability. *Neuroscience* **2016**, *320*, 205–209. [CrossRef]

Journal of
Clinical Medicine

Article

Transcutaneous Electrical Neuromodulation of the Cervical Spinal Cord Depends Both on the Stimulation Intensity and the Degree of Voluntary Activity for Training. A Pilot Study

Hatice Kumru [1,2,3,*,†], María Rodríguez-Cañón [1,4,†], Victor R. Edgerton [1,5], Loreto García [1,2,3], África Flores [4], Ignasi Soriano [1,2,3], Eloy Opisso [1,2,3], Yury Gerasimenko [6,7,8], Xavier Navarro [1,4], Guillermo García-Alías [1,4,‡] and Joan Vidal [1,2,3,‡]

1   Fundación Institut Guttmann, Institut Universitari de Neurorehabilitació Adscrit a la UAB, 08916 Badalona, Spain; rodriguezcanonmaria@gmail.com (M.R.-C.); vre@ucla.edu (V.R.E.); loretogarcia@guttmann.com (L.G.); isoriano@guttmann.com (I.S.); eopisso@guttmann.com (E.O.); xavier.navarro@uab.cat (X.N.); guillermo.garcia@uab.cat (G.G.-A.); jvidal@guttmann.com (J.V.)
2   Universitat Autònoma de Barcelona, 08193 Barcelona, Spain
3   Fundació Institut d'Investigació en Ciències de la Salut Germans Trias i Pujol, 08916 Badalona, Spain
4   Departament de Biologia Cel·lular, Fisiologia i Immunologia & Insititute of Neuroscience, Universitat Autònoma de Barcelona, 08193 Bellaterra, Spain; africa.flores@uab.cat
5   Department of Neurobiology, University of California Los Angeles, Los Angeles, CA 90095, USA
6   Pavlov Institute of Physiology, 199034 St. Petersburg, Russia; yury.gerasimenko@louisville.edu
7   Department of Physiology and Biophysics, University of Louisville, Louisville, KY 40292, USA
8   Kentucky Spinal Cord Injury Research Center, University of Louisville, Louisville, KY 40292, USA
*   Correspondence: hkumru@guttmann.com
†   Equal role as first author.
‡   Equal role as senior PI.

Citation: Kumru, H.; Rodríguez-Cañón, M.; Edgerton, V.R.; García, L.; Flores, Á.; Soriano, I.; Opisso, E.; Gerasimenko, Y.; Navarro, X.; García-Alías, G.; et al. Transcutaneous Electrical Neuromodulation of the Cervical Spinal Cord Depends Both on the Stimulation Intensity and the Degree of Voluntary Activity for Training. A Pilot Study. *J. Clin. Med.* **2021**, *10*, 3278. https://doi.org/10.3390/jcm10153278

Academic Editors: Ursula S. Hofstoetter and Karen Minassian

Received: 2 June 2021
Accepted: 22 July 2021
Published: 25 July 2021

Publisher's Note: MDPI stays neutral with regard to jurisdictional claims in published maps and institutional affiliations.

Abstract: Electrical enabling motor control (eEmc) through transcutaneous spinal cord stimulation offers promise in improving hand function. However, it is still unknown which stimulus intensity or which muscle force level could be better for this improvement. Nine healthy individuals received the following interventions: (i) eEmc intensities at 80%, 90% and 110% of abductor pollicis brevis motor threshold combined with hand training consisting in 100% handgrip strength; (ii) hand training consisting in 100% and 50% of maximal handgrip strength combined with 90% eEmc intensity. The evaluations included box and blocks test (BBT), maximal voluntary contraction (MVC), F wave persistency, F/M ratio, spinal and cortical motor evoked potentials (MEP), recruitment curves of spinal MEP and cortical MEP and short-interval intracortical inhibition. The results showed that: (i) 90% eEmc intensity increased BBT, MVC, F wave persistency, F/M ratio and cortical MEP recruitment curve; 110% eEmc intensity increased BBT, F wave persistency and cortical MEP and recruitment curve of cortical MEP; (ii) 100% handgrip strength training significantly modulated MVC, F wave persistency, F/M wave and cortical MEP recruitment curve in comparison to 50% handgrip strength. In conclusion, eEmc intensity and muscle strength during training both influence the results for neuromodulation at the cervical level.

Keywords: transcutaneous spinal cord stimulation; intensity effect; muscle strength effect; hand training; neuromodulation; cervical spinal cord

1. Introduction

Transcutaneous spinal cord stimulation (tSCS) used as a method of electrical enabling motor control (eEmc) is a novel, non-invasive method and alone or combined with hand training offers promise in improving hand function since it can modify the functional state of the sensory-motor system [1,2]. eEmc consists of low intensity electrical stimulation applied for changing the physiological states of spinal networks to a level that it enables the spinal network to better respond to voluntary commands and proprioceptive inputs.

In the literature, the different terminologies have been used for transcutaneous spinal cord stimulation: "transcutaneous spinal stimulation (TSS) [3], painless transcutaneous electrical enabling motor control (pcEmc) [4], transcutaneous electrical stimulation of the spinal cord (TESS)" [5] or "transcutaneous enabling motor control (tEmc)" [6]. Recent developments in therapeutic approaches for spinal cord injury (SCI) showed that tSCS alone [3] or a combination of spinal cord stimulation with pharmacological treatment [7] or with hand training [6,8,9] allowed improving hand motor function in individuals with SCI. It has been hypothesized that spinal electrical stimulation can change the excitability of spinal circuitry and potentially neuromodulate the spinal network to facilitate and enhance the restoration of paralyzed limb function [5–9]. Plastic changes have been also reported in the control of upper limb function in healthy subjects following eEmc alone [5] and combined with hand training [10]. It has been suggested that the mechanisms recruited by eEmc combined with physical training, although partially overlapping, may involve different and perhaps synergistic processes leading to more effective reorganization of neural circuits [10]. It was recently reported that tSCS was capable of facilitating cortically evoked muscle responses and the degree of facilitation progressively increased during the 1 s stimulation training, and was still evident 0.5 s after the end of the training in monkeys [11]. The most likely mechanism of eEmc occurs via transcutaneous tonic spinal activation by elevating spinal networks excitability [12] and may affect interneuronal pathways that generate action potentials on motoneurons within a motor pool in a more normal stochastic time frame [13]. As such, eEmc is hypothesized to potentiate the generation of postsynaptic excitatory potentials and, thus, shift the spinal motor network excitability closer to the excitation threshold. In addition, activation of back musculature under the electrodes of tSCS [11] and of sensory afferents at the level of dorsal roots or via the spinal pathways can contribute to elevating neural excitability [14,15].

According to reports in the literature, there was great variability in the intensity of eEmc applied, though most of the studies used high intensities, close to the participants' tolerance threshold. In cervical SCI, the stimulus intensity for eEmc varied from below the resting spinal motor threshold (RMT) to an intensity adjusted to enable maximal grip strength or adjusted based on the participant's functional task performance [5–9]. When eEmc was combined with hand training, variable protocols were used: either maximal handgrip to submaximal isometric hand movement such as squeezing/grasping or standard stretching, active assistive range of motion exercises, intensive gross and fine motor skill training or the intensity that made the task easiest [6,8,9]. All these studies reported significant functional improvement of hand muscle strength and/or neurophysiological changes. We recently showed that even one single session of cervical eEmc is able to modify the excitability of neural networks controlling upper limb function in healthy subjects [10]. These changes strongly depend on the combination of eEmc with hand training, since improved hand grip force and increased spinal and corticospinal output were found in comparison to each intervention tested alone [10]. The intervention consisted in eEmc at an intensity of 90% the RMT at the cervical spinal cord and handgrip training at maximum voluntary contraction (MVC) [10]. However, it remains unknown if the intensity of electrical stimulation for eEmc and the level of hand grip strength in such combined strategy can affect motor strength and/or functional outcome, and if plastic changes occur at spinal or cortical level.

Thus, the objectives of this study were to test: (1) the effect of different electrical stimulation intensities for eEmc combined with the maximum force of hand grip during training, and (2) the effect of different hand grip strength during training combined with 90% of spinal RMT electrical stimulation for eEmc on hand function and spinal cord and cortical excitability. We hypothesized that eEmc at higher stimulus intensity applied at two sites of cervical spinal cord combined with a higher level of hand grip strength can enhance motor strength and functional outcome and eventually modulate spinal or cortical neural circuits more than lower stimulus intensity and lower hand grip strength during training.

2. Experimental Section

2.1. Study Design

Nine healthy volunteer subjects (age range 25–60 years; mean age 39.8 ± 11.1 years; 3 females and 6 males) participated in the study (Table 1). The inclusion criteria were: age between 18 and 65 years, without any neurological disorder and uncontrolled disease, which could limit the experiment, and given written informed consent. Exclusion criteria were any implanted metallic or electrical devices, and pregnancy. The study protocol was approved by the Research Ethics Committee of the Institute Guttmann and was conducted in accordance with the Declaration of Helsinki.

Table 1. Demographic data of healthy individuals and current intensities used for cervical stimulation at C3-C4 and at C6-C7 level.

				eEmc Intensity Applied during Stimulation (mA)							
				C3-C4				C6-C7			
				% eEmc + 100% MVC			90%eEmc + 50% MVC	%eEmc + 100%MVC			90%eEmc + 50% MVC
Subject	Sex	Age	Hand	80%	90%	110%		80%	90%	110%	
1	M	44	R	32	29	40	29	34	38	44	38
2	M	60	R	69	59	77	63	72	72	90	72
3	F	25	R	27	30	33	29	32	30	40	32
4	F	27	L	32	34	44	34	36	36	46	41
5	F	33	R	37	54	53	47	48	63	62	58
6	M	41	R	27	36	37	31	38	45	42	34
7	M	51	R	51	52	75	63	61	63	90	81
8	M	39	R	48	59	77	49	56	72	88	59
9	M	38	B	45	67	73	61	53	77	75	85

M: male; F: female; L: left; R: right; B: both.

This study was performed in 2 parts. In the first part, we studied the effect of different electrical stimulation intensities for eEmc calculated from the spinal RMT of abductor pollicis brevis (APB) muscle combined with the maximum hand grip strength during training. In the second part, we studied the effect of different handgrip strength levels during training combined with 90% spinal RMT of stimulus intensity for eEmc (Figure 1).

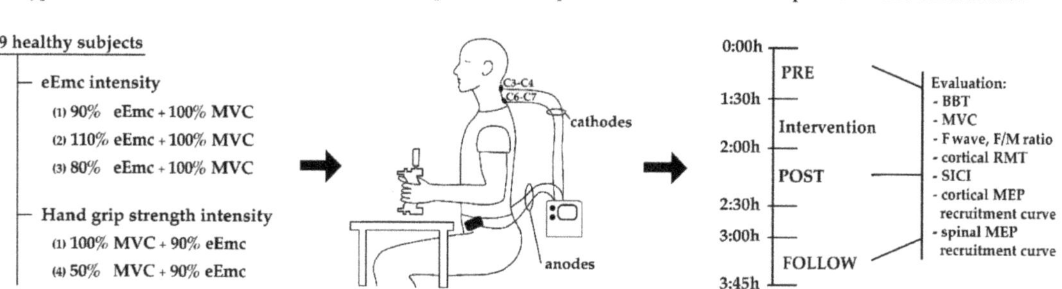

Figure 1. Schematic representation of the experiment conditions, order of intervention and the evaluation time points during each experiment, and the functional and motor strength assessment and neurophysiological assessments performed.

2.2. Interventions

The study included a total of four interventions, tested in two parts separated at least one week apart. All interventions combined eEmc with hand training. In the first part, we studied the effect of three electrical eEmc stimulation at intensity 80%, 90% and 110% of spinal RMT of the APB muscle, combined with the maximum hand grip strength (100% of

MVC). In the second part, we studied two different levels of hand grip strength for training, 50% and 100% of MVC, combined with eEmc at stimulus intensity of 90% of spinal RMT.

Each intervention consisted of trains of 20 s of concomitant eEmc stimulation and hand training, followed by 80 s of rest, for 30 min. The subject alternated the two hands during eEmc, resulting in nine hand-training attempts for each hand and 18 in total during the whole intervention [10].

The eEmc was delivered through two circular hydrogel adhesive electrodes (2 cm diameter, Axion GmbH, Leonberg, Germany) placed along the midline over spinous processes C3-C4 and C6-C7. eEmc was delivered using biphasic rectangular 1 ms pulses at a frequency of 30 Hz, with each pulse filled with a carrier frequency of 10 kHz [10]. For eEmc, we used two channels of a five-channel current-controlled stimulator of Biostim-5 stimulator (Cosyma Inc., Moscow, Russia) with two pairs of anode and cathode. Each channel was set up independently. Stimulation of the first channel was delivered by a cathode at C3-C4 level and stimulation of the second channel by a cathode at C6-C7, with anodes placed at the iliac crests.

The hand training consisted of holding the hand grip dynamometer and maintaining a hand grip strength of 100% of MVC for all stimulus intensities of eEmc in the first part. In the second part, either 100% (100% MVC) or 50% (50% MVC) of hand grip strength during hand training was combined with eEmc at 90% of stimulus intensity.

2.3. Functional and Motor Strength Assesment of the Healthy Participants

The evaluation protocol consisted of two hand functional outcomes, the Box and Block test (BBT) and the measurement of hand grip strength during a MVC with a dynamometer. Neurophysiological assessment was made on the dominant hand and arm muscles. The dominant hand was determined according to the Edinburgh inventory [16]. F-wave and spinal motor evoked potentials (spinal MEPs) in response to single-pulse C3-C4 and C6-C7 tSCS were recorded in the APB muscle to study the spinal cord excitability. At the cortical level, using transcranial magnetic stimulation (TMS), we determined the cortical RMT of APB muscle, and recorded cortical motor evoked potentials (cortical MEPs), short-interval intracortical inhibition (SICI) and recruitment curve of cortical MEPs. The neurophysiological studies were performed with an EMG machine (Medelec Synergy, Oxford Instruments; Surrey, England). The evaluation protocol was repeated at three time points: before (PRE), just after (POST) and one hour after finishing the intervention (FOLLOW) in each subject and experimental study (Figure 1). The total duration of each experiment was around four hours.

2.3.1. Box and Block Test (BBT)

The BBT was used to assess hand dexterity. The participant was seated in front of a box split in two halves, the one on the preferred side full of small cube-shaped blocks [17]. The subject had to move the blocks one by one to the other side, crossing the middle line, during 60 s. The total number of moved blocks was the BBT score.

2.3.2. Maximum Voluntary Contraction (MVC) during Hand Grip

The participant was seated in front of a table, with the dominant forearm resting on it in neutral position, holding a dynamometer (Jamar Model 5030J1, Sammons Preston, NJ, USA) with the hand. For trigger signal, we used a mild electric stimulus (3 mA intensity with pulse width 0.5 ms) delivered by means of ring electrodes around the fifth finger. Following the trigger signal, the participant had to perform a MVC and maintain it during 4 s; the strength was measured in kilograms (kg) in three consecutive trials, with at least one minute rest between them.

2.4. Neurophysiological Assesment

Disposable adhesive surface electrodes (outer diameter 20 mm; Technomed, Maastricht Airport, Netherlands) were placed over the muscle belly of the APB, abductor digiti minimi (ADM), flexor carpi radialis (FCR), extensor digitorum (ED) and biceps brachii (BB) muscles of the dominant arm, with the cathode proximal and the anode approximately 2 cm distally after standard skin preparation. The EMG signal was amplified, filtered using band-pass of 30 Hz–10 kHz, and epochs of 100 ms sweep duration at amplitude sensitivity of 0.1–0.5 mV recorded at a sampling rate of 50 kHz. EMG activity was visualized online, with the subject relaxing her/his arm and hand. If any background EMG activity was observed after stimulus delivery, the recording was discarded, and the stimulation was repeated. Offline analysis was performed with MATLAB.

2.4.1. F Wave Persistency and F/M Wave Ratio

Suprathreshold electrical stimulation (pulse width 0.5 ms) of the median nerve at the wrist was used to elicit the maximum M wave and F wave in the APB muscle for at least 15 recordings. We assessed spinal motoneuron excitability using F-wave persistency and the Fmax/Mmax ratio (F/M ratio) in the APB muscle.

2.4.2. Spinal MEP

Stimulation was delivered using two hydrogel adhesive disk electrodes (20 mm diameter; Axion GmbH, Germany) as cathodes along the midline between spinous processes C3-C4 and C6-C7. The anodes were two adhesive rectangular electrodes (5×12 cm) placed symmetrically over the iliac crest. Monophasic rectangular single pulses of 1 ms duration were delivered from a Biostim-5 stimulator (Cosyma Inc., Moscow, Russia). The spinal RMT was calculated as the lowest intensity of electrical stimulation applied first at C3-C4 and then at C6-C7 that evoked a spinal MEP of $\geq$50 µV peak-to-peak amplitude in the APB muscle in at least 5 of 10 consecutive trials. Then, the recruitment of spinal MEPs on all the muscles tested was recorded for each stimulation site and time point. Recruitment curves of spinal MEPs were derived from responses to gradually increasing stimulation intensity from 90% to 150% of the baseline spinal RMT at 10% increments, with three repetitions at each intensity.

2.4.3. Transcranial Magnetic Stimulation (TMS)

A Magstim® BiStim^2 TMS (Magstim Company, Whitland, Wales, UK) apparatus was used. A figure-of-eight coil was held tangentially to the scalp over the motor area of the dominant hand in the optimal position for activating the APB in a posterior-anterior current direction. The hot point was defined as a point over the scalp where the largest amplitude cortical MEP in APB muscle was recorded. Subjects were seated in a chair, resting their pronated forearms on a desk in front of them and were asked to stay relaxed but awake throughout the test.

2.4.4. Neurophysiological Parameters

We recorded the following parameters pre and post intervention (1) the BBT score to evaluate hand dexterity and (2) the MVC for handgrip strength. To evaluate spinal cord excitability changes: (1) F wave persistency and F/M wave ratio; (2) recruitment curve of spinal MEP. To evaluate cortical excitability changes: (1) cortical RMT, defined as the lowest intensity of TMS that evoked a cortical MEP of $\geq$50 µV peak-to-peak amplitude in the APB muscle in at least 5 of 10 consecutive trials; (2) mean amplitude of cortical MEPs using single-pulse TMS at 120% of cortical RMT of ABP in 5 recordings; (3) short intracortical inhibition (SICI) using paired-pulse TMS with a subthreshold conditioning stimulus (80% of cortical RMT) and a suprathreshold test stimulus (120% of cortical RMT) at interstimulus interval of 2 ms in 5 recordings without background activity [18]; (4) recruitment curves of cortical MEP obtained at increasing intensities from 90% to 150% of cortical RMT of APB, at 10% increments (three recordings at each intensity). The absence of baseline EMG activity was verified before carrying out each of the recordings; between 1 to 6 recordings for each subject were rejected because of background activity.

2.5. Data Analysis and Statistics

All data are expressed as the mean ± standard error of the mean (SEM), at baseline (PRE), just after (POST) and sixty minutes after finishing the intervention (FOLLOW) in each condition. We calculated the % changes with respect to baseline for each time point evaluation when appropriate. In the analysis of the recruitment curves, differences between pre vs. post intervention, and vs. follow were calculated. BBT was calculated just from one trial and MVC (hand muscle strength) was calculated as the mean of 3 trials.

For the F wave, the peak-to-peak amplitude and for Mmax, the maximum amplitude of the M wave was measured. F wave was considered if peak-to-peak amplitude was at least 1% of M wave (Mmax) amplitude. F wave persistence was calculated by dividing the number of F responses by the number of stimuli from 15 recordings. The maximum amplitude of F wave was normalized to Mmax amplitude to obtain Fmax/Mmax ratio.

For recruitment curves of spinal MEPs and cortical MEPs, we measured the peak-to-peak MEP amplitude in each recording, and then calculated the mean amplitude of MEPs from 3 recordings for each intensity and for each subject and for each experimental condition. Recruitment curves obtained at post and follow-up were normalized calculating the amplitude difference according to pre intervention recruitment curves values.

For SICI, averaged peak-to-peak amplitude of the conditioned MEP (obtained after the conditioning stimulus of 80% cortical RMT) was expressed as a percentage of the average amplitude of the test MEP (obtained at supramaximal 120% cortical RMT stimulus), according to % = (conditioned MEP/test MEP) * 100.

Shapiro–Wilk test was used to assess if data were normally distributed, and sphericity was evaluated with Mauchly's test. Greenhouse–Geisser correction was used when the assumption of sphericity was violated ($p < 0.05$). All data was normally distributed, and a two-way repeated measures ANOVA (RM-ANOVA) was used for all statistical analyses to study the main effects of intervention and time on the functional and motor strength assessment and neurophysiological assessments for the raw and the normalized data. Interaction effect was also studied to determine if the interaction between intervention and time factors could affect the dependent variable. For significant interaction in the normalized data, time point and intervention simple effects were calculated and shown in results and figures. Time point multiple comparisons were performed with the Dunnett test (comparison between baseline vs. post and baseline vs. follow-up). The intervention multiple comparisons were compared with the Tukey test for the first part of the study (three interventions) and with Šidák test for the second part (two interventions). Significance level was set at $p < 0.05$ in all cases. Posterior estimation effect size η^2 was calculated (0.01, 0.06, >0.14 as small, medium and large effect respectively).

3. Results

Demographic data of healthy individuals and used intensities for eEmc are given in Table 1. All volunteers finished the four experimental interventions with some discomfort sensation subjectively reported but not quantified, especially for the higher intensities during cervical stimulation, on the neck around the electrodes and slight extension of the neck because of cervical paravertebral muscle contraction.

3.1. Effects of eEmc Intensity

Raw data of functional and motor strength and neurophysiological assessment during the different electrical stimulus intensity for eEmc, in the first part of the study, are shown in Table 2.

Table 2. Raw data and statistics of functional and motor strength and neurophysiological assessment during the electrical stimulus intensity for eEmc at 80%, 90% and 110% combined with maximum (100%) handgrip strength during hand training.

	Time	80%	90%	110%	F (DFn, DFd)	*p* Value	η^2
BBT (blocks number)	Pre	74.1 ± 3.0	72.8 ± 3.1	71.2 ± 2.7	Ftime (2, 16) = 17.32	<0.0001	0.033
	Post	74.6 ± 3.1	75.7 ± 3.5	74.4 ± 2.9	Finterven. (2, 16) = 0.16	0.855	0.002
	Foll.	74.6 ± 3.2	77.4 ± 3.4	78.0 ± 2.7	Finterac. (4, 32) = 3.49	0.018	0.014
MVC Grip Force (kg)	Pre	37.8 ± 8.3	31.3 ± 11.3	36.9 ± 7.6	Ftime (2, 16) = 5.87	0.012	0.002
	Post	35.1 ± 7.6	32.7 ± 11.0	35.5 ± 8.7	Finterven. (2, 16) = 3.59	0.052	0.040
	Foll.	35.8 ± 9.2	33.6 ± 9.8	36.4 ± 7.5	Finterac. (4, 32) = 2.71	0.048	0.009
F wave Persistency	Pre	65.4 ± 7.9	48.8 ± 9.1	71.5 ± 6.5	Ftime (2, 16) = 0.33	0.725	0.002
	Post	59.3 ± 8.5	60.2 ± 8.2	59.7 ± 8.0	Finterven. (2, 16) = 3.33	0.062	0.019
	Foll.	59.9 ± 9.5	57.9 ± 9.2	60.2 ± 10.1	Finterac. (4, 32) = 2.92	0.031	0.031
Mmax wave (microV)	Pre	13,296.3 ± 1754.9	16,829.0 ± 1863.2	16,844.5 ± 1204.9	Ftime (2, 16) = 1.77	0.202	0.018
	Post	12,752.8 ± 1427.5	15,485.3 ± 1937.6	16,234.6 ± 1579.3	Finterven. (2, 16) = 4.85	0.023	0.119
	Foll.	13,697.6 ± 1726.9	17,153.6 ± 1889.8	18,452.1 ± 970.3	Finterac. (4, 32) = 0.29	0.882	0.003
Ratio Fmax/Mmax	Pre	5.5 ± 0.8	3.7 ± 0.8	4.2 ± 0.6	Ftime (1.92, 15.33) = 0.15	0.851	0.001
	Post	4.6 ± 0.7	5.0 ± 0.8	4.5 ± 0.5	Finterven. (1.38, 11.03) = 0.55	0.530	0.008
	Foll.	4.2 ± 0.6	4.3 ± 0.7	5.4 ± 0.6	Finterac. (1.83, 14.64) = 1.93	0.082	0.069
RMT TMS	Pre	38.7 ± 2.3	39.1 ± 2.6	39.1 ± 2.0	Ftime (1.85, 14.81) = 1.46	0.263	0.002
	Post	37.8 ± 2.4	37.8 ± 2.4	39.3 ± 2.5	Finterven. (1.89, 15.14) = 0.21	0.805	0.002
	Foll.	38.2 ± 2.4	39.1 ± 2.6	38.4 ± 2.0	Finterac. (1.91,15.31) = 1.46	0.263	0.003
SICI in APB (%)	Pre	13.1 ± 3.7	13.8 ± 2.9	8.3 ± 1.2	Ftime (1.13, 9.05) = 0.93	0.374	0.021
	Post	10.8 ± 2.6	12.4 ± 4.2	7.1 ± 2.6	Finterven. (1.51, 12.07) = 0.14	0.816	0.003
	Follow	13.2 ± 4.2	16.7 ± 4.7	11.2 ± 4.1	Finterac. (1.90, 15.21) = 0.34	0.708	0.010
Cortical MEP at 120% RMT in APB (mV)	Pre	0.6 ± 0.1	0.8 ± 0.1	0.8 ± 0.1	Ftime (2, 16) = 3.43	0.058	0.091
	Post	1.1 ± 0.2	0.9 ± 0.1	1.1 ± 0.2	Finterven. (2, 16) = 5.25	0.018	0.056
	Foll.	0.8 ± 0.1	0.7 ± 0.1	1.6 ± 0.3	Finterac. (4, 32) = 4.99	0.003	0.113
Stimulus Intensity		**80%**	**90%**	**110%**			
TMS recruitment: diff post-pre (mV)	0.9	−0.002 ± 0.08	0.08 ± 0.11	0.04 ± 0.08			
	1	0.06 ± 0.20	0.13 ± 0.37	−0.04 ± 0.47			
	1.1	−0.08 ± 0.68	0.14 ± 0.80	0.16 ± 0.77	$F_{intensity\ (6,\ 48)}$ = 10.69	<0.001	0.121
	1.2	0.004 ± 0.71	0.37 ± 1.20	0.68 ± 1.14	$F_{interven.\ (2,\ 16)}$ = 0.74	0.491	0.030
	1.3	0.62 ± 0.96	0.07 ± 1.02	0.12 ± 1.00	$F_{interac.\ (12,\ 96)}$ = 2.17	0.019	0.079
	1.4	0.124 ± 0.90	0.91 ± 1.21	1.03 ± 0.71			
	1.5	0.14 ± 0.77	1.17 ± 1.48	1.34 ± 0.64			
TMS recruitment: diff foll-pre (mV)	0.9	−0.01 ± 0.06	−0.02 ± 0.03	0.03 ± 0.09			
	1	0.05 ± 0.28	−0.01 ± 0.25	0.17 ± 0.80			
	1.1	−0.16 ± 0.46	−0.10 ± 0.43	0.43 ± 1.18	$F_{intensity\ (6,\ 48)}$ = 4.09	0.002	0.078
	1.2	0.17 ± 0.61	0.20 ± 0.59	0.88 ± 1.15	$F_{interven.\ (2,\ 16)}$ = 3.61	0.051	0.102
	1.3	−0.20 ± 0.66	0.39 ± 0.65	−0.01 ± 1.03	$F_{interac.\ (12,\ 96)}$ = 3.73	<0.001	0.109
	1.4	0.13 ± 1.14	0.34 ± 0.72	1.24 ± 0.70			
	1.5	−0.47 ± 0.88	0.76 ± 0.63	1.05 ± 0.57			
C3-C4 recruitment: diff post-pre (mV)	0.9	0.02 ± 0.07	−0.004 ± 0.01	0.01 ± 0.02			
	1	0.01 ± 0.09	−0.01 ± 0.02	0.02 ± 0.05			
	1.1	0.01 ± 0.04	−0.03 ± 0.11	0.001 ± 0.03	$F_{intensity\ (6,\ 48)}$ = 0.96	0.460	0.023
	1.2	0.01 ± 0.03	−0.04 ± 0.11	0.004 ± 0.02	$F_{interven.\ (2,\ 16)}$ = 0.99	0.395	0.057
	1.3	−0.03 ± 0.09	−0.04 ± 0.11	−0.01 ± 0.03	$F_{interac.\ (12,\ 96)}$ = 1.69	0.081	0.037
	1.4	0.03 ± 0.09	−0.06 ± 0.11	0.01 ± 0.04			
	1.5	0.04 ± 0.10	−0.003 ± 0.09	−0.02 ± 0.04			
C3-C4 recruitment: diff foll-pre (mV)	0.9	0.03 ± 0.09	0.02 ± 0.05	0.01 ± 0.02			
	1	0.02 ± 0.09	−0.01 ± 0.03	0.04 ± 0.10			
	1.1	0.01 ± 0.05	−0.002 ± 0.11	0.01 ± 0.02	$F_{intensity\ (6,\ 48)}$ = 0.31	0.929	0.008
	1.2	0.02 ± 0.05	0.004 ± 0.02	0.02 ± 0.03	$F_{interven.\ (2,\ 16)}$ = 0.78	0.474	0.036
	1.3	0.03 ± 0.04	0.01 ± 0.02	0.001 ± 0.03	$F_{interac.\ (12,\ 96)}$ = 1.10	0.373	0.035
	1.4	0.04 ± 0.08	−0.01 ± 0.05	0.01 ± 0.03			
	1.5	0.06 ± 0.08	0.001 ± 0.04	0.01 ± 0.03			
C6-C7 recruitment: diff post-pre (mV)	0.9	0.003 ± 0.03	−0.001 ± 0.03	0.01 ± 0.02			
	1	−0.001 ± 0.03	0.04 ± 0.10	0.01 ± 0.03			
	1.1	0.004 ± 0.02	−0.01 ± 0.05	0.01 ± 0.03	$F_{intensity\ (6,\ 48)}$ = 2.14	0.066	0.041
	1.2	−0.03 ± 0.09	−0.01 ± 0.06	−0.01 ± 0.03	$F_{interven.\ (2,\ 16)}$ = 0.67	0.525	0.023
	1.3	−0.04 ± 0.11	−0.02 ± 0.04	0.001 ± 0.05	$F_{interac.\ (12,\ 96)}$ = 0.51	0.906	0.023
	1.4	−0.03 ± 0.08	−0.01 ± 0.05	−0.004 ± 0.05			
	1.5	−0.02 ± 0.12	0.01 ± 0.04	0.01 ± 0.06			
C6-C7 recruitment: diff foll-pre (mV)	0.9	0.01 ± 0.02	−0.004 ± 0.03	0.02 ± 0.02			
	1	−0.0004 ± 0.03	0.03 ± 0.09	0.02 ± 0.02			
	1.1	0.01 ± 0.02	−0.001 ± 0.04	0.02 ± 0.04	$F_{intensity\ (6,\ 48)}$ = 0.40	0.878	0.010
	1.2	−0.02 ± 0.10	0.01 ± 0.02	0.003 ± 0.03	$F_{interven.\ (2,\ 16)}$ = 0.16	0.850	0.005
	1.3	0.02 ± 0.04	0.01 ± 0.02	−0.02 ± 0.08	$F_{interac.\ (12,\ 96)}$ = 0.72	0.725	0.039
	1.4	0.01 ± 0.09	0.004 ± 0.03	0.004 ± 0.04			
	1.5	−0.01 ± 0.14	0.002 ± 0.02	0.03 ± 0.06			

diff post-pre: differences between post-pre evaluation; diff foll-pre: differences between baseline-follow up evaluation.

3.1.1. Functional and Motor Strength Assessment

Box and Block test. There was a significant effect of time and interaction in the raw data (Table 2) and an effect of time, intervention and interaction in the normalized data. The percentage change of number of boxes during follow-up was significantly higher for 90% and 110% intensity of eEmc with respect to 80% (Figure 2A).

Grip force during MVC. RM-ANOVA showed significant effect of the time and interaction in the raw data; and effect of time, intervention and interaction in the normalized data. Intervention simple effect in percentage changes of MVC showed higher maximum muscle grip strength with 90% intensity of eEmc than with 110% and 80% at post intervention and follow-up time points (Figure 2B).

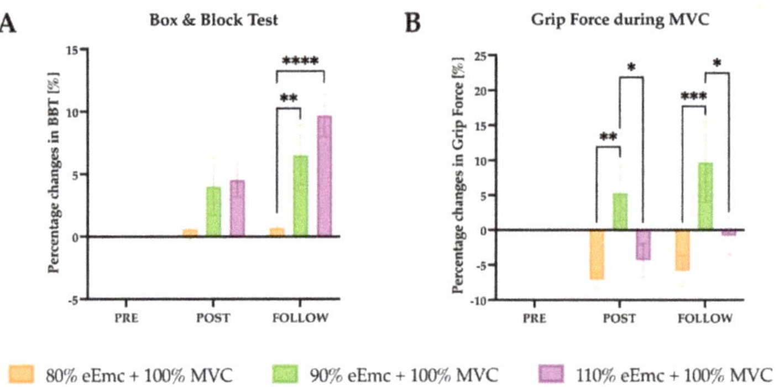

Figure 2. Functional and motor outcomes assessed by the Box and Block test and MVC. (**A**) percentage changes in the number of blocks moved in the Box and Block test with respect to baseline. Intervention simple effect shows significant differences at follow of 90% (** $p = 0.004$) and 110% eEmc (**** $p < 0.0001$) with respect to 80% eEmc; (**B**) percentage changes in MVC with respect to baseline. Intervention simple effect showed differences of 90% eEmc compared with 80% (** $p = 0.0026$) and 110% eEmc at post (* $p = 0.0211$) and follow (*** $p = 0.0108$; * $p = 0.0002$) time points.

3.1.2. Neurophysiological Assessment

F-wave. There was a significant effect of interaction on the persistence of the F wave in the raw data. The percentage change of F wave persistency was significantly higher for 90% of eEmc intensity at post intervention in comparison with 80% and 110%, and at follow-up in both 90% and 110% in comparison with 80% (Figure 3A).

For Mmax wave, RM-ANOVA showed significant effect of intervention in the raw data, but the percentage change of Mmax amplitude did not show any significant effect. On the other hand, the normalized Fmax/Mmax ratio showed significant effect of interaction. It was significantly higher with 90% of eEmc in comparison with 80% at post intervention; and with 90% and 110% in comparison with 80% of eEmc at follow-up evaluation (Figure 3B).

Spinal MEPs recruitment curve. RM-ANOVA showed no significant differences in the raw or the normalized data regarding results of the recruitment curve of spinal MEPs at C3-C4 or at C6-C7 level.

Cortical RMT, TMS-induced cortical MEPs and SICI. Cortical RMT and SICI did not change significantly at any eEmc stimulation intensity ($p > 0.05$) neither in the raw nor in the normalized data.

For the TMS-induced cortical MEPs at 120% of RMT, RM-ANOVA showed significant effect of intervention and interaction in the raw data. However, there was significant effect of time, intervention and interaction in the normalized data. According to intervention simple effect, the percentage change of MEP amplitude was significantly higher at 110% of eEmc with respect to 80% and to 90% at follow-up (Figure 4A).

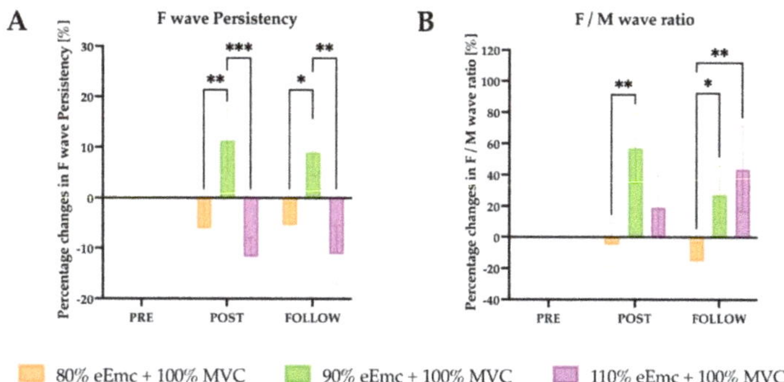

Figure 3. F wave persistency and F/M wave ratio. (**A**) percentage changes in F wave persistency with respect to baseline. Intervention simple effect showed significant differences of 90% eEmc with respect to 80% (** $p = 0.0079$) and 110% eEmc (*** $p = 0.0005$) at post and at follow (* $p = 0.0304$; ** $p = 0.0020$); (**B**) percentage changes in F/M ratio with respect to baseline. Intervention simple effect showed significant differences at post of 90% and 80% eEmc (** $p = 0.0032$); and at follow of 80% eEmc and 90% (* $p = 0.0491$) and 110% eEmc (** $p = 0.0051$).

For recruitment curves of cortical MEP, there was a significant effect of intensity and interaction for difference between pre and post, and an effect of interaction for difference between pre and follow-up. We found a significantly higher MEP amplitude with 90% and 110% of eEmc compared with 80% at post intervention (Figure 4B). The MEP amplitude was significantly higher at follow-up testing in 90% and 110% of eEmc with respect to 80% for 1.4× and 1.5 × RMT (Figure 4C), and in 110% with respect to 80% and 90% of eEmc for 1.1 × RMT (Figure 4C).

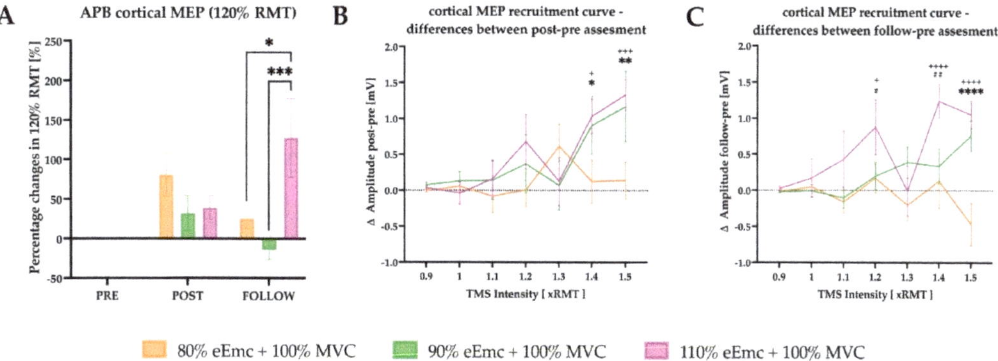

Figure 4. Cortical excitability outcomes. (**A**) percentage changes in cortical MEP evoked by TMS 120% of RMT with respect to baseline. Intervention simple effect showed significant differences at follow of 110% eEmc with respect to 80% (* $p = 0.0126$) and 90% eEmc (*** $p = 0.0005$); (**B**) difference in cortical MEP amplitude between post and pre time points. Intensity simple effects showed significant differences of 80% eEmc with respect 90% and 110% of eEmc at 1.4 (* $p = 0.0390$, + $p = 0.0133$) and 1.5 (** $p = 0.0042$, +++ $p = 0.0008$) RMT cortical MEP; (**C**) difference in cortical MEP amplitude between follow and pre time points. Intensity simple effects showed significant differences of 80% and 110% eEmc at 1.2 (+ $p = 0.0162$), 1.4 (++++ $p < 0.0001$) and 1.5 RMT (++++ $p < 0.0001$); and of 90% and 110% eEmc at 1.2 (# $p < 0.0225$) and 1.4 RMT multiple (## $p = 0.0015$); and of 80% and 90% eEmc at 1.5 (**** $p < 0.0001$) RMT multiple.

3.2. Effects of Hand Grip Force during Training

Raw data of functional and motor strength assessment and neurophysiological assessments during the second part of the study, evaluating different hand grip strength levels for training, combined with a fixed eEMC intensity, are shown in Table 3.

Table 3. Raw data and statistics of functional and motor strength and neurophysiological assessment during 100% or 50% of hand grip strength during hand training combined with the 90% of electrical stimulus intensity for eEmc.

	Time	80%	90%	110%	F (DFn, DFd)	*p* Value	η^2
BBT (blocks number)	Pre	74.1 ± 3.0	72.8 ± 3.1	71.2 ± 2.7	Ftime $(2, 16) = 17.32$	**<0.0001**	0.033
	Post	74.6 ± 3.1	75.7 ± 3.5	74.4 ± 2.9	Finterven. $(2, 16) = 0.16$	**0.855**	0.002
	Foll.	74.6 ± 3.2	77.4 ± 3.4	78.0 ± 2.7	Finterac. $(4, 32) = 3.49$	**0.018**	0.014
MVC Grip Force (kg)	Pre	37.8 ± 8.3	31.3 ± 11.3	36.9 ± 7.6	Ftime $(2, 16) = 5.87$	**0.012**	0.002
	Post	35.1 ± 7.6	32.7 ± 11.0	35.5 ± 8.7	Finterven. $(2, 16) = 3.59$	**0.052**	0.040
	Foll.	35.8 ± 9.2	33.6 ± 9.8	36.4 ± 7.5	Finterac. $(4, 32) = 2.71$	**0.048**	0.009
F wave Persistency	Pre	65.4 ± 7.9	48.8 ± 9.1	71.5 ± 6.5	Ftime $(2, 16) = 0.33$	**0.725**	0.002
	Post	59.3 ± 8.5	60.2 ± 8.2	59.7 ± 8.0	Finterven. $(2, 16) = 3.33$	0.062	0.019
	Foll.	59.9 ± 9.5	57.9 ± 9.2	60.2 ± 10.1	Finterac. $(4, 32) = 2.92$	0.031	0.031
Mmax wave (microV)	Pre	$13,296.3 \pm 1754.9$	$16,829.0 \pm 1863.2$	$16,844.5 \pm 1204.9$	Ftime $(2, 16) = 1.77$	0.202	0.018
	Post	$12,752.8 \pm 1427.5$	$15,485.3 \pm 1937.6$	$16,234.6 \pm 1579.3$	Finterven. $(2, 16) = 4.85$	0.023	0.119
	Foll.	$13,697.6 \pm 1726.9$	$17,153.6 \pm 1889.8$	$18,452.1 \pm 970.3$	Finterac. $(4, 32) = 0.29$	0.882	0.003
Ratio Fmax/Mmax	Pre	5.5 ± 0.8	3.7 ± 0.8	4.2 ± 0.6	Ftime $(1.92, 15.33) = 0.15$	0.851	0.001
	Post	4.6 ± 0.7	5.0 ± 0.8	4.5 ± 0.5	Finterven. $(1.38, 11.03) = 0.55$	0.530	0.008
	Foll.	4.2 ± 0.6	4.3 ± 0.7	5.4 ± 0.6	Finterac. $(1.83, 14.64) = 1.93$	0.082	0.069
RMT TMS	Pre	38.7 ± 2.3	39.1 ± 2.6	39.1 ± 2.0	Ftime $(1.85, 14.81) = 1.46$	0.263	0.002
	Post	37.8 ± 2.4	37.8 ± 2.4	39.3 ± 2.5	Finterven. $(1.89, 15.14) = 0.21$	0.805	0.002
	Foll.	38.2 ± 2.4	39.1 ± 2.6	38.4 ± 2.0	Finterac. $(1.91, 15.31) = 1.46$	0.263	0.003
SICI in APB (%)	Pre	13.1 ± 3.7	13.8 ± 2.9	8.3 ± 1.2	Ftime $(1.13, 9.05) = 0.93$	0.374	0.021
	Post	10.8 ± 2.6	12.4 ± 4.2	7.1 ± 2.6	Finterven. $(1.51, 12.07) = 0.14$	0.816	0.003
	Follow	13.2 ± 4.2	16.7 ± 4.7	11.2 ± 4.1	Finterac. $(1.90, 15.21) = 0.34$	0.708	0.010
Cortical MEP at 120% RMT in APB (mV)	Pre	0.6 ± 0.1	0.8 ± 0.1	0.8 ± 0.1	Ftime $(2, 16) = 3.43$	0.058	0.091
	Post	1.1 ± 0.2	0.9 ± 0.1	1.1 ± 0.2	Finterven. $(2, 16) = 5.25$	0.018	0.056
	Foll.	0.8 ± 0.1	0.7 ± 0.1	1.6 ± 0.3	Finterac. $(4, 32) = 4.99$	0.003	0.113
Stimulus Intensity		**80%**	**90%**	**110%**			
TMS recruitment: diff post-pre (mV)	0.9	-0.002 ± 0.08	0.08 ± 0.11	0.04 ± 0.08			
	1	0.06 ± 0.20	0.13 ± 0.37	-0.04 ± 0.47			
	1.1	-0.08 ± 0.68	0.14 ± 0.80	0.16 ± 0.77	$F_{intensity\ (6,\ 48)} = 10.69$	**<0.001**	0.121
	1.2	0.004 ± 0.71	0.37 ± 1.20	0.68 ± 1.14	$F_{interven.\ (2,\ 16)} = 0.74$	0.491	0.030
	1.3	0.62 ± 0.96	0.07 ± 1.02	0.12 ± 1.00	$F_{interac.\ (12,\ 96)} = 2.17$	**0.019**	0.079
	1.4	0.124 ± 0.90	0.91 ± 1.21	1.03 ± 0.71			
	1.5	0.14 ± 0.77	1.17 ± 1.48	1.34 ± 0.64			
TMS recruitment: diff foll-pre (mV)	0.9	-0.01 ± 0.06	-0.02 ± 0.03	0.03 ± 0.09			
	1	0.05 ± 0.28	-0.01 ± 0.25	0.17 ± 0.80			
	1.1	-0.16 ± 0.46	-0.10 ± 0.43	0.43 ± 1.18	$F_{intensity\ (6,\ 48)} = 4.09$	**0.002**	0.078
	1.2	0.17 ± 0.61	0.20 ± 0.59	0.88 ± 1.15	$F_{interven.\ (2,\ 16)} = 3.61$	0.051	0.102
	1.3	-0.20 ± 0.66	0.39 ± 0.65	-0.01 ± 1.03	$F_{interac.\ (12,\ 96)} = 3.73$	**<0.001**	0.109
	1.4	0.13 ± 1.14	0.34 ± 0.72	1.24 ± 0.70			
	1.5	-0.47 ± 0.88	0.76 ± 0.63	1.05 ± 0.57			
C3-C4 recruitment: diff post-pre (mV)	0.9	0.02 ± 0.07	-0.004 ± 0.01	0.01 ± 0.02			
	1	0.01 ± 0.09	-0.01 ± 0.02	0.02 ± 0.05			
	1.1	0.01 ± 0.04	-0.03 ± 0.11	0.001 ± 0.03	$F_{intensity\ (6,\ 48)} = 0.96$	0.460	0.023
	1.2	0.01 ± 0.03	-0.04 ± 0.11	0.004 ± 0.02	$F_{interven.\ (2,\ 16)} = 0.99$	0.395	0.057
	1.3	-0.03 ± 0.09	-0.04 ± 0.11	-0.01 ± 0.03	$F_{interac.\ (12,\ 96)} = 1.69$	0.081	0.037
	1.4	0.03 ± 0.09	-0.06 ± 0.11	0.01 ± 0.04			
	1.5	0.04 ± 0.10	-0.003 ± 0.09	-0.02 ± 0.04			
C3-C4 recruitment: diff foll-pre (mV)	0.9	0.03 ± 0.09	0.02 ± 0.05	0.01 ± 0.02			
	1	0.02 ± 0.09	-0.01 ± 0.03	0.04 ± 0.10			
	1.1	0.01 ± 0.05	-0.002 ± 0.11	0.01 ± 0.02	$F_{intensity\ (6,\ 48)} = 0.31$	0.929	0.008
	1.2	0.02 ± 0.05	0.004 ± 0.02	0.02 ± 0.03	$F_{interven.\ (2,\ 16)} = 0.78$	0.474	0.036
	1.3	0.03 ± 0.04	0.01 ± 0.02	0.001 ± 0.03	$F_{interac.\ (12,\ 96)} = 1.10$	0.373	0.035
	1.4	0.04 ± 0.08	-0.01 ± 0.05	0.01 ± 0.03			
	1.5	0.06 ± 0.08	0.001 ± 0.04	0.01 ± 0.03			

Table 3. *Cont.*

	Time	80%	90%	110%	F (DFn, DFd)	*p* Value	η^2
C6-C7 recruitment: diff post-pre (mV)	0.9	0.003 ± 0.03	-0.001 ± 0.03	0.01 ± 0.02			
	1	-0.001 ± 0.03	0.04 ± 0.10	0.01 ± 0.03			
	1.1	0.004 ± 0.02	-0.01 ± 0.05	0.01 ± 0.03	$F_{\text{intensity }(6, 48)} = 2.14$	0.066	0.041
	1.2	-0.03 ± 0.09	-0.01 ± 0.06	-0.01 ± 0.03	$F_{\text{interven. }(2, 16)} = 0.67$	0.525	0.023
	1.3	-0.04 ± 0.11	-0.02 ± 0.04	0.001 ± 0.05	$F_{\text{interac. }(12, 96)} = 0.51$	0.906	0.023
	1.4	-0.03 ± 0.08	-0.01 ± 0.05	-0.004 ± 0.05			
	1.5	-0.02 ± 0.12	0.01 ± 0.04	0.01 ± 0.06			
C6-C7 recruitment: diff foll-pre (mV)	0.9	0.01 ± 0.02	-0.004 ± 0.03	0.02 ± 0.02			
	1	-0.0004 ± 0.03	0.03 ± 0.09	0.02 ± 0.02			
	1.1	0.01 ± 0.02	-0.001 ± 0.04	0.02 ± 0.04	$F_{\text{intensity }(6, 48)} = 0.40$	0.878	0.010
	1.2	-0.02 ± 0.10	0.01 ± 0.02	0.003 ± 0.03	$F_{\text{interven. }(2, 16)} = 0.16$	0.850	0.005
	1.3	0.02 ± 0.04	0.01 ± 0.02	-0.02 ± 0.08	$F_{\text{interac. }(12, 96)} = 0.72$	0.725	0.039
	1.4	0.01 ± 0.09	0.004 ± 0.03	0.004 ± 0.04			
	1.5	-0.01 ± 0.14	0.002 ± 0.02	0.03 ± 0.06			

diff post-pre: differences between post-pre evaluation; diff foll-pre: differences between baseline-follow up evaluation.

3.2.1. Functional and Motor Strength Assessment

Box and Block test. RM-ANOVA showed an effect of time and interaction in the raw data (Table 3), and only of time in the normalized data. The percentage of BBT changes was not significant between 50% vs. 100% handgrip strength condition.

Grip force during MVC. RM-ANOVA showed an effect of intervention and interaction in the raw and in the normalized data (Table 3). The percentage of MVC increased significantly in 100% handgrip strength condition compared with 50% at post intervention and at follow-up (Figure 5A).

3.2.2. Neurophysiological Assessment

F-wave persistence. RM-ANOVA showed effect of interaction in the raw data, and effect of intervention and interaction in the normalized data. According to intervention simple effects, the percentage of F wave persistency increased significantly with 100% hand grip strength compared to 50% at post intervention and follow-up (Figure 5B).

F/M wave ratio. RM-ANOVA showed no effect in the raw data but showed significant effect of intervention and interaction in the normalized data. There was a significantly higher ratio with 100% of handgrip strength intervention than with 50% at post intervention (Figure 5C).

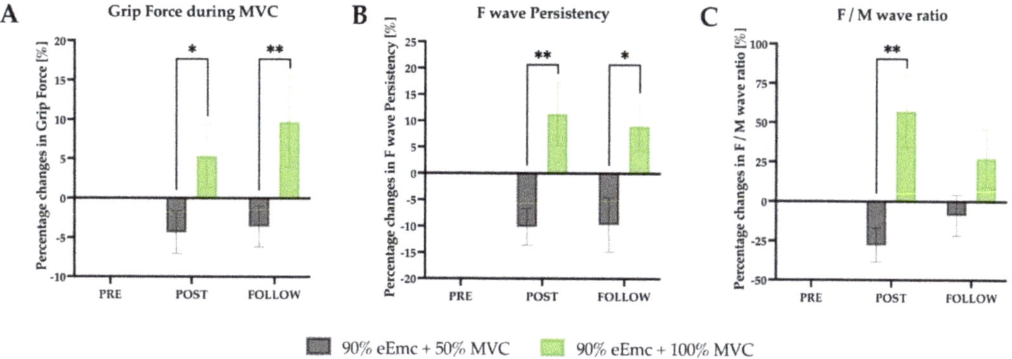

Figure 5. Changes in MVC, F wave persistency and F/M wave ratio. (**A**) percentage changes in grip force during MVC according to baseline. Intervention simple effect showed differences at post (* *p* = 0.0144) and follow (** *p* = 0.0011); (**B**) percentage changes in F wave persistency according to baseline. Intervention simple effects showed significant differences at post (** *p* = 0.0073) and follow (* *p* = 0.0191); (**C**) percentage changes in F/M ratio according to baseline. Intervention simple effect showed significant differences at post (** *p* = 0.0016).

Spinal MEPs recruitment curve. RM-ANOVA did not show any significant differences in raw and normalized data of the recruitment curve of spinal MEP evoked with stimulation either at C3-C4 or C6-C7 ($p > 0.05$).

Cortical RMT, TMS-induced cortical MEPs and SICI. RM-ANOVA showed significant effect of time in the raw and in the normalized data for cortical RMT, but not for SICI. On the other hand, TMS-induced cortical MEPs at 120% RMT showed significant effect of interaction in the raw and in the normalized data. The cortical MEP amplitude was higher in 50% than in 100% handgrip strength intervention at follow-up (Figure 6A).

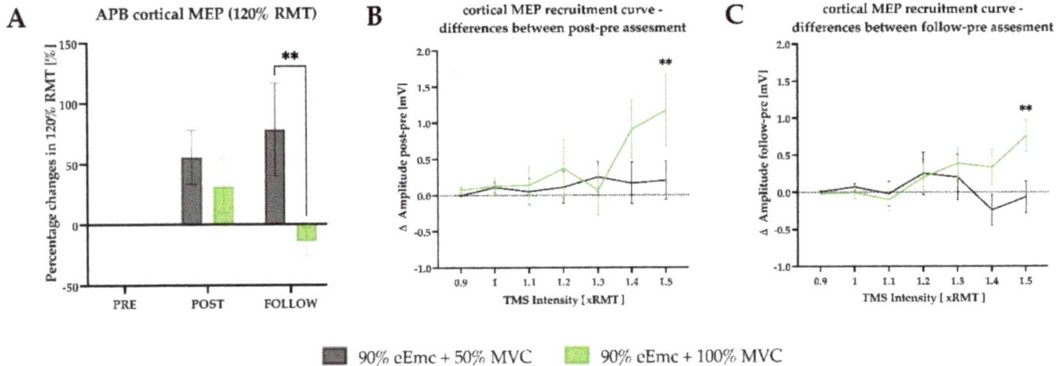

Figure 6. Cortical excitability outcomes. (**A**) percentage changes in cortical MEP from 120% of RMT according to baseline. Intervention simple effect showed significant differences at follow; (**B**) difference amplitude between post and pre cortical MEP recruitment. Intensity simple effects showed significant differences at 1.5 RMT multiple (** $p = 0.0059$); (**C**) difference amplitude between follow and pre MEP recruitment. Intensity simple effects showed significant differences at 1.5 RMT multiple (** $p = 0.0063$).

In the recruitment curve of cortical MEPs, RM-ANOVA showed significant effect of intensity and interaction between pre and post, and effect of interaction between pre and follow-up. Additionally, 100% of handgrip strength intervention significantly increased the cortical MEP amplitude at 1.5 × RMT with respect to 50% of MVC at post intervention and at follow-up (Figure 6B,C).

4. Discussion

In a previous study we reported that eEmc combined with hand training improves hand grip force and increases spinal and corticospinal excitability in comparison to each intervention tested alone [8]. In the present study, we further investigated if the levels of both eEmc intensity and muscle strength during training could influence the results for neuromodulation at the cervical spinal cord. The results of this study show that eEmc at 90% intensity of spinal RMT of the APB muscle combined with maximal handgrip strength induced better hand function and muscle strength than the other conditions assayed. The results of F wave persistency and F/M ratio and of cortical MEP recruitment curves reflected stronger plastic changes in the spinal and cortical pathways of eEmc at 90% intensity compared to eEmc at intensities of 80% or 110%. Most of these effects were maintained at least for one hour following the intervention. On the other hand, eEmc at 110% intensity induced better hand function and higher F/M ratio and cortical plasticity changes with respect to eEmc at 80%. In contrast, eEmc at 80% of intensity did not show any noticeable effect on hand function, spinal and cortical excitability. In the second part of the study, eEmc at 90% intensity was combined with training at the maximal (100%) or at half (50%) MVC in hand grip strength. With 100% MVC, there was higher muscle strength and more plastic changes at spinal cord, measured by F wave persistency and F/M ratio, and at cortical level, measured by cortical MEP recruitment curve, than with

50% MVC. Most of these effects were also maintained at least one hour during follow-up testing. In contrast, with the 50% MVC of grip strength there was only a significant increase in cortical excitability measured by cortical MEP.

4.1. Effects of eEmc Intensity

There is no consensus in the literature about which intensity of electrical stimulation may modulate more effectively the functional outputs of the cervical spinal cord. Thus, previous studies have used different methods to select the eEmc intensity. In our previous study in healthy individuals, we empirically used a stimulation intensity at 90% of spinal RMT of the APB muscle [10], whereas Benavides et al. [5] had used 100% of RMT of the Biceps brachii muscle in SCI and healthy individuals. Gad et al. [6] adjusted the eEmc to enable maximal grip strength or submaximal isometric hand movement without causing discomfort in SCI subjects. Freverty et al. [7] used varying combinations of eEmc parameters to obtain optimal facilitation of voluntary hand grip by identification of the relative activation levels of the motor pools, and combined eEmc with medication (Buspirone) without hand training. Studying cervical SCI subjects, other authors used different strategies, such as beginning eEmc at 50 mA and adjusting the intensity based on the functional task performance and subjective feedback during the intervention [9], increasing stimulus intensity in 10 mA intervals from 10 to 120 mA [8], or from 5 to 68 mA [19].

Despite the different stimulus intensity used for eEmc, the most likely direct mechanism of eEmc occurs via transcutaneous tonic spinal activation by elevating spinal networks excitability [12] and may activate interneuronal pathways that generate action potentials on motoneurons within a motor pool in a more normal stochastic time frame [13]. As such, eEmc is hypothesized to potentiate the generation of postsynaptic excitatory potentials and thus shift the spinal motor network excitability closer to the excitation threshold. The constantly changing postsynaptic potentials intrinsic to spinal networks [20] can contribute to a constantly changing population of interneurons being asynchronously activated in a random pattern [21]. According to our results, it can be interpreted that eEmc, at higher levels when delivered at 90% stimulus intensity, brings interneurons and motoneurons closer to motor threshold and, therefore, more likely to respond to descending drive, and elevates the excitability of spinal cord networks and possibly also of motor cortex. In addition, sensory afferents at the level of dorsal roots and/or the dorsal column spinal pathways may be affected by the tSCS and thus contribute differentially to elevating or suppressing spinal and/or supraspinal neural excitability [14,15,21]. Despite we cannot preclude this mechanism, considering that the stimulation was applied at C3 level, far cranial from the APB muscle root innervation segment (C8-T1), we hypothesize that the effect on muscle response is more likely due to stimulation of descending pathways and premotoneuronal network than to the stimulation of proprioceptive fibers within posterior roots.

From our results, it appears that there is a window of intensity for the neuromodulatory effects of eEmc. The effects of an eEmc intensity of 90% of spinal RMT of the target muscle exceeded the effects of lower (80%) and higher (110%) stimulus intensities. It is unclear, however, why only the 90% level of intensity enabled higher voluntarily controlled drive to produce stronger contractions of hand training, which subsequently strengthens the neuromuscular network. It is, noteworthy that eEmc at 110% intensity augmented hand muscle strength less than 90% intensity but induced more plastic changes at cortical level than 90% and 80% intensities. Perhaps the higher eEmc intensity can elicit impulses that are conducted antidromically to the cortex or induce a greater facilitatory effect among brain networks, inducing more effectively plastic changes at the cortical than the spinal level. Thus, our data suggest that eEmc at suprathreshold intensity modulates more effectively the cortical excitability than at subthreshold intensities, but hand function, muscle strength and spinal cord excitability were modulated more effectively by subthreshold intensity of 90% spinal RMT.

4.2. Effects of Handgrip Strength during Training

Our results indicate that the combination of maximal (100% of MVC) handgrip strength with eEmc further increases the hand muscle strength and induces more effectively plastic changes at spinal cord and cortical levels in comparison to lower (50%) handgrip strength.

There is no consensus either about what kind of hand training could help to modulate, more effectively, spinal cord motor function in healthy individuals or following SCI. Indeed, only a few studies on eEmc at cervical cord level combined tSCS with hand training. In the study by Gad et al. [6] with cervical SCI subjects, two tasks were performed: isometric maximum handgrip contraction or voluntary rhythmic efforts of submaximal contraction (opening and closing the hand). In another study on a single SCI subject, a functional task combined with eEmc aimed to address hand grasp and pinch bilaterally [9]. In healthy individuals, we previously used maximal handgrip force for hand training combined with eEmc [10]. Several observations made after a single training or stimulation session suggest that short-term plastic changes occur [5,10,22]. Performing the maximum handgrip strength combined with eEmc at 90% of stimulus intensity modulates more effectively hand motor function and induces more effectively neurophysiological changes at spinal cord and cortical levels than when performing lower hand grip strength.

According to our results, activity-based plasticity depends on the level of hand grip strength for hand training and the eEmc intensity for neuromodulation. Activity-based neural plasticity mechanisms involve both functional modifications of existing synapses and structural changes that alter the synaptic connectivity of neurons (formation, removal and remodeling of synaptic buttons and even dendrites) [22]. A single exposure to activity-dependent plasticity by physical training combined with electrical stimulation may induce these short-term changes that in turn could promote occurrence of long-term plasticity under repeated exposure that involves mechanisms such as long-term potentiation and depression, morphological changes of dendrites, synaptogenesis and axonal branching/regeneration [23].

4.3. Limitations of the Study

It has to be noted that the sample size in this study was relatively small, limiting statistical power and interpretation of the results. Four days of experiments and the long duration of each experiment may result in fatigue factors, limiting the number of replicates that could be recorded for neurophysiological parameters, resulting in more variability. The only follow-up assessment period was for one hour after the end of session, thus providing no comparison for other longer timeframes. Nevertheless, the obtained results open a work frame for further studies refining the conditions of combinatory neuromodulation for SCI patients.

5. Conclusions

Taken together, this study demonstrates an approach to assess two modulatory interventions, eEmc and activity-dependent intervention at the supraspinal, and spinal and peripheral levels. It also provides basic physiological responses in healthy subjects, which will help the design of future studies involving other dysfunctional conditions. The findings of the study indicate that the stimulus intensity for eEmc is relevant for improving hand function and/or muscle strength, as well as hand muscle strength level and recruitment of motor units within a motor pool during training in conjunction with electrical neuromodulation. The data also demonstrate both the immediacy and short-term persistence of at least one hour after intervention. Interestingly, our results also allow us to hypothesize that spinal and cortical excitability may be modulated differently depending on the eEmc intensity. Therefore, both the electrical stimulus intensity of eEmc and the intensity of the voluntary effort during training should be considered when optimizing rehabilitation protocols for restoring motor performance.

Author Contributions: Conceptualization, H.K., M.R.-C., Á.F., G.G.-A. and J.V.; methodology, H.K., M.R.-C. and J.V.; software, M.R.-C.; formal analysis, H.K. and M.R.-C.; validation, H.K., M.R.-C., G.G.-A. and J.V.; investigation, H.K., M.R.-C., Á.F., L.G., I.S., G.G.-A. and J.V.; resources, H.K., M.R.-C., E.O., G.G.-A. and J.V.; data curation, H.K., M.R.-C. and L.G.; writing—original draft preparation, H.K., M.R.-C. and X.N.; writing—review and editing, H.K., M.R.-C., V.R.E., L.G., E.O., I.S., J.V., Y.G., Á.F., G.G.-A. and X.N.; visualization, H.K. and M.R.-C.; supervision, H.K. and G.G.-A.; project administration, H.K., G.G.-A. and J.V.; funding acquisition, H.K., G.G.-A. and J.V. All authors have read and agreed to the published version of the manuscript.

Funding: This research was funded by H2020-ERA-NET Neuron (AC16/00034) to G.G.-A. and J.V.; AES 2019 ISCIII (PI19/01680) to J.V.; Fundació La Marató de TV3 2017 (201713.31) to G.G.-A.; Premi Beca "Mike Lane" 2019-Castellers de la Vila de Gràcia to H.K.; and National Institutes of Health Grant 1R01 NS102920-01A1 to Y.G.

Institutional Review Board Statement: The study was conducted according to the guidelines of the Declaration of Helsinki, and approved by the Ethics Committee of Institut Guttmann (protocol code 2017255 and date of approval: 29 April 2017).

Informed Consent Statement: Informed consent was obtained from all subjects involved in the study.

Data Availability Statement: The data presented in this study are available on request from the corresponding author. The data are not publicly available due to privacy restrictions.

Conflicts of Interest: V.R.E. and Y.G. hold shareholder interest in NeuroRecovery Technologies and hold inventorship rights on intellectual property licensed by The Regents of the University of California to NeuroRecovery Technologies and its subsidiaries. Y.G. also holds shareholder interest in Cosyma Inc., St. Petersburg, Russia. The funders had no role in the design of the study, in the collection, analyses, or interpretation of data, in the writing of the manuscript, or in the decision to publish the results.

References

1. Gerasimenko, Y.P.; Lu, D.C.; Modaber, M.; Zdunowski, S.; Gad, P.; Sayenko, D.G.; Morikawa, E.; Haakana, P.; Ferguson, A.R.; Roy, R.R.; et al. Noninvasive Reactivation of Motor Descending Control after Paralysis. *J. Neurotrauma* **2015**, *32*, 1968–1980. [CrossRef] [PubMed]
2. Megía García, A.; Serrano-Muñoz, D.; Taylor, J.; Avendaño-Coy, J.; Gómez-Soriano, J. Transcutaneous Spinal Cord Stimulation and Motor Rehabilitation in Spinal Cord Injury: A Systematic Review. *Neurorehabil. Neural Repair* **2020**, *34*, 3–12. [CrossRef] [PubMed]
3. Calvert, J.S.; Manson, G.A.; Grahn, P.J.; Sayenko, D.G. Preferential activation of spinal sensorimotor networks via lateralized transcutaneous spinal stimulation in neurologically intact humans. *J. Neurophysiol.* **2019**, *122*, 2111–2118. [CrossRef] [PubMed]
4. Gerasimenko, Y.; Gorodnichev, R.; Moshonkina, T.; Sayenko, D.; Gad, P.; Edgerton, V.R. Transcutaneous electrical spinal-cord stimulation in humans. *Ann. Phys. Rehabil. Med.* **2015**, *58*, 225–231. [CrossRef]
5. Benavides, F.D.; Jo, H.J.; Lundell, H.; Edgerton, V.R.; Gerasimenko, Y.; Perez, M.A. Cortical and subcortical effects of transcutaneous spinal cord stimulation in humans with tetraplegia. *J. Neurosci.* **2020**, *40*, 2633–2643. [CrossRef] [PubMed]
6. Gad, P.; Lee, S.; Terrafranca, N.; Zhong, H.; Turner, A.; Gerasimenko, Y.; Edgerton, V.R. Non-Invasive Activation of Cervical Spinal Networks after Severe Paralysis. *J. Neurotrauma* **2018**, *35*, 2145–2158. [CrossRef]
7. Freyvert, Y.; Yong, N.A.; Morikawa, E.; Zdunowski, S.; Sarino, M.E.; Gerasimenko, Y.; Edgerton, V.R.; Lu, D.C. Engaging cervical spinal circuitry with non-invasive spinal stimulation and buspirone to restore hand function in chronic motor complete patients. *Sci. Rep.* **2018**, *8*, 15546. [CrossRef]
8. Inanici, F.; Samejima, S.; Gad, P.; Edgerton, V.R.; Hofstetter, C.P.; Moritz, C.T. Transcutaneous Electrical Spinal Stimulation Promotes Long-Term Recovery of Upper Extremity Function in Chronic Tetraplegia. *IEEE Trans. Neural Syst. Rehabil. Eng.* **2018**, *26*, 1272–1278. [CrossRef] [PubMed]
9. Zhang, F.; Momeni, K.; Ramanujam, A.; Ravi, M.; Carnahan, J.; Kirshblum, S.; Forrest, G.F. Cervical Spinal Cord Transcutaneous Stimulation Improves Upper Extremity and Hand Function in People with Complete Tetraplegia: A Case Study. *IEEE Trans. Neural Syst. Rehabil. Eng.* **2020**, *28*, 3167–3174. [CrossRef] [PubMed]
10. Kumru, H.; Flores, Á.; Rodríguez-Cañón, M.; Edgerton, V.R.; García, L.; Benito-Penalva, J.; Navarro, X.; Gerasimenko, Y.; García-Alías, G.; Vidal, J. Cervical Electrical Neuromodulation Effectively Enhances Hand Motor Output in Healthy Subjects by Engaging a Use-Dependent Intervention. *J. Clin. Med.* **2021**, *10*, 195. [CrossRef] [PubMed]
11. Guiho, T.; Baker, S.N.; Jackson, A. Epidural and transcutaneous spinal cord stimulation facilitates descending inputs to upper-limb motoneurons in monkeys. *J. Neural. Eng.* **2021**, *18*, 046011. [CrossRef]
12. Hofstoetter, U.S.; Freundl, B.; Binder, H.; Minassian, K. Common neural structures activated by epidural and transcutaneous lumbar spinal cord stimulation: Elicitation of posterior root-muscle reflexes. *PLoS ONE* **2018**, *13*, e0192013. [CrossRef]

13. Stein, R.B.; Gossen, E.R.; Jones, K.E. Neuronal variability: Noise or part of the signal? *Nat. Rev. Neurosci.* **2005**, *6*, 389–397. [CrossRef]

14. Gomes-Osman, J.; Field-Fote, E.C. Cortical vs. afferent stimulation as an adjunct to functional task practice training: A randomized, comparative pilot study in people with cervical spinal cord injury. *Clin. Rehabil.* **2015**, *29*, 771–782. [CrossRef] [PubMed]

15. Gomes-Osman, J.; Tibbett, J.A.; Poe, B.P.; Field-Fote, E.C. Priming for Improved Hand Strength in Persons with Chronic Tetraplegia: A Comparison of Priming-Augmented Functional Task Practice, Priming Alone, and Conventional Exercise Training. *Front. Neurol.* **2017**, *7*, 7–242. [CrossRef] [PubMed]

16. Oldfield, R.C. The assessment and analysis of handedness: The Edinburgh inventory. *Neuropsychologia* **1971**, *9*, 97–113. [CrossRef]

17. Mathiowetz, V.; Volland, G.; Kashman, N.; Weber, K. Adult Norms for the Box and Block Test of Manual Dexterity. *Am. J. Occup. Ther.* **1985**, *39*, 386–391. [CrossRef] [PubMed]

18. Kujirai, T.; Caramia, M.D.; Rothwell, J.C.; Day, B.L.; Thompson, P.D.; Ferbert, A.; Wroe, S.; Asselman, P.; Marsden, C.D. Corticocortical inhibition in human motor cortex. *J. Physiol.* **1993**, *471*, 501–519. [CrossRef] [PubMed]

19. Murray, L.M.; Knikou, M. Remodeling Brain Activity by Repetitive Cervicothoracic Transspinal Stimulation after Human Spinal Cord Injury. *Front. Neurol.* **2017**, *20*, 8–50. [CrossRef] [PubMed]

20. Taccola, G.; Sayenko, D.; Gad, P.; Gerasimenko, Y.; Edgerton, V.R. And yet it moves: Recovery of volitional control after spinal cord injury. *Prog. Neurobiol.* **2018**, *160*, 64–81. [CrossRef]

21. Angeli, C.A.; Edgerton, V.R.; Gerasimenko, Y.P.; Harkema, S.J. Altering spinal cord excitability enables voluntary movements after chronic complete paralysis in humans. *Brain* **2014**, *137*, 1394–1409. [CrossRef] [PubMed]

22. Smith, A.C.; Knikou, M. A Review on Locomotor Training after Spinal Cord Injury: Reorganization of Spinal Neuronal Circuits and Recovery of Motor Function. *Neural Plast.* **2016**, *2016*. [CrossRef] [PubMed]

23. Knikou, M. Neural control of locomotion and training-induced plasticity after spinal and cerebral lesions. *Clin. Neurophysiol.* **2010**, *121*, 1655–1668. [CrossRef] [PubMed]

Journal of
Clinical Medicine

Article

Low-Intensity and Short-Duration Continuous Cervical Transcutaneous Spinal Cord Stimulation Intervention Does Not Prime the Corticospinal and Spinal Reflex Pathways in Able-Bodied Subjects

Atsushi Sasaki [1,2], Roberto M. de Freitas [3], Dimitry G. Sayenko [4], Yohei Masugi [1,5], Taishin Nomura [3], Kimitaka Nakazawa [1] and Matija Milosevic [3,*]

[1] Department of Life Sciences, Graduate School of Arts and Sciences, The University of Tokyo, Tokyo 153-8902, Japan; atsushi-sasaki@g.ecc.u-tokyo.ac.jp (A.S.); ymasugi@tiu.ac.jp (Y.M.); nakazawa@idaten.c.u-tokyo.ac.jp (K.N.)

[2] Japan Society for the Promotion of Science, Tokyo 102-0083, Japan

[3] Department of Mechanical Science and Bioengineering, Graduate School of Engineering Science, Osaka University, Osaka 560-8531, Japan; r.freitas@bpe.es.osaka-u.ac.jp (R.M.d.F.); taishin@bpe.es.osaka-u.ac.jp (T.N.)

[4] Center for Neuroregeneration, Department of Neurosurgery, Houston Methodist Research Institute, Houston, TX 77030, USA; dgsayenko@houstonmethodist.org

[5] School of Health Sciences, Tokyo International University, Saitama 350-1197, Japan

* Correspondence: matija@bpe.es.osaka-u.ac.jp

Citation: Sasaki, A.; de Freitas, R.M.; Sayenko, D.G.; Masugi, Y.; Nomura, T.; Nakazawa, K.; Milosevic, M. Low-Intensity and Short-Duration Continuous Cervical Transcutaneous Spinal Cord Stimulation Intervention Does Not Prime the Corticospinal and Spinal Reflex Pathways in Able-Bodied Subjects. *J. Clin. Med.* **2021**, *10*, 3633. https://doi.org/10.3390/jcm10163633

Academic Editors: Ursula S. Hofstoetter, Karen Minassian and Hiroyuki Katoh

Received: 1 July 2021
Accepted: 13 August 2021
Published: 17 August 2021

Publisher's Note: MDPI stays neutral with regard to jurisdictional claims in published maps and institutional affiliations.

Abstract: Cervical transcutaneous spinal cord stimulation (tSCS) has been utilized in applications for improving upper-limb sensory and motor function in patients with spinal cord injury. Although therapeutic effects of continuous cervical tSCS interventions have been reported, neurophysiological mechanisms remain largely unexplored. Specifically, it is not clear whether sub-threshold intensity and 10-min duration continuous cervical tSCS intervention can affect the central nervous system excitability. Therefore, the purpose of this study was to investigate effects of sub-motor-threshold 10-min continuous cervical tSCS applied at rest on the corticospinal and spinal reflex circuit in ten able-bodied individuals. Neurophysiological assessments were conducted to investigate (1) corticospinal excitability via transcranial magnetic stimulation applied on the primary motor cortex to evoke motor-evoked potentials (MEPs) and (2) spinal reflex excitability via single-pulse tSCS applied at the cervical level to evoke posterior root muscle (PRM) reflexes. Measurements were recorded from multiple upper-limb muscles before, during, and after the intervention. Our results showed that low-intensity and short-duration continuous cervical tSCS intervention applied at rest did not significantly affect corticospinal and spinal reflex excitability. The stimulation duration and/or intensity, as well as other stimulating parameters selection, may therefore be critical for inducing neuromodulatory effects during cervical tSCS.

Keywords: cervical; transcutaneous spinal cord stimulation; corticospinal pathway; spinal reflex; neuromodulation

1. Introduction

Spinal cord stimulation (SCS) has been applied in rehabilitation as a neuromodulation application to assist recovery of sensory and motor function after spinal cord injury (SCI), with remarkable achievements in clinical research [1]. Specifically, it has been demonstrated that lumbar epidural stimulation can contribute to restoration of voluntary motor control of lower-limb muscles, postural stability, and gait function in chronic SCI individuals [2–4]. Similarly, cervical epidural SCS has been used for promoting volitional hand function in individuals with chronic tetraplegia [5]. Consistent with epidural SCS approaches, non-invasive transcutaneous spinal cord stimulation (tSCS) has also been utilized successfully in applications for improving sensory and motor function during lower-limb voluntary movement [6] and walking using lumbar stimulation [7–12], trunk stability and standing with lower thoracic and lumbar stimulation [13,14], and gripping and upper-limb function with cervical stimulation [15–19]. Computer simulations and experimental studies using animal models as well as human studies have shown compelling evidence that electric impulses induced by either implanted epidural or surface non-invasive electrodes can primarily activate the afferent fibers in the posterior roots of the spinal cord [20–25]. Since the stimulated afferent fibers activated by tSCS have synaptic connections to the spinal interneurons and motoneurons, tSCS application can therefore be utilized to modulate spinal motor excitability.

Due to the importance of walking in humans, lumbar tSCS applications for modulation of spinal locomotor circuits have been the main focus of many previous studies [7–12]. Although cervical tSCS is also expected to have similar neuromodulatory effects on the upper-limb muscle function, it has been unexplored relative to lumbar tSCS application until recently. Recent reports have shown impressive therapeutic effects of continuous cervical tSCS [16–18,26]. However, little is known about the neural mechanisms and how continuous cervical tSCS affects central nervous system (CNS) excitability and the innervated upper-limb muscles.

Several recent studies have reported effects of continuous cervical tSCS intervention on neural excitability of upper-limb muscles [15,19]. Specifically, Benavides et al. [15] demonstrated that upper-limb subcortical but not cortical motor-evoked potentials, and intracortical inhibition can be facilitated after 20 min of continuous cervical tSCS, suggesting that cervical tSCS has an excitatory effect on the spinal networks and an inhibitory effect on the cortical networks. Moreover, it was shown that these effects were similar for able-bodied and SCI patients [15], indicating that studies with able-bodied participants can also be translated for rehabilitation protocols in SCI patients. Similarly, another study also demonstrated that continuous cervical tSCS applied intermittently (20 s ON/80 s OFF) at rest in able-bodied participants for 30 min can affect the F-wave amplitude, which indicates the excitability of anterior horn motoneurons in the spinal cord [19].

Taken together, these studies revealed that relatively long (20–30 min), continuous cervical tSCS interventions can modulate spinal and/or cortical networks controlling the upper-limb muscles after the intervention duration. However, the effect of a single, short intervention (<10 min) and understanding of how neural activity is modulated by the intervention remain unclear. Shortening the intervention duration and using low-intensity stimulation may improve therapy adherence for patients in an effort to minimize discomforts in actual clinical therapy since it was reported that high-intensity cervical tSCS could induce discomfort [27].

Therefore, the aim of this study was to investigate how low-intensity and short-duration continuous cervical tSCS applied at rest could affect corticospinal and spinal reflex excitability before, during, and after the intervention. We hypothesized that short-term application of low-intensity continuous tSCS would affect corticospinal and spinal circuits of upper-limb muscles [10,15,19,28]. To test our hypotheses, we used transcranial magnetic stimulation (TMS) to investigate motor-evoked potentials (MEPs), which reflect corticospinal excitability [29]. Moreover, single-pulse cervical tSCS was used to investigate spinal reflex responses, which reflect the excitability of the spinal reflex circuits in multiple upper-limb muscles.

2. Materials and Methods

2.1. Participants

Ten able-bodied individuals were recruited in our current study (25.8 ± 2.6 years, 68.8 ± 8.3 kg, and 173.4 ± 5.1 cm (mean ± SD)). None of the participants had a history of neurological and musculoskeletal impairments. All participants gave written informed consent in accordance with the Declaration of Helsinki. The experimental procedures were approved by the local institutional ethics committee at the Graduate School of Arts and Sciences at The University of Tokyo.

2.2. Experimental Procedures

During the experiments, participants remained in the supine position (Figure 1A). Assessments consisted of evaluating corticospinal excitability through motor-evoked potential (MEP) elicited by single-pulse TMS (see Section 2.3.2) and spinal reflex circuits excitability through posterior root muscle (PRM) reflexes elicited by single-pulse cervical tSCS (see Section 2.3.3). Assessments were performed (i) at baseline before the intervention (Pre); (ii) at mid-point during the intervention (5 min after the start of the intervention; Figure 1B) (During); and (iii) immediately after the end of the 10-min intervention (Post). MEP responses were assessed in Pre, During, and Post, while PRM reflex responses were assessed in Pre and Post since same electrodes were used for eliciting PRM reflex responses and for delivering the intervention. Eight MEP (TMS) and PRM reflex (tSCS) responses were elicited in each assessment phase. Both assessments were performed in the same session, while the order of TMS or tSCS assessments was pseudorandomized between participants (Figure 1B).

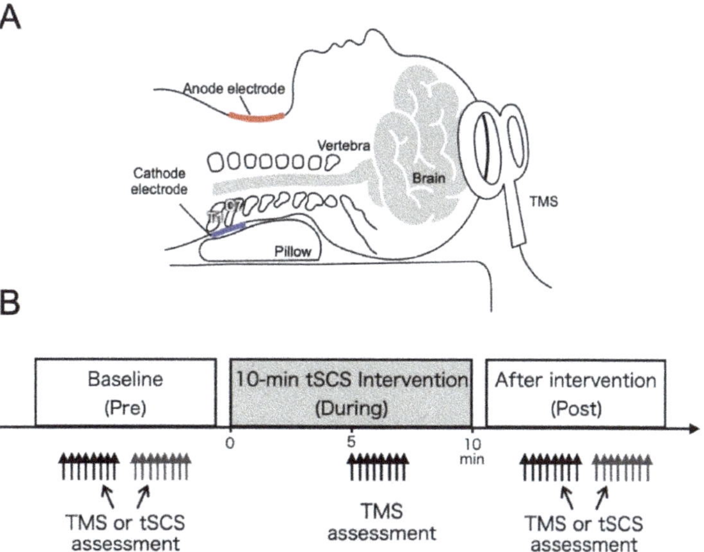

Figure 1. (**A**) Experimental setup showing head posture of participants during the experiments, including the cervical transcutaneous spinal cord stimulation (tSCS) setup and transcranial magnetic stimulation (TMS) setup. (**B**) TMS assessments were conducted before (Pre), during continuous cervical tSCS (During), and after (Post) the intervention, while tSCS assessments were conducted at Pre and Post intervention time points. At each assessment point, TMS or tSCS stimuli consisted of 8 elicited responses. In the Pre and Post assessment, the order of TMS and tSCS assessments was randomized between participants.

2.3. Assessments

2.3.1. Electromyography (EMG) Activity

Electromyographic (EMG) activity was recorded unilaterally from the right (dominant in all the subjects) hand: (1) biceps brachii (BB), (2) triceps brachii (TB), (3) flexor carpi radialis (FCR), (4) extensor carpi radialis (ECR), (5) first dorsal interosseous (FDI), and (6) abductor pollicis brevis (APB) muscles. Surface EMG electrodes (Ag/AgCl; Vitrode F-150S; Nihon Kohden, Tokyo, Japan) were placed on the muscle belly of the right upper limb with an approximate interelectrode distance of 20 mm, whereas APB electrodes were placed over the muscle belly and first metacarpophalangeal joint, and FDI electrodes were placed over the muscle belly and the second metacarpophalangeal joint [23]. A reference electrode was placed around the lateral epicondyle. Prior to application of electrodes, skin was cleaned using alcohol swabs to reduce skin impedance. Signals were band-pass filtered (5–1000 Hz) and amplified ($\times$1000) using a multi-channel EMG amplifier (MEG-6108, Nihon Kohden, Tokyo, Japan). All data were digitized at a sampling frequency of 4000 Hz using an analog-to-digital converter (PowerLab/16SP, AD Instruments, Castle Hill, Australia) and stored on a computer for post processing.

2.3.2. Transcranial Magnetic Stimulation (TMS)

TMS was delivered over the primary motor cortex using a mono-phasic magnetic stimulator (Magstim 200, Magstim Co., Whitland, UK) through a figure-of-eight coil (Magstim Co., Whitland, UK). The optimal stimulation spot ("hot spot") was searched over the left cortex, where MEPs could be recorded from the right FDI muscle. The motor threshold (MT) was determined while the participants remained in the supine position and relaxed. MT was defined as the minimum TMS intensity for which FDI MEPs had a peak-to-peak amplitudes larger than 50 μV and were evoked in at least five out of ten successive trials [30]. The stimulation intensity for experiment was set at 120% of the MT level (1.2MT) (62.3 $\pm$ 11.3 % of maximal stimulator output). MEPs were simultaneously recorded from the FDI, APB, FCR, and ECR muscles, although stimulation "hot spot" was optimized for the FDI muscle. Specifically, the mean amplitudes of the MEPs of the FDI, APB, FCR, and ECR muscles during Pre phase were 0.12 $\pm$ 0.13, 0.16 $\pm$ 0.12, 0.97 $\pm$ 0.70, and 0.53 $\pm$ 0.57 mV (mean $\pm$ SD), respectively. It should be noted that responses for some participants were notably small (4 out of 10 subjects for FCR and 1 out of 10 for ECR were <50 μV) during Pre phase, but they were kept in the analysis.

2.3.3. Transcutaneous Spinal Cord Stimulation (tSCS)

Single- (or double-) pulse tSCS can be used to consistently elicit spinal reflexes through the activation of monosynaptic connection between Ia sensory fibers and motoneurons at multiple spinal levels innervating upper-limb muscles [21,23,27]. To evaluate excitability of the spinal reflex circuits in multiple upper-limb muscles simultaneously, a constant current electrical stimulator (DS7A, Digitimer Ltd., Welwyn Garden City, UK) was used to apply a single monophasic square pulse with a 2-ms pulse width [23,27]. As during the conditioning intervention, the cathode electrode (50 $\times$ 50 mm) was placed on the spine between C6 and T1 cervical spine processes on the posterior side of the neck (Figure 1), and the anode electrode (75 $\times$ 100 mm) was placed along the midline of the anterior side of the neck [23]. Prior to the experiments, the cathode was adjusted to determine the optimal stimulation location. Specifically, tSCS-evoked responses were tested when the cathode was positioned on the C6, C7, or T1 levels with the same stimulus amplitude. The location that induced largest responses in all tested muscles was chosen as the stimulation site (C6: $n = 0$; C7: $n = 2$; T1: $n = 8$). Electrodes were fixed with adhesive tape to prevent their movement during the experiments. Next, to determine the stimulus intensity, the recruitment curves of the responses of all tested upper-limb muscles were obtained for each participant by gradually increasing the tSCS stimulation amplitude. To eliminate the ceiling effect of the evoked responses, the stimulus intensity was adjusted to evoke responses between the lower and middle portions of the ascending part of the recruitment curve in all muscles [21,31] and was kept constant for the duration of the experiment (57.0 $\pm$ 4.0 mA). Prior to starting the experiments, a paired-pulse stimulation protocol (50-ms inter-stimulus interval) was applied to confirm whether the evoked responses were initiated in the afferent fibers to evoke PRM reflex responses [23]. Suppression of the second evoked response demonstrated post-activation depression, confirming that PRM reflexes were evoked by activating the afferent roots [23,24,32–37]. Notably, the second responses were significantly depressed by the first stimulus activation in all muscles (Section 3.2.1). Specifically, the peak-to-peak amplitudes of the second responses for all recorded muscles were significantly depressed by the first stimulus activation (Wilcoxon signed-rank test, $p < 0.01$; Section 3.2.1), which confirmed the reflex nature of the evoked responses [38].

2.4. Conditioning Intervention Using Continuous Cervical tSCS

Continuous cervical tSCS intervention was delivered over the course of 10 min using the same electrode configurations (anode: midline of the anterior side of the neck; cathode: C7 or T1) that were determined for single-pulse tSCS (see Section 2.3.3). Continuous stimulation was applied using a constant current electrical stimulator (Rehab, Chattanooga, DJO Global, Vista, CA, USA) to apply biphasic rectangular pulses with a 400-µs pulse width at a frequency of 30 Hz. Stimulus intensity for each participant was adjusted to evoke paresthesias of the arm muscles by gradually increasing the stimulation pulse amplitude from 0 mA in 1 mA increments until the participants self-reported paresthesias consistent with previous lumbar tSCS study [10,28]. Specifically, participants were asked to identity the abnormal sensation tingling/pricking in the muscles distal to the stimulus delivery. The stimulation amplitudes ranged between 15 and 28 mA, with an average of 20.9 ± 3.4 mA (mean $\pm$ SD) between participants. It should be noted that pre assessments were performed just prior to adjusting the stimulating amplitudes to avoid possible effects.

2.5. Data Analysis

Peak-to-peak amplitudes were calculated for the TMS-induced MEP responses and tSCS-induced PRM reflex responses of each muscle and for each trial using a custom written script in MATLAB (2017a, The MathWorks Inc., Natick, MA, USA). Eight repeated trials were averaged for each phase (i.e., Pre, During, and Post for MEPs and Pre and Post for PRM reflexes). MEP and PRM reflex amplitudes were then normalized as a percentage of the Pre phase amplitude.

2.6. Statistics

Normalized MEP peak-to-peak responses were compared between Pre, During, and Post assessments using the Friedman test, a non-parametric repeated measure one-way analysis of variance (ANOVA). For paired-pulse tSCS protocol, PRM reflex peak-to-peak amplitudes were compared between first and second responses using the Wilcoxon signed-rank test, a non-parametric paired *t*-test. Normalized PRM reflex peak-to-peak responses were compared between Pre and Post assessments using the Wilcoxon signed-rank test. Non-parametric tests were chosen because the Shapiro–Wilk test showed that some identified measures were not normally distributed, and the sample size remains relatively small. Significance level was set to $p < 0.05$. All statistical analyses were performed using SPSS Statistics ver.25 (IBM Corp., Chicago, IL, USA).

3. Results

3.1. MEP Amplitude

The results of the MEP responses are shown in Figure 2. Averaged waveforms obtained from one representative subject are shown in Figure 2A. Box plots indicate the peak-to-peak amplitudes (% Pre) of MEP responses in During and Post phase (Figure 2B). The Friedman test showed that MEP peak-to-peak amplitudes were not statistically significantly different in any of the recorded muscles (FCR: $\chi^2(2) = 1.38$, $p = 0.500$; ECR: $\chi^2(2) = 4.20$, $p = 0.122$; FDI: $\chi^2(2) = 5.60$, $p = 0.061$; APB: $\chi^2(2) = 0.60$, $p = 0.741$). We also confirmed that comparison of area under the curve of MEP responses in each assessment point were consistent with the results of peak-to-peak amplitudes.

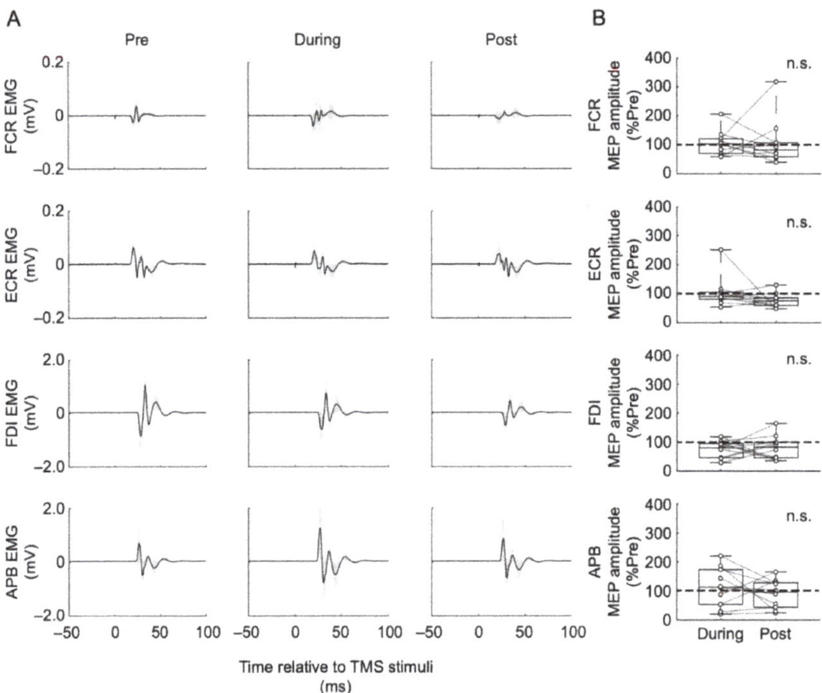

Figure 2. (**A**) Averaged motor evoked potentials (MEPs) for one representative participant at before (Pre), during continuous cervical tSCS (During), and after (Post) assessment points. Shadows represent standard deviation. (**B**) Box plots show MEP amplitude group data in each assessment point. MEP amplitudes were normalized with respect to the MEP amplitude during the Pre assessment (% of Pre) for each participant. Data are shown for the flexor carpi radialis (FCR), extensor carpi radialis (ECR), first dorsal interosseous (FDI), and abductor pollicis brevis (APB) muscles. The horizontal lines in the box plots indicate median values. The ends of the boxes represent the 25th and 75th percentiles. The whiskers on the box plots illustrate the minimum and maximum values. Legend: n.s., non-significant.

3.2. PRM Reflex

3.2.1. Paired-Pulse Protocol

The results of the paired-pulse protocol are shown in Figure 3. Averaged waveforms obtained from one representative participant are shown in Figure 3A. Box plots indicate the peak-to-peak amplitude (mV) of first and second responses during the paired-pulse stimulation protocol (Figure 3B). The Wilcoxon signed-rank test showed that the second responses were significantly smaller than first responses in all recorded muscles (BB: Z = 2.803, p = 0.005; TB: Z = 2.803, p = 0.005; FCR: 2.803, p = 0.005; ECR: Z = 2.803, p = 0.005; FDI: Z = 2.701, p = 0.007; APB: Z = 2.803, p = 0.005).

3.2.2. PRM Reflex Amplitude

The results of the PRM reflex responses are shown in Figure 4. Average waveforms obtained from one representative participant are shown in Figure 4A. Box plots indicate the peak-to-peak amplitudes (% of Pre) of spinal reflex responses during Post phase (Figure 4B). The Wilcoxon signed-rank test showed that the spinal reflex peak-to-peak amplitudes during Post phase were not statistically significantly different from Pre phase in all recorded muscles (BB: Z = 0.889, p = 0.374; TB: Z = 0.415, p = 0.678; FCR: Z = 1.01, p = 0.314; ECR: Z = 1.01, p = 0.314; FDI: Z = 1.24, p = 0.214; APB: Z = 0.533, p = 0.564).

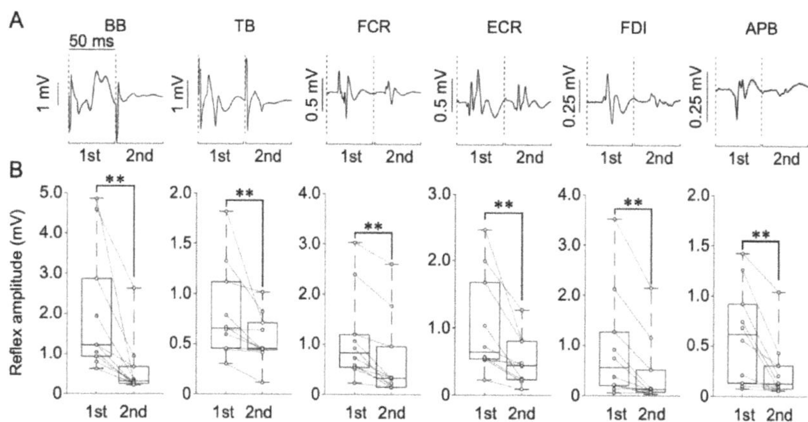

Figure 3. (**A**) Averaged spinal reflex responses for one representative participant during the paired-pulse stimulation protocol. First and second responses were separated by 50 ms. Shadows represent standard deviation. (**B**) Group data for the first and second responses. Data represent the biceps brachii (BB), triceps brachii (TB), flexor carpi radialis (FCR), extensor carpi radialis (ECR), first dorsal interosseous (FDI), and abductor pollicis brevis (APB) upper-limb muscles. The horizontal lines in the box plots indicate median values. The ends of the boxes represent the 25th and 75th percentiles. The whiskers on the box plots illustrate the minimum and maximum values. Legend: ** $p < 0.01$.

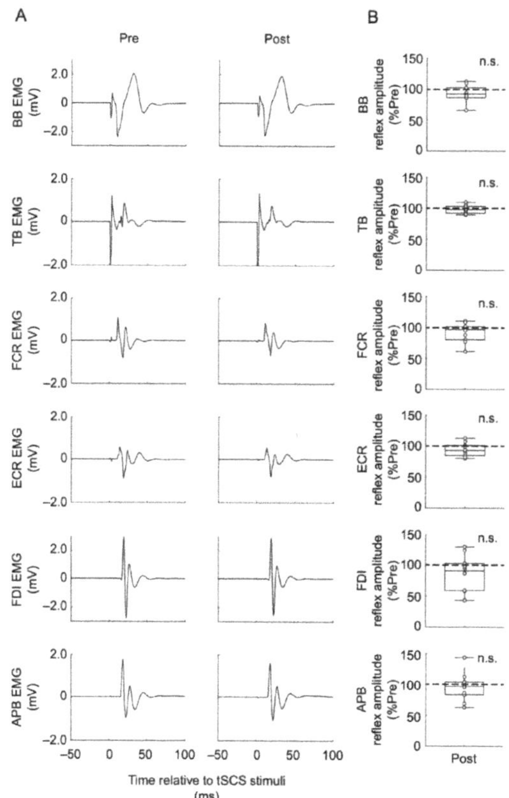

Figure 4. (**A**) Averaged spinal reflex responses for one representative participant at before (Pre) and

after (Post) assessment points. Shadows represent standard deviation. (**B**) Box plots show posterior root muscle (PRM) reflex group data in the Post assessment point. PRM reflex amplitudes were normalized with respect to the PRM reflex amplitude during the Pre assessment (% of Pre) for each participant. Data are shown for the biceps brachii (BB), triceps brachii (TB), flexor carpi radialis (FCR), extensor carpi radialis (ECR), first dorsal interosseous (FDI), and abductor pollicis brevis (APB) muscles. The horizontal lines in the box plots indicate median values. The ends of the boxes represent the 25th and 75th percentiles. The whiskers on the box plots illustrate the minimum and maximum values. Legend: n.s., non-significant.

4. Discussion

In the present study, we investigated the effects of relatively short-duration (10 min) and low-intensity (sub-motor-threshold) continuous cervical tSCS applied at rest on the upper-limb muscle MEPs elicited by TMS as well as PRM reflexes elicited by single-pulse cervical tSCS. Our results showed that the intervention did not significantly affect corticospinal (Figure 2) and spinal reflex excitabilities (Figure 4), suggesting that the utilized intervention did not effectively modulate the excitability of spinal monosynaptic connections between Ia sensory afferents and motoneurons and corticospinal pathways in able-bodied participants.

4.1. Continuous Cervical tSCS Parameters for Inducing Neuromodulatory Effect

In this study, continuous cervical tSCS was delivered at 30 Hz with biphasic rectangular pulses of 400-µs width and stimulation intensities at a level to elicit paresthesias in the upper-limb muscles for a total duration of 10 min while participants remained in the supine position at rest. Many previous studies reporting therapeutic effects of tSCS used stimulation frequencies between 15 and 50 Hz [6,10–12,15,26,28,39,40], while we used 30 Hz in the current study. We also used 400-µs pulse width, which was within the range previously utilized in neuromodulation applications. Specifically, 100-µs pulses were delivered as 10-kHz carrier frequency, within 1-ms bursts at 30 Hz [16–19]. Additionally, 200 µs pulses were delivered as single pulses or as a train of pulses using a 5-kHz carrier frequency in a recent study by Benavides et al. [15]. Notably, Benavides et al. [15] demonstrated that 200-µs biphasic pulses delivered at 30 Hz can increase corticospinal and spinal excitability in both SCI patients and able-bodied participants without the use of a 5-kHz carrier frequency. While the carrier frequency was reported as beneficial for suppressing pain [41,42], which may reduce discomfort during tSCS [43,44], the choice of stimulating frequency parameters used herein (i.e., 30 Hz without the carrier frequency) is unlikely to be related to the lack of neuromodulation. Moreover, the biphasic rectangular pulse waveform adopted in the present study is also in accordance with other previous neuromodulatory tSCS applications [10,11,16–19,28]. Additionally, the sub-motor-threshold stimulation intensity applied to elicit paresthesias is consistent with the considerations adopted by Hofstoetter et al. [10,28] for lumbar tSCS interventions in SCI patients. Although the overall stimulation configurations used in the present study are consistent with previous interventions that demonstrated neuromodulation effects in the corticospinal and spinal levels in able-bodied and SCI participants, our intervention did not induce any effects on the studied outcomes, contrary to our hypothesis.

While it may not be possible to attribute a single-stimulation parameter setting to the lack of effects observed herein, next, we discuss the tSCS parameter settings in our study. In particular, our stimulation intervention is generally consistent with the study by Benavides et al. [15], while we adopted longer pulse-width stimulating pulses, lower stimulation amplitude intensities (sub-motor-threshold), and shorter intervention duration. Specifically, by using a twice-longer pulse width (i.e., 400 µs vs. 200 µs used by Benavides et al. [15]) and twice-shorter intervention duration (i.e., 10 min vs. 20 min used by Benavides et al. [15]), the corticospinal and spinal excitability neuromodulation effects were not demonstrated in our study. It is possible that the intervention duration might have influenced our results. Specifically, it was previously demonstrated that 4-min interventions proposed by Parhizi et al. [45] also did not produce neuromodulatory effects. This suggests that the intervention duration may be important when delivering cervical tSCS in the rest conditions. Moreover, it was previously shown that spinal reflex excitability can be modulated using sub-motor-threshold stimulation amplitudes when the intervention was delivered over longer durations, i.e., 30 min for lumbar tSCS [28]. Therefore, while the pulse width and intervention duration trade-off normalized the conditions between our current study and that of Benavides et al. [15], the lower stimulation amplitude (sub-motor-threshold vs. motor-threshold intensity used by Benavides et al. [15]) may suggest that the energy delivery yielded lower efficacy in our current intervention [46]. Taken together, a compensation between stimulation amplitudes and intervention duration during continuous cervical tSCS intervention protocol may have influenced our current results.

It should also be noted that FDI muscle MEPs (target muscle for TMS assessments) during the intervention showed an inhibitory trend (8 of 10 participants had reduced MEP responses during the intervention compared to Pre, as shown in Figure 2B), but there were no statistically significant effects ($p = 0.061$). Additionally, it should also be pointed out that despite the changes in F-wave amplitude and persistency, which indicated change in excitability of cervical spinal circuits in a study by Kumru et al. [19], PRM reflex responses to single-pulse tSCS were not affected after the intervention in their study. On the other hand, cervical tSCS applied at supra-threshold but not at sub-threshold intensities produced lower-limb spinal (H-reflex) facilitation [47]. In addition to the intervention duration, this also points out that intensity of cervical tSCS intervention can be among the important factors underlying spinal reflex excitability. Therefore, future studies are warranted to compare different stimulation amplitude intensities and pulse widths as well as different intervention duration on the neuromodulatory effectiveness.

4.2. Voluntary Involvement Combined with Continuous Cervical tSCS May Be Required for Effective Neuromodulation

In addition to the stimulation parameters, voluntary engagement during cervical tSCS may be important to induce neuromodulatory effects. In our current study, continuous cervical tSCS intervention was applied when participants remained at rest in the supine position. On the other hand, many previous studies reporting therapeutic effect of continuous cervical tSCS on upper-limb motor function have delivered stimulation combined with functional task performance [16–18,26]. Voluntary engagement combined with afferent recruitment may elicit a form of Hebbian plasticity in the CNS during application of electrical stimulation [48,49]. A recent study reported that continuous cervical tSCS combined with hand training enhanced hand motor outputs, increased F-wave amplitudes and persistency, reduced TMS-induced resting motor thresholds, and facilitated MEP amplitudes after the intervention [19]. In contrast, cervical tSCS without hand training only increased the F-wave amplitudes [19]. Therefore, a neuromodulatory effect resulting from short-duration or single-session continuous cervical tSCS may depend on the brain state enhanced by voluntary drive when stimulation is delivered. Taken together, voluntary engagement may be essential to maximize neuromodulatory effects of cervical tSCS during short-term interventions.

4.3. Electrode Configuration Considerations for Cervical tSCS

In our current study, the cathode electrode was placed on the spine between the C7 and T1 cervical spinal processes on the posterior side of the neck, and the anode electrode was placed along the midline of the anterior side of the neck (Figure 1) [21,23,27]. Previous studies investigating the neuromodulatory effects of cervical tSCS placed anode electrodes bilaterally over the iliac crests [15,16,19]. Recent findings revealed that anode configuration over the anterior neck may elicit larger spinal reflex responses compared to when anode electrode were placed over iliac crests bilaterally, while there were no differences in the discomfort experienced between these different configurations at similar stimulation intensities [21]. Although it is also plausible that the anode location can affect the effectiveness of cervical tSCS intervention, it is unlikely that it was a critical factor because our previous work demonstrated that the anode placement over the anterior neck effectively targets the spinal circuits [21]. Future studies should also consider optimizing the placement of the anode electrode during continuous cervical tSCS so as to minimize discomfort.

4.4. Limitations

Our work has several limitations that should be noted. First, although neurophysiological effect of stimulation parameters used in our current study (i.e., sub-threshold intensity continuous tSCS) has not been investigated in previous cervical tSCS applications, the parameters used herein are not completely novel since similar approaches have been used with lumbar tSCS [28]. Therefore, future studies should explore more optimal stimulation settings for cervical tSCS. Second, we did not investigate neurophysiological effects in patients with neurological injury, such as spinal cord injury. For the clinical application of cervical tSCS, neurophysiological effect in patients should also be investigated using various stimulation parameters in the future. Third, we assessed corticospinal excitability using TMS during delivery of cervical tSCS, while spinal reflex excitability could not be assessed during cervical tSCS application since same electrodes were used for eliciting PRM reflex responses and for delivering the intervention. Therefore, spinal excitability during delivery of tSCS should be investigated using other assessment methods, such as H-reflex or F-wave, although it is well known that these methods can be applied to assess only a limited number of upper-limb muscles [50]. Finally, our current study was not able to identify the minimum time required for inducing neurophysiological effectiveness of cervical tSCS. This is very important for optimizing cervical tSCS stimulation-parameter settings. Therefore, future studies should consider the dose-response relationship to identify a minimum time required for effective application of cervical tSCS.

5. Conclusions

We investigated effects of continuous cervical tSCS on corticospinal and spinal reflex excitability. Our results showed that low-intensity and short-duration continuous cervical tSCS intervention applied at rest did not significantly affect corticospinal and spinal reflex excitability in able-bodied subjects. In addition to the amplitude and the duration of the intervention; the stimulating pulse and/or the pulse width, including the carrier frequency of the stimulating pulses; as well as voluntary engagement during the intervention may be important for inducing short-term neuromodulatory effect of cervical tSCS. Therefore, future studies should consider stimulation-parameters settings while minimizing stimulation-induced discomforts during cervical tSCS in actual clinical applications.

Author Contributions: A.S., Y.M., K.N. and M.M. designed the experiment; A.S. and M.M. performed experiments; A.S. analyzed data; A.S., R.M.d.F., K.N. and M.M. interpreted results of experiments; A.S. prepared figures; A.S. drafted the manuscript; A.S., R.M.d.F., D.G.S., Y.M., T.N., K.N. and M.M. edited and revised the manuscript. All authors have read and agreed to the published version of the manuscript.

Funding: This project was supported by Grant-in-Aid (KAKENHI) from the Japan Society for the Promotion of Science (JSPS) for Fellows awarded to A.S. (#19J22927), Grant-in-Aid for Early-Career Scientists and Grant-in-Aid for Research Activity Start-up awarded to M.M. (#20K19412 and #19K23606), and the Grants-in-Aid for Scientific Research (A) awarded to K.N. (#18H04082).

Institutional Review Board Statement: The study was conducted according to the guidelines of the Declaration of Helsinki and approved by the local Ethics Committee of the Graduate School of Arts and Science at the University of Tokyo (protocol code: 533-2, and date of approval: 31 January 2018).

Informed Consent Statement: Informed consent was obtained from all subjects involved in the study.

Data Availability Statement: The data related to the findings of this study are available from the corresponding author, upon reasonable request.

Conflicts of Interest: The authors declare no conflict of interest.

References

1. Moritz, C. A giant step for spinal cord injury research. *Nat. Neurosci.* **2018**, *21*, 1645–1646. [CrossRef]
2. Gill, M.L.; Grahn, P.J.; Calvert, J.S.; Linde, M.B.; Lavrov, I.A.; Strommen, J.A.; Beck, L.A.; Sayenko, D.G.; Van Straaten, M.G.; Drubach, D.I.; et al. Neuromodulation of lumbosacral spinal networks enables independent stepping after complete paraplegia. *Nat. Med.* **2018**, *24*, 1677–1682. [CrossRef]
3. Harkema, S.; Gerasimenko, Y.; Hodes, J.; Burdick, J.; Angeli, C.; Chen, Y.; Ferreira, C.; Willhite, A.; Rejc, E.; Grossman, R.G.; et al. Effect of epidural stimulation of the lumbosacral spinal cord on voluntary movement, standing, and assisted stepping after motor complete paraplegia: A case study. *Lancet* **2011**, *377*, 1938–1947. [CrossRef]
4. Wagner, F.B.; Mignardot, J.-B.; Le Goff-Mignardot, C.G.; Demesmaeker, R.; Komi, S.; Capogrosso, M.; Rowald, A.; Seáñez, I.; Caban, M.; Pirondini, E.; et al. Targeted neurotechnology restores walking in humans with spinal cord injury. *Nature* **2018**, *563*, 65–71. [CrossRef]
5. Lu, D.C.; Edgerton, V.R.; Modaber, M.; AuYong, N.; Morikawa, E.; Zdunowski, S.; Sarino, M.E.; Sarrafzadeh, M.; Nuwer, M.R.; Roy, R.R.; et al. Engaging Cervical Spinal Cord Networks to Reenable Volitional Control of Hand Function in Tetraplegic Patients. *Neurorehabil. Neural Repair* **2016**, *30*, 951–962. [CrossRef]
6. Meyer, C.; Hofstoetter, U.S.; Hubli, M.; Hassani, R.H.; Rinaldo, C.; Curt, A.; Bolliger, M. Immediate Effects of Transcutaneous Spinal Cord Stimulation on Motor Function in Chronic, Sensorimotor Incomplete Spinal Cord Injury. *J. Clin. Med.* **2020**, *9*, 3541. [CrossRef] [PubMed]
7. Gad, P.; Gerasimenko, Y.; Zdunowski, S.; Turner, A.; Sayenko, D.; Lu, D.C.; Edgerton, V.R. Weight Bearing Over-ground Stepping in an Exoskeleton with Non-invasive Spinal Cord Neuromodulation after Motor Complete Paraplegia. *Front. Neurosci.* **2017**, *11*, 333. [CrossRef]
8. Gerasimenko, Y.; Gorodnichev, R.; Moshonkina, T.; Sayenko, D.; Gad, P.; Reggie Edgerton, V. Transcutaneous electrical spinal-cord stimulation in humans. *Ann. Phys. Rehabil. Med.* **2015**, *58*, 225–231. [CrossRef] [PubMed]
9. Hofstoetter, U.S.; Freundl, B.; Lackner, P.; Binder, H. Transcutaneous Spinal Cord Stimulation Enhances Walking Performance and Reduces Spasticity in Individuals with Multiple Sclerosis. *Brain Sci.* **2021**, *11*, 472. [CrossRef] [PubMed]
10. Hofstoetter, U.S.; Krenn, M.; Danner, S.M.; Hofer, C.; Kern, H.; McKay, W.B.; Mayr, W.; Minassian, K. Augmentation of Voluntary Locomotor Activity by Transcutaneous Spinal Cord Stimulation in Motor-Incomplete Spinal Cord-Injured Individuals: Augmentation of Locomotion by tSCS in Incomplete SCI. *Artif. Organs* **2015**, *39*, E176–E186. [CrossRef] [PubMed]
11. Hofstoetter, U.S.; Hofer, C.; Kern, H.; Danner, S.M.; Mayr, W.; Dimitrijevic, M.R.; Minassian, K. Effects of transcutaneous spinal cord stimulation on voluntary locomotor activity in an incomplete spinal cord injured individual. *Biomed. Eng. Biomed. Tech.* **2013**, *58*. [CrossRef]
12. Minassian, K.; Hofstoetter, U.S.; Danner, S.M.; Mayr, W.; Bruce, J.A.; McKay, W.B.; Tansey, K.E. Spinal Rhythm Generation by Step-Induced Feedback and Transcutaneous Posterior Root Stimulation in Complete Spinal Cord–Injured Individuals. *Neurorehabil. Neural Repair* **2016**, *30*, 233–243. [CrossRef]
13. Rath, M.; Vette, A.H.; Ramasubramaniam, S.; Li, K.; Burdick, J.; Edgerton, V.R.; Gerasimenko, Y.P.; Sayenko, D.G. Trunk Stability Enabled by Noninvasive Spinal Electrical Stimulation after Spinal Cord Injury. *J. Neurotrauma* **2018**, *35*, 2540–2553. [CrossRef]
14. Sayenko, D.G.; Rath, M.; Ferguson, A.R.; Burdick, J.W.; Havton, L.A.; Edgerton, V.R.; Gerasimenko, Y.P. Self-Assisted Standing Enabled by Non-Invasive Spinal Stimulation after Spinal Cord Injury. *J. Neurotrauma* **2019**, *36*, 1435–1450. [CrossRef]
15. Benavides, F.D.; Jo, H.J.; Lundell, H.; Edgerton, V.R.; Gerasimenko, Y.; Perez, M.A. Cortical and Subcortical Effects of Transcutaneous Spinal Cord Stimulation in Humans with Tetraplegia. *J. Neurosci.* **2020**, *40*, 2633–2643. [CrossRef]
16. Gad, P.; Lee, S.; Terrafranca, N.; Zhong, H.; Turner, A.; Gerasimenko, Y.; Edgerton, V.R. Non-Invasive Activation of Cervical Spinal Networks after Severe Paralysis. *J. Neurotrauma* **2018**, *35*, 2145–2158. [CrossRef] [PubMed]
17. Inanici, F.; Brighton, L.N.; Samejima, S.; Hofstetter, C.P.; Moritz, C.T. Transcutaneous Spinal Cord Stimulation Restores Hand and Arm Function After Spinal Cord Injury. *IEEE Trans. Neural Syst. Rehabil. Eng.* **2021**, *29*, 310–319. [CrossRef] [PubMed]
18. Inanici, F.; Samejima, S.; Gad, P.; Edgerton, V.R.; Hofstetter, C.P.; Moritz, C.T. Transcutaneous Electrical Spinal Stimulation Promotes Long-Term Recovery of Upper Extremity Function in Chronic Tetraplegia. *IEEE Trans. Neural Syst. Rehabil. Eng.* **2018**, *26*, 1272–1278. [CrossRef]
19. Kumru, H.; Flores, Á.; Rodríguez-Cañón, M.; Edgerton, V.R.; García, L.; Benito-Penalva, J.; Navarro, X.; Gerasimenko, Y.; García-Alías, G.; Vidal, J. Cervical Electrical Neuromodulation Effectively Enhances Hand Motor Output in Healthy Subjects by Engaging a Use-Dependent Intervention. *J. Clin. Med.* **2021**, *10*, 195. [CrossRef]
20. Danner, S.M.; Hofstoetter, U.S.; Ladenbauer, J.; Rattay, F.; Minassian, K. Can the Human Lumbar Posterior Columns Be Stimulated by Transcutaneous Spinal Cord Stimulation? A Modeling Study: Modeling transcutaneous posterior columns stimulation. *Artif. Organs* **2011**, *35*, 257–262. [CrossRef] [PubMed]
21. De Freitas, R.M.; Sasaki, A.; Sayenko, D.G.; Masugi, Y.; Nomura, T.; Nakazawa, K.; Milosevic, M. Selectivity and excitability of upper-limb muscle activation during cervical transcutaneous spinal cord stimulation in humans. *J. Appl. Physiol.* **2021**, *131*, 746–759. [CrossRef]
22. Greiner, N.; Barra, B.; Schiavone, G.; Lorach, H.; James, N.; Conti, S.; Kaeser, M.; Fallegger, F.; Borgognon, S.; Lacour, S.; et al. Recruitment of upper-limb motoneurons with epidural electrical stimulation of the cervical spinal cord. *Nat. Commun.* **2021**, *12*, 435. [CrossRef]

23. Milosevic, M.; Masugi, Y.; Sasaki, A.; Sayenko, D.G.; Nakazawa, K. On the reflex mechanisms of cervical transcutaneous spinal cord stimulation in human subjects. *J. Neurophysiol.* **2019**, *121*, 1672–1679. [CrossRef] [PubMed]
24. Minassian, K.; Persy, I.; Rattay, F.; Dimitrijevic, M.R.; Hofer, C.; Kern, H. Posterior root-muscle reflexes elicited by transcutaneous stimulation of the human lumbosacral cord. *Muscle Nerve* **2007**, *35*, 327–336. [CrossRef] [PubMed]
25. Rattay, F.; Minassian, K.; Dimitrijevic, M. Epidural electrical stimulation of posterior structures of the human lumbosacral cord: 2. quantitative analysis by computer modeling. *Spinal Cord* **2000**, *38*, 473–489. [CrossRef]
26. Qian, Q.; Ling, Y.T.; Zhong, H.; Zheng, Y.-P.; Alam, M. Restoration of arm and hand functions via noninvasive cervical cord neuromodulation after traumatic brain injury: A case study. *Brain Inj.* **2020**, *34*, 1771–1780. [CrossRef]
27. Wu, Y.-K.; Levine, J.M.; Wecht, J.R.; Maher, M.T.; LiMonta, J.M.; Saeed, S.; Santiago, T.M.; Bailey, E.; Kastuar, S.; Guber, K.S.; et al. Posteroanterior cervical transcutaneous spinal stimulation targets ventral and dorsal nerve roots. *Clin. Neurophysiol.* **2020**, *131*, 451–460. [CrossRef]
28. Hofstoetter, U.S.; McKay, W.B.; Tansey, K.E.; Mayr, W.; Kern, H.; Minassian, K. Modification of spasticity by transcutaneous spinal cord stimulation in individuals with incomplete spinal cord injury. *J. Spinal Cord Med.* **2014**, *37*, 202–211. [CrossRef]
29. Rothwell, J.; Thompson, P.; Day, B.; Boyd, S.; Marsden, C. Stimulation of the human motor cortex through the scalp. *Exp. Physiol.* **1991**, *76*, 159–200. [CrossRef]
30. Rossini, P.M.; Burke, D.; Chen, R.; Cohen, L.G.; Daskalakis, Z.; Di Iorio, R.; Di Lazzaro, V.; Ferreri, F.; Fitzgerald, P.B.; George, M.S.; et al. Non-invasive electrical and magnetic stimulation of the brain, spinal cord, roots and peripheral nerves: Basic principles and procedures for routine clinical and research application: An updated report from an I.F.C.N. Committee. *Clin. Neurophysiol.* **2015**, *126*, 1071–1107. [CrossRef]
31. Masugi, Y.; Sasaki, A.; Kaneko, N.; Nakazawa, K. Remote muscle contraction enhances spinal reflexes in multiple lower-limb muscles elicited by transcutaneous spinal cord stimulation. *Exp. Brain Res.* **2019**, *237*, 1793–1803. [CrossRef]
32. Courtine, G.; Harkema, S.J.; Dy, C.J.; Gerasimenko, Y.P.; Dyhre-Poulsen, P. Modulation of multisegmental monosynaptic responses in a variety of leg muscles during walking and running in humans. *J. Physiol.* **2007**, *582*, 1125–1139. [CrossRef]
33. Dy, C.J.; Gerasimenko, Y.P.; Edgerton, V.R.; Dyhre-Poulsen, P.; Courtine, G.; Harkema, S.J. Phase-dependent modulation of percutaneously elicited multisegmental muscle responses after spinal cord injury. *J. Neurophysiol.* **2010**, *103*, 2808–2820. [CrossRef] [PubMed]
34. Masugi, Y.; Obata, H.; Inoue, D.; Kawashima, N.; Nakazawa, K. Neural effects of muscle stretching on the spinal reflexes in multiple lower-limb muscles. *PLoS ONE* **2017**, *12*, e0180275. [CrossRef]
35. Masugi, Y.; Kawashima, N.; Inoue, D.; Nakazawa, K. Effects of movement-related afferent inputs on spinal reflexes evoked by transcutaneous spinal cord stimulation during robot-assisted passive stepping. *Neurosci. Lett.* **2016**, *627*, 100–106. [CrossRef]
36. Roy, F.D.; Gibson, G.; Stein, R.B. Effect of percutaneous stimulation at different spinal levels on the activation of sensory and motor roots. *Exp. Brain Res.* **2012**, *223*, 281–289. [CrossRef]
37. Sasaki, A.; Kaneko, N.; Masugi, Y.; Milosevic, M.; Nakazawa, K. Interlimb neural interactions in corticospinal and spinal reflex circuits during preparation and execution of isometric elbow flexion. *J. Neurophysiol.* **2020**, *124*, 652–667. [CrossRef]
38. Hofstoetter, U.S.; Freundl, B.; Binder, H.; Minassian, K. Recovery cycles of posterior root-muscle reflexes evoked by transcutaneous spinal cord stimulation and of the H reflex in individuals with intact and injured spinal cord. *PLoS ONE* **2019**, *14*, e0227057. [CrossRef]
39. Estes, S.P.; Iddings, J.A.; Field-Fote, E.C. Priming Neural Circuits to Modulate Spinal Reflex Excitability. *Front. Neurol.* **2017**, *8*, 17. [CrossRef]
40. Megía-García, Á.; Serrano-Muñoz, D.; Taylor, J.; Avendaño-Coy, J.; Comino-Suárez, N.; Gómez-Soriano, J. Transcutaneous Spinal Cord Stimulation Enhances Quadriceps Motor Evoked Potential in Healthy Participants: A Double-Blind Randomized Controlled Study. *J. Clin. Med.* **2020**, *9*, 3275. [CrossRef]
41. Ward, A.R.; Robertson, V.J. Variation in torque production with frequency using medium frequency alternating current. *Arch. Phys. Med. Rehabil.* **1998**, *79*, 1399–1404. [CrossRef]
42. Ward, A.R.; Robertson, V.J. Sensory, motor, and pain thresholds for stimulation with medium frequency alternating current. *Arch. Phys. Med. Rehabil.* **1998**, *79*, 273–278. [CrossRef]
43. Gerasimenko, Y.; Gorodnichev, R.; Puhov, A.; Moshonkina, T.; Savochin, A.; Selionov, V.; Roy, R.R.; Lu, D.C.; Edgerton, V.R. Initiation and modulation of locomotor circuitry output with multisite transcutaneous electrical stimulation of the spinal cord in noninjured humans. *J. Neurophysiol.* **2015**, *113*, 834–842. [CrossRef]
44. Ward, A.R. Electrical Stimulation Using Kilohertz-Frequency Alternating Current. *Phys. Ther.* **2009**, *89*, 181–190. [CrossRef]
45. Parhizi, B.; Barss, T.S.; Mushahwar, V.K. Simultaneous Cervical and Lumbar Spinal Cord Stimulation Induces Facilitation of Both Spinal and Corticospinal Circuitry in Humans. *Front. Neurosci.* **2021**, *15*, 615103. [CrossRef] [PubMed]
46. Miller, J.P.; Eldabe, S.; Buchser, E.; Johanek, L.M.; Guan, Y.; Linderoth, B. Parameters of Spinal Cord Stimulation and Their Role in Electrical Charge Delivery: A Review: SCS Parameters and Charge Delivery. *Neuromodul. Technol. Neural Interface* **2016**, *19*, 373–384. [CrossRef]
47. Islam, M.A.; Zaaya, M.; Comiskey, E.; Demetrio, J.; O'Keefe, A.; Palazzo, N.; Pulverenti, T.S.; Knikou, M. Modulation of soleus H-reflex excitability following cervical transspinal conditioning stimulation in humans. *Neurosci. Lett.* **2020**, *732*, 135052. [CrossRef] [PubMed]

48. Milosevic, M.; Marquez-Chin, C.; Masani, K.; Hirata, M.; Nomura, T.; Popovic, M.R.; Nakazawa, K. Why brain-controlled neuroprosthetics matter: Mechanisms underlying electrical stimulation of muscles and nerves in rehabilitation. *Biomed. Eng. OnLine* **2020**, *19*, 81. [CrossRef]
49. Stein, R.B.; Everaert, D.G.; Roy, F.D.; Chong, S.; Soleimani, M. Facilitation of Corticospinal Connections in Able-bodied People and People With Central Nervous System Disorders Using Eight Interventions. *J. Clin. Neurophysiol.* **2013**, *30*, 66–78. [CrossRef]
50. McNeil, C.J.; Butler, J.E.; Taylor, J.L.; Gandevia, S.C. Testing the excitability of human motoneurons. *Front. Hum. Neurosci.* **2013**, *7*, 152. [CrossRef]

Journal of
Clinical Medicine

Article

Effect of Cervical Transcutaneous Spinal Cord Stimulation on Sensorimotor Cortical Activity during Upper-Limb Movements in Healthy Individuals

Ciarán McGeady [1], Monzurul Alam [2,*,†], Yong-Ping Zheng [2] and Aleksandra Vučković [1,*,†]

[1] Centre for Rehabilitation Engineering, University of Glasgow, Glasgow G12 8QQ, UK; c.mcgeady.1@research.gla.ac.uk

[2] Department of Biomedical Engineering, The Hong Kong Polytechnic University, Hung Hom, Kowloon, Hong Kong; yongping.zheng@polyu.edu.hk

* Correspondence: md.malam@connect.polyu.hk (M.A.); aleksandra.vuckovic@glasgow.ac.uk (A.V.)

† These authors contributed equally to this work.

Abstract: Transcutaneous spinal cord stimulation (tSCS) can improve upper-limb motor function after spinal cord injury. A number of studies have attempted to deduce the corticospinal mechanisms which are modulated following tSCS, with many relying on transcranial magnetic stimulation to provide measures of corticospinal excitability. Other metrics, such as cortical oscillations, may provide an alternative and complementary perspective on the physiological effect of tSCS. Hence, the present study recorded EEG from 30 healthy volunteers to investigate if and how cortical oscillatory dynamics are altered by 10 min of continuous cervical tSCS. Participants performed repetitive upper-limb movements and resting-state tasks while tSCS was delivered to the posterior side of the neck as EEG was recorded simultaneously. The intensity of tSCS was tailored to each participant based on their maximum tolerance (mean: 50 ± 20 mA). A control session was conducted without tSCS. Changes to sensorimotor cortical activity during movement were quantified in terms of event-related (de)synchronisation (ERD/ERS). Our analysis revealed that, on a group level, there was no consistency in terms of the direction of ERD modulation during tSCS, nor was there a dose-effect between tSCS and ERD/ERS. Resting-state oscillatory power was compared before and after tSCS but no statistically significant difference was found in terms of alpha peak frequency or alpha power. However, participants who received the highest stimulation intensities had significantly weakened ERD/ERS (10% ERS) compared to when tSCS was not applied (25% ERD; $p = 0.016$), suggestive of cortical inhibition. Overall, our results demonstrated that a single 10 min session of tSCS delivered to the cervical region of the spine was not sufficient to induce consistent changes in sensorimotor cortical activity among the entire cohort. However, under high intensities there may be an inhibitory effect at the cortical level. Future work should investigate, with a larger sample size, the effect of session duration and tSCS intensity on cortical oscillations.

Keywords: neuromodulation; transcutaneous spinal cord stimulation; electroencephalography; event-related desynchronisation; rehabilitation; posterior root muscle reflex

Citation: McGeady, C.; Alam, M.; Zheng, Y.-P.; Vučković, A. Effect of Cervical Transcutaneous Spinal Cord Stimulation on Sensorimotor Cortical Activity during Upper-Limb Movements in Healthy Individuals. *J. Clin. Med.* **2022**, *11*, 1043. https://doi.org/10.3390/jcm11041043

Academic Editor: Karen Minassian

Received: 1 December 2021
Accepted: 15 February 2022
Published: 17 February 2022

Publisher's Note: MDPI stays neutral with regard to jurisdictional claims in published maps and institutional affiliations.

1. Introduction

Transcutaneous spinal cord stimulation is a non-invasive neuromodulatory technique that has shown potential in reversing upper-limb paralysis in spinal cord injury (SCI) patients [1,2]. The technique often involves placing one or more cathode electrodes at and around the spinal level of injury to deliver high-frequency currents at sub-threshold intensities. It has been postulated that electrical interaction with a combination of structures, such as dorsal column fibres, the dorsal horn and posterior/ventral roots, decreases the motor threshold, making voluntary motor control easier through residual descending pathways [3–5]. When combined with conventional rehabilitative therapies such as physical practice, tSCS has led to lasting functional improvements [1,2,6]. The extent to which tSCS modulates corticospinal pathways, however, is still a matter of contention.

Numerous studies have investigated tSCS modulation at both the cortical and spinal level [7–13]. Benavides et al., for example, investigated cortical modulation by comparing motor evoked potentials (MEPs) induced by transcranial magnetic stimulation (TMS) before and after 20 min of tSCS. They found that MEP amplitudes tended to increase following stimulation, implying facilitation of the corticospinal tract. Ambiguities still exist surrounding tSCS-based neuromodulation, however. In a similar study, Sasaki et al. reported a null effect of tSCS on MEP amplitude, albeit with sessions of a shorter duration [10]. Both studies, and indeed the majority of similar studies, used MEP amplitudes to provide a metric of cortical excitability. Other measures, such as cortical oscillations, offer an alternative perspective on the physiological effects of tSCS. Although MEP and oscillation amplitudes have both been associated with motor cortical excitability, they are not strongly correlated, and likely reflect different neural processes [14,15]. Where cortical oscillations tend to reflect the induced excitability of large populations of cortical neurons, MEPs are affected by the global excitability of corticospinal pathways [14,16]. An understanding of how each measure is affected by tSCS will build a stronger foundation in which to guide future tSCS-based neurorehabilitation strategies. A further benefit of understanding the influence of tSCS on cortical oscillations concerns the use of brain–computer interfaces, which are increasingly being used in neurorehabilitation, often when combined with stimulation-based therapies [17,18]. Such BCI paradigms rely on distinct and consistent modulation of sensorimotor oscillations during imagined or attempted movement. Facilitated expression of sensorimotor oscillations may improve the performance of such systems [19,20].

As far as we are aware, no studies have yet considered tSCS-based neuromodulation in terms of sensorimotor cortical oscillations as measured from the electroencephalogram (EEG). Given reports of enhanced excitability of motoneuron and cortico-motoneuronal synapses through spinal stimulation, we would expect an expression of neuromodulation in terms of cortical oscillations, as is the case with other stimulation-based modalities such as functional electrical stimulation (FES) [21], and transcutaneous electrical nerves stimulation (TENS) [22]. The variety of modulation is a matter of conjecture, however. On the one hand, we may expect sensorimotor cortical excitation, as sensory afferent volleys may be amplified resulting in stronger activation of the somatosensory cortex. On the other hand, we may expect cortical inhibition given high-frequency spinal cord stimulation has been linked to suppression of nociceptive transmission [23]. At the very least, we would expect a quantifiable difference in sensorimotor cortical activity when tSCS is applied compared to when stimulation is not present. Therefore, the aim of the present study was to investigate if sensorimotor cortical activity during upper-limb movement could be modulated by short duration continuous tSCS.

To test this hypothesis, we had healthy volunteers perform upper-limb movements as continuous tSCS was delivered to the posterior region of the neck, using typical clinical stimulation parameters [1,2]. EEG was recorded simultaneously and sensorimotor dynamics were extracted in an offline analysis. The alpha frequency is the most dominant EEG feature during the resting state, and its event-related (de)synchronisation (ERD/ERS) has been associated with cortical activation during sensorimotor tasks, reflecting asynchronous neural firing [16,24,25]. We performed a side-by-side comparison of ERD/ERS with and without tSCS. A further hypothesis was that sensorimotor neuromodulation by tSCS would be subject to a dose effect where the modulation would be facilitated or attenuated as a function of time. We tested this by considering the ERD/ERS of alpha and beta frequency bands across movement repetitions. In addition to ERD during movement, we compared resting-state EEG before and after tSCS.

2. Materials and Methods

2.1. Participants

Thirty able-bodied volunteers (9 females, 21 males; 26.7 ± 3.0 years old) participated in this study. Exclusion criteria included musculoskeletal pathology of the upper limbs, metal or electronic implants, medications that influenced neural excitability (antiepileptic, antipsychotics, or antidepressants), allergy to the electrode material, epilepsy, and pregnancy.

Sessions were conducted at the same time of day to minimise baseline EEG variances and subjects were allowed to take breaks in between recording runs. Written informed consent was obtained from all participants. This study was approved by the Human Subjects Ethics Sub-committee of the Hong Kong Polytechnic University and conducted according to the principles and guidelines of the Declaration of Helsinki.

2.2. Experimental Protocol

Based on a two-day crossover design, participants underwent two sessions on different days. Both sessions had participants perform a 10 min upper-limb movement task as EEG and EMG were recorded from the sensorimotor region of the scalp and forearms respectively (see Figure 1A for an illustration of the experimental setup). Continuous tSCS was applied concurrently to the cervical region of the neck during only one of these sessions (Figure 2A). The order in which participants received both sessions was pseudo-randomised.

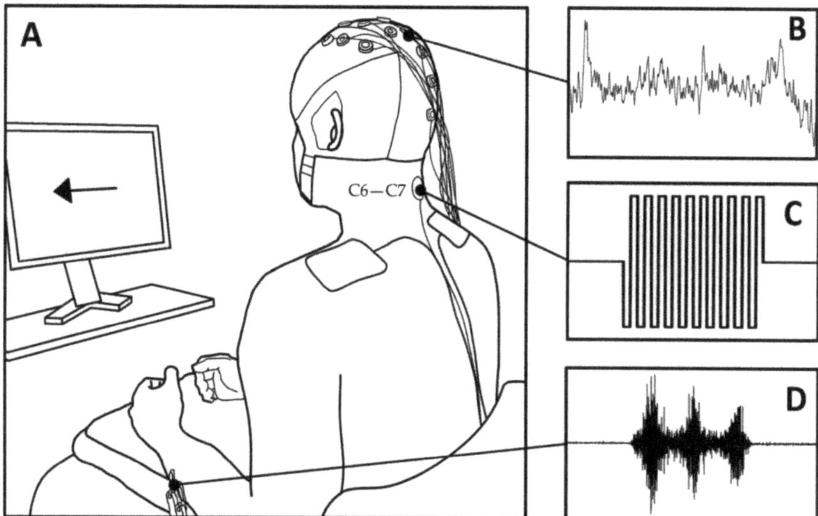

Figure 1. Experimental setup showing recording and stimulation modalities. (**A**) Participant receives cues from a computer screen to perform upper-limb movements. (**B**) EEG is recorded from the central area of the scalp. (**C**) One millisecond long burst containing 10 biphasic pulses is delivered at 30 Hz to the posterior region of the neck during continuous tSCS. (**D**) EMG during left-hand rhythmic finger flexion/extension over the extensor carpi radialis. The same setup was used on the right side.

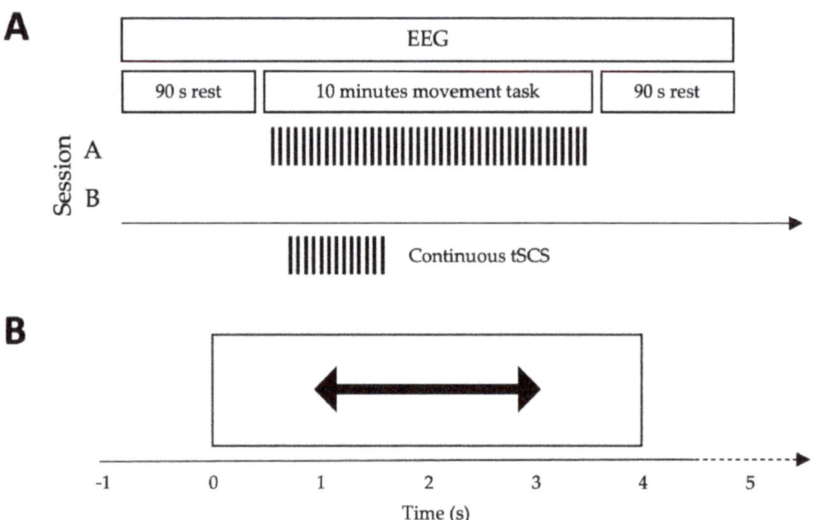

Figure 2. Experimental session protocol and movement task timing scheme. (**A**) Outline of the experimental sessions, carried out on different days. Both sessions began and ended with the recording of resting-state EEG with eyes closed. An upper-limb movement task lasted 10 min while EEG was recorded simultaneously. Only during session A was continuous tSCS applied to the cervical region of the spine. (**B**) The timing scheme of a single trial from the movement task. At t = 0 s an arrow appeared onscreen prompting the participant to perform either left, right, or bimanual finger flexion. The movement was sustained for four seconds. This was followed by a randomised 1.5–2.5 s inter-trial interval. There were 30 repetitions of each movement, totalling 90 trials.

There were two parts to a session: (1) resting-state EEG recording, and (2) a movement execution task. Part (1) was performed before and after the movement task to investigate potential modulation of physiological markers. While recording, participants were required to sit still in an upright position, minimising all body and eye movements. Resting-state EEG was recorded for 90 s with eyes closed. The movement execution task was performed in an upright, seated position and had participants perform rhythmic right-hand, left-hand and bimanual finger flexion, as cued by an interface on a computer screen (Figure 1A). A rightwards-pointing arrow cued right-hand movement, a leftwards-pointing arrow cued left-hand movement, and a double arrow pointing both left and right cued bimanual movements. We included a bimanual condition as SCI patients often use both hands during tSCS training, and most activities of daily living include coordination of both hands [2,26,27]. Each movement was performed and sustained for four seconds and repeated 30 times, with a randomised 1.5 to 2.5 s inter-trial interval. The timing scheme is illustrated in Figure 2B. EMG was recorded from the forearm muscles to measure movement onset.

2.3. Electroencephalography (EEG)

Two g.USBamp biosignal amplifiers (g.tec, Schiedlberg, Austria) recorded EEG at 1200 Hz from 19 passive electrodes: Fz, FC3, FC1, FCz, FC2, FC4, C3, C1, Cz, C2, C4, CP3, CP1, CPz, CP2, CP4, Pz, POz, and Oz, according to the international 10–20 system (See Figure 1A,B) [28]. Electrode AFz was used as ground and the reference electrode was placed on the right earlobe. EEG was filtered with a band-pass (0.01–100 Hz) and a notch filter (50 Hz). Electrode impedances were kept below 5 kΩ throughout the recording session, and participants were instructed to minimise head and eye movements in order to ensure high fidelity recordings. Given the considerable artefacts produced by concurrent tSCS, conventional data-cleaning techniques were unsuitable [29]. For example, the high amplitude stimulation component meant that applying rejection thresholds on peak-to-peak amplitudes would eliminate segments of otherwise meaningful EEG. Hence, rejection thresholds were not used during pre-processing and instead strict adherence to the protocol outlined above was followed.

2.4. Electromyography (EMG)

To determine the onset of upper-limb movement, electromyography (EMG) was used to measure the activity of the extensor carpi radialis (ERC) muscles (See Figure 1D). Two electrodes (Ag/AgCl; F-301, Skintact, Innsbruck, Austria) were positioned on the dorsal side of each forearm, above the belly of the ERC, with a 20 mm inter-electrode distance. Ground electrodes were attached to the lateral epicondyles. EMG was recorded with the same biosignal amplifier outlined above (band-pass filter: 5–1200 Hz; notch filter: 50 Hz) to ensure synchronisation with EEG. Movement onset was defined as the moment EMG activity exceeded the mean of the resting phase plus two times its standard deviation for at least 100 ms [30].

2.5. Transcutaneous Spinal Cord Stimulation (tSCS)

Using a DS8R Biphasic Constant Current Stimulator (Digitimer, Hertfordshire, UK), spinal cord stimulation was delivered in bursts of ten 100 µs long biphasic rectangular pulses at a frequency of 30 Hz (see Figure 1B for an illustration of a single burst) [31]. A round 3.2 cm cathode electrode (Axelgaard Manufacturing Co., Fallbrook, CA, USA) was placed between the C5–C6 intervertebral space, placement reflective of upper-limb rehabilitation in clinical practice. Rectangular inter-connected anode electrodes (8.9×5.0 cm) were placed symmetrically on the shoulders, above the acromion (see Figure 1A for an illustration) [32]. We used feedback from the participant to determine the current intensity. Starting at 0 mA, the current was gradually increased in 2.5 mA increments until the participant verbally communicated their wish to stop increasing. Participants were asked before each incremental increase whether they would be able to tolerate the sensation for at least 30 s. If they were unable to tolerate the intensity, the current was reduced by one increment and was used for the remainder of the movement task. The area of discomfort varied across participants. Some participants reported that discomfort was focused under the cathode electrode; others found the contraction of back and neck muscles intolerable; some reported a combination of both. Across all participants, tSCS current intensity was on average 50 ± 20 mA, with a minimum and maximum current of 10 and 85 mA respectively.

2.6. Quantifying Sensorimotor Cortical Activity during tSCS

EEG was pre-processed offline with a 3rd-order Butterworth band-pass filter (1–40 Hz) and notch filter (50 Hz). Next, continuous EEG was segmented into epochs from −2 to 6 s relative to movement onset. The power spectral density across time and frequency was found using the multitaper method (1–25 Hz) with a resolution of 0.5 Hz. This analysis was performed with channels C3, C4, and the mean of C3 and C4, for right, left, and bimanual movements respectively. Time-frequency power was normalised with respect to a pre-movement baseline, defined as −1.25 to −0.25 s before movement and the average time-frequency powers were averaged across all subjects for each movement type. Statistical masking was added to time-frequency plots to display only power values which deviated significantly ($p < 0.05$) from baseline, as determined by a cluster-based permutation test.

We then separately considered the mean alpha (7–13 Hz) and beta (14–25 Hz) band ERD/ERS during two phases of movement: (1) movement initiation (0.5–1.5 s), and (2) sustained movement (1.5–3.0 s). We expected tSCS would strengthen ERD during the sustained movement phase, reflecting similar results observed using FES during motor imagery [19]. ERD values were averaged across movement phases and compared between stimulation conditions with a Wilcoxon signed-rank test, where $p < 0.05$ indicated a statistically significant difference in cortical activity.

A topographical analysis was performed by averaging the movement phases outlined above in the alpha and beta frequency bands for each recorded channel. The spatial distributions of cortical activation were used in a cluster-based permutation test to compare the ERD patterns while tSCS was on compared to when tSCS was off. A significance threshold of 0.05 was used to identify significant differences in topographical distributions between the two stimulation conditions.

Finally, in order to investigate a dose-effect of tSCS on cortical activity, we considered the correlation between ERD during each trial and sequence of trials by calculating Pearson's correlation coefficient. A Wilcoxon signed-rank test was used to determine if the participant-wise average correlation coefficients significantly differed between stimulation conditions.

2.7. Neuromodulation of Resting-State EEG

We explored whether tSCS exerted a neuromodulatory effect on resting-state EEG by comparing individual alpha frequency before and after the movement task. We used resting state, eyes closed EEG and segmented it into one-second epochs with a 0.1 s overlap. Each epoch was windowed using a Hamming window and the periodograms (2^{15} point FFT) were averaged to estimate the power spectral density (PSD). The alpha peak frequency was defined as the frequency with the maximum power in the 7–13 Hz range. The alpha peak frequency after the intervention was expressed as a percentage change from the alpha peak before the intervention. We also considered the power of the alpha peak and similarly normalised this with respect to pre-intervention power. A Wilcoxon signed-rank test was performed to determine if there was a significant difference in the change of alpha peak frequency and power between the tSCS-off and tSCS-on conditions.

3. Results

3.1. Event-Related (De)synchronisation (ERD/ERS)

To investigate the effect of tSCS on sensorimotor activity during movement we calculated alpha and beta band power differences with respect to rest. Figure 3 shows time-frequency power values averaged across all participants for left, right, and bimanual finger flexion. The plots only display power values that significantly differed ($p < 0.05$) from baseline, as determined by a cluster-based permutation test. Each movement type showed significant broadband (8–25 Hz) ERD with particular power suppression in the alpha band (8–12 Hz). Right and bimanual movements tend to show similar patters of ERD regardless of whether tSCS had been applied or not. Left-hand movements appeared to have deeper and more sustained alpha desynchronisation when tSCS was applied.

To test for a significant difference of ERD between conditions we divided each movement into two phases: (1) movement initiation (0.5–1.5 s after movement onset), and (2) sustained movement (1.5–3.0 s after movement onset). Figure 4 shows the average ERD during movement initiation for each movement type and stimulation condition in the alpha and beta bands. There were no significant differences detected in the alpha band (Figure 4A: Left: $p = 0.15$; Right: $p = 0.14$; Bimanual: $p = 0.90$), nor in the beta band (Figure 4B: Left: $p = 0.77$; Right: $p = 0.60$; Bimanual: $p = 0.75$). Although ERD shows variability, the variance is inline with other studies reporting ERD within participants and across sessions [33]. On average, however, there was a lack of consistency in the direction of modulation with some participants having stronger ERD with stimulation, and some having suppressed ERD.

Similar results are seen in Figure 5 which presents ERD values during sustained movement (Alpha: Left: $p = 0.19$; Right: $p = 0.12$; Bimanual: $p = 0.40$; Beta: Left: $p = 0.4$; Right: $p = 0.90$; Bimanual: $p = 0.94$).

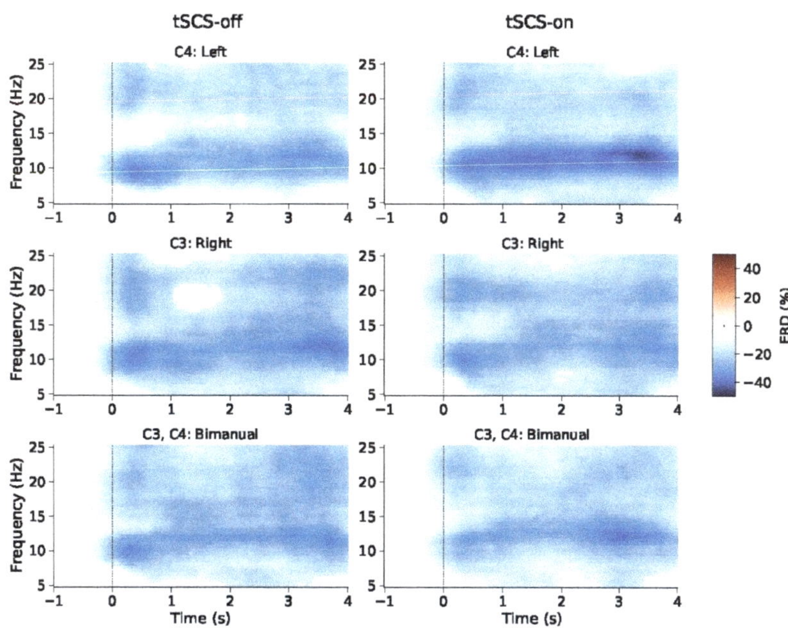

Figure 3. Time-frequency plots of event-related desynchronisation (ERD) during repetitive left, right, and bimanual finger flexion with and without tSCS. Only values significantly different from 0% ERD ($p < 0.05$) are shown, as determined by a cluster-based permutation test.

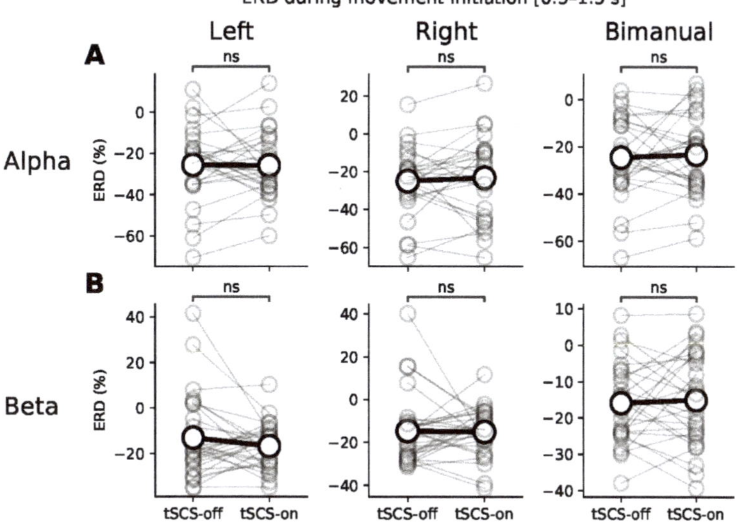

Figure 4. Average ERD during movement initiation (0.5–1.5 s) for each type of upper-limb movement (left, right, and bimanual finger flexion). (**A**,**B**) show ERD in the alpha and beta bands respectively. Grey markers show ERD of individual participants and the black markers show the participant-wise average. A Wilcoxon signed-rank test explored statistically significance differences between the tSCS-off and tSCS-on conditions for each movement and frequency band ('ns' denotes no significant difference).

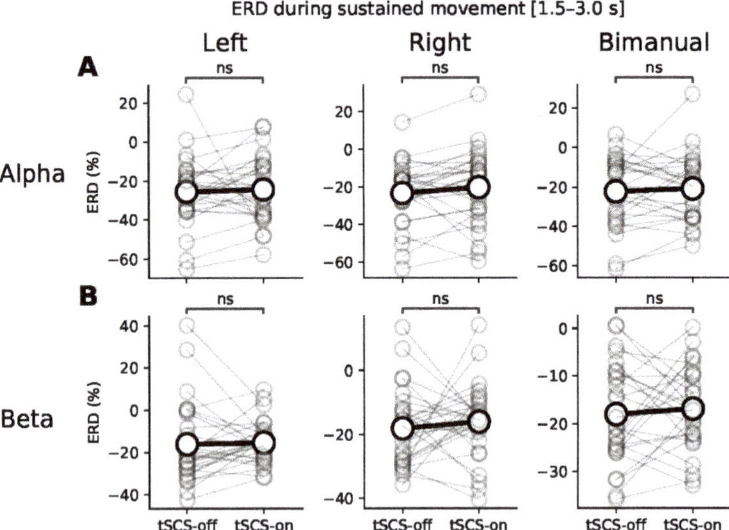

Figure 5. Average ERD during sustained movement (1.5–3.0 s) for each type of upper-limb movement (left, right, and bimanual finger flexion). (**A**,**B**) show ERD in the alpha and beta bands respectively. Grey markers show ERD of individual participants and the black markers show the participant-wise average. A Wilcoxon signed-rank test explored statistically significance differences between the tSCS-off and tSCS-on conditions for each movement and frequency band ('ns' denotes no significant difference).

3.2. Topographic Analysis of ERD

The ERD topographic patterns during movement initiation and sustained movement are illustrated in Figures 6 and 7 respectively. It can be seen that there is desynchronisation present at all the electrodes in the alpha and beta frequency bands in both stimulation conditions. Figure 7A shows bilateral alpha ERD when tSCS is off. When tSCS is on the pattern appears more contralaterally dominant over C4 electrodes (Figure 7C). However, a cluster-based permutation test showed that there were no regions of the topographical distributions that significantly differed between conditions. This was the case for the beta band and for sustained movement shown in Figure 6.

3.3. Dose Effect of Event-Related Desynchronisation

We found that on average there was no dose effect of tSCS on alpha or beta ERD (Figure 8). Taken as a group, the average correlation coefficients were close to zero with or without the presence of tSCS. A Wilcoxon signed-rank test corrected for multiple comparisons found no significant difference between conditions in either frequency band (Alpha: $p = 0.16$; Beta: $p = 0.75$).

3.4. Resting State Modulation

We found that resting state individual alpha peak frequency was not significantly altered by tSCS ($p = 0.67$), showing an approximately 0% change from pre-intervention alpha for both stimulation conditions (Figure 9A). Furthermore, the change in alpha power was also unaffected by tSCS ($p = 0.20$), shown in Figure 9B.

ERD during left hand movement initiation [0.5–1.5 s]

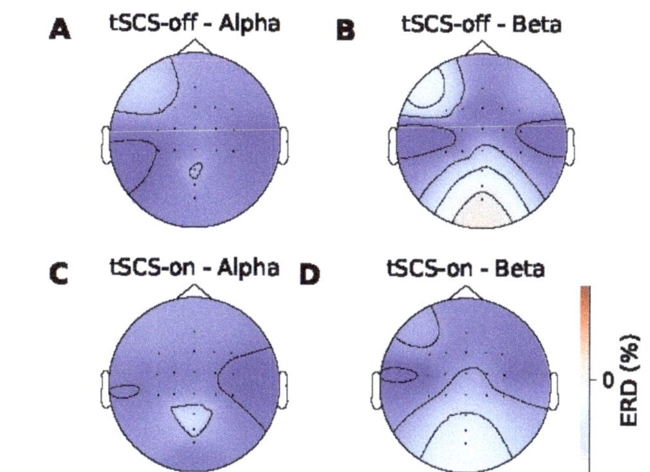

Figure 6. Topographic ERD/ERS distribution during left handed movement initiation (0.5–1.5 s after movement onset). (**A,B**) show spatial distribution of ERD/ERS in the alpha and beta bands without tSCS. (**C,D**) show the spatial distribution in the alpha and beta bands during with tSCS.

ERD during sustained left movement [1.5–3.0 s]

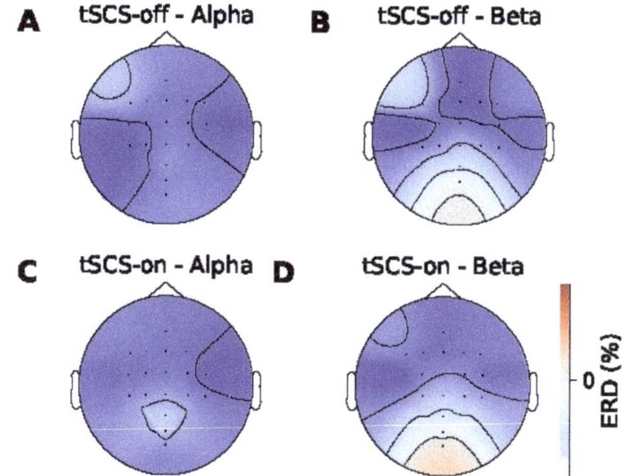

Figure 7. Topographic ERD/ERS distribution during sustained left handed movement (1.5–3.0 s after movement onset). (**A,B**) show spatial distribution of ERD/ERS in the alpha and beta bands without tSCS. (**C,D**) show the spatial distribution in the alpha and beta bands during tSCS.

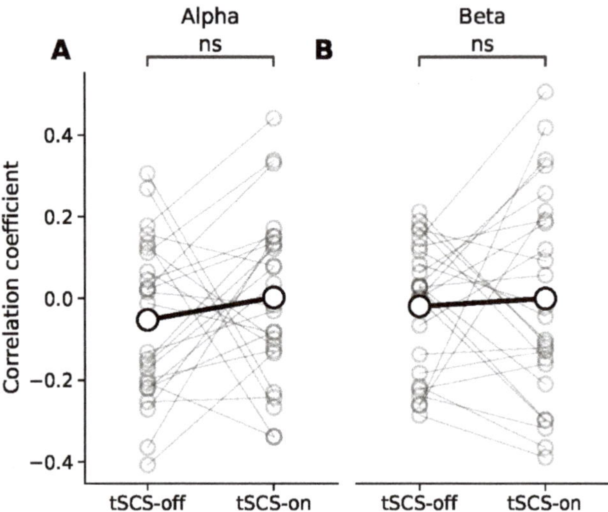

Figure 8. Correlation coefficient between event-related desynchronisation (ERD) in the alpha (**A**) and beta (**B**) bands during repetitive bimanual finger flexion and the sequence of trials in with and without tSCS. The grey markers represent correlation coefficient values for individual participants and the black markers represent the across-participant session average. 'ns' where there was no significant difference.

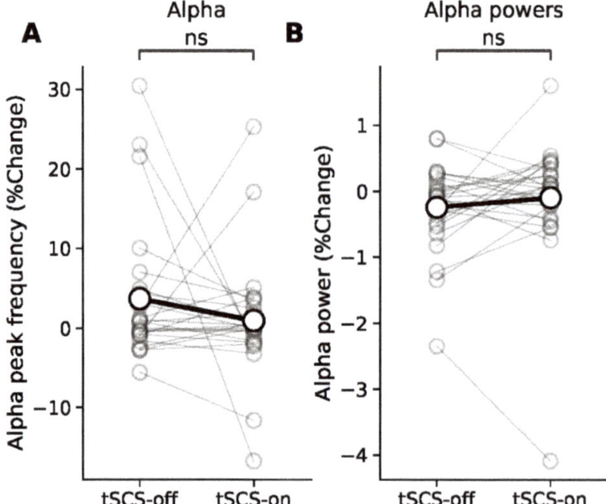

Figure 9. Resting state EEG. (**A**) shows the change (%) in peak alpha frequencies from baseline during resting state with eyes closed with and without tSCS. (**B**) shows the change in power of the peak frequencies. The grey markers represent individual participants and the black markers represent the across-participant session averages. Non-significance, as determined by a Wilcoxon signed-rank test, is expressed as 'ns'.

3.5. Effect of tSCS Intensity

Given the variability across sessions shown in Figures 4 and 5, we investigated whether the variance could partially be explained by stimulation current intensity, given intensity was tailored to the individual. Figure 10 shows that ERD/ERS appears similarly distributed between conditions at around 20% ERD for intensities between 10 and 60 mA. Intensities above around 65 mA, however, tended to result in suppressed ERD, or even ERS, relative to the tSCS-off condition. A linear regression found that ERD/ERS and tSCS intensity were indeed positively, and significantly, correlated ($r = 0.409$, $p = 0.025$).

The discomfort felt by participants tended to grow as a function of tSCS intensity. It may have been the case, therefore, that relative alpha power was being suppressed by the uncomfortable sensation, resulting in less desynchronisation during movement, a known consequence of pain on the alpha rhythm [34,35]. Suppression would likely have been more prominent in participants who received the highest intensities. To test this, we found the correlation between intensity and pre-movement relative alpha power (-1.5 s to -0.5 s relative to movement onset): $r = -0.062$, $p = 0.75$. Although the correlation was not significant, the participants who received the highest intensities tended to have reduced alpha power during rest.

Interestingly, when two sub-groups were formed from participants from the lower and upper 25% of the intensity distribution, ERD/ERS become significantly altered by tSCS in the high-intensity group only. Figure 11 shows that in the early phase of movement, ERD/ERS is significantly elevated, ($p = 0.016$) from around -25% without tSCS to around 10% during tSCS, reflective of (event-related) synchronisation rather than desynchronisation. This is seen also in the beta band ($p = 0.015$) and the trend is seen during sustained movement but without significance ($p > 0.05$).

Resting-state alpha frequency and power were also reevaluated in terms of current intensity but no altered effect was found.

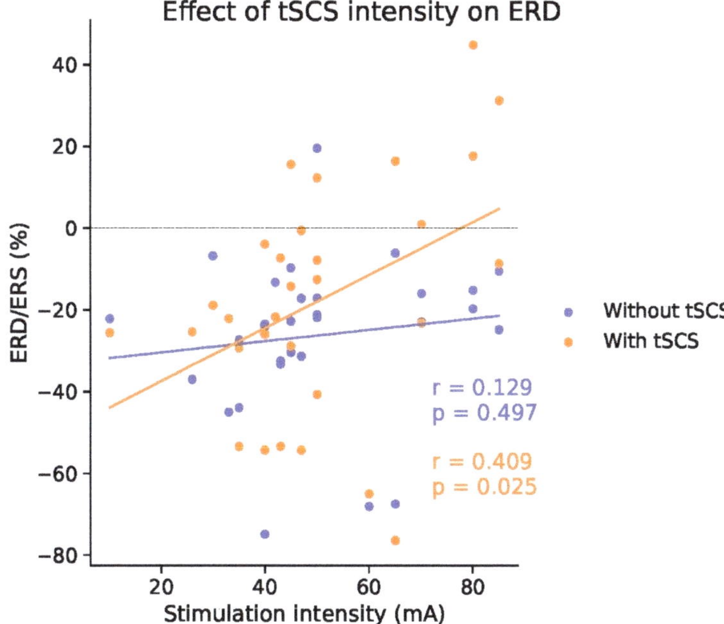

Figure 10. Subject-wise ERD/ERS (%) against tSCS intensity. To aid comparison, ERD/ERS values from the tSCS-off condition are also shown. Stimulation intensity relates to the tSCS-on condition only. Statistical outcomes from a linear regression are given in terms of *r* and *p* values.

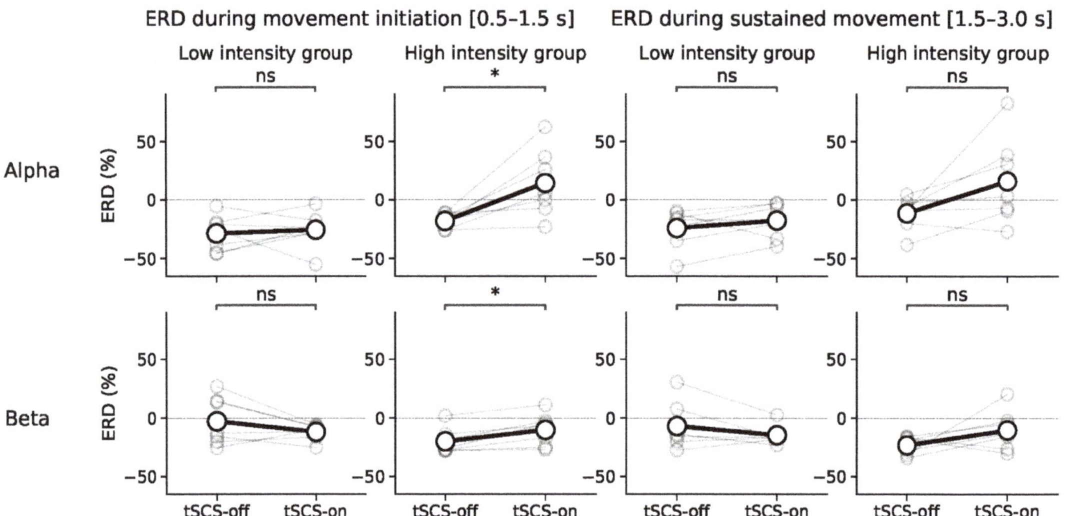

Figure 11. Event-related desynchronisation during movement with participants divided into two groups depending stimulation intensity. Low-intensity participants received tSCS at currents between 10 and 40 mA. High-intensity participants received tSCS at currents between 60 and 85 mA. A Wilcoxon signed-rank rest was used to determine statistically significant differences in ERD between experimental sessions. * $p < 0.05$, ns denotes non-significance.

3.6. Stimulation Adherence

Continuous tSCS was well tolerated by the majority of participants. In two cases, upon receiving tSCS at the beginning of the session, the sensation was considered overwhelming and the participants opted not to continue with the experiment. Both reported that, although not painful, stimulation was uncomfortable and made sitting still difficult.

4. Discussion

The present study showed that a 10 min session of tSCS did not significantly modulate sensorimotor brain rhythms during repetitive upper-limb movements. Similarly, resting-state EEG, as characterised by alpha-band peak frequency and power, was unaffected by continuous tSCS. An investigation of tSCS intensity revealed, however, that cortical activity may have been suppressed among participants who received the highest stimulation intensities, given ERD/ERS was significantly altered for these participants. This work suggests that tSCS intensity may be an important factor to elicit consistent modulation at the cortical level. However, as this high-intensity group is a subset of the overall participant sample, the sample number is small and must be verified on a larger scale.

The inter-participant and inter-session variability in measures such as ERD and alpha power, tended to reflect the inherent variances associated with these measures, as they are in line with other research [10,33]. However, the variance may partially be attributable to current intensity, which was individualised for each participant based on their maximum tolerance. This choice of protocol was based on typical clinical procedures for determining current intensity [2,36,37]. The alpha and beta ERD/ERS of participants who received the highest intensity stimulation tended to be weaker compared to when tSCS was not present. This may imply cortical inhibition following tSCS, which would echo similar claims made by Benavides et al. [7]. Conversely, this reduction in ERD/ERD may have been a consequence of the discomfort associated with high-intensity currents as reduced resting-state alpha power has been associated with exposure to painful sensations [34,35], and lower alpha often correlates with weaker ERD during movement [38]. It is difficult to speculate on the role tSCS intensity played on the individual as each participant received only one level of tSCS intensity. Future analyses should have each participant receive multiple current intensities in order to discern if an intensity-dependent effect exists.

Transcutaneous spinal cord stimulation must penetrate deep into spinal structures, passing multiple layers of skin, fat, muscle, and vertebrae, in order to exert a neuromodulatory effect [39]. Stimulation intensity must, therefore, be strong enough to overcome the impedance of the medium between electrodes. High-intensity stimulation, however, can result in intense discomfort or pain following the contraction of neck and back muscles, and activation of cutaneous pain receptors [40]. In this study, stimulation was set to the participants' maximum tolerance. Maximum stimulation tolerance was shown by Manson et al. to constitute approximately 56% of the intensity required to induce a motor response [40]. This sub-threshold intensity is within the range that clinical studies have reported functional improvements following cervical tSCS [2,36,37]. It is possible, however, that participants with relatively poor stimulation tolerances did not receive activation of posterior-root afferents. This may explain the fact that neuromodulation of cortical oscillations was only observed in the subset of high-intensity participants. It may be the case, therefore, that tSCS, by its very nature, is unsuitable for a portion of a given sample. Future studies may need to consider exclusion criteria that eliminate participants who cannot tolerate stimulation intensities capable of spinal cord interaction.

Although this study is the first to investigate the effects of transcutaneous spinal cord stimulation on cortical oscillations, other studies have reported neuromodulation through electrical stimulation of peripheral musculature and nerves [21,41,42]. For instance, Insausti-Delgado et al. reported enhanced alpha and beta ERD during high intensity neuromuscular electrical stimulation of the wrist extensors [21]. They attributed this effect to the activation of muscle spindles and joint afferents which recruited proprioceptive fibres in the spinal cord, which in turn affected the motor cortex. Indeed, tSCS has been reported to also recruit large-to-medium proprioceptive fibres within posterior roots [39]. Yet the present study found that participants who underwent high-intensity tSCS displayed suppressed alpha and beta band ERD during movement, suggestive of inhibited cortical activity. It may be the case that high-frequency stimulation interfered with the conduction of sensory information to the somatosensory cortex, reducing cortical area activated during movement, which in turn resulted in decreased expression of alpha and beta ERD. Benavides et al. also noted cortical inhibition following cervical tSCS with a 5 kHz carrier frequency, and the effect was even more pronounced in SCI patients [7]. They attributed this inhibition to the activation of inhibitory cortical circuits which influenced motor cortical activity. It is unclear whether this inhibition is related to the reduction of cortical activity in the present study.

Some EEG-based investigations featuring electrical stimulation are challenging or impossible without applying artefact-attenuation techniques [43]. However, stimulation artefact contamination was not considered a confounding factor here as previous work by our group showed that, so long as the spectral region of interest does not overlap with the stimulation frequency, resulting EEG bares statistically similar characteristics to that of normal EEG [29]. Therefore, any differences found in spectral power would be attributable to endogenous neuromodulation and not signal corruption.

The lack of sham condition in this study may constitute a limitation given that the placebo effect has been shown to impact EEG-based metrics [44]. However, implementing a sham control with tSCS is non-trivial as the intensity range at which tSCS exerts a non-therapeutic effect is currently unknown. Similarly, non-therapeutic duration is also unknown, hence, protocols that ramp down after a brief period of stimulation were considered unsuitable. Further, the intense, non-painful sensation associated with tSCS, even at low currents, makes the ambiguity required for establishing an effective sham control difficult. Indeed, as Turner et al. showed using transcranial direct current stimulation, participants were aware of whether they were or were not receiving active stimulation throughout the experimental procedure [45]. We expect that placebo effect contamination to be low, however, as the procedure and equipment were identical in both sessions, and the outcome measures (ERD/ERS during movement) were not known by participants. Effective sham-blinding protocols should be verified in the future, perhaps by stimulating a spinal level that does not project to the motor pools under investigation.

A significant limitation of this study is that it lacks a clinical population. We note that studies that include a patient cohort in addition to healthy controls often reported more marked modulation in the SCI group [7]. It may be the case that, in healthy participants, a ceiling effect limits the recruitment of additional fibres as the cortical–spinal network is already being used to its fullest extent during movement. Additionally, an SCI cohort would allow for higher currents to be explored, owing to reduced sensitivity at and below the spinal level of injury. This would likely minimise the effect noted here whereby individuals receiving the highest intensities of tSCS exhibited reduced resting-state alpha power due to discomfort.

5. Conclusions

This study, for the first time, investigated cervical tSCS neuromodulation in terms of sensorimotor oscillations as measured by EEG. Our results showed that, on a group level, there was no consistent excitatory or inhibitory effect in terms of cortical activity during upper-limb movement. However, consistency appeared to emerge among participants who received the highest stimulation intensities. ERD, a measure of sensorimotor cortical activity, was diminished in these participants, potentially implying an inhibitory effect of tSCS at the cortical level. However, this sub-set of participants constitutes a small population size. Future work should, therefore, specifically investigate the effects of tSCS intensity on cortical oscillations. Additionally, future work should endeavour to determine the critical duration required for cervical tSCS to exert a measurable effect on sensorimotor cortical activity.

Author Contributions: Conceptualisation, C.M. and A.V.; methodology, C.M.; software, C.M.; validation, C.M. and A.V.; formal analysis, C.M.; investigation, C.M.; resources, Y.-P.Z.; data curation, C.M.; writing—original draft preparation, C.M.; writing—review and editing, A.V., Y.-P.Z. and M.A.; visualisation, C.M.; project administration, M.A.; funding acquisition, M.A. All authors have read and agreed to the published version of the manuscript.

Funding: This work was supported by RCUK PhD scholarship EP/N509668/1, the University of Glasgow Graduate School Mobility Scholarship, the Hong Kong Polytechnic University (UAKB), and the Telefield Charitable Fund (83D1).

Institutional Review Board Statement: The study was conducted according to the guidelines of the Declaration of Helsinki and approved by the Human Subjects Ethics Sub-committee of the Hong Kong Polytechnic University (HSEARS20201105003, 26 November 2020).

Informed Consent Statement: Informed consent was obtained from all subjects involved in the study.

Data Availability Statement: Data will be made available upon reasonable request to the authors.

Acknowledgments: We sincerely thank the research participants for their patience and commitment to our study.

Conflicts of Interest: The authors declare no conflict of interest.

References

1. Zhang, F.; Momeni, K.; Ramanujam, A.; Ravi, M.; Carnahan, J.; Kirshblum, S.; Forrest, G.F. Cervical Spinal Cord Transcutaneous Stimulation Improves Upper Extremity and Hand Function in People With Complete Tetraplegia: A Case Study. *IEEE Trans. Neural Syst. Rehabil. Eng.* **2020**, *28*, 3167–3174.
2. Inanici, F.; Brighton, L.N.; Samejima, S.; Hofstetter, C.P.; Moritz, C.T. Transcutaneous spinal cord stimulation restores hand and arm function after spinal cord injury. *IEEE Trans. Neural Syst. Rehabil. Eng.* **2021**, *29*, 310–319.
3. Hofstoetter, U.S.; Freundl, B.; Binder, H.; Minassian, K. Recovery cycles of posterior root-muscle reflexes evoked by transcutaneous spinal cord stimulation and of the H reflex in individuals with intact and injured spinal cord. *PLoS ONE* **2019**, *14*, e0227057. [CrossRef] [PubMed]

4. Hofstoetter, U.S.; Freundl, B.; Danner, S.M.; Krenn, M.J.; Mayr, W.; Binder, H.; Minassian, K. Transcutaneous Spinal Cord Stimulation Induces Temporary Attenuation of Spasticity in Individuals with Spinal Cord Injury. *J. Neurotrauma* **2020**, *37*, 481–493. [CrossRef]

5. Duffell, L.D.; Donaldson, N.D.N. A Comparison of FES and SCS for Neuroplastic Recovery After SCI: Historical Perspectives and Future Directions. *Front. Neurol.* **2020**, *11*, 607. [CrossRef] [PubMed]

6. Taylor, C.; McHugh, C.; Mockler, D.; Minogue, C.; Reilly, R.B.; Fleming, N. Transcutaneous spinal cord stimulation and motor responses in individuals with spinal cord injury: A methodological review. *PLoS ONE* **2021**, *16*, e0260166. [CrossRef]

7. Benavides, F.D.; Jo, H.J.; Lundell, H.; Edgerton, V.R.; Gerasimenko, Y.; Perez, M.A. Cortical and Subcortical Effects of Transcutaneous Spinal Cord Stimulation in Humans with Tetraplegia. *J. Neurosci.* **2020**, *40*, 2633–2643. [CrossRef]

8. Kumru, H.; Rodríguez-Cañón, M.; Edgerton, V.R.; García, L.; Flores, Á.; Soriano, I.; Opisso, E.; Gerasimenko, Y.; Navarro, X.; García-Alías, G.; et al. Transcutaneous Electrical Neuromodulation of the Cervical Spinal Cord Depends Both on the Stimulation Intensity and the Degree of Voluntary Activity for Training. A Pilot Study. *J. Clin. Med.* **2021**, *10*, 3278. [CrossRef]

9. Kumru, H.; Flores, Á.; Rodríguez-Cañón, M.; Edgerton, V.R.; García, L.; Benito-Penalva, J.; Navarro, X.; Gerasimenko, Y.; García-Alías, G.; Vidal, J. Cervical Electrical Neuromodulation Effectively Enhances Hand Motor Output in Healthy Subjects by Engaging a Use-Dependent Intervention. *J. Clin. Med.* **2021**, *10*, 195. [CrossRef]

10. Sasaki, A.; de Freitas, R.M.; Sayenko, D.G.; Masugi, Y.; Nomura, T.; Nakazawa, K.; Milosevic, M. Low-Intensity and Short-Duration Continuous Cervical Transcutaneous Spinal Cord Stimulation Intervention Does Not Prime the Corticospinal and Spinal Reflex Pathways in Able-Bodied Subjects. *J. Clin. Med.* **2021**, *10*, 3633.

11. Megía-García, Á.; Serrano-Muñoz, D.; Taylor, J.; Avendaño-Coy, J.; Comino-Suárez, N.; Gómez-Soriano, J. Transcutaneous Spinal Cord Stimulation Enhances Quadriceps Motor Evoked Potential in Healthy Participants: A Double-Blind Randomized Controlled Study. *J. Clin. Med.* **2020**, *9*, 3275.

12. Gerasimenko, Y.; Gorodnichev, R.; Moshonkina, T.; Sayenko, D.; Gad, P.; Reggie Edgerton, V. Transcutaneous electrical spinal-cord stimulation in humans. *Ann. Phys. Rehabil. Med.* **2015**, *58*, 225–231. [CrossRef]

13. Al'joboori, Y.; Hannah, R.; Lenham, F.; Borgas, P.; Kremers, C.J.P.; Bunday, K.L.; Rothwell, J.; Duffell, L.D. The Immediate and Short-Term Effects of Transcutaneous Spinal Cord Stimulation and Peripheral Nerve Stimulation on Corticospinal Excitability. *Front. Neurosci.* **2021**, *15*, 749042. [CrossRef]

14. Rau, C.; Plewnia, C.; Hummel, F.; Gerloff, C. Event-related desynchronization and excitability of the ipsilateral motor cortex during simple self-paced finger movements. *Clin. Neurophysiol.* **2003**, *114*, 1819–1826. [CrossRef]

15. Mäki, H.; Ilmoniemi, R.J. EEG oscillations and magnetically evoked motor potentials reflect motor system excitability in overlapping neuronal populations. *Clin. Neurophysiol.* **2010**, *121*, 492–501. [CrossRef] [PubMed]

16. Pfurtscheller, G.; Lopes da Silva, F.H. Event-related EEG/MEG synchronization and desynchronization: Basic principles. *Clin. Neurophysiol.* **1999**, *110*, 1842–1857. [CrossRef]

17. Salisbury, D.B.; Parsons, T.D.; Monden, K.R.; Trost, Z.; Driver, S.J. Brain–computer interface for individuals after spinal cord injury. *Rehabil. Psychol.* **2016**, *61*, 435–441. [CrossRef]

18. Daly, J.J.; Huggins, J.E. Brain-Computer Interface: Current and Emerging Rehabilitation Applications. *Arch. Phys. Med. Rehabil.* **2015**, *96*, S1–S7. [CrossRef] [PubMed]

19. Reynolds, C.; Osuagwu, B.A.; Vuckovic, A. Influence of motor imagination on cortical activation during functional electrical stimulation. *Clin. Neurophysiol.* **2015**, *126*, 1360–1369. [CrossRef]

20. Lotte, F.; Bougrain, L.; Cichocki, A.; Clerc, M.; Congedo, M.; Rakotomamonjy, A.; Yger, F. A review of classification algorithms for EEG-based brain–computer interfaces: A 10 year update. *J. Neural Eng.* **2018**, *15*, 031005. [CrossRef] [PubMed]

21. Insausti-Delgado, A.; López-Larraz, E.; Omedes, J.; Ramos-Murguialday, A. Intensity and Dose of Neuromuscular Electrical Stimulation Influence Sensorimotor Cortical Excitability. *Front. Neurosci.* **2021**, *14*, 1359. [CrossRef]

22. Sharon, O.; Fahoum, F.; Nir, Y. Transcutaneous Vagus Nerve Stimulation in Humans Induces Pupil Dilation and Attenuates Alpha Oscillations. *J. Neurosci.* **2021**, *41*, 320–330.

23. Fürst, S. Transmitters involved in antinociception in the spinal cord. *Brain Res. Bull.* **1999**, *48*, 129–141. [CrossRef]

24. Ramos-Murguialday, A.; Birbaumer, N. Brain oscillatory signatures of motor tasks. *J. Neurophysiol.* **2015**, *113*, 3663–3682.

25. Graimann, B.; Pfurtscheller, G. Quantification and visualization of event-related changes in oscillatory brain activity in the time–frequency domain. In *Progress in Brain Research*; Elsevier: Amsterdam, The Netherlands, 2006; Volume 159, pp. 79–97. [CrossRef]

26. Kantak, S.; Jax, S.; Wittenberg, G. Bimanual coordination: A missing piece of arm rehabilitation after stroke. *Restor. Neurol. Neurosci.* **2017**, *35*, 347–364. [CrossRef] [PubMed]

27. Hoffman, L.R.; Field-Fote, E.C. Cortical Reorganization Following Bimanual Training and Somatosensory Stimulation in Cervical Spinal Cord Injury: A Case Report. *Phys. Ther.* **2007**, *87*, 208–223. [CrossRef] [PubMed]

28. Kasashima-Shindo, Y.; Fujiwara, T.; Ushiba, J.; Matsushika, Y.; Kamatani, D.; Oto, M.; Ono, T.; Nishimoto, A.; Shindo, K.; Kawakami, M.; et al. Brain-computer interface training combined with transcranial direct current stimulation in patients with chronic severe hemiparesis: Proof of concept study. *J. Rehabil. Med.* **2015**, *47*, 318–324. [CrossRef]

29. McGeady, C.; Vučković, A.; Zheng, Y.P.; Alam, M. EEG Monitoring Is Feasible and Reliable during Simultaneous Transcutaneous Electrical Spinal Cord Stimulation. *Sensors* **2021**, *21*, 6593.

30. Merletti, R.; Farina, D. *Surface Electromyography: Physiology, Engineering, and Applications*; John Wiley & Sons: Hoboken, NJ, USA, 2016.
31. Wecht, J.R.; Savage, W.M.; Famodimu, G.O.; Mendez, G.A.; Levine, J.M.; Maher, M.T.; Weir, J.P.; Wecht, J.M.; Carmel, J.B.; Wu, Y.K.; et al. Posteroanterior Cervical Transcutaneous Spinal Cord Stimulation: Interactions with Cortical and Peripheral Nerve Stimulation. *J. Clin. Med.* **2021**, *10*, 5304.
32. Barss, T.S.; Parhizi, B.; Mushahwar, V.K. Transcutaneous spinal cord stimulation of the cervical cord modulates lumbar networks. *J. Neurophysiol.* **2020**, *123*, 158–166.
33. Espenhahn, S.; de Berker, A.O.; van Wijk, B.C.M.; Rossiter, H.E.; Ward, N.S. Movement-related beta oscillations show high intra-individual reliability. *NeuroImage* **2017**, *147*, 175–185. [CrossRef] [PubMed]
34. Peng, W.; Hu, L.; Zhang, Z.; Hu, Y. Changes of spontaneous oscillatory activity to tonic heat pain. *PLoS ONE* **2014**, *9*, e91052. [CrossRef] [PubMed]
35. Tan, L.L.; Oswald, M.J.; Kuner, R. Neurobiology of brain oscillations in acute and chronic pain. *Trends Neurosci.* **2021**, *44*, 629–642. [CrossRef]
36. Zhang, H.; Liu, Y.; Zhou, K.; Wei, W.; Liu, Y. Restoring Sensorimotor Function Through Neuromodulation After Spinal Cord Injury: Progress and Remaining Challenges. *Front. Neurosci.* **2021**, *15*, 1312. [CrossRef] [PubMed]
37. Gad, P.; Lee, S.; Terrafranca, N.; Zhong, H.; Turner, A.; Gerasimenko, Y.; Edgerton, V.R. Non-Invasive Activation of Cervical Spinal Networks after Severe Paralysis. *J. Neurotrauma* **2018**, *35*, 2145–2158. [CrossRef]
38. López-Larraz, E.; Ray, A.M.; Birbaumer, N.; Ramos-Murguialday, A. Sensorimotor rhythm modulation depends on resting-state oscillations and cortex integrity in severely paralyzed stroke patients. In Proceedings of the 2019 9th International IEEE/EMBS Conference on Neural Engineering (NER), San Francisco, CA, USA, 20–23 March 2019; pp. 37–40; ISSN 1948-3546. [CrossRef]
39. Hofstoetter, U.S.; Freundl, B.; Binder, H.; Minassian, K. Common neural structures activated by epidural and transcutaneous lumbar spinal cord stimulation: Elicitation of posterior root-muscle reflexes. *PLoS ONE* **2018**, *13*, e0192013.
40. Manson, G.A.; Calvert, J.S.; Ling, J.; Tychhon, B.; Ali, A.; Sayenko, D.G. The relationship between maximum tolerance and motor activation during transcutaneous spinal stimulation is unaffected by the carrier frequency or vibration. *Physiol. Rep.* **2020**, *8*, e14397.
41. Corbet, T.; Iturrate, I.; Pereira, M.; Perdikis, S.; Millán, J.D.R. Sensory threshold neuromuscular electrical stimulation fosters motor imagery performance. *NeuroImage* **2018**, *176*, 268–276. [CrossRef]
42. Takemi, M.; Masakado, Y.; Liu, M.; Ushiba, J. Sensorimotor event-related desynchronization represents the excitability of human spinal motoneurons. *Neuroscience* **2015**, *297*, 58–67. [CrossRef]
43. Kohli, S.; Casson, A.J. Removal of Gross Artifacts of Transcranial Alternating Current Stimulation in Simultaneous EEG Monitoring. *Sensors* **2019**, *19*, 190.
44. Li, L.; Wang, H.; Ke, X.; Liu, X.; Yuan, Y.; Zhang, D.; Xiong, D.; Qiu, Y. Placebo Analgesia Changes Alpha Oscillations Induced by Tonic Muscle Pain: EEG Frequency Analysis Including Data during Pain Evaluation. *Front. Comput. Neurosci.* **2016**, *10*, 45. [PubMed]
45. Turner, C.; Jackson, C.; Learmonth, G. Is the "end-of-study guess" a valid measure of sham blinding during transcranial direct current stimulation? *Eur. J. Neurosci.* **2021**, *53*, 1592–1604.

Journal of
Clinical Medicine

Article

Influence of Spine Curvature on the Efficacy of Transcutaneous Lumbar Spinal Cord Stimulation

Veronika E. Binder [1,2], Ursula S. Hofstoetter [2,*], Anna Rienmüller [3], Zoltán Száva [4], Matthias J. Krenn [5,6], Karen Minassian [2,†] and Simon M. Danner [7,*,†]

[1] Institute for Analysis and Scientific Computing, Vienna University of Technology, 1040 Vienna, Austria; VeronikaBinder@gmx.at

[2] Center for Medical Physics and Biomedical Engineering, Medical University of Vienna, 1090 Vienna, Austria; karen.minassian@meduniwien.ac.at

[3] Department of Orthopedic and Trauma Surgery, Medical University of Vienna, 1090 Vienna, Austria; anna.rienmueller@meduniwien.ac.at

[4] Competence Center Medizintechnik, Austrian Workers' Compensation Board, 1200 Vienna, Austria; z.szava@gmail.com

[5] Department of Neurobiology and Anatomical Sciences, University of Mississippi Medical Center, Jackson, MS 39216, USA; mkrenn@umc.edu

[6] Center for Neuroscience and Neurological Recovery, Methodist Rehabilitation Center, Jackson, MS 39216, USA

[7] Department of Neurobiology and Anatomy, College of Medicine, Drexel University, Philadelphia, PA 19129, USA

* Correspondence: ursula.hofstoetter@meduniwien.ac.at (U.S.H.); smd395@drexel.edu (S.M.D.)

† Equal contribution.

Citation: Binder, V.E.; Hofstoetter, U.S.; Rienmüller, A.; Száva, Z.; Krenn, M.J.; Minassian, K.; Danner, S.M. Influence of Spine Curvature on the Efficacy of Transcutaneous Lumbar Spinal Cord Stimulation. *J. Clin. Med.* **2021**, *10*, 5543. https://doi.org/10.3390/jcm10235543

Academic Editor: Moussa Antoine Chalah

Received: 13 August 2021
Accepted: 24 November 2021
Published: 26 November 2021

Publisher's Note: MDPI stays neutral with regard to jurisdictional claims in published maps and institutional affiliations.

Abstract: Transcutaneous spinal cord stimulation is a non-invasive method for neuromodulation of sensorimotor function. Its main mechanism of action results from the activation of afferent fibers in the posterior roots—the same structures as targeted by epidural stimulation. Here, we investigated the influence of sagittal spine alignment on the capacity of the surface-electrode-based stimulation to activate these neural structures. We evaluated electromyographic responses evoked in the lower limbs of ten healthy individuals during extension, flexion, and neutral alignment of the thoracolumbar spine. To control for position-specific effects, stimulation in these spine alignment conditions was performed in four different body positions. In comparison to neutral and extended spine alignment, flexion of the spine resulted in a strong reduction of the response amplitudes. There was no such effect on tibial-nerve evoked H reflexes. Further, there was a reduction of post-activation depression of the responses to transcutaneous spinal cord stimulation evoked in spinal flexion. Thus, afferent fibers were reliably activated with neutral and extended spine alignment. Spinal flexion, however, reduced the capacity of the stimulation to activate afferent fibers and led to the co-activation of motor fibers in the anterior roots. This change of action was due to biophysical rather than neurophysiological influences. We recommend applying transcutaneous spinal cord stimulation in body positions that allow individuals to maintain a neutral or extended spine.

Keywords: biophysics; H reflex; human; M wave; neuromodulation; posterior root-muscle reflex; posterior root stimulation; spine alignment; spinal cord; spinal cord stimulation; spine; transcutaneous

1. Introduction

Transcutaneous lumbar spinal cord stimulation was designed to activate large-diameter posterior root afferent fibers through the use of skin-surface stimulation electrodes placed over the spine at the thoracolumbar junction, overlying the lumbosacral spinal cord, and indifferent electrodes over the anterior lower trunk [1]. Single stimuli were shown to evoke posterior root-muscle (PRM) reflexes (short-latency spinal reflexes with physiological similarities to the soleus H reflex) in many lower limb muscles [1–5]. Trains of stimuli can therefore provide tonic multisegmental afferent input to the spinal cord, comparable with epidural spinal cord stimulation [6]. As a result, transcutaneous lumbar spinal cord stimulation has been used for neurophysiological studies by investigating reflex modulation in multiple lower limb muscles simultaneously [7–9] as well as for neuromodulation of sensorimotor function after spinal cord injury [10–12] and multiple sclerosis [13,14].

To reliably apply transcutaneous lumbar spinal cord stimulation, posterior root afferent fibers must be activated selectively and consistently. Although the stimulation seems unspecific, the biophysical properties of the lumbosacral spinal cord and its surrounding spinal roots enable a rather selective recruitment of large-diameter posterior root fibers [15–18]. Yet, under certain conditions, anterior roots containing the axons of the spinal motoneurons can be (co-)activated [15,16]. In neurophysiological studies, this (co-)activation would evoke direct M wave-like responses [1,2] that bypass the spinal sensorimotor circuits; in neuromodulation approaches, continuous contractions of the respectively innervated muscles would be generated, impeding the intended effects. We have previously shown that selective posterior root recruitment can be achieved when stimulation is applied in supine or normal upright standing positions [1,3,6,10,19], but when applied in the prone position, the thresholds of anterior root motor fibers relative to those of the posterior roots are decreased [19].

Which neural structures are being activated depends on the interplay of several biophysical and anatomical factors that influence the electric field acting on the target neural structures [15–19]. The electric current crosses the spine mainly through the soft tissues between the bony vertebrae [15,20]. With flexion and extension of the spine, the vertebrae undergo relative movements and rotations, resulting in changes in the distance between bony structures, in the dimensions of the intervertebral foramina, and in deformation of the soft tissues [21].

We here hypothesized that these changes in the properties of the volume conductor in-between the surface electrodes would influence the current flow through the soft tissues of the spine and, in turn, the current acting on the anterior and posterior root fibers. We applied transcutaneous spinal cord stimulation in individuals with intact nervous systems who assumed three spine alignment conditions (neutral, extension, flexion), each within four body positions (supine, lateral recumbent, sitting, standing). We show that PRM reflexes of similar size are elicited in the extended and neutral spine alignment condition, while they diminish in the flexed spine alignment condition and co-activation of motor fibers in the anterior roots occurs. By concomitantly monitoring the H reflex, we demonstrate that these effects are not caused by neurophysiological modulation but rather by biophysical influences on the stimulation conditions. Such influences are important to consider because a neutral spinal curvature cannot always be maintained by individuals with sensorimotor disorders, and spinal curvature can change during dynamic motor tasks.

2. Materials and Methods

2.1. Participants

Ten neurologically intact individuals (21–33 years; three females; body mass index: 18.8–25.7, mean: 21.8) participated in the study. Exclusion criteria were previous spinal, cranial, or abdominal surgeries, meningitis, primary muscular diseases, pronounced postural anomalies or restricted flexibility of the lower limbs, and pregnancy. The study was approved by the institutional review board of the Medical University of Vienna, Austria (EK 1686/2014), and conducted in accordance with the Declaration of Helsinki. All participants signed written informed consent prior to their enrollment into the study.

2.2. Investigated Body Positions and Spine Alignments

Lower-limb muscle responses to tibial nerve stimulation and transcutaneous spinal cord stimulation were studied in the supine, lateral recumbent, sitting, and standing positions. In each of these body positions, three static alignments of the thoracolumbar spine in the sagittal plane were studied: maximum extension, neutral, and maximum flexion. Maximum extension and flexion were defined as the maximum displacements that the participants felt comfortable to assume. Maximum extension, neutral, and maximum flexion angles were not controlled to be the same across body positions. The experiments were designed to investigate the effects of relative changes in spine alignment. The spine was straight across positions in the frontal plane, with the head and neck aligned in a neutral position. In the supine and standing positions, the arms were relaxed, parallel to the torso; in the lateral recumbent position, the elbows were flexed with the forearms in a comfortable position. In the sitting position, the forearms rested on the thighs. In the supine position, the examination table was adjusted, and therapeutic pillows were used to support and stabilize the different spine alignments. In the lateral recumbent position, participants lay on their left sides, a support ensured a straight spine in the frontal plane, and stabilization was provided to support the pelvis and shoulder girdle against rotations. In the sitting position, the participants sat on the examination table with their feet placed on a height-adjusted footrest and with hip, knee, and ankle flexed at 90°. In the standing position, the participants assumed an upright standing position next to a wall for reference, with no external support and with the shoulders aligned above the hips across spine alignment conditions. An orthopedic surgeon (A.R.) monitored proper positioning of the study participants throughout the recordings. An ankle orthosis (Malleo Sprint, Otto Bock HealthCare GmbH, Duderstadt, Germany) was used to maintain equal muscle length of the left soleus across recordings.

2.3. Electromyographic Recordings

Surface electromyographic (EMG) activity was acquired bilaterally from rectus femoris, the hamstrings muscle group, tibialis anterior, and soleus with pairs of silver-silver chloride electrodes (Intec Medizintechnik GmbH, Klagenfurt, Austria) placed with an interelectrode distance of 3 cm (Figure 1) [6,22]. Ground electrodes were placed over the lateral malleolus of each leg. An abrasive paste (Nuprep, Weaver and Company, Aurora, CO, USA) was used to reduce EMG electrode resistance below 10 kΩ. EMG signals were amplified with a gain set to 602, filtered to a bandwidth of 10–600 Hz, digitized at 10,000 samples per second and channel using a USB-NI 6261 data acquisition card (National Instruments, Inc., Austin, TX, USA), and recorded using DasyLab 12.0 (Measurement Computing Corporation, Norton, MA, USA).

Stimulation and recording setups

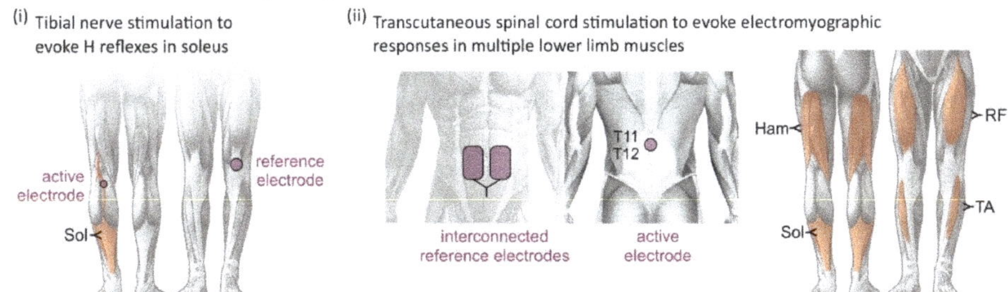

Figure 1. Stimulation and recording setups. (**i**) H reflexes in soleus (Sol) were evoked with an active electrode placed over the posterior tibial nerve in the popliteal fossa and a reference electrode placed proximal to the patella. (**ii**) Posterior root-muscle (PRM) reflexes were elicited bilaterally in the rectus femoris (RF), the hamstrings (Ham) muscle group, tibialis anterior (TA), and Sol by transcutaneous spinal cord stimulation with an active electrode placed over the spine between the T11 and T12 spinous processes and a pair of interconnected reference electrodes over the lower abdomen.

2.4. Stimulation Procedures

For all stimulation procedures, a current-controlled stimulator (Stimulette r2x-S1, Dr. Schuhfried Medizintechnik GmbH, Moedling, Austria) was used and set to deliver charge-balanced, symmetrical, biphasic rectangular pulses of 1 ms width per phase.

Stimulation of the posterior tibial nerve of the left leg was carried out using a self-adhesive hydrogel surface electrode (Ø = 1 cm; Leonhard Lang GmbH, Innsbruck, Austria) placed in the popliteal fossa and a reference electrode (Ø = 5 cm; Schwamedico GmbH, Ehringhausen, Germany) placed over the anterior aspect of the knee (Figure 1i). The electrode position in the popliteal fossa was adjusted to minimize the stimulation amplitude required to elicit an H reflex in the soleus while producing an isolated ankle plantar-flexion movement when increasing the stimulation amplitude. Maximum M waves (Mmax) were determined with the participants lying supine with neutral spine alignment.

The H reflex was used to assess potential neurophysiological influences of the different static spine alignment conditions on reflex excitability. To this end, tibial-nerve stimulation amplitudes were first separately adjusted for a neutral spine alignment in each of the four body positions to elicit H reflexes with peak-to-peak amplitudes corresponding to 25% of Mmax on the ascending limb of the recruitment curve [23,24] with present M waves. For each body position, the stimulation amplitude was then kept constant across the extension, neutral, and flexion conditions. With the designated stimulation amplitude, six stimulation pulses were applied per body position and spine alignment condition, separated by a minimum of 8 s between consecutive stimuli.

Stimulation of the lumbar and upper sacral spinal roots was carried out through a self-adhesive hydrogel surface electrode (Ø = 5 cm; Schwamedico GmbH, Ehringhausen, Germany) placed medially over the spine between the T11 and T12 spinous processes and a pair of interconnected electrodes (each 8 × 13 cm) on the lower abdomen (Figure 1ii). With reference to the abdominal electrodes, the paraspinal electrode acted as the anode for the first phase of the biphasic stimulation pulses and as the cathode for the second phase [6].

To study the dependence of transcutaneous lumbar spinal cord stimulation on spine alignment, stimulation amplitudes were adjusted in each body position with a neutral spine alignment to elicit EMG responses in the left soleus with peak-to-peak amplitudes that best matched those of the H reflexes in this muscle as well as to concomitantly elicit responses in all other muscles studied. For each body position, stimulation amplitudes were kept constant across the extension, neutral, and flexion conditions. With the designated stimulation amplitude, three single stimulation pulses and three double stimulation pulses with an interstimulus interval of 35 ms were applied, separated by a minimum of 8 s between consecutive single or double stimuli. Correct adhesion of all stimulation electrodes was regularly checked in each of the positions and spine alignment conditions.

2.5. Study Protocol

Each subject assumed body positions in the following order: supine, lateral recumbent, standing, and lastly, sitting. For each body position, the stimulation intensities for tibial nerve stimulation and transcutaneous lumbar spinal cord stimulation were first adjusted (as described above) with neutral spine alignment. Then, both types of stimulation were applied in neutral, followed by extended and flexed spine alignment. For each body position and spine alignment condition, first tibial nerve stimulation and then transcutaneous spinal cord stimulation was applied.

2.6. Data Analysis

Before the analysis, all recordings were visually inspected. For each body position and spine alignment condition investigated, EMG peak-to-peak amplitudes were calculated for the H reflexes and M waves elicited in the left soleus. Peak-to-peak amplitudes were also calculated for EMG responses elicited by single-pulse and paired-pulse transcutaneous spinal cord stimulation in rectus femoris, hamstrings, tibialis anterior, and soleus of both legs. For the responses to double-pulse transcutaneous spinal cord stimulation, the peak-to-peak amplitudes of the responses to the second stimulus of the pair were normalized to the respective first ones. Normalized second-to-first response amplitudes were only included in the statistical model if the corresponding first response sizes were > 100 μV.

Statistical analysis was performed with R 4.0.0 (R Foundation for Statistical Computing, Vienna, Austria) [25]. Each statistical model included a full factorial dispersion model, fit using Template Model Builder [26,27] interfaced through the glmmTMB package [28]. Backward elimination was performed on the random effects. Random effects were removed from the model if the likelihood-ratio test was not significant. Model assumptions were tested using the DHARMa package for R with distribution, dispersion, outliers, and quantile deviation tests performed. Q-Q plots and plots of residuals against the predicted values were inspected. If any assumptions were violated, other error distributions and link functions were tested. To control for multiple comparisons, Bonferroni–Holm correction was applied to all post hoc tests [29]. An alpha error of $p < 0.05$ was regarded as significant.

Separate generalized linear mixed models with spine alignment condition (extended, neutral, flexed) as fixed effect and body position (supine, lateral recumbent, sitting, standing) as random effect were run for the EMG peak-to-peak amplitudes of H reflexes, M waves, and responses evoked by transcutaneous spinal cord stimulation in the left soleus. For the statistical model of the M wave, a Gamma distribution with a log link function was used to satisfy model assumptions.

For the responses to transcutaneous lumbar spinal cord stimulation in rectus femoris, hamstrings, tibialis anterior, and soleus, results derived from the left and right lower limbs were considered as separate cases. Separate generalized linear mixed models with spine alignment condition, muscle group, and their interaction as fixed effects and body position as a random effect were performed for the EMG peak-to-peak amplitudes of responses elicited by single (or respective first stimulation pulses of a pair) as well as for the normalized second-to-first response amplitudes. A Gamma distribution with a log link function was used for both models to ensure all model assumptions were met.

Recordings of one individual in the lateral recumbent and sitting body positions were excluded from analysis because the subject was not able to hold these positions steadily without generating EMG activity in the lower limb muscles.

3. Results

3.1. Influence of Spine Alignment Condition on H Reflexes and PRM Reflexes in Soleus

Soleus EMG responses evoked by tibial nerve stimulation as well as by transcutaneous spinal cord stimulation in an individual in a lateral recumbent position with extension, neutral alignment, and flexion of the thoracolumbar spine are displayed in Figure 2. M waves and H reflexes with similar EMG peak-to-peak amplitudes were evoked across spine alignment conditions (Figure 2A). In the extension and neutral spine alignment conditions, transcutaneous spinal cord stimulation evoked PRM reflexes—as documented by the post-stimulation depression of the second responses when double-stimuli were applied [1,6] (Figure 2B). The EMG responses were comparable in the extension and neutral spine alignment conditions but diminished in size when stimulation with the same amplitude was applied in the flexed condition. This reduction of the response size with spinal flexion was likely caused by biophysical influences because no depression of the H reflex was observed (Figure 2A).

The generalized linear mixed model on the peak-to-peak amplitude of the soleus H reflex revealed a significant fixed effect of spine alignment condition on the attained peak-to-peak amplitudes ($F_{2,97} = 4.355$, $p = 0.0154$, $NG_p^2 = 0.082$; Figure 3A). Post hoc tests demonstrated larger peak-to-peak amplitudes with flexed compared to neutral spine conditions ($t_{97} = 2.859$, $p = 0.0143$), with no further significant differences detected (extension vs. flexion, $t_{97} = 2.275$, $p = 0.0642$; extension vs. neutral alignment, $t_{97} = 0.655$, $p = 0.7902$). Further in-depth post hoc analyses unveiled significantly larger peak-to-peak amplitudes with flexed compared to neutral spine alignment only in the lateral recumbent position ($t_{88} = -2.805$, $p = 0.0169$; all other $p > 0.100$). In the case of the M wave, no significant effect of spine alignment condition on the attainable peak-to-peak amplitudes was detected ($F_{2;105} = 1.816$, $p = 0.1678$, $NG_p^2 < 0.0001$), signifying consistent tibial nerve stimulation conditions.

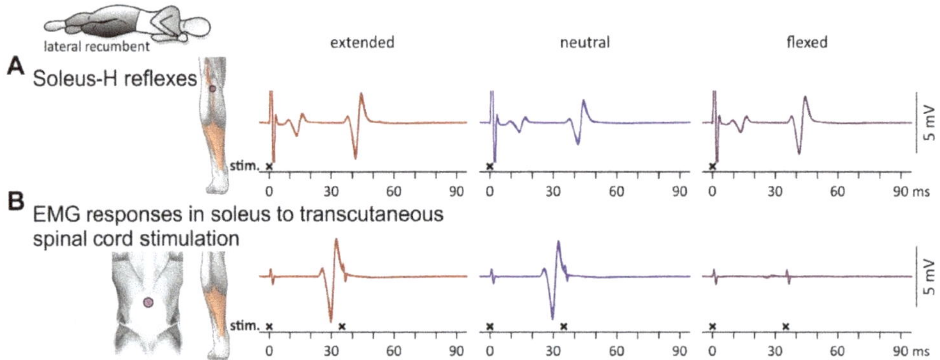

Figure 2. Influence of the spine alignment condition on M waves and H reflexes as well as on electromyographic (EMG) responses to transcutaneous spinal cord stimulation in left soleus: exemplary results. (**A**) EMG responses evoked by tibial nerve stimulation; six stimulus-triggered repetitions displayed superimposed. Traces show stimulation artifacts followed by M waves and (at > 30 ms) H reflexes. (**B**) EMG responses evoked by transcutaneous spinal cord stimulation. Superimposed representation of responses to three single stimuli as well as to three double stimuli with an interstimulus interval of 35 ms. EMG responses to transcutaneous spinal cord stimulation diminished in size during static flexion of the thoracolumbar spine, while M waves and H reflexes did not demonstrate such changes. Black crosses mark times of stimulus application. All recordings derived from the left soleus of one subject in lateral recumbent position, with the spine in extended, neutral, and flexed conditions. Note that amplitudes of tibial nerve and spinal cord stimulation were unchanged across the different spine alignment conditions.

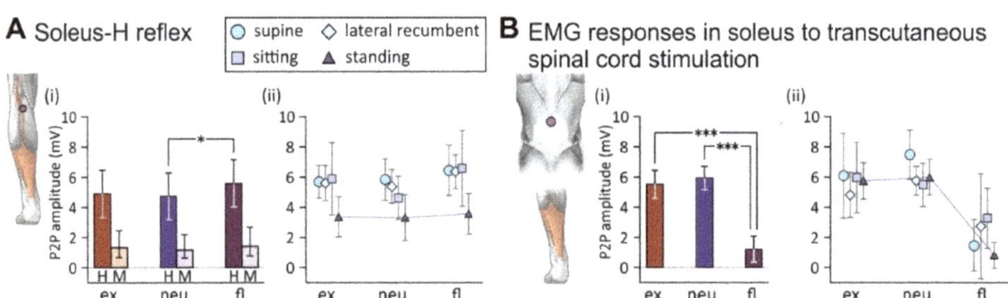

Figure 3. Influence of the spine alignment condition on M waves and H reflexes as well as electromyographic (EMG) responses to transcutaneous spinal cord stimulation in soleus: group results. (**A**) (**i**) Marginal mean EMG peak-to-peak (P2P) amplitudes of H reflexes (H) and M waves (M) elicited across body positions with extended (ex.), neutral (neu.), and flexed (fl.) spine. P2P amplitudes of H reflexes were significantly larger with a flexed than with a neutral spine. (**ii**) P2P amplitudes of H reflexes elicited with different body positions and spine alignment conditions. (**B**) (**i**) Marginal mean P2P amplitudes of soleus responses to transcutaneous spinal cord stimulation across body positions elicited in different spine alignment conditions. P2P amplitudes were significantly smaller with flexed spine alignment compared to extended and neutral alignments. (**ii**) P2P amplitudes of EMG responses elicited with different body positions and spine alignment conditions. Whiskers extend from the lower to the upper limits of the respective 95% confidence intervals. Asterisks denote significant results of pairwise contrasts (*, $p < 0.05$; ***, $p < 0.0001$).

The statistical model for the soleus peak-to-peak amplitudes of the EMG responses to transcutaneous spinal cord stimulation revealed a highly significant fixed effect of spine alignment condition on the response size ($F_{2;98} = 47.879$, $p < 0.0001$, $NG_p^2 = 0.494$; Figure 3B).

Post hoc tests revealed significantly smaller peak-to-peak amplitudes with flexed compared to the neutral ($t_{98} = 9.410$, $p < 0.0001$), as well as to the extended spine alignment conditions ($t_{98} = 7.335$, $p < 0.0001$). No differences were found between extended and neutral spine alignments ($t_{98} = 0.781$, $p = 0.7153$).

3.2. Influence of Spine Alignment Condition on Responses in Thigh and Leg Muscles Evoked by Transcutaneous Spinal Cord Stimulation

EMG responses of multiple lower-limb muscles to transcutaneous spinal cord stimulation elicited in an individual standing upright and with extension, neutral alignment, and flexion of the thoracolumbar spine are displayed in Figure 4. Stimulation applied in the spinal extension and neutral spine alignment conditions evoked PRM reflexes in hamstrings, tibialis anterior, and soleus, as demonstrated by the post-activation depression of the second responses when double-stimuli were applied. With flexion of the spine, however, response sizes diminished in these muscles, despite the unchanged stimulation amplitude. In rectus femoris, double stimuli evoked two responses of similar EMG peak-to-peak amplitudes, revealing direct M wave-like response-components elicited by anterior root stimulation [2,15]. Indeed, the shape of the EMG responses in rectus femoris to the first and second pulses of the double stimuli indicate mixed stimulation of upper lumbar posterior and anterior roots with spine extension and neutral spine alignments, but only anterior root (without posterior root) stimulation with spine flexion (Appendix A Figure A1). Direct anterior root recruitment can be the dominating effect of stimulation when applied in a sitting position (Figure A2).

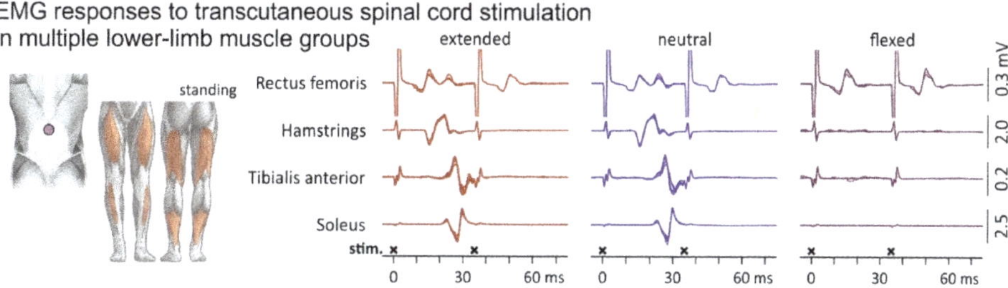

Figure 4. Influence of the spine alignment condition on lower-limb muscle responses evoked by transcutaneous spinal cord stimulation: exemplary results. Electromyographic (EMG) recordings from rectus femoris, hamstrings, tibialis anterior, and soleus. Superimposed representations of three responses to single stimuli (cropped at 35 ms) as well as three responses to double stimuli with an interstimulus interval of 35 ms. Black crosses mark times of stimulus application. All recordings derived from one subject while standing, with the spine in extended, neutral, and flexed alignments. Note that stimulation amplitudes were kept constant across spine alignment conditions.

The statistical model run for the responses to the single or first stimulation pulses of a pair revealed a significant fixed-effect of the spine alignment condition on the attainable EMG peak-to-peak amplitudes of responses across muscles ($F_{2;720} = 23.660$, $p < 0.0001$, $NG_p^2 = 0.062$) with smaller responses evoked with flexed than extended ($t_{720} = 5.691$, $p < 0.0001$) and neutral spine alignments ($t_{720} = 6.831$, $p < 0.0001$; Figure 5i). No differences in response size were found between extended and neutral spine alignments ($t_{720} = 1.257$, $p = 0.4200$). Across spine alignment conditions, the fixed effect of muscle was significant ($F_{3;720} = 187.233$, $p < 0.0001$, $NG_p^2 = 0.438$), indicating that response amplitudes differed between muscles (Figure 5ii). All post hoc pairwise comparisons were significant (all $p < 0.001$). EMG peak-to-peak amplitudes were smallest in rectus femoris and largest in soleus. Finally, the interaction between spine alignment condition and muscle was significant, ($F_{6;730} = 3.193$, $p = 0.0040$, $NG_p^2 = 0.026$), indicating that the influence of spine alignment on the stimulation differed between muscles (Figure 5iii). Planned pairwise contrasts revealed significantly lower EMG peak-to-peak amplitudes of the responses in hamstrings, tibialis anterior, and soleus when evoked with flexed compared to both extended as well as neutral spine alignments (all $p < 0.05$). No such differences were found in the case of rectus femoris (extended vs. flexed spine, $p = 0.9693$; neutral vs. flexed spine alignment, $p = 0.9762$).

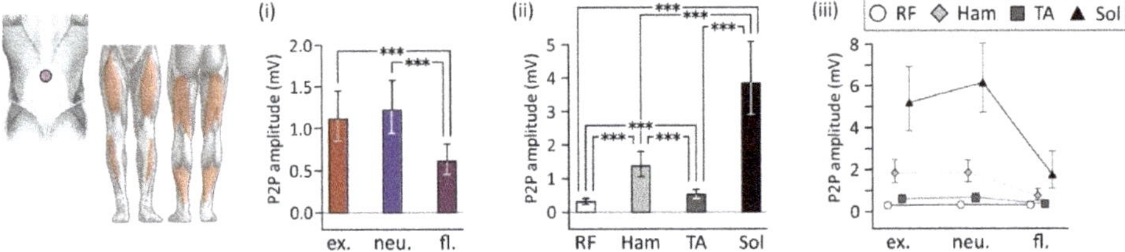

Figure 5. Influence of the spine alignment condition on lower-limb muscle responses evoked by transcutaneous spinal cord stimulation: group results. (i) Marginal mean peak-to-peak (P2P) amplitudes of electromyographic (EMG) responses evoked with extended (ex.), neutral (neu.), and flexed (fl.) spine alignment conditions, across body positions (supine, lateral recumbent, sitting, standing) and muscle groups (rectus femoris, RF; hamstrings; Ham; tibialis anterior, TA; soleus, Sol). Response sizes were significantly smaller with spinal flexion than with extension and with neutral spine alignment. (ii) Marginal mean P2P amplitudes of responses evoked in RF, Ham, TA, and Sol, across body positions and spine alignment conditions. (iii) P2P amplitudes of responses evoked in muscles and with spine alignment conditions. Whiskers extend from the lower to the upper limits of the respective 95% confidence intervals. Asterisks denote significant results of pairwise contrasts (***, $p < 0.0001$).

Double stimuli applied with an interstimulus interval of 35 ms provided essential information on the influence of spine alignment condition on the recruitment of afferent fibers in the posterior roots and potential concomitant activation of motor fibers in the anterior roots—larger normalized second-to-first response amplitudes without changes in voluntary descending inputs indicate an increased likelihood of motor fiber co-activation [1,2,19]. The statistical models run for the normalized second-to-first response amplitudes revealed a significant effect of the spine alignment condition across muscles ($F_{2;569} = 61.288$, $p < 0.0001$, $NG_p^2 = 0.177$), with significantly larger normalized second-to-first response amplitudes obtained with flexed than with extended ($t_{569} = 8.886$, $p < 0.0001$) or neutral spine alignments ($t_{569} = 10.551$, $p < 0.0001$; Figure 6i). No differences were found between extended and neutral spine alignments ($t_{569} = 1.539$, $p = 0.2736$). Across spine alignment conditions, the fixed effect of muscle was significant ($F_{3;569} = 125.374$, $p < 0.0001$, $NG_p^2 = 0.398$), with the largest normalized second-to-first response amplitudes found in rectus femoris and the smallest in soleus (Figure 6ii). All pairwise post hoc comparisons were significant (all $p < 0.0001$), except for the one between hamstrings and tibialis anterior ($p = 0.1631$). Finally, the interaction between spine alignment condition and muscle was significant ($F_{6;569} = 8.759$, $p = 0.0040$, $NG_p^2 = 0.085$), indicating that the effect of spine alignment on normalized second-to-first response amplitudes differed between muscles (Figure 6iii). Pairwise contrasts revealed significantly larger normalized second-to-first response amplitudes in hamstrings, tibialis anterior, and soleus when evoked with flexed compared to both extended as well as neutral spine alignments (all $p < 0.002$). No such differences were found in the case of rectus femoris (extended vs. flexed, $p = 0.3276$; neutral vs. flexed, $p = 0.2423$).

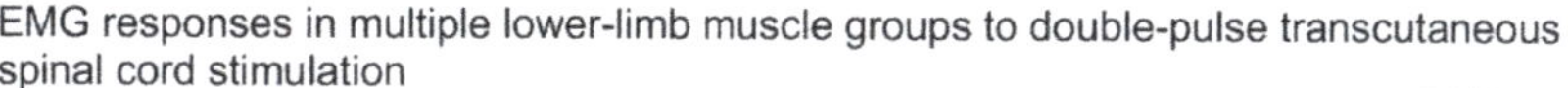

EMG responses in multiple lower-limb muscle groups to double-pulse transcutaneous spinal cord stimulation

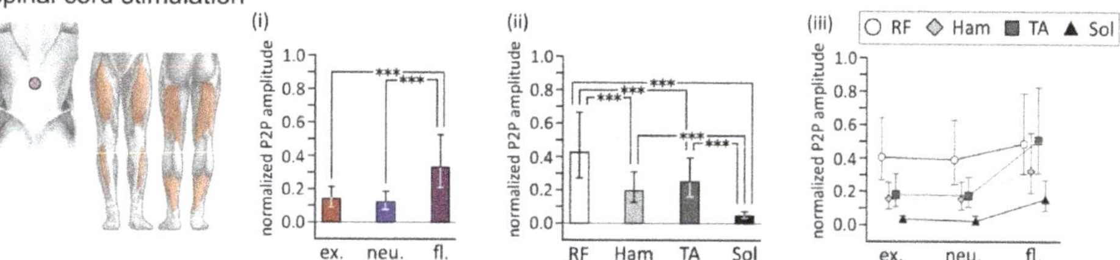

Figure 6. Influence of spine alignment condition on lower-limb muscle responses to double-pulse transcutaneous spinal cord stimulation: group results. (**i**) Marginal mean normalized peak-to-peak (P2P) amplitudes of second-to-first electromyographic (EMG) responses elicited with extended (ex.), neutral (neu.), and flexed (fl.) spine alignment, across body positions and muscle groups. Normalized response sizes were significantly larger with flexed than with extended or neutral spine alignments. (**ii**) Marginal mean normalized P2P amplitudes of responses elicited in rectus femoris (RF), hamstrings (Ham), tibialis anterior (TA), and soleus (Sol), across body positions and spine alignment conditions. (**iii**) Normalized P2P amplitudes of responses elicited in muscle groups and with spine alignments as indicated. Whiskers extend from the lower to the upper limits of the respective 95% confidence intervals. Asterisks denote significant results of pairwise contrasts (***, $p < 0.0001$).

4. Discussion

We demonstrated that the alignment of the spine in the sagittal plane affects the capacity of transcutaneous spinal cord stimulation to recruit posterior root afferents. While PRM reflexes could be elicited in neutral and extended spine alignment conditions, flexed spine alignment resulted in a strong reduction of the response amplitudes and in (co-)activation of motor fibers. The underlying mechanisms were of biophysical rather than neurophysiological nature as suggested by the lack of depression of the tibial nerve-evoked soleus H reflex in the same condition. To improve the generalizability of the results, we performed the study in four different body positions and controlled for position-specific effects in the statistical models.

Similarly to epidural electrical stimulation [30–33], the target neural structures of transcutaneous spinal cord stimulation are afferent fibers within the posterior rootlets of several spinal cord segments, as corroborated by computer simulations [15,17,18,34] and neurophysiological studies [1,6,35]. Additionally, computer simulations of transcutaneous spinal cord stimulation identified low-threshold sites of motor axons within the anterior roots at their exits from the vertebral canal in the intervertebral foramina, where they join the corresponding posterior roots to form the spinal nerves [15,17]. Stimulation at such sites would lead to a mixed activation of afferent and efferent axons.

Mixed activation of sensory posterior root and motor anterior root fibers would be disadvantageous for both major applications of transcutaneous spinal cord stimulation, i.e., neurophysiological and interventional studies. In neurophysiological studies, single-stimulus evoked PRM reflexes are utilized to probe the spinal sensorimotor circuits using specific conditioning-test paradigms [3,8,9,36,37]. The concomitant activation of anterior root efferents would lead to direct M wave-like responses superimposed on the EMG signals of the PRM reflexes, owing to their similar onset latencies [1,2,35]. In interventional studies using tonic transcutaneous spinal cord stimulation [10,11,13,38–43], electrical activation of anterior roots would bypass the target spinal circuits and generate continuous contractions of the respectively innervated lower-limb muscles.

We here identified different types of responses evoked by transcutaneous spinal cord stimulation based on their refractory behavior tested by double stimuli. PRM reflexes were characterized by their clear suppression when evoked 35 ms following a preceding activation (e.g., soleus in Figure 2B; hamstrings, tibialis anterior, and soleus in Figure 4) [1,4,6,44]. Other responses had major EMG components that could be evoked in close succession with little to no attenuation (e.g., rectus femoris in Figure 4; cf. Figure 4 in [45]). These responses likely reflected a mixed activation of motor axons in the anterior roots and afferent axons in the posterior roots (Figures A1 and A2) [19,35,46].

The present results suggest that spine alignment affects the preferential site of neural stimulation: in the extended and neutral spine alignment conditions, elicitation of PRM reflexes suggests activation of sensory afferent fibers in the posterior rootlets only; in the flexed spine alignment condition, the decreased suppression of the second response to the double stimuli suggests (co-)activation of motor and sensory fibers of the anterior and posterior roots in the intervertebral foramina. An explanation for this phenomenon could be that the flexed spine alignment condition profoundly alters current flow through the spine and consequently the current acting upon the different neural structures.

Computer simulations showed that the relatively low conductivity of the vertebrae [20] impede the conduction of electrical current across the spine and major current flow directions develop across the ligaments, the dural sac, the intervertebral discs, as well as the intervertebral foramina (Figures A3 and A4) [15,17].

The relative rotations and translations of the rigid vertebrae involved in the flexion and extension of the spine are associated with the deformation of their connecting intervertebral discs and longitudinal and intervertebral ligaments [21]. This results in changes of the volumes occupied by the soft tissues and fluid in the vertebral canal as well as the size of the intervertebral foramina: with sagittal flexion of the spine, the length of the vertebral canal increases because the axes of rotation shift anteriorly towards the intervertebral discs (Figure A5). In addition, all ligaments bordering the canal are stretched [21], the cross-sectional area of the spinal canal at the level of the intervertebral discs increases [47,48], and the vertical and horizontal dimensions of the intervertebral foramina are maximized (Figure A5B). Conversely, with spinal extension, there is a reduction of the cross-sectional area of the spinal canal caused by bulging of the intervertebral discs and an increase in thickness of the ligamenta flava, which are pushed anteriorly by the superior articular processes of the underlying vertebrae [47–49]. Additionally, the diameters of the intervertebral foramina decrease with spinal extension as the pedicles come closer together [48,50].

The increased volume of the vertebral canal and increased diameters of the intervertebral foramina in the flexed spine alignment condition could explain the associated reduction in EMG response sizes and increased second-to-first response ratios: stimulation of afferent fibers in the posterior rootlets might have been impeded by decreased current densities in the spinal canal and dural sac, and stimulation of motor fibers might have been augmented by increased channeling of current flow through the intervertebral foramina.

If this theory is true, then the sitting position is the least preferential body position for posterior root stimulation because the pelvis is rotated backwards, resulting in a flattening of the lumbar spine and a strong flexion bias compared to the other studied body positions [51,52]. Even the extended spine condition in sitting shows more relative flexion than the neutral spine condition in the standing and supine body positions [51]. Indeed, the example recording shown in Figure A2 indicates ubiquitous activation of motor fibers in the sitting position. Note that by modeling the body position as random effects, we controlled for such position-specific differences.

The capacity of transcutaneous spinal cord stimulation to activate posterior roots could have been additionally influenced by relative changes in paraspinal electrode location with respect to the spinal cord and roots due to movement of the skin or the spinal cord accompanying sagittal flexion of the spine. However, skin movement was shown to agree with the movement of the underlying spinous processes within about 10% difference [53], and the flexibility of the spinal cord allows it to largely follow the changes of length of the vertebral canal [21]. The influence of potential changes in trajectories and orientations of anterior and posterior roots within the generated electric field remains to be elucidated. The proposed underlying mechanisms remain hypothetical, and further investigations are needed. Specifically, computational studies could help delineate the impact of the various biophysical and anatomical consequences of sagittal spinal flexion on current flow and neural excitation. Detailed computer simulations including digital twin generation may lead to the development of adaptive stimulation methods for reliable afferent fiber stimulation independent from spine alignment conditions.

5. Conclusions

Transcutaneous spinal cord stimulation for neurophysiological studies as well as neuromodulative applications relies on the activation of sensory afferent fibers. Here, we showed that afferent fibers can be reliably stimulated with neutral and extended spine alignment conditions in various body positions. In contrast, sagittal flexion of the spine detrimentally impacted the activation of afferent fibers and could result in co-activation of efferent fibers. Thus, transcutaneous spinal cord stimulation should be applied in a body position that allows for stable extended or neutral sagittal spine alignment. Further, the capacity of transcutaneous spinal cord stimulation to recruit afferent fibers should be confirmed in the same body position and spine alignment condition prior to its intended scientific or interventional application.

Author Contributions: Conceptualization, S.M.D.; methodology, U.S.H., A.R., M.J.K. and K.M.; data curation, V.E.B., M.J.K. and S.M.D.; formal analysis, V.E.B. and S.M.D.; funding acquisition, U.S.H. and S.M.D.; investigation, V.E.B., A.R., M.J.K. and Z.S.; resources, U.S.H. and K.M.; data curation, V.E.B., M.J.K. and S.M.D.; writing—original draft preparation, V.E.B., U.S.H., K.M. and S.M.D.; writing—review and editing, V.E.B., U.S.H., M.J.K., A.R., Z.S., K.M. and S.M.D.; visualization, U.S.H., Z.S. and K.M.; supervision, A.R. and S.M.D.; project administration, S.M.D. All authors have read and agreed to the published version of the manuscript.

Funding: This research was partially funded by the Austrian Science Fund (FWF), grant number I 3837-B34, and by the National Institutes of Health (NIH) grant numbers R01 NS115900 and R01 NS112304. Open access funding by the Austrian Science Fund (FWF), grant number I 3837-B34.

Institutional Review Board Statement: The study was approved by the institutional review board of the Medical University of Vienna, Austria (EK 1686/2014).

Informed Consent Statement: All participants signed written informed consent prior to their enrollment into the study.

Data Availability Statement: The data of this study are available in the manuscript.

Conflicts of Interest: The authors declare no conflict of interest.

Appendix A

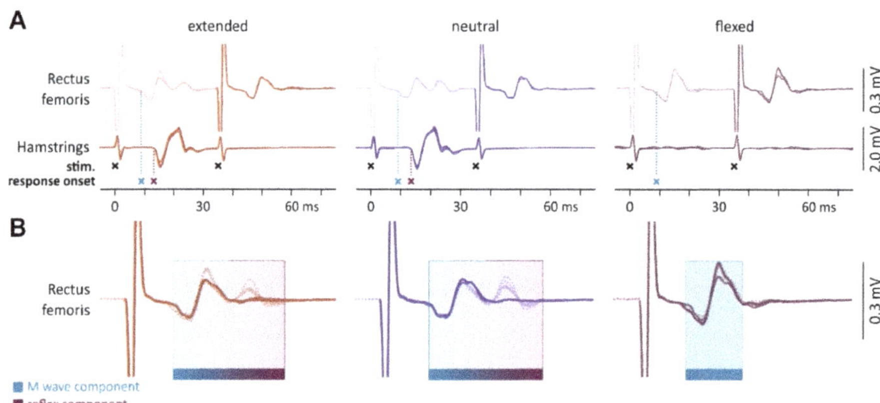

Figure A1. Electromyographic (EMG) waveforms of responses to double-pulse transcutaneous spinal cord stimulation reveal posterior and anterior root stimulation in rectus femoris. (**A**) Hamstrings responses in spine extension and neutral spine alignment conditions demonstrate complete post-stimulation depression when evoked with an interstimulus interval of 35 ms, a hallmark behavior of PRM reflexes [1–4]. Black crosses mark times of stimulus application. In the same spine alignment conditions, EMG waveforms of the rectus femoris show a more complex behavior. Stimulus-triggered superposition of the first and second responses shown in (**B**) reveal a complete suppression of the late potentials of the EMG waveform without suppression of the early potentials. These early EMG potentials reflect the direct electrical stimulation of motor axons in the anterior roots [2,35,54,55]. With spinal flexion, no responses were evoked in hamstrings and no late EMG components were evoked in rectus femoris. In rectus femoris, the early EMG component was consistently evoked as seen in the other spine alignment conditions. Thus, spinal flexion impeded posterior root but not anterior root stimulation. Furthermore, the onset latency of rectus femoris responses were considerably shorter than those of hamstrings (blue and purple crosses in A). The differences in susceptibility to changes in spine alignment and the different latencies support the theory proposed by computer simulations of different anatomical sites of posterior and anterior root fiber stimulation, i.e., along the most proximal portions of the posterior rootlets within the dural sac and the most distal portions of the anterior root fibers within the intervertebral foramina, respectively [15–17].

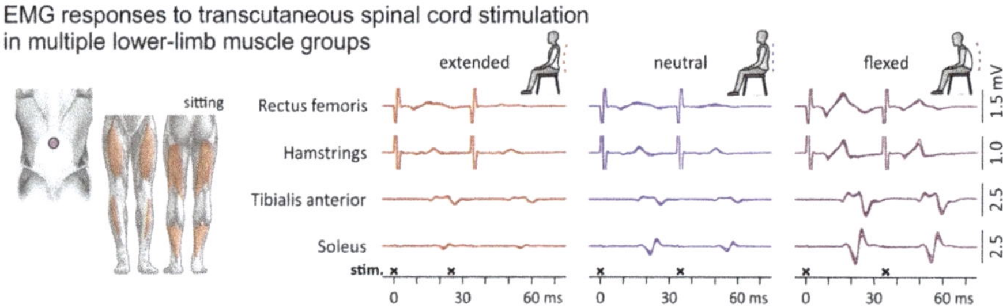

Figure A2. Influence of the spine alignment condition on electromyographic (EMG) responses evoked by transcutaneous spinal cord stimulation in multiple lower-limb muscles: exemplary results in the sitting position. Superimposed representations of three responses to single stimuli (cropped at 35 ms) as well as three responses to double stimuli with an interstimulus interval of 35 ms in rectus femoris, hamstrings, tibialis anterior, and soleus. Black crosses mark times of stimulus application. All recordings derived from one subject while sitting, with the spine in extended, neutral, and flexed conditions. Note that stimulation amplitudes were kept constant across spine alignment conditions. With double-stimuli, the second responses evoked at an interval of 35 ms following the first responses demonstrate little to no suppression across muscles and spine alignment conditions. Thus, the detected responses evoked by transcutaneous spinal cord stimulation in the sitting position were largely direct muscle responses evoked in the anterior roots, bypassing the spinal cord.

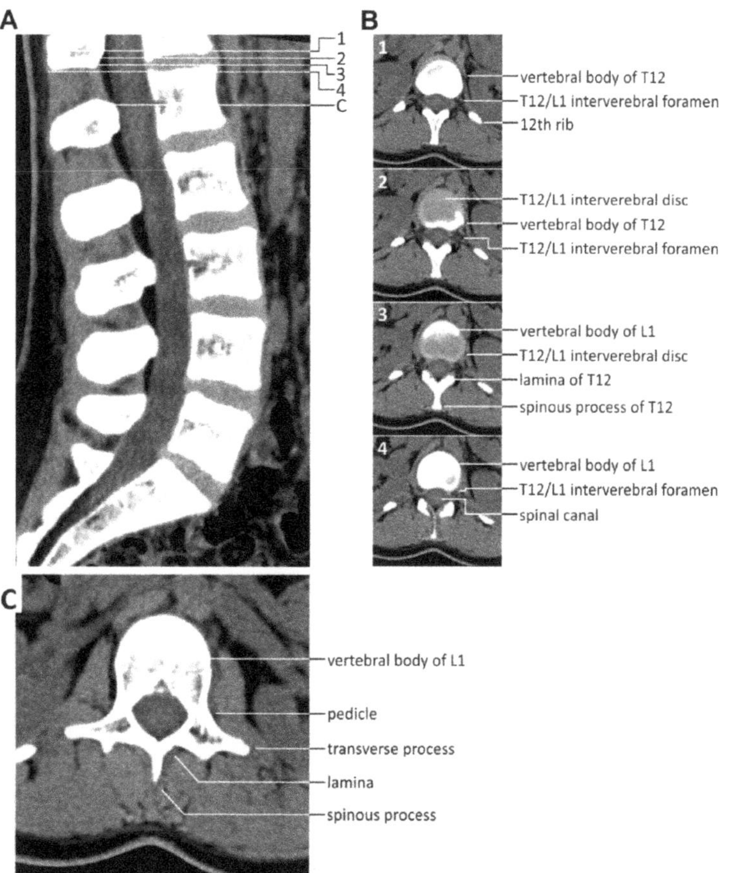

Figure A3. Vertebrae and soft tissues of the spine. (**A**) Sagittal CT scan of the normal spine demonstrating the substantial volumes in-between the vertebrae T12 to L5 that are occupied by soft tissues and liquids, all of which have better electrical conductivities than bone. (**B**) Axial CT scans of the thoracolumbar junction as indicated (1–4) showing the spinal canal, the T12/L1 intervertebral foramen, and the T12/L1 intervertebral disc. (**C**) Bony spinal canal at the level of the L1 pedicles. Case courtesy of Dr Bruno Di Muzio, Radiopaedia.org, rID: 46533.

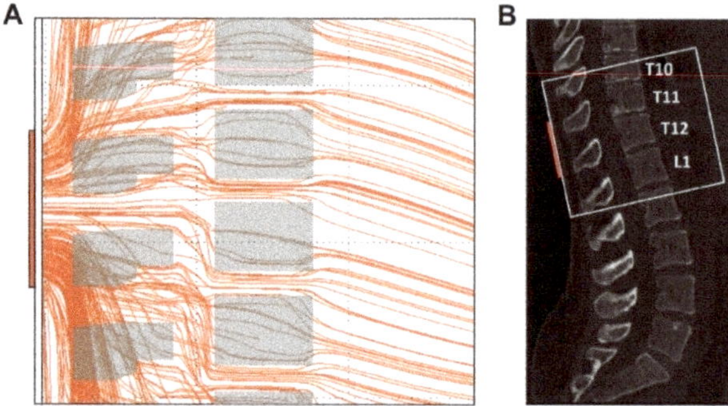

Figure A4. Current flow through the spine generated by transcutaneous spinal cord stimulation with the paraspinal electrode placed over the T11 and T12 spinal processes and reference electrodes over the lower abdomen (not shown). (**A**) Computer simulation of the current flow in the sagittal plane at the level of the stimulating paraspinal electrode. Shown are the results of a section of the three-dimensional model in a midsagittal layer with a thickness of 2 mm. Various anatomical structures with tissue-specific electrical conductivities were considered in the model, including paraspinal muscles, vertebral bones, the vertebral canal with the dural sac containing the spinal cord, and intervertebral discs. Only the mid-sagittal cross-sectional areas of the vertebrae are displayed. The current largely crosses the spine via the soft tissues in-between the vertebrae. The density of the lines is proportional to the local current densities. Scaling: the paraspinal electrode (vertical red rectangle on the far left side) represents 5 cm. Computer simulation model from [15,17]. (**B**) Midsagittal bone window CT scan, case courtesy of Assoc Prof Frank Gaillard, Radiopaedia.org, rID: 43822. The white box indicates the approximate anatomic region shown in (**A**).

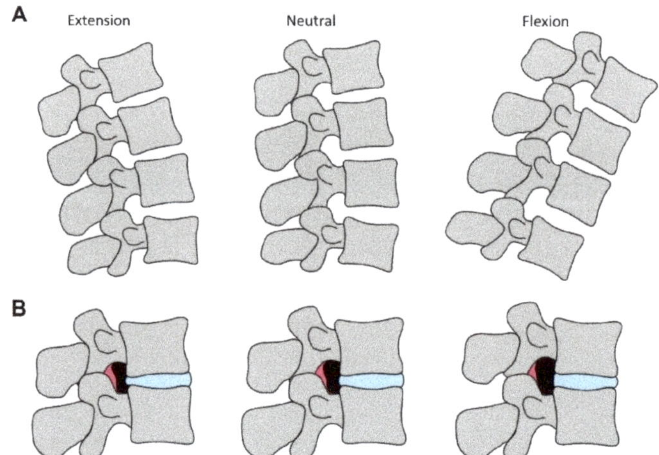

Figure A5. Anatomical changes with flexed, neutral, and extended spine alignment conditions. (**A**) Sketches depict changes in vertebral alignment when the spine goes from extension to a neutral alignment and to flexion. During flexion, the spinal canal is lengthened because of the location of the rotation axis anterior to the canal, leading to increased distances between spinous processes as well as between pedicles of neighboring vertebrae. (**B**) Sketches depict changes in vertical and horizontal diameters of the intervertebral foramina (black areas) with different spine alignment conditions. With a change from extension to flexion of the spine, the intervertebral foramina where the afferent posterior and efferent anterior roots join to exit the spine are subjected to geometrical changes due to relative movements of adjacent pedicles and soft tissue bulge (magenta, ligamentum flavum; blue, intervertebral disc).

References

1. Minassian, K.; Persy, I.; Rattay, F.; Dimitrijevic, M.R.; Hofer, C.; Kern, H. Posterior root-muscle reflexes elicited by transcutaneous stimulation of the human lumbosacral cord. *Muscle Nerve* **2007**, *35*, 327–336. [CrossRef]
2. Minassian, K.; Freundl, B.; Hofstoetter, U.S. The posterior root-muscle reflex. In *Neurophysiology in Neurosurgery*; Elsevier: Amsterdam, The Netherlands, 2020; pp. 239–253.
3. Hofstoetter, U.S.; Freundl, B.; Binder, H.; Minassian, K. Recovery cycles of posterior root-muscle reflexes evoked by transcutaneous spinal cord stimulation and of the H reflex in individuals with intact and injured spinal cord. *PLoS ONE* **2019**, *14*, e0227057. [CrossRef] [PubMed]
4. Andrews, J.C.; Stein, R.B.; Roy, F.D. Post-activation depression in the human soleus muscle using peripheral nerve and transcutaneous spinal stimulation. *Neurosci. Lett.* **2015**, *589*, 144–149. [CrossRef] [PubMed]
5. Kitano, K.; Koceja, D.M. Spinal reflex in human lower leg muscles evoked by transcutaneous spinal cord stimulation. *J. Neurosci. Methods* **2009**, *180*, 111–115. [CrossRef]
6. Hofstoetter, U.S.; Freundl, B.; Binder, H.; Minassian, K. Common neural structures activated by epidural and transcutaneous lumbar spinal cord stimulation: Elicitation of posterior root-muscle reflexes. *PLoS ONE* **2018**, *13*, e0192013. [CrossRef] [PubMed]
7. Dy, C.J.; Gerasimenko, Y.P.; Edgerton, V.R.; Dyhre-Poulsen, P.; Courtine, G.; Harkema, S.J. Phase-dependent modulation of percutaneously elicited multisegmental muscle responses after spinal cord injury. *J. Neurophysiol.* **2010**, *103*, 2808–2820. [CrossRef]
8. Saito, A.; Nakagawa, K.; Masugi, Y.; Nakazawa, K. Inter-muscle differences in modulation of motor evoked potentials and posterior root-muscle reflexes evoked from lower-limb muscles during agonist and antagonist muscle contractions. *Exp. Brain Res.* **2021**, *239*, 463–474. [CrossRef] [PubMed]
9. Hofstoetter, U.S.; Minassian, K.; Hofer, C.; Mayr, W.; Rattay, F.; Dimitrijevic, M.R. Modification of reflex responses to lumbar posterior root stimulation by motor tasks in healthy subjects. *Artif. Organs* **2008**, *32*, 644–648. [CrossRef] [PubMed]
10. Minassian, K.; Hofstoetter, U.S.; Danner, S.M.; Mayr, W.; Bruce, J.A.; McKay, W.B.; Tansey, K.E. Spinal Rhythm Generation by Step-Induced Feedback and Transcutaneous Posterior Root Stimulation in Complete Spinal Cord-Injured Individuals. *Neurorehabil. Neural Repair* **2016**, *30*, 233–243. [CrossRef] [PubMed]
11. Estes, S.; Zarkou, A.; Hope, J.M.; Suri, C.; Field-Fote, E.C. Combined Transcutaneous Spinal Stimulation and Locomotor Training to Improve Walking Function and Reduce Spasticity in Subacute Spinal Cord Injury: A Randomized Study of Clinical Feasibility and Efficacy. *J. Clin. Med.* **2021**, *10*, 1167. [CrossRef]
12. Hofstoetter, U.S.; Freundl, B.; Danner, S.M.; Krenn, M.J.; Mayr, W.; Binder, H.; Minassian, K. Transcutaneous Spinal Cord Stimulation Induces Temporary Attenuation of Spasticity in Individuals with Spinal Cord Injury. *J. Neurotrauma* **2020**, *37*, 481–493. [CrossRef]
13. Hofstoetter, U.S.; Freundl, B.; Lackner, P.; Binder, H. Transcutaneous Spinal Cord Stimulation Enhances Walking Performance and Reduces Spasticity in Individuals with Multiple Sclerosis. *Brain Sci.* **2021**, *11*, 472. [CrossRef]
14. Roberts, B.W.R.; Atkinson, D.A.; Manson, G.A.; Markley, R.; Kaldis, T.; Britz, G.W.; Horner, P.J.; Vette, A.H.; Sayenko, D.G. Transcutaneous spinal cord stimulation improves postural stability in individuals with multiple sclerosis. *Mult. Scler. Relat. Disord.* **2021**, *52*, 103009. [CrossRef]
15. Ladenbauer, J.; Minassian, K.; Hofstoetter, U.S.; Dimitrijevic, M.R.; Rattay, F. Stimulation of the human lumbar spinal cord with implanted and surface electrodes: A computer simulation study. *IEEE Trans. Neural Syst. Rehabil. Eng.* **2010**, *18*, 637–645. [CrossRef] [PubMed]
16. Danner, S.M.; Hofstoetter, U.S.; Ladenbauer, J.; Rattay, F.; Minassian, K. Can the human lumbar posterior columns be stimulated by transcutaneous spinal cord stimulation? A modeling study. *Artif. Organs* **2011**, *35*, 257–262. [CrossRef]
17. Száva, Z.; Danner, S.M.; Minassian, K. *Transcutaneous Electrical Spinal Cord Stimulation: Biophysics of a New Rehabilitation Method after Spinal Cord Injury*; VDM Verlag Dr. Müller: Saarbrücken, Germany, 2011.
18. Hofstoetter, U.S.; Danner, S.M.; Minassian, K. Paraspinal Magnetic and Transcutaneous Electrical Stimulation. In *Encyclopedia of Computational Neuroscience*; Jaeger, D., Jung, R., Eds.; Springer: New York, NY, USA, 2015; pp. 2194–2212.
19. Danner, S.M.; Krenn, M.; Hofstoetter, U.S.; Toth, A.; Mayr, W.; Minassian, K. Body Position Influences Which Neural Structures Are Recruited by Lumbar Transcutaneous Spinal Cord Stimulation. *PLoS ONE* **2016**, *11*, e0147479. [CrossRef] [PubMed]
20. Geddes, L.A.; Baker, L.E. The specific resistance of biological material—A compendium of data for the biomedical engineer and physiologist. *Med. Biol. Eng.* **1967**, *5*, 271–293. [CrossRef]
21. White, A.A.; Panjabi, M.M. *Clinical Biomechanics of the Spine*; JB Lippincott: Philadelphia, PA, USA, 1990.
22. Sherwood, A.M.; McKay, W.B.; Dimitrijević, M.R. Motor control after spinal cord injury: Assessment using surface EMG. *Muscle Nerve* **1996**, *19*, 966–979. [CrossRef]
23. Pierrot-Deseilligny, E.; Mazevet, D. The monosynaptic reflex: A tool to investigate motor control in humans. Interest and limits. *Clin. Neurophysiol.* **2000**, *30*, 67–80. [CrossRef]
24. Burke, D. Clinical uses of H reflexes of upper and lower limb muscles. *Clin. Neurophysiol. Pract.* **2016**, *1*, 9–17. [CrossRef]
25. R Core Team. *R: A Language and Environment for Statistical Computing*; R Foundation for Statistical Computing: Vienna, Austria, 2020.
26. Kristensen, K.; Nielsen, A.; Berg, C.; Skaug, H.; Bell, B. TMB: Automatic Differentiation and Laplace Approximation. *J. Stat. Softw.* **2016**, *70*, 1–21. [CrossRef]

27. Thygesen, U.; Albertsen, C.; Berg, C.; Kristensen, K.; Nielsen, A. Validation of Ecological State Space Models Using the Laplace Approximation. *Environ. Ecol. Stat.* **2017**, *24*, 317–339. [CrossRef]
28. Brooks, M.E.; Kristensen, K.; Benthem, K.J.V.; Magnusson, A.; Berg, C.W.; Nielsen, A.; Skaug, H.J.; Mächler, M.; Bolker, B.M. glmmTMB Balances Speed and Flexibility Among Packages for Zero-inflated Generalized Linear Mixed Modeling. *R Journal* **2017**, *9*, 378–400. [CrossRef]
29. Holm, S. A Simple Sequentially Rejective Multiple Test Procedure. *Scand. J. Stat.* **1979**, *6*, 65–70.
30. Rattay, F.; Minassian, K.; Dimitrijevic, M.R. Epidural electrical stimulation of posterior structures of the human lumbosacral cord: 2. quantitative analysis by computer modeling. *Spinal Cord* **2000**, *38*, 473–489. [CrossRef] [PubMed]
31. Formento, E.; Minassian, K.; Wagner, F.; Mignardot, J.B.; Le Goff-Mignardot, C.G.; Rowald, A.; Bloch, J.; Micera, S.; Capogrosso, M.; Courtine, G. Electrical spinal cord stimulation must preserve proprioception to enable locomotion in humans with spinal cord injury. *Nat. Neurosci.* **2018**, *21*, 1728–1741. [CrossRef] [PubMed]
32. Capogrosso, M.; Wenger, N.; Raspopovic, S.; Musienko, P.; Beauparlant, J.; Bassi Luciani, L.; Courtine, G.; Micera, S. A computational model for epidural electrical stimulation of spinal sensorimotor circuits. *J. Neurosci. Off. J. Soc. Neurosci.* **2013**, *33*, 19326–19340. [CrossRef]
33. Minassian, K.; Jilge, B.; Rattay, F.; Pinter, M.M.; Binder, H.; Gerstenbrand, F.; Dimitrijevic, M.R. Stepping-like movements in humans with complete spinal cord injury induced by epidural stimulation of the lumbar cord: Electromyographic study of compound muscle action potentials. *Spinal Cord* **2004**, *42*, 401–416. [CrossRef]
34. Danner, S.M.; Hofstoetter, U.S.; Minassian, K. Finite Element Models of Transcutaneous Spinal Cord Stimulation. In *Encyclopedia of Computational Neuroscience*; Jaeger, D., Jung, R., Eds.; Springer: New York, NY, USA, 2015; pp. 1197–1202.
35. Zhu, Y.; Starr, A.; Haldeman, S.; Chu, J.K.; Sugerman, R.A. Soleus H-reflex to S1 nerve root stimulation. *Electroencephalogr. Clin. Neurophysiol.* **1998**, *109*, 10–14. [CrossRef]
36. Calvert, J.S.; Gill, M.L.; Linde, M.B.; Veith, D.D.; Thoreson, A.R.; Lopez, C.; Lee, K.H.; Gerasimenko, Y.P.; Edgerton, V.R.; Lavrov, I.A.; et al. Voluntary Modulation of Evoked Responses Generated by Epidural and Transcutaneous Spinal Stimulation in Humans with Spinal Cord Injury. *J. Clin. Med.* **2021**, *10*, 4898. [CrossRef]
37. Wecht, J.; Savage, W.; Famodimu, G.; Mendez, G.; Levine, J.; Maher, M.; Weir, J.; Wecht, J.; Carmel, J.; Wu, Y.; et al. Posteroanterior Cervical Transcutaneous Spinal Cord Stimulation: Interactions with Cortical and Peripheral Nerve Stimulation. *J. Clin. Med.* **2021**, *10*, 5304. [CrossRef]
38. Hofstoetter, U.S.; Perret, I.; Bayart, A.; Lackner, P.; Binder, H.; Freundl, B.; Minassian, K. Spinal motor mapping by epidural stimulation of lumbosacral posterior roots in humans. *iScience* **2020**, *24*, 101930. [CrossRef]
39. Meyer, C.; Hofstoetter, U.S.; Hubli, M.; Hassani, R.H.; Rinaldo, C.; Curt, A.; Bolliger, M. Immediate Effects of Transcutaneous Spinal Cord Stimulation on Motor Function in Chronic, Sensorimotor Incomplete Spinal Cord Injury. *J. Clin. Med.* **2020**, *9*, 3541. [CrossRef]
40. Estes, S.P.; Iddings, J.A.; Field-Fote, E.C. Priming Neural Circuits to Modulate Spinal Reflex Excitability. *Front. Neurol.* **2017**, *8*, 17. [CrossRef]
41. Al'joboori, Y.; Massey, S.J.; Knight, S.L.; Donaldson, N.d.N.; Duffell, L.D. The Effects of Adding Transcutaneous Spinal Cord Stimulation (tSCS) to Sit-To-Stand Training in People with Spinal Cord Injury: A Pilot Study. *J. Clin. Med.* **2020**, *9*, 2765. [CrossRef] [PubMed]
42. Kumru, H.; Flores, Á.; Rodríguez-Cañón, M.; Edgerton, V.R.; García, L.; Benito-Penalva, J.; Navarro, X.; Gerasimenko, Y.; García-Alías, G.; Vidal, J. Cervical Electrical Neuromodulation Effectively Enhances Hand Motor Output in Healthy Subjects by Engaging a Use-Dependent Intervention. *J. Clin. Med.* **2021**, *10*, 195. [CrossRef]
43. Sandler, E.B.; Condon, K.; Field-Fote, E.C. Efficacy of Transcutaneous Spinal Stimulation versus Whole Body Vibration for Spasticity Reduction in Persons with Spinal Cord Injury. *J. Clin. Med.* **2021**, *10*, 3267. [CrossRef] [PubMed]
44. Kagamihara, Y.; Hayashi, A.; Okuma, Y.; Nagaoka, M.; Nakajima, Y.; Tanaka, R. Reassessment of H-reflex recovery curve using the double stimulation procedure. *Muscle Nerve* **1998**, *21*, 352–360. [CrossRef]
45. Halter, J.; Dolenc, V.; Dimitrijevic, M.R.; Sharkey, P.C. Neurophysiological assessment of electrode placement in the spinal cord. *Appl. Neurophysiol.* **1983**, *46*, 124–128. [CrossRef]
46. Minassian, K.; Hofstoetter, U.S.; Rattay, F. Transcutaneous lumbar posterior root stimulation for motor control studies and modification of motor activity after spinal cord injury. In *Restorative Neurology of Spinal Cord Injury*; Dimitrijevic, M., Kakulas, B., McKay, W., Vrbova, G., Eds.; Oxford University Press: New York, NY, USA, 2011; pp. 226–255.
47. Schönström, N.; Lindahl, S.; Willén, J.; Hansson, T. Dynamic changes in the dimensions of the lumbar spinal canal: An experimental study in vitro. *J. Orthop. Res.* **1989**, *7*, 115–121. [CrossRef]
48. Schmid, M.R.; Stucki, G.; Duewell, S.; Wildermuth, S.; Romanowski, B.; Hodler, J. Changes in cross-sectional measurements of the spinal canal and intervertebral foramina as a function of body position: In vivo studies on an open-configuration MR system. *Am. J. Roentgenol.* **1999**, *172*, 1095–1102. [CrossRef] [PubMed]
49. Weishaupt, D.; Boxheimer, L. Magnetic Resonance Imaging of the Weight-Bearing Spine. *Semin. Musculoskelet. Radiol.* **2003**, *7*, 277–286.
50. Mayoux-Benhamou, M.A.; Revel, M.; Aaron, C.; Chomette, G.; Amor, B. A morphometric study of the lumbar foramen. *Surg. Radiol. Anat.* **1989**, *11*, 97–102. [CrossRef] [PubMed]

51. Hirasawa, Y.; Bashir, W.A.; Smith, F.W.; Magnusson, M.L.; Pope, M.H.; Takahashi, K. Postural Changes of the Dural Sac in the Lumbar Spines of Asymptomatic Individuals Using Positional Stand-Up Magnetic Resonance Imaging. *Spine* **2007**, *32*, E136–E140. [CrossRef]
52. Hey, H.W.D.; Lau, E.T.-C.; Tan, K.-A.; Lim, J.L.; Choong, D.; Lau, L.-L.; Liu, K.-P.G.; Wong, H.-K. Lumbar Spine Alignment in Six Common Postures. *Spine (Phila. PA, 1976).* **2017**, *42*, 1447–1455. [CrossRef]
53. Hindle, R.J.; Pearcy, M.J.; Cross, A.T.; Miller, D.H. Three-dimensional kinematics of the human back. *Clin. Biomech.* **1990**, *5*, 218–228. [CrossRef]
54. Dimitrijevic, M.R.; Faganel, J.; Sherwood, A.M. Spinal cord stimulation as a tool for physiological research. *Appl. Neurophysiol.* **1983**, *46*, 245–253. [CrossRef] [PubMed]
55. Minassian, K.; Persy, I.; Rattay, F.; Pinter, M.M.; Kern, H.; Dimitrijevic, M.R. Human lumbar cord circuitries can be activated by extrinsic tonic input to generate locomotor-like activity. *Hum. Mov. Sci.* **2007**, *26*, 275–295. [CrossRef]

Journal of
Clinical Medicine

Article

Algorithms for Automated Calibration of Transcutaneous Spinal Cord Stimulation to Facilitate Clinical Applications

Christina Salchow-Hömmen [1,†], Thomas Schauer [2,*,†], Philipp Müller [2], Andrea A. Kühn [1], Ursula S. Hofstoetter [3,†] and Nikolaus Wenger [1,†]

1   Department of Neurology, Charité–Universitätsmedizin Berlin, 10117 Berlin, Germany; christina.salchow@charite.de (C.S.-H.); andrea.kuehn@charite.de (A.A.K.); nikolaus.wenger@charite.de (N.W.)
2   Control Systems Group, Technische Universität Berlin, 10587 Berlin, Germany; mueller@control.tu-berlin.de
3   Center for Medical Physics and Biomedical Engineering, Medical University of Vienna, 1090 Vienna, Austria; ursula.hofstoetter@meduniwien.ac.at
*   Correspondence: schauer@control.tu-berlin.de; Tel.: +49-(0)30-314-24404
†   These authors contributed equally to this work.

Citation: Salchow-Hömmen, C.; Schauer, T.; Müller, P.; Kühn, A.A.; Hofstoetter, U.S.; Wenger, N. Algorithms for Automated Calibration of Transcutaneous Spinal Cord Stimulation to Facilitate Clinical Applications. *J. Clin. Med.* **2021**, *10*, 5464. https://doi.org/10.3390/jcm10225464

Academic Editor: Mark A. Korsten

Received: 11 October 2021
Accepted: 13 November 2021
Published: 22 November 2021

Publisher's Note: MDPI stays neutral with regard to jurisdictional claims in published maps and institutional affiliations.

Abstract: Transcutaneous spinal cord stimulation (tSCS) is a promising intervention that can benefit spasticity control and augment voluntary movement in spinal cord injury (SCI) and multiple sclerosis. Current applications require expert knowledge and rely on the thorough visual analysis of electromyographic (EMG) responses from lower-limb muscles to optimize attainable treatment effects. Here, we devised an automated tSCS setup by combining an electrode array placed over low-thoracic to mid-lumbar vertebrae, synchronized EMG recordings, and a self-operating stimulation protocol to systematically test various stimulation sites and amplitudes. A built-in calibration procedure classifies the evoked responses as reflexes or direct motor responses and identifies stimulation thresholds as recommendations for tSCS therapy. We tested our setup in 15 individuals (five neurologically intact, five SCI, and five Parkinson's disease) and validated the results against blinded ratings from two clinical experts. Congruent results were obtained in 13 cases for electrode positions and in eight for tSCS amplitudes, with deviations of a maximum of one position and 5 to 10 mA in amplitude in the remaining cases. Despite these minor deviations, the calibration found clinically suitable tSCS settings in 13 individuals. In the remaining two cases, the automatic setup and both experts agreed that no reflex responses could be detected. The presented technological developments may facilitate the dissemination of tSCS into non-academic environments and broaden its use for diagnostic and therapeutic purposes.

Keywords: automation; electromyography; noninvasive; Parkinson's disease; posterior root-muscle reflexes; spasticity; spinal cord injury; spinal cord stimulation; transcutaneous

1. Introduction

Epidural electrical stimulation (EES) of the lumbar spinal cord has recently experienced a surge of interest because of its potential to restore voluntary control of locomotion in individuals after severe spinal cord injury (SCI) [1–3]. Through the recruitment of large-to-medium diameter proprioceptive and cutaneous afferents within lumbar and upper sacral posterior roots [4,5], EES can facilitate the alleviation of severe lower limb spasticity [6] and the generation or augmentation of rhythmic and locomotor-like lower limb activity in otherwise paralyzed legs in individuals with SCI [1,4,7,8]. Moreover, in other neurological disorders, treatment effects of EES are under active investigation. For example, in advanced Parkinson's disease (PD), EES was shown to ameliorate motor symptoms such as impaired gait function and postural stability [9–12], yet a recent study could not reproduce these outcomes [13]. A general problem of EES is the lack of clinical or physiological markers for identifying treatment responders in advance [14].

The target neural structures of lumbar EES [4,5] can also be recruited noninvasively by using transcutaneous spinal cord stimulation (tSCS) [15–17]. Transcutaneous SCS uses surface electrodes placed on the paravertebral and abdominal skin to generate a current flow through the lower trunk, partially crossing the dural sac [5,17] (see Figure 1). Independent studies have shown the efficacy of tSCS to ameliorate spasticity and augment voluntary motor control, including locomotion in individuals with SCI [18–24] as well as multiple sclerosis [25,26]. As a clinically accessible and noninvasive approach, tSCS was suggested to hold the potential to develop into a widely used neurorehabilitation technique and to serve as a screening tool to estimate individually attainable therapeutic outcomes of EES [23].

The application of tSCS has so far been restricted to specialized research centers, in part owing to the required expert knowledge and constraints in clinical time management. For example, a necessary prerequisite to induce neuromodulatory effects in the lower limbs by EES or tSCS is the specific placement of the epidural or surface electrodes, respectively, so as to overlie the spinal cord segments innervating lower extremity muscles [6,8,17,20,23,27]. Such placement can be validated via the electromyographic (EMG) recording of evoked responses, i.e., short-latency reflexes initiated within the posterior roots, so-called posterior root-muscle (PRM) reflexes [1,7,8,16,18,22,28–31]. These reflexes are generally thought to result from the activation of proprioceptive fibers within the posterior roots that cause synchronized responses of motoneurons in the spinal cord [32,33]. With respect to tSCS specifically, the placement of the paravertebral stimulating electrode typically follows a multi-step procedure. First, after rough palpation of the spinal column, a self-adhesive surface electrode is placed over the spine at a level estimated to correspond to the T11/T12 spinous processes [29]. Single stimulation pulses are then applied at incremental amplitudes with the aim to elicit PRM reflexes in L2–S2 innervated lower limb muscles [16,29]. In order to achieve this aim, the electrode may have to be relocated by several centimeters in either the rostral or caudal direction, and with each new position, the procedure of single stimulation pulse application is repeated. With the electrode placement eventually designated for therapy, double stimuli are applied, normally with interstimulus intervals of 30–100 ms, to test for the presence of post-stimulation attenuation of the evoked responses and, hence, identify them as reflexes [18,29]. In contrast, post-stimulation attenuation is not observed for EMG responses that result from the direct stimulation of efferent motor pathways (M-waves) in the anterior roots or peripheral nerves [18,29]. Finally, reflex response thresholds are determined based on available EMG recordings. In tSCS applications for spasticity control, the stimulation amplitude is then set to approximately 90% of the lowest PRM reflex threshold [18,23,25]. This time-consuming procedure likely imposes an impediment to the wider use of tSCS in real-world clinical practice. Furthermore, stimulation devices normally used in neurorehabilitation do not allow for the application of single and double stimuli, let alone for the synchronous recording of EMG activity from several muscles.

The aim of the present study was to cast this expert knowledge into an automated tSCS setup that identifies appropriate stimulation sites over the lumbosacral spinal cord and determines clinically suitable stimulation amplitudes for the application of tSCS in spasticity control [18,20,23,25]. To this end, we combined an electrode-array configuration, a self-operating stimulation protocol, synchronous recordings of EMG responses from several lower-limb muscles, and online evaluation algorithms. We tested our approach in neurologically intact individuals as well as in individuals with SCI and PD. The accuracies of the automated results were validated by two independent experts. To ease clinical decision making, we propose a graphical user interface that indicates optimal stimulation configurations with an intuitive color code. Our results may aid the dissemination of tSCS technologies into non-academic environments and broaden the use of tSCS for both diagnostic and therapeutic applications in the near future.

2. Materials and Methods

2.1. Participants

The automated tSCS setup was tested in five neurologically intact volunteers (mean age ± SD, 32.8 ± 2.77 years), five individuals with chronic SCI (47.0 ± 11.51 years), and five individuals with idiopathic PD (70.2 ± 6.42 years; Table 1). Among the exclusion criteria were any active implants or passive implants at vertebral level T8 or caudally. All procedures were approved by the Ethics Committee of the Berlin Chamber of Physicians (ETH-28/17; 2017 and 2019; SCI) and the Ethics Committee of the Charité–Universitätsmedizin Berlin (EA2/118/18, 2018; PD and healthy controls). All participants signed written informed consent forms prior to their enrollment into the study.

Table 1. Clinical characteristics of study participants.

Group	Subject	Sex	Age	Height	Weight	SCI		PD	
			Years	cm	kg	Level	AIS	H and Y	FD
Control	C1	m	36	184	82	-	-	-	-
(*n* = 5)	C2	f	30	177	65	-	-	-	-
	C3	m	35	190	78	-	-	-	-
	C4	m	33	181	70	-	-	-	-
	C5	f	30	168	56	-	-	-	-
SCI	S1	m	40	182	80	T5-6	A	-	-
(*n* = 5)	S2	m	57	180	74	T5-6	A	-	-
	S3	f	49	179	76	T5-6	B	-	-
	S4	m	58	185	75	C7	B	-	-
	S5	f	31	165	49	T1-3	B	-	-
PD	P1	m	60	180	70	-	-	2	12
(*n* = 5)	P2	f	77	175	68	-	-	2	9
	P3	f	74	172	80	-	-	2	5
	P4	m	70	176	92	-	-	2	14
	P5	m	70	178	72	-	-	2	2

AIS, American Spinal Cord Injury Association Impairment Scale; *FD*, years since first diagnosis; *H and Y*, Hoehn and Yahr disease severity classification [34]; *level*, neurological level of spinal cord injury; *PD*, Parkinson's disease; *SCI*, spinal cord injury.

2.2. Stimulation and Recording Setup

Four separate hydrogel surface electrodes (each 5 × 5 cm; axion, Leonberg, Germany) were placed over the spine in a rostrocaudal arrangement [35], starting with the most caudal electrode that was positioned between the L3/L4 spinous processes (position 1) based on palpation (Figure 1A(i)). The remaining three electrodes were then placed rostrally along the spine (positions 2–4), with an inter-electrode distance of 1 cm. Such setup ensured coverage of the targeted lumbosacral posterior roots by at least one of the electrodes. In individuals C5, S5, and P5, only three electrodes were used due to their comparatively smaller body heights. Two interconnected hydrogel electrodes (each 7 × 12 cm; axion, Leonberg, Germany) were placed on the lower abdomen symmetrically to the umbilicus [17,18]. Each of the paraspinal electrodes could be separately selected as stimulating electrode, with the abdominal electrodes acting as the common indifferent electrode. Stimulation was applied via a four-channel electrical stimulator (RehaMove3; HASOMED, Magdeburg, Germany) set to deliver charge-balanced, symmetric, and biphasic rectangular pulses of 2 ms width (1 ms per phase) [18]. With respect to the abdominal electrodes, the selected stimulating electrode served as anode for the first phase and the cathode for the second phase of the biphasic pulses.

EMG activity was acquired from the right and left L2–L4 innervated quadriceps muscle groups (RQ and LQ) as well as the L5–S2 innervated triceps surae muscle groups (RTS and LTS) [36] by using pairs of surface electrodes placed centrally over the muscle bellies with a distance of 1–2 cm (Figure 1A(ii)). For each lower limb, a wireless two-channel EMG sensor was used to sample the EMG activity of Q and TS at 1 kHz (MuscleLab; Ergotest, Porsgrunn, Norway). Common reference electrodes for both EMG channels of a limb were placed bilaterally on the outer edge of the patella. The skin was cleaned prior to EMG electrode placement to minimize signal noise.

All hardware was controlled by a customized program, developed in Matlab/Simulink (MathWorks, Natick, MA, USA), by using a modified Linux ERT target [37] and a specially created stimulation interface (Python, Kivy). Post-processing analysis was performed in MATLAB (R2021a).

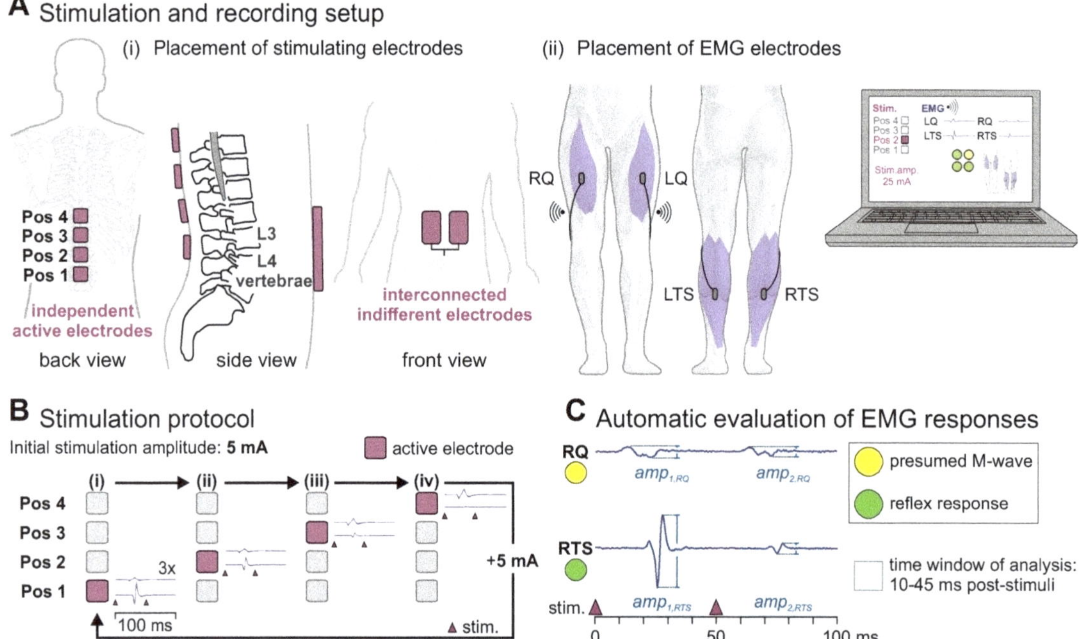

Figure 1. Automated stimulation and recording setup for lumbosacral posterior root stimulation. (**A**) (i) Separate active electrodes were placed at four rostrocaudal positions over the spine (Pos 1–4), with the most caudal electrode located between the L3 and L4 spinous processes. Two interconnected electrodes on the lower abdomen acted as common indifferent electrode. (ii) Surface electromyographic (EMG) recordings were acquired from right (R) and left (L) quadriceps (Q) and triceps surae (TS) muscle groups by using a two-channel sensor per lower limb. Information on the selected active electrode and the recorded EMG activity was transmitted to a laptop and displayed in a custom user interface. (**B**) The automated stimulation protocol systematically tested electrode positions 1–4 using three double stimuli per stimulation amplitude, starting with 5 mA. In each iteration of the protocol, the stimulation amplitude was increased by 5 mA up to a maximum of 75 mA or the individually maximally tolerated amplitude. (**C**) The custom user interface allowed for the visualization of averaged EMG waveforms of muscle responses elicited by the three double stimuli with a given amplitude, here exemplarily shown for RQ and RTS. Peak-to-peak amplitudes of the responses to the respective first and second stimuli were automatically computed ($amp_{1,RQ}$, $amp_{2,RQ}$, and $amp_{1,RTS}$ and $amp_{2,RTS}$). By applying the rules of the defined rating light system (Table 2), the responses were automatically classified as being a presumed M-wave or reflex response.

2.3. Automatic Determination of Electrodes and Parameters

2.3.1. Automated Stimulation Protocol

Stimulation was applied with the participants lying in the supine position. Starting with the most caudal electrode (position 1), each of the four paraspinal electrodes was selected, one by one, as the stimulating electrode (Figure 1B). From each position, three double stimuli with an inter-pulse-interval of 50 ms were administered with a stimulating amplitude initially set at 5 mA, with 5 s between repetitions. Subsequently, the stimulation amplitude was increased by 5 mA. The same procedure was repeated for stimulation amplitudes up to 75 mA or the individually maximum tolerated amplitude. For each double stimulus, raw EMG activity was recorded from the four muscle groups for time windows of 600 ms starting shortly before the first stimulus (see below).

The automated stimulation protocol could be monitored by the custom user interface, which provided information on the active electrode position at each time as well as a countdown to the application of the next double stimulus. The lower and upper limits of the stimulation amplitude could be manually adjusted to differ from the default values of 5 mA and 75 mA, respectively. The custom user interface additionally provided an online view of the raw EMG data derived from the four muscle groups studied as well as, optionally, the visualization of stimulus-triggered muscle responses and the averaged EMG waveforms derived from the three double stimuli at a given stimulation amplitude (cf. Figure 1C). In addition, a color-coded rating of the EMG responses can be displayed (cf. Section 2.3.3).

2.3.2. EMG Pre-Processing
EMG Synchronization

The wirelessly transmitted EMG data, acquired with the two sensors placed on the left and right lower limbs, were first synchronized to the times of double stimulus application. Data from Q and TS of a single limb were acquired by the same sensor and, hence, required no further synchronization. For the synchronization of the EMG data to the double stimuli, stimulation artifacts detected in the raw EMG signals of RQ and LQ were utilized, and time windows from 30 ms before to 300 ms after the first stimulus were extracted. Specifically, the stimulation artifacts were detected by performing non-causal double differentiation of the EMG signals of RQ and LQ by using a discrete approximation of the Laplace's differential operator Δ:

$$emg_{i,\Delta}(n, I, j, \tilde{t}) = 4\Delta(emg_{i,\text{raw}}(n, I, j, \tilde{t})), \quad i \in \{\text{LQ}, \text{RQ}\}, \tilde{t} = 1, \ldots, 600 \qquad (1)$$

where $emg_{i,\text{raw}}(n, k, j, \tilde{t}))$ is the raw EMG of the muscle $i \in \{LQ, RQ\}$ with the corresponding sample index $\tilde{t}$ for the position $n \in \{1, 2, 3, 4\}$ of the active tSCS electrode and applied stimulation amplitude $I = 5, 10, \ldots \leq 75$ mA. The index $j \in \{1, 2, 3\}$ describes the repetition number of the double stimuli.

Samples $\tilde{t}$ were marked as potential stimulation artifacts instances $\tilde{t}^*$, caused by the first stimulus, if they fulfilled the following four conditions.

$$(\mathcal{C}1) \qquad |emg_{i,\Delta}(n, I, j, \tilde{t}^*)| > 3\text{SD}(\{emg_{i,\Delta}(n, I, j, \tilde{t})\}_{\tilde{t}=300,\cdots,600}), \qquad (2)$$

$$(\mathcal{C}2) \qquad |emg_{i,\Delta}(n, I, j, \tilde{t}^* + 50)| > 3\text{SD}(\{emg_{i,\Delta}(n, I, j, \tilde{t})\}_{\tilde{t}=300,\cdots,600}), \qquad (3)$$

$$(\mathcal{C}3) \qquad |emg_{i,\Delta}(n, I, j, \tilde{t}^* + 100)| \leq 3\text{SD}(\{emg_{i,\Delta}(n, I, j, \tilde{t})\}_{\tilde{t}=300,\cdots,600}), \qquad (4)$$

$$(\mathcal{C}4) \qquad (\text{sign}(emg_{i,\text{raw,mf}}(n, I, j, \tilde{t}^*)) \neq \text{sign}(emg_{i,\text{raw,mf}}(n, I, j, \tilde{t}^* - 1))) \vee \qquad (5)$$
$$(\text{sign}(emg_{i,\text{raw,mf}}(n, I, j, \tilde{t}^*)) \neq \text{sign}(emg_{i,\text{raw,mf}}(n, I, j, \tilde{t}^* + 1))).$$

Here, the signal $emg_{i,\text{raw,mf}}$ was obtained by subtracting its median from the raw EMG signal in order to remove any present offset levels. Hence, condition $(\mathcal{C}4)$ checks for a sign change in the median-free raw EMG signal. The earliest candidate $\tilde{t}^*$ in the interval $\tilde{t} \in [10, 200]$ was chosen as new time instance ($t = 0$) of the first stimulus and is used to time-align the EMG recordings of the two sensors. Afterwards, the EMG signals were cropped to the new time range -30 to 300 ms for t.

If the synchronization failed, i.e., if no time instant could be found at which the conditions $\mathcal{C}1$–$\mathcal{C}4$ were fulfilled, then the corresponding measurement data were discarded, except for the data containing only baseline EMG (noise). Such baseline recordings can occur, e.g., in the case of low stimulation amplitudes (normally $<20\,\text{mA}$) and for the most rostral electrode position 4, as stimulation artifacts are barely present. In this case, the first 331 EMG samples were extracted for displaying at the user interface and to capture the noise characteristics. The measurement was declared as baseline EMG when condition ($\mathcal{C}1$) was not fulfilled for any single sample.

EMG Pre-Filtering

The median was removed from the cropped and synchronized raw EMG signals for offset correction, and detected stimulation artifacts were blanked (set to 0) for intervals of -2–$2\,\text{ms}$ and 48–52 ms. A second-order notch filter (Butterworth bandstop filter for the frequency range 43–47 Hz) as well as a first-order low-pass filter (Butterworth filter design with cutoff frequency of 300 Hz) were subsequently non-causally applied to the pre-processed EMG signals. The blanked intervals of the stimulation artifacts were restored in the filtered EMG signals to enable the synchronization and averaging of EMG responses to the three repetitions of double stimuli applied with a given stimulation amplitude and from a given stimulation electrode position.

Muscle-Specific Base Noise Level

For further analysis, the noise level $\overline{emg}_i$ of each muscle i at rest was defined by the standard deviation of all EMG samples $emg_i(n, I, j, t)$ for all n, I, j, and $100 \leq t \leq 300$.

Similarity Check and Averaging

For each muscle studied, the filtered EMG signals recorded during the three repetitions of double stimulus application from a given electrode position and with a given stimulation amplitude were checked for similarity by analyzing the root mean square error (RMSE) between the repetitions in the time interval 5–45 ms plus 55–95 ms. Signals were declared as similar if their RMSE was less than 16 times the muscle-specific base noise level $\overline{emg}_i$. The threshold parameter was determined empirically. Similar EMG signals were then cropped to a time range of -1 to 200 ms relative to the first of the double stimuli of each repetition and averaged to obtain the mean EMG signal $emg_{i,\text{avg}}(n, I, t)$, with n being the electrode position, I being the stimulation amplitude, and t being the sample index. Non-similar repetitions were discarded. If no similarity was found in at least two repetitions, the respective recording was marked as invalid. Such cases could occur, e.g., when stimulating during unintended movements, or when the initial artifact detection yielded incorrect results. The similarity check prevented such data from falsifying the results of the subsequent automatic evaluation.

2.3.3. Automatic Evaluation of EMG Responses

The averaged EMG signals $emg_{i,\text{avg}}(n, I, t)$, obtained for a given electrode position n and stimulation amplitude I, were further analyzed by separately calculating the EMG peak-to-peak amplitudes of the responses to the first and the second of the double stimuli, i.e., $amp_{1,i}(n, I)$ and $amp_{2,i}(n, I)$, within time intervals of 10–45 ms following the respective stimuli. The suppression level of the second to the first response was then calculated as follows:

$$sup_i(n, I) = 1 - amp_{2,i}(n, I) / amp_{1,i}(n, I) \tag{6}$$

with $i \in \{\text{LQ}, \text{LTS}, \text{RQ}, \text{RTS}\}$. The level was limited to the range of [0,1]. A value of one would signify complete suppression, while a value of zero would indicate two responses of same size, i.e., the lack of post-stimulation depression and, hence, the presence of M-waves [29].

The averaged EMG signals of a given muscle were then classified into four cases (cf. Table 2): "no response" if $amp_{1,i}(n, I)$ was $\leq 50\,\mu\text{V}$ or did not differ from the baseline EMG by at least six standard deviations; "reflex response" if $amp_{1,i}(n, I)$ was $> 50\,\mu\text{V}$ and differed from the baseline EMG by at least six standard deviations with a suppression level $sup_i(n, I) > 60\%$; "presumed M-wave" if previous conditions on $amp_{1,i}(n, I)$ were valid but suppression level was $\leq 60\%$; and "invalid" if no similar EMG signals were obtained for stimulation from a given electrode position and with given stimulation amplitude. The assigned color codes in Table 2 enable a rating light system in which green indicates the desired reflex responses.

Table 2. Rating light system for the automatic evaluation of EMG responses.

Color Code	Description	Condition
gray	No response	$(amp_{1,i}(n, I) \leq 6\,\overline{emg}_i) \vee (amp_{1,i}(n, I) \leq 50\mu\text{V})$
green	Reflex response	$(amp_{1,i}(n, I) > 6\,\overline{emg}_i) \wedge (amp_{1,i}(n, I) > 50\mu\text{V}) \wedge sup_i(n, I) > 60\%$
yellow	Presumed M-wave	$(amp_{1,i}(n, I) > 6\,\overline{emg}_i) \wedge (amp_{1,i}(n, I) > 50\mu\text{V}) \wedge sup_i(n, I) \leq 60\%$
red	Invalid	No similar EMG signals obtained for stimulation from a given electrode position and with given stimulation amplitude

2.3.4. Automatic Delineation of Stimulation Position and Amplitude for tSCS Therapy

Two independent approaches were implemented to identify a suitable electrode position and stimulation amplitude for tSCS therapy, the *ranking approach* considering the color code classification of responses only, and the *cost function approach* additionally considering the obtained EMG peak-to-peak amplitudes and suppression levels.

Ranking Approach

Based on the color code classification, a ranking of pairs (*n*—electrode position and *I*—stimulation amplitude) was performed by utilizing a nested search in the following priority (order):

1. At least two green labels present (i.e., reflex responses elicited in at least two of the four muscle groups studied; minimal requirement);
2. Largest number of green labels (i.e., reflex responses);
3. Smallest current level difference between stimulation amplitude I and I', at which the first green label was obtained (amplitude threshold);
4. Lowest stimulation amplitude I.

All electrodes fulfilling the minimal requirement were ranked, with rank "1" indicating best fulfillment of criteria 2–4 within the available search space. For tSCS therapy in spasticity, the best ranked electrode position n and a stimulation amplitude corresponding to 90% of the respective stimulation amplitude threshold I' [23] were suggested as therapy parameters.

Cost Function Approach

In addition to the color code classification, the cost function approach considered the attainable peak-to-peak amplitudes of the EMG responses and the precise suppression levels by using the following cost function.

$$J(n, I) = \frac{1}{4} \sum_{i \in \{\text{LQ,LTS,RQ,RTS}\}} \frac{amp_{1,i}(n, I)}{\overline{amp}_{1,i}} sup_i(n, I). \tag{7}$$

Here, the response amplitudes of each muscle were normalized with respect to their maximal amplitude $\overline{amp}_{1,i}$ found in the corresponding muscle and scaled with the observed suppression ratio.

The electrode position with the largest cost function value J was selected for tSCS therapy, favoring electrode positions yielding reflex responses of large peak-to-peak amplitudes. The stimulation amplitude for tSCS for spasticity control was the same as suggested by the ranking approach, i.e., 90% of the amplitude threshold of the selected electrode position.

Graphical Visualization

The color coded results of the automatic evaluation of the EMG responses $emg_{i,avg}(n, I)$ are presented as a rating light matrix (Figure 2) for all muscles i, electrode positions n, and applied stimulation amplitudes I. After termination of the stimulation protocol, the electrode position and stimulation amplitude pairs with the highest rank (ranking approach) were marked, and the suggested stimulation amplitude for the tSCS intervention was provided. For a more detailed overview, the rating details matrix was additionally developed, including the color codes as well as details obtained with both the ranking and the cost function approaches (see Figure 3 for an example). The rating detail matrix provided information on the five best ranked electrode positions along with the respective cost function values.

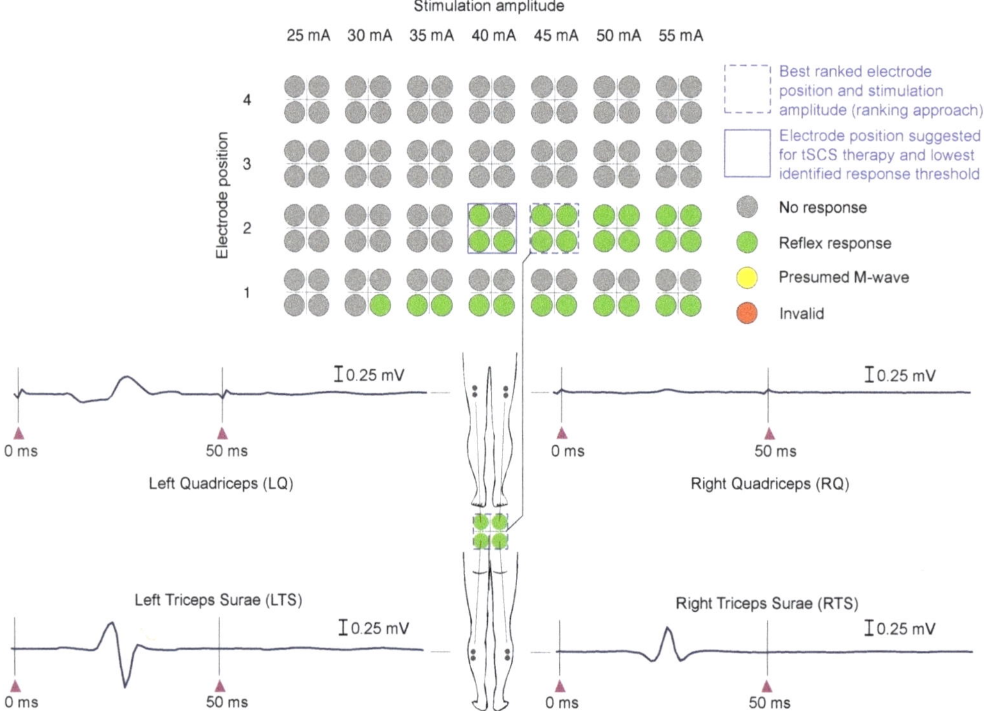

Figure 2. Rating light matrix derived for a neurologically intact individual, C1, along with EMG signals averaged from three repetitions of double stimulus application from electrode position 2 with a stimulation amplitude of 45 mA (best ranked parameter combination). The four color-coded circles represent the results obtained for the four studied muscle groups per electrode position n and tested stimulation amplitude I (cf. Table 2). In favor of a compact overview, results are only displayed for the stimulation amplitude before the first detected response, in this case from 25 mA instead of from 5 mA, until the highest applied intensity.

2.4. Validation

In order to validate the proposed algorithms for automatic tSCS calibration, all preprocessed averaged EMG signals $emg_{i,avg}(n, I, t)$ of our data set were presented to two experts in the field of tSCS (U.S.H. and N.W.). For each participant, the experts annotated the electrode position and stimulation amplitude that they would choose for a sub-threshold tSCS therapy setup (e.g., for treating spasticity). The experts were blinded to the participant's condition. The annotations were then compared with the results of the automatic determination by the ranking approach and the cost function approach.

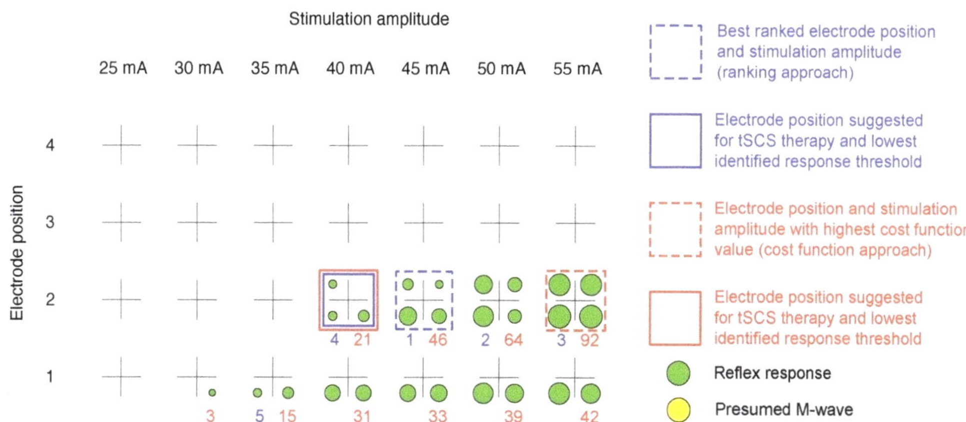

Figure 3. Rating details matrix derived for a neurologically intact individual, C1. In addition to the color codes, reflex peak-to-peak amplitudes are reflected by the size of the colored circles representing the four muscle groups studied. If no response had been elicited, the circle shrank to a dot. Blue numbers are ranks 1–5 as derived from the ranking approach, and red numbers are cost function values *J* calculated in the cost function approach, with higher values signifying favorable results.

3. Results

The stimulation and recording setups as well as the algorithms for the calibration of tSCS settings for therapeutic application were successfully and safely applied in all 15 study participants. No adverse events were encountered. Across participants, the complete execution of the stimulation protocol and the determination of stimulation parameters took on average ten min, ranging from 7 to 15 min.

Exemplary pre-processed averaged EMG signals $emg_{i,\mathrm{avg}}(n, I, t)$ derived during the automatic execution of the stimulation protocol in one neurologically intact individual and used for expert validation are displayed in Figure 4. The corresponding rating light matrix using color codes for the classification of electrode positions and stimulation amplitudes along with the averaged representation of EMG signals obtained with the best ranked setting (n, I) is shown in Figure 2. In this case, electrode position $n = 2$ was selected since it was the only electrode position resulting in the elicitation of reflex responses in all four studied muscle groups. Furthermore, a stimulation amplitude of 35 mA was suggested for therapy (90% of I', amplitude threshold, here 40 mA). The rating details matrix (Figure 3) exhibits that both the ranking approach as well as the cost function approach resulted in the suggestion of the same parameter settings for therapy despite differences in the rankings (stimulation with 45 mA in the case of the ranking approach vs. 55 mA in the case of the cost function approach; see also Table 3).

Across participants, validation revealed a high accuracy of the automatic ranking of stimulation site and amplitude relative to the experts' classification (Table 3). Specifically, the same electrode position was selected by both experts as well as both evaluation approaches in 11 out of 15 participants. In another two participants in whom no EMG responses were detected even at the maximum applied stimulation amplitudes (participants P3 and P4 with PD), no electrode position was suggested, neither by the experts nor the automatic approaches. In one of the remaining participants (S4), expert #1 selected an electrode that was one position more caudal than the one automatically detected as well as chosen by expert #2. In the other remaining participant (S1), the two experts selected different but neighboring electrode positions, $n = 2$ and $n = 1$, as did the two automatic approaches, also $n = 2$ and $n = 1$. This was the only case in which the ranking approach and the cost function approach did not select the same electrode position.

Table 3. Comparison of expert's selection with the results of the automatic tSCS parameter determination.

| Subject | Expert's Selection | | | | Automatic Selection | | | |
| | Expert #1 | | Expert #2 | | Cost Function | | Ranking | |
	n	*I′*	*n*	*I′*	*n*	*I′*	*n*	*I′*
C1	2	40 mA	2	35 mA	2	40 mA	2	40 mA
C2	2	25 mA	2	25 mA	2	25 mA	2	25 mA
C3	3	25 mA	3	25 mA	3	25 mA	3	25 mA
C4	2	20 mA	2	20 mA	2	20 mA	2	20 mA
C5	3	35 mA	3	35 mA	3	35 mA	3	35 mA
S1	2	25 mA	1	35 mA	2	25 mA	1	35 mA
S2	3	30 mA	3	30 mA	3	20 mA	3	20 mA
S3	2	45 mA	2	40 mA	2	35 mA	2	35 mA
S4	2	25 mA	3	25 mA	3	25 mA	3	25 mA
S5	2	25 mA	2	20 mA	2	20 mA	2	20 mA
P1	2	45 mA	2	40 mA	2	40 mA	2	40 mA
P2	1	45 mA	1	45 mA	1	45 mA	1	45 mA
P3	-	-	-	-	-	-	-	-
P4	-	-	-	-	-	-	-	-
P5	1	35 mA	1	35 mA	1	40 mA	1	40 mA

C, neurologically intact control; *I′*, stimulation amplitude threshold; *n*, number of electrode position (cf. Figure 1A); *P*, individual with Parkinson's disease; *S*, individual with SCI; -, no response elicited with stimulation amplitudes up to 75 mA; hence, no electrode and stimulation amplitude were suggested. The stimulation amplitude for tSCS therapy is then set to 90% of the identified thresholds *I′*.

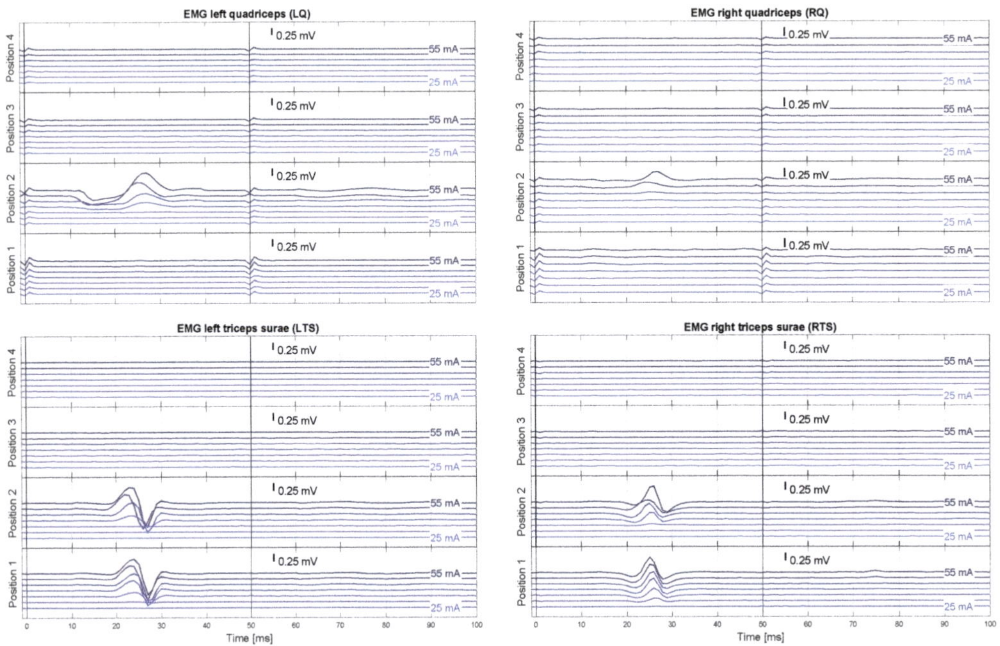

Figure 4. EMG recordings derived from bilateral quadriceps and triceps surae muscle groups during tSCS applied from different rostrocaudal electrode positions with double-stimuli of incremental amplitudes. Exemplary traces derived from a neurologically intact individual, C1, as they were shown to experts. A step-wise increase in stimulation amplitude resulted in the elicitation of responses in bilateral triceps surae by the first pulse of the double stimuli, yet only when applied from the two caudal electrode positions 1 and 2. Bilateral quadriceps responses were present only with stimulation delivered from position 2, hence the best ranked position. Response suppression when the second pulses of the double stimuli were applied after 50 ms identified the responses as reflexes. Each trace averaged from up to three double stimuli per stimulation amplitude; blue-shaded values are stimulation amplitudes.

With respect to the stimulation amplitude suggested for therapy, congruent values between experts and algorithms were found in six out of the fifteen participants, with another two participants for whom no stimulation amplitudes were suggested (P3 and P4; see above). In participant S1, expert #1 chose the same stimulation amplitude as the cost function approach (25 mA, $n = 2$), and expert #2 agreed with the ranking approach (35 mA, $n = 1$). In four participants (C1, S5, P1, and P5), there was a difference of one increment (i.e., 5 mA) between the four suggested stimulation amplitudes for therapy, and in two participants (S2 and S3), there was a maximum difference of 10 mA. The rating light matrix and the rating details matrix derived for participant S2 are shown in Figure 5, and the corresponding EMG signals are provided in Figure A1 in the Appendix A. Exemplary EMG responses and rating matrices of a PD individual (P1) are shown in Figures A2 and A3. Notably, across participants, only 18 out of 601 evaluated EMG responses were classified as being invalid, corresponding to an error ratio of 3%.

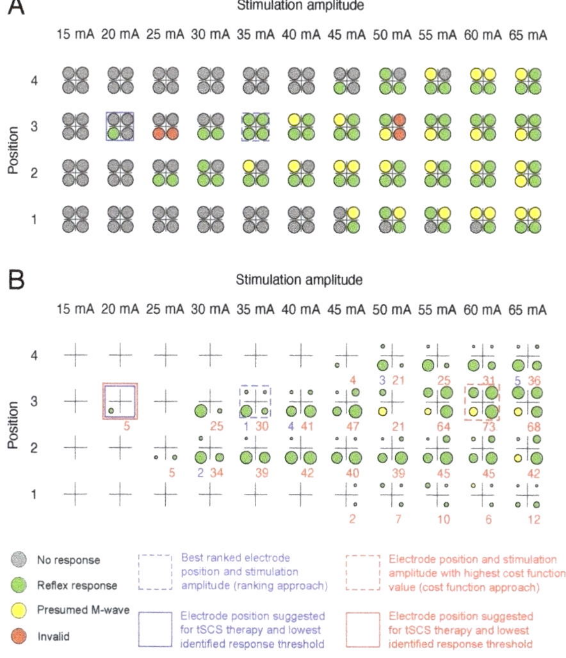

Figure 5. Comparison of ranking based on the two automatic response evaluation approaches applied. (**A**) Rating light matrix and (**B**) rating details matrix derived for participant S2 with SCI. Both approaches resulted in the suggestion of electrode position 3 and identified 20 mA as the lowest response threshold. The four color-coded circles represent the results obtained for the four muscle groups studied (cf. Table 2). In the rating details matrix, the cycle size additionally provides information on the attainable response sizes. If no response had been elicited, the circle shrank to a dot. Blue numbers are ranks 1–5 as derived from the ranking approach, and red numbers are cost function values J calculated in the cost function approach, with higher values signifying favorable results.

4. Discussion

In this article, we presented a novel method that maps expert knowledge on the application of tSCS as a neurorehabilitative method into an automated evaluation algorithm. The developed combined stimulation and evaluation protocol allows for real-time classification of responses evoked by double-stimuli tSCS in several lower limb muscles bilaterally, identifying them either as direct motor responses (M-waves) or as PRM reflexes initiated within afferent fibers of the lumbosacral posterior roots [16,29]. This information is further processed into individual treatment recommendations for electrode position and stimulation amplitude for the application of tSCS in therapeutic settings, e.g., for the control of spinal spasticity [18,20,23,25]. Blinded validations by two independent experts confirmed the accuracy of the automatically derived recommendations. With respect to the electrode position specifically, congruent results were obtained by both evaluation approaches implemented as well as both experts in 13 out of 15 cases (87%).

Our novel approach with an array of four surface electrodes placed longitudinally over the spine to cover mid-lumbar to low-thoracic vertebral levels allowed us to simultaneously evaluate several stimulation sites, rather than having to relocate a single stimulating electrode as in traditional tSCS setups until the required PRM reflex distribution in the lower limb muscles is obtained [18]. This is of specific relevance when treating individuals with severely limited mobility and further helps minimize the time needed to find individually tailored stimulation settings. Other studies of tSCS for neurorehabilitation have used a pragmatic design, placing a single active electrode over vertebral levels T11/T12 in all participants [20,38]. However, previous studies of EES in lower-limb motor control have pointed at the distinctive importance of the specific placement of active electrodes over lumbar and upper sacral spinal cord segments [1,6,27,28]. Consequently, EES parameter settings are thoroughly tested and individually optimized prior to the full implantation of the system for chronic stimulation [14]. Our automated calibration of tSCS may, hence, present a first step towards facilitating individually tailored stimulation setups and maximizing the attainable therapeutic outcomes.

In the present study, the duration for the execution of the automated protocols was approximately 10 min on average. This duration could likely be further reduced by roughly two min by omitting the lower range of stimulation amplitudes (5 mA and 10 mA) at which no responses were evoked in any of our study participants. Furthermore, using a setup with three instead of four rostrocaudally arranged stimulation electrodes, as in participants C5, S5, and P5, would shorten the protocol by another estimated 2 min, resulting in an overall duration of 6 min for the entire procedure. Notably, the most rostral electrode position was not recommended for therapy for any of the individuals tested with four electrodes, neither by the automated calibration nor the two experts. The repeated use of our tSCS setup in an individual patient could additionally limit the search space, i.e., by using less electrodes and test even fewer stimulation amplitudes.

We implemented two evaluation approaches emphasizing different aspects of the detected EMG responses, i.e., the occurrence of PRM reflexes in the majority of studied muscles or the precise amplitude and suppression characteristics, respectively. Both approaches resulted in the same recommendations of electrode positions and the same detection of response thresholds across all but one participant. Both algorithms further provided information on tSCS configurations in addition to the best ranked results that could be alternatively employed should they be perceived as more comfortable by the individual treated. The patient's perception is a criterion that is not reflected by our automated evaluation approach, but plays a particular role in the tolerance of tSCS therapy, especially in individuals with intact sensory perception in the stimulation area (here, controls and PD patients). By providing the rating light and rating details matrix, we want to enable the user to select other highly ranked tSCS configurations based on individual preferences and criteria, for example, the patient's perception of continuous stimulation (50 Hz).

Although PD patients often had higher baseline muscle activity than controls and SCI patients, mostly due to dyskinesia as a side effect of dopaminergic medication (e.g., in patient P1), the algorithms were able to robustly detect PRM reflexes and gave reliable recommendations on stimulation parameters in three PD participants. The proposed methods take the baseline muscle activity into account in the used threshold parameters. However, in two individuals with PD, no responses were elicited by tSCS applied from either of the four electrode positions with stimulation amplitudes of up to 65 mA and 75 mA, respectively. A previous study has shown a significant increase in Hoffmann-reflex thresholds in individuals with PD compared to age-matched controls [39], and it can be assumed that similar changes would also affect the PRM reflex. Notably, the individuals with PD in the present study were collectively older than the neurologically intact controls and the individuals with SCI, and studies have shown a lower excitability of the Hoffmann-reflex in elderly adults [40]. These observations suggest that screening tSCS responses, for example, before EES implantation, could add valuable information for the prospective characterization of patients in future clinical trials.

Our algorithms are currently designed to provide recommendations for the application of tSCS at amplitudes below the threshold for the elicitation of PRM reflexes in the lower limb muscles, as utilized in spasticity treatment [23] or for the augmentation of voluntary locomotor activity in incomplete SCI [19]. Future applications could be extended to the recommendation of parameter settings for locomotion therapy or standing and balance training using higher stimulation amplitudes [14,21,26,41,42]. Furthermore, our present approach was optimized for application of tSCS in the supine position. It is known that the body position influences which neural structures are recruited by lumbar tSCS [43]. An adaptation of our method for other body positions such as the upright position in locomotor training seems useful and well possible in the future. Another improvement in the future could be to allow individual adjustments in the selection criteria, which could facilitate more experienced users to set or fine-tune the rules for specific therapy goals (e.g., prioritizing behavior in specific muscles).

The proposed algorithms can be implemented in interpreted higher programming languages such as Python or MATLAB and applied to EMG data from other clinical recording systems. Artifact detection and signal alignment (synchronization) steps can be omitted if no wireless transmission is involved and the stimulator unit provides a trigger output for synchronization with the measurement system. Both features are usually available in present recording and stimulation systems at clinical research centers. In addition, we are currently developing a tablet-controlled setup with an iOS app that uses Bluetooth low energy to communicate wirelessly with EMG sensors and a stimulator for future broader use in rehabilitation centers and physiotherapy practices. This system will directly exploit the algorithms presented in this article.

5. Conclusions

In this study, we developed an automated calibration method for tSCS that accurately identified suitable electrode positions and stimulation amplitudes for therapeutic applications. The setup proved to be applicable by non-specialized health professionals, allowing them to individually calibrate tSCS by the use of a comprehensive graphical user interface. Our approach may, hence, provide an easy-to-use and time-effective solution for clinical decision making. These developments may aid the dissemination of tSCS technologies into non-academic environments and broaden the use of tSCS for diagnostic and therapeutic applications.

Author Contributions: Conceptualization, T.S., P.M., A.A.K., U.S.H. and N.W.; methodology, C.S.-H., T.S., P.M., U.S.H. and N.W.; software, C.S.-H., T.S. and P.M.; validation, C.S.-H., T.S., U.S.H. and N.W.; formal analysis, T.S.; investigation, C.S.-H. and N.W.; resources, C.S.-H., T.S., A.A.K. and N.W.; data curation, C.S.-H., U.S.H. and N.W.; writing—original draft preparation, C.S.-H. and T.S.; writing—review and editing, C.S.-H., T.S., U.S.H. and N.W.; visualization, C.S.-H., T.S. and U.S.H.; supervision, T.S., U.S.H. and N.W.; project administration, C.S.-H., T.S., A.A.K. and N.W.; funding acquisition, T.S., A.A.K. and N.W. All authors have read and agreed to the published version of the manuscript.

Funding: This research was funded by the German Federal Ministry of Economics and Energy (ZF4171703TS7). We acknowledge support by the German Research Foundation and the Open Access Publication Fund of TU Berlin. N.W. is a Freigeist-Fellow supported by the Volkswagen Foundation and a participant in the BIH Charité Clinician Scientist Program. U.S.H. is supported by the Austrian Science Fund (FWF), project nr. I 3837-B34. A.A.K. and N.W. are supported by funding by the Deutsche Forschungsgemeinschaft (DFG, German Research Foundation) project ID 424778381-TRR 295.

Institutional Review Board Statement: The study was conducted according to the guidelines of the Declaration of Helsinki and approved by the Ethics Committee of the Berlin Chamber of Physicians (ETH-28/17; 2017 and 2019; SCI) and the Ethics Committee of the Charité–Universitätsmedizin Berlin (EA2/118/18, 2018; PD and healthy controls).

Informed Consent Statement: Informed consent was obtained from all subjects involved in the study.

Data Availability Statement: The data presented in this study are available upon reasonable request from the corresponding author.

Acknowledgments: We like to express our appreciation to Christian Dikau for his support in the technical realization of the tSCS setup.

Conflicts of Interest: TS is co-funder of the SensorStim Neurotechnology GmbH, which is a company developing sensor-based stimulation devices. The other authors declare no conflict of interest.

Abbreviations

The following abbreviations are used in this manuscript:

AIS	American Spinal Cord Injury Association Impairment Scale;
EES	Epidural electrical stimulation;
EMG	Electromyography;
FD	Years since first diagnosis of PD;
H and Y	Hoehn and Yahr scale;
LQ	Left quadriceps;
LTS	Left tricpes surae;
PD	Parkinson's disease;
PRM reflex	Posterior root-muscle reflex;
RQ	Right quadriceps;
RTS	Right tricpes surae;
SCI	Spinal cord injury;
tSCS	Transcutaneous spinal cord stimulation.

Appendix A

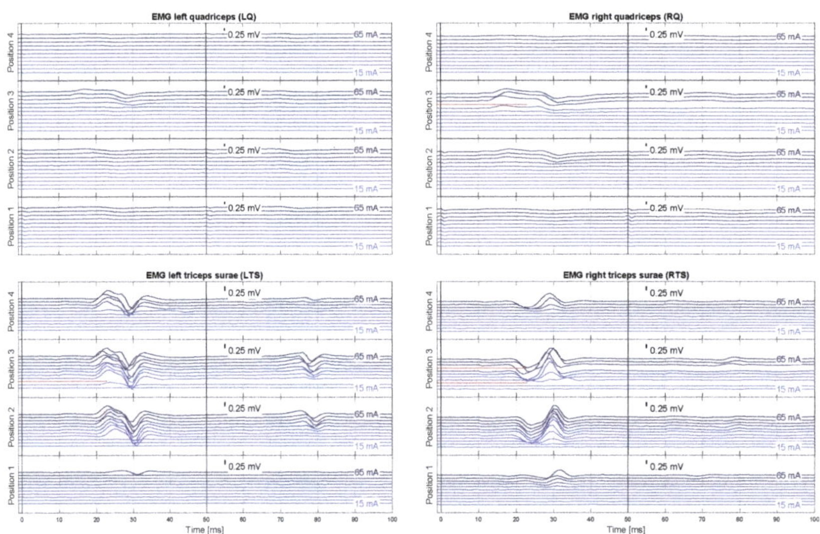

Figure A1. EMG recordings derived from bilateral quadriceps and triceps surae muscle groups during tSCS applied from different rostrocaudal electrode positions with double-stimuli of incremental amplitudes. Exemplary results derived from an individual with SCI, S2. Each trace averaged from up to three double stimuli per stimulation amplitude; blue-shaded values are stimulation amplitudes.

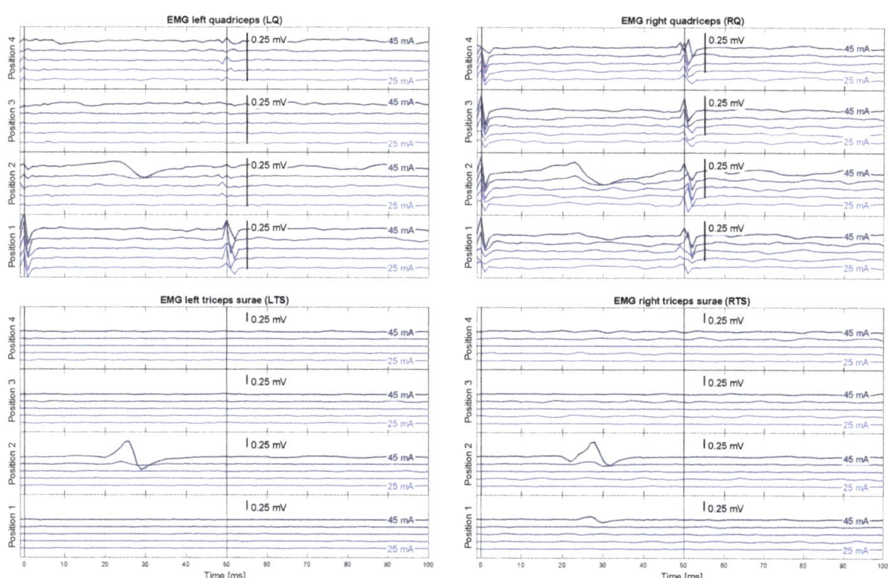

Figure A2. EMG recordings derived from bilateral quadriceps and triceps surae muscle groups during tSCS applied from different rostrocaudal electrode positions with double-stimuli of incremental amplitudes. Exemplary results derived from an individual with PD, P1. Each trace averaged from up to three double stimuli per stimulation amplitude; blue-shaded values are stimulation amplitudes.

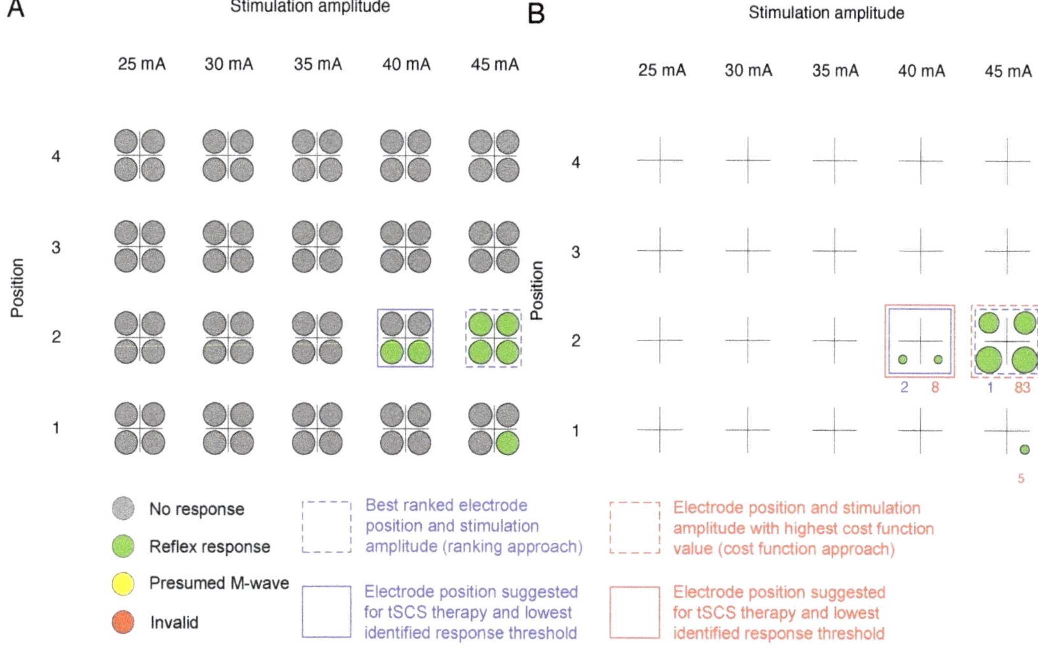

Figure A3. Comparison of ranking based on the two automatic response evaluation approaches applied. (**A**) Rating light matrix and (**B**) rating details matrix derived for participant P1 with PD. Both approaches resulted in the suggestion of electrode position 2 for therapy and identified 40 mA as the lowest response threshold (suggested stimulation amplitude for therapy: 0.9×40 mA = 36 mA). The four color-coded circles represent the results obtained for the four muscle groups studied (green, reflex response; cf. Table 2)). In the rating details matrix, the circle size additionally provided information on the attainable response amplitudes. If no response had been elicited, the circle shrank to a dot. Blue numbers are ranks as derived from the ranking approach; red numbers are cost function values J calculated in the cost function approach, with higher values signifying favorable results.

References

1. Wagner, F.B.; Mignardot, J.B.; Le Goff-Mignardot, C.G.; Demesmaeker, R.; Komi, S.; Capogrosso, M.; Rowald, A.; Seáñez, I.; Caban, M.; Pirondini, E.; et al. Targeted neurotechnology restores walking in humans with spinal cord injury. *Nature* **2018**, *563*, 65–71. [CrossRef] [PubMed]
2. Angeli, C.A.; Boakye, M.; Morton, R.A.; Vogt, J.; Benton, K.; Chen, Y.; Ferreira, C.K.; Harkema, S.J. Recovery of over-ground walking after chronic motor complete spinal cord injury. *N. Engl. J. Med.* **2018**, *379*, 1244–1250. [CrossRef] [PubMed]
3. Gill, M.L.; Grahn, P.J.; Calvert, J.S.; Linde, M.B.; Lavrov, I.A.; Strommen, J.A.; Beck, L.A.; Sayenko, D.G.; Van Straaten, M.G.; Drubach, D.I.; et al. Neuromodulation of lumbosacral spinal networks enables independent stepping after complete paraplegia. *Nat. Med.* **2018**, *24*, 1677–1682. [CrossRef] [PubMed]
4. Minassian, K.; Jilge, B.; Rattay, F.; Pinter, M.; Binder, H.; Gerstenbrand, F.; Dimitrijevic, M.R. Stepping-like movements in humans with complete spinal cord injury induced by epidural stimulation of the lumbar cord: Electromyographic study of compound muscle action potentials. *Spinal Cord.* **2004**, *42*, 401–416. [CrossRef]
5. Ladenbauer, J.; Minassian, K.; Hofstoetter, U.; Dimitrijevic, M.R.; Rattay, F. Stimulation of the human lumbar spinal cord with implanted and surface electrodes: A computer simulation study. *IEEE Trans. Neural Syst. Rehabil. Eng.* **2010**, *18*, 637–645. [CrossRef]
6. Pinter, M.; Gerstenbrand, F.; Dimitrijevic, M. Epidural electrical stimulation of posterior structures of the human lumbosacral cord: 3. Control of spasticity. *Spinal Cord.* **2000**, *38*, 524–531. [CrossRef]
7. Angeli, C.; Edgerton, V.; Gerasimenko, Y.; Harkema, S. Altering spinal cord excitability enables voluntary movements after chronic complete paralysis in humans. *Brain* **2014**, *137*, 1394–1409. [CrossRef]
8. Minassian, K.; McKay, W.; Binder, H.; Hofstoetter, U. Targeting Lumbar Spinal Neural Circuitry by Epidural Stimulation to Restore Motor Function After Spinal Cord Injury. *Neurotherapeutics* **2016**, *13*, 284–294. [CrossRef]
9. de Andrade, E.M.; Ghilardi, M.G.; Cury, R.G.; Barbosa, E.R.; Fuentes, R.; Teixeira, M.J.; Fonoff, E.T. Spinal cord stimulation for Parkinson's disease: A systematic review. *Neurosurg. Rev.* **2015**, *39*, 27–35. [CrossRef]
10. Pinto de Souza, C.; Hamani, C.; Oliveira Souza, C.; Lopez Contreras, W.O.; dos Santos Ghilardi, M.G.; Cury, R.G.; Reis Barbosa, E.; Jacobsen Teixeira, M.; Talamoni Fonoff, E. Spinal cord stimulation improves gait in patients with Parkinson's disease previously treated with deep brain stimulation. *Mov. Disord.* **2017**, *32*, 278–282. [CrossRef]
11. de Lima-Pardini, A.; Coelho, D.; Pinto de Souza, C.; Souza, C.O.; dos Santos Ghilardi, M.G.; Garcia, T.; Voos, M.; Milosevic, M.; Hamani, C.; Teixeira, L.A.; et al. Effects of spinal cord stimulation on postural control in Parkinson's disease patients with freezing of gait. *Elife* **2018**, *7*, e37727. [CrossRef]
12. Fonoff, E.; de Lima-Pardini, A.; Coelho, D.; Monaco, B.; Machado, B.; Pinto de Souza, C.; dos Santos Ghilardi, M.; Hamani, C. Spinal cord stimulation for freezing of gait: From bench to bedside. *Front. Neurol.* **2019**, *10*, 905. [CrossRef]
13. Prasad, S.; Aguirre-Padilla, D.; Poon, Y.; Kalsi-Ryan, S.; Lozano, A.M.; Fasano, A. Spinal Cord Stimulation for Very Advanced Parkinson's Disease: A 1-Year Prospective Trial. *Mov. Disord.* **2020**, *35*, 1082–1083. [CrossRef]
14. Hofstoetter, U.S.; Freundl, B.; Binder, H.; Minassian, K. Spinal cord stimulation as a neuromodulatory intervention for altered motor control following spinal cord injury. In *Advanced Technologies for the Rehabilitation of Gait and Balance Disorders*; Biosystems & Biorobotics; Springer: Cham, Switzerland, 2018; Volume 19, pp. 501–521. [CrossRef]
15. Courtine, G.; Harkema, S.; Dy, C.; Gerasimenko, Y.P.; Dyhre-Poulsen, P. Modulation of multisegmental monosynaptic responses in a variety of leg muscles during walking and running in humans. *J. Physiol.* **2007**, *582*, 1125–1139. [CrossRef]
16. Minassian, K.; Hofstoetter, U.; Frank, R. Transcutaneous lumbar posterior root stimulation for motor control studies and modification of motor activity after spinal cord injury. In *Restorative Neurology of Spinal Cord Injury*; Dimitrijevic, M., Kakulas, B., McKay, W., Vrbová, G., Eds.; Oxford University Press: New York, NY, USA, 2011; pp. 226–255.
17. Hofstoetter, U.; Freundl, B.; Binder, H.; Minassian, K. Common neural structures activated by epidural and transcutaneous lumbar spinal cord stimulation: Elicitation of posterior root-muscle reflexes. *PLoS ONE* **2018**, *13*, e0192013. [CrossRef]
18. Hofstoetter, U.; McKay, W.; Tansey, K.; Mayr, W.; Kern, H.; Minassian, K. Modification of spasticity by transcutaneous spinal cord stimulation in individuals with incomplete spinal cord injury. *J. Spinal Cord. Med.* **2014**, *37*, 202–211. [CrossRef]
19. Hofstoetter, U.; Krenn, M.; Danner, S.; Hofer, C.; Kern, H.; McKay, W.B.; Mayr, W.; Minassian, K. Augmentation of Voluntary Locomotor Activity by Transcutaneous Spinal Cord Stimulation in Motor-Incomplete Spinal Cord-Injured Individuals. *Artif. Organs* **2015**, *39*, E176–E186. [CrossRef]
20. Estes, S.P.; Iddings, J.A.; Field-Fote, E.C. Priming neural circuits to modulate spinal reflex excitability. *Front. Neurol.* **2017**, *8*, 17. [CrossRef]

21. Gad, P.; Gerasimenko, Y.; Zdunowski, S.; Turner, A.; Sayenko, D.; Lu, D.; Edgerton, V. Weight Bearing Over-ground Stepping in an Exoskeleton with Non-invasive Spinal Cord Neuromodulation after Motor Complete Paraplegia. *Front. Neurosci.* **2017**, *11*, 333. [CrossRef]
22. Sayenko, D.G.; Rath, M.; Ferguson, A.R.; Burdick, J.W.; Havton, L.A.; Edgerton, V.R.; Gerasimenko, Y.P. Self-Assisted Standing Enabled by Non-Invasive Spinal Stimulation after Spinal Cord Injury. *J. Neurotrauma* **2019**, *36*, 1435–1450. [CrossRef]
23. Hofstoetter, U.S.; Freundl, B.; Danner, S.M.; Krenn, M.J.; Mayr, W.; Binder, H.; Minassian, K. Transcutaneous Spinal Cord Stimulation Induces Temporary Attenuation of Spasticity in Individuals with Spinal Cord Injury. *J. Neurotrauma* **2020**, *37*, 481–493. [CrossRef] [PubMed]
24. Meyer, C.; Hofstoetter, U.S.; Hubli, M.; Hassani, R.H.; Rinaldo, C.; Curt, A.; Bolliger, M. Immediate Effects of Transcutaneous Spinal Cord Stimulation on Motor Function in Chronic, Sensorimotor Incomplete Spinal Cord Injury. *J. Clin. Med.* **2020**, *9*, 3541. [CrossRef] [PubMed]
25. Hofstoetter, U.; Freundl, B.; Lackner, P.; Binder, H. Transcutaneous Spinal Cord Stimulation Enhances Walking Performance and Reduces Spasticity in Individuals with Multiple Sclerosis. *Brain Sci.* **2021**, *11*, 472. [CrossRef] [PubMed]
26. Roberts, B.; Atkinson, D.; Manson, G.; Markley, R.; Kaldis, T.; Britz, G.; Horner, P.; Vette, A.; Sayenko, D. Transcutaneous spinal cord stimulation improves postural stability in individuals with multiple sclerosis. *Mult. Scler. Relat. Disord.* **2021**, *52*, 103009. [CrossRef]
27. Hofstoetter, U.S.; Perret, I.; Bayart, A.; Lackner, P.; Binder, H.; Freundl, B.; Minassian, K. Spinal motor mapping by epidural stimulation of lumbosacral posterior roots in humans. *iScience* **2021**, *24*, 101930. [CrossRef]
28. Murg, M.; Binder, H.; Dimitrijevic, M. Epidural electric stimulation of posterior structures of the human lumbar spinal cord: 1. muscle twitches—A functional method to define the site of stimulation. *Spinal Cord.* **2000**, *38*, 394–402. [CrossRef]
29. Minassian, K.; Persy, I.; Rattay, F.; Dimitrijevic, M.R.; Hofer, C.; Kern, H. Posterior root-muscle reflexes elicited by transcutaneous stimulation of the human lumbosacral cord. *Muscle Nerv.* **2007**, *35*, 327–336. [CrossRef]
30. Hofstoetter, U.S.; Minassian, K.; Hofer, C.; Mayr, W.; Rattay, F.; Dimitrijevic, M.R. Modification of reflex responses to lumbar posterior root stimulation by motor tasks in healthy subjects. *Artif. Organs* **2008**, *32*, 644–648. [CrossRef]
31. Calvert, J.S.; Grahn, P.J.; Strommen, J.A.; Lavrov, I.A.; Beck, L.A.; Gill, M.L.; Linde, M.B.; Brown, D.A.; Van Straaten, M.G.; Veith, D.D.; et al. Electrophysiological Guidance of Epidural Electrode Array Implantation over the Human Lumbosacral Spinal Cord to Enable Motor Function after Chronic Paralysis. *J. Neurotrauma* **2019**, *36*, 1451–1460. [CrossRef]
32. Rattay, F.; Minassian, K.; Dimitrijevic, M. Epidural electrical stimulation of posterior structures of the human lumbosacral cord: 2. quantitative analysis by computer modeling. *Spinal Cord.* **2000**, *38*, 473–489. [CrossRef]
33. Capogrosso, M.; Wenger, N.; Raspopovic, S.; Musienko, P.; Beauparlant, J.; Luciani, L.B.; Courtine, G.; Micera, S. A computational model for epidural electrical stimulation of spinal sensorimotor circuits. *J. Neurosci.* **2013**, *33*, 19326–19340. [CrossRef]
34. Hoehn, M.M.; Yahr, D. Parkinsonism: Onset, progression, and mortality. *Neurology* **1967**, *17*, 427–442. [CrossRef]
35. Salchow-Hömmen, C.; Dikau, C.; Mueller, P.; Hofstoetter, U.; Kühn, A.; Schauer, T.; Wenger, N. Characterization of optimal electrode configurations for transcutaneous spinal cord stimulation. In Proceedings of the IFESS Conference at Rehab Week, Toronto, ON, Canada, 24–28 June 2019.
36. Sayenko, D.; Atkinson, D.; Dy, C.; Gurley, K.; Smith, V.; Angeli, C.; Harkema, S.; Edgerton, V.; Gerasimenko, Y. Spinal segment-specific transcutaneous stimulation differentially shapes activation pattern among motor pools in humans. *J. Appl. Physiol.* **2015**, *118*, 1364–1374. [CrossRef]
37. Sojka, M.; Píša, P. Usable Simulink Embedded Coder Target for Linux. In Proceedings of the 16th Real Time Linux Workshop, Düsseldorf, Germany, 12–13 October 2014.
38. Estes, S.; Zarkou, A.; Hope, J.M.; Suri, C.; Field-Fote, E.C. Combined Transcutaneous Spinal Stimulation and Locomotor Training to Improve Walking Function and Reduce Spasticity in Subacute Spinal Cord Injury: A Randomized Study of Clinical Feasibility and Efficacy. *J. Clin. Med.* **2021**, *10*, 1167. [CrossRef]
39. Pierantozzi, M.; Palmieri, M.; Galati, S.; Stanzione, P.; Peppe, A.; Tropepi, D.; Brusa, L.; Pisani, A.; Moschella, V.; Marciani, M.G.; et al. Pedunculopontine nucleus deep brain stimulation changes spinal cord excitability in Parkinson's disease patients. *J. Neural. Transm.* **2008**, *115*, 731–735. [CrossRef]
40. Baudry, S.; Penzer, F.; Duchateau, J. Input-output characteristics of soleus homonymous Ia afferents and corticospinal pathways during upright standing differ between young and elderly adults. *Acta Physiol.* **2014**, *210*, 667–677. [CrossRef]
41. Minassian, K.; Hofstoetter, U.S.; Danner, S.M.; Mayr, W.; Bruce, J.A.; McKay, W.B.; Tansey, K.E. Spinal rhythm generation by step-induced feedback and transcutaneous posterior root stimulation in complete spinal cord–injured individuals. *Neurorehabil. Neural. Repair.* **2016**, *30*, 233–243. [CrossRef]
42. McHugh, L.; Miller, A.; Leech, K.; Salorio, C.; Martin, R. Feasibility and utility of transcutaneous spinal cord stimulation combined with walking-based therapy for people with motor incomplete spinal cord injury. *Spinal Cord. Ser. Cases* **2020**, *6*, 104. [CrossRef]
43. Danner, S.; Krenn, M.; Hofstoetter, U.; Toth, A.; Mayr, W.; Minassian, K. Body position influences which neural structures are recruited by lumbar transcutaneous spinal cord stimulation. *PLoS ONE* **2016**, *11*, e0147479. [CrossRef]

Journal of
Clinical Medicine

MDPI

Article

Efficacy of Transcutaneous Spinal Stimulation versus Whole Body Vibration for Spasticity Reduction in Persons with Spinal Cord Injury

Evan B. Sandler [1,2], Kyle Condon [1] and Edelle C. Field-Fote [1,2,3,*]

1 Shepherd Center, Crawford Research Institute, Atlanta, GA 30309, USA; evan.sandler@shepherd.org (E.B.S.); kyle.condon@shepherd.org (K.C.)
2 Program in Biological Sciences, Georgia Institute of Technology, Atlanta, GA 30332, USA
3 Division of Physical Therapy, Emory University School of Medicine, Atlanta, GA 30322, USA
* Correspondence: Edelle.Field-Fote@Shepherd.org; Tel.: +1-404-603-4274

Citation: Sandler, E.B.; Condon, K.; Field-Fote, E.C. Efficacy of Transcutaneous Spinal Stimulation versus Whole Body Vibration for Spasticity Reduction in Persons with Spinal Cord Injury. *J. Clin. Med.* **2021**, *10*, 3267. https://doi.org/10.3390/jcm10153267

Academic Editors: Ursula S. Hofstoetter and Karen Minassian

Received: 26 June 2021
Accepted: 22 July 2021
Published: 24 July 2021

Publisher's Note: MDPI stays neutral with regard to jurisdictional claims in published maps and institutional affiliations.

Abstract: Transcutaneous spinal stimulation (TSS) and whole-body vibration (WBV) each have a robust ability to activate spinal afferents. Both forms of stimulation have been shown to influence spasticity in persons with spinal cord injury (SCI), and may be viable non-pharmacological approaches to spasticity management. In thirty-two individuals with motor-incomplete SCI, we used a randomized crossover design to compare single-session effects of TSS versus WBV on quadriceps spasticity, as measured by the pendulum test. TSS (50 Hz, 400 µs, 15 min) was delivered in supine through a cathode placed over the thoracic spine (T11-T12) and an anode over the abdomen. WBV (50 Hz; eight 45-s bouts) was delivered with the participants standing on a vibration platform. Pendulum test first swing excursion (FSE) was measured at baseline, immediately post-intervention, and 15 and 45 min post-intervention. In the whole-group analysis, there were no between- or within-group differences of TSS and WBV in the change from baseline FSE to any post-intervention timepoints. Significant correlations between baseline FSE and change in FSE were associated with TSS at all timepoints. In the subgroup analysis, participants with more pronounced spasticity showed significant decreases in spasticity immediately post-TSS and 45 min post-TSS. TSS and WBV are feasible physical therapeutic interventions for the reduction of spasticity, with persistent effects.

Keywords: antispasmodic; electrical stimulation; neuromodulation; paraplegia; pendulum test; tetraplegia

1. Introduction

At discharge from inpatient rehabilitation after spinal cord injury (SCI), more than half of all individuals report experiencing spasticity; a large proportion continue to report that spasticity interferes with function 5 years post-injury [1]. Comprehensively described as "disordered sensori-motor control presenting as intermittent or sustained involuntary activation of muscles," [2] spasticity impacts the ability to perform functional movements such as transfers, and can lead to contractures and pain [3]. Spasticity is often difficult to manage, and while antispasmodics are the most common management approach, the evidence supporting their value is weak [4]. Moreover, in a survey that acquired responses from 1076 individuals with SCI, only 38% reported that their spasticity was improved by prescribed antispasmodics [5]. By comparison, physical therapeutic interventions such as stretching and exercise were reported to improve spasticity in 48% and 45% of respondents, respectively. In this survey study, spasticity was defined for the respondents in an inclusive way, to encompass characteristics associated with the experience of spasticity, including involuntary spasms, spasms triggered by stimuli, and stiffness.

The common element among all physical therapeutic interventions directed at reducing spasticity is that they activate sensory afferents. Afferent input activates inhibitory spinal interneurons [6–10], and this effect likely underlies the reduction in spasticity in persons with SCI, associated with various forms of afferent stimulation [9,11–14]. Transcutaneous spinal stimulation (TSS) and whole-body vibration (WBV) are among the approaches that appear to have value for spasticity management [7,15–17]; both TSS and WBV activate large-diameter afferent fibers [18–20]. Evidence suggests that both electrical (TSS) and mechanical (vibration) approaches to the activation of large-diameter afferents have neuromodulatory effects arising from the activation of inhibitory mechanisms, which likely underlie the observed reduction in spasticity [7,9,12–17,21].

Beyond their influence on inhibitory mechanisms, TSS and WBV are well-suited to therapeutic applications because they have modifiable dosing parameters. Recent studies show that 50 Hz TSS, administered for 30 min over the thoracic spine at an intensity below motor threshold, reduces quadriceps spasticity, with effects persisting for up to 2 h post-intervention [7,15,22]. WBV parameters of higher frequency (50 Hz) and longer duration reduced quadriceps up to 45 min post-intervention [16]. Similar to TSS, WBV has been demonstrated to reduce the excitation of the Ia reflex arc through the activation of inhibitory interneurons [20,23]. However, unlike the direct electrical activation of dorsal roots by TSS, WBV repeatedly activates muscle spindles, which provide Ia afferent input to the spinal cord. The more direct effect of TSS on spinal circuits may have a greater impact on the reduction in spasticity.

Clinical feasibility and ease of use are strong determinants of the utility of an intervention in physical rehabilitation and/or the home environment. WBV can present economic limitations, and treatments require an individual to stand on the vibration platform, which is not possible for some who are affected by spasticity. These factors could limit the utility of WBV in the home setting compared to TSS' ease of electrode application, and the potential for lower-cost stimulation units. TSS may, therefore, be able to benefit a larger group of individuals experiencing lower extremity spasticity. Although both TSS and WBV have been studied separately, no studies have directly compared the effects of TSS and WBV on spasticity. Therefore, our primary aim was to compare the effects of these two approaches on quadriceps spasticity.

2. Materials and Methods

This study was conducted with ethical approval from the Shepherd Center Research Review Committee. All participants gave their written informed consent prior to study enrollment, in accordance with the Declaration of Helsinki. This study was registered with clinicaltrials.gov (accessed on 7 October 2014) (NCT02340910).

2.1. Subjects

Individuals with SCI were eligible for participation if they met the following inclusion criteria: injury level at or above T12, self-report of at least mild spasticity affecting the lower extremity muscles, ability to stand for ≥ 1 min using upper extremities for balance only, ability to take a step with at least one leg with or without an assistive device, and ability to rise to a standing position, requiring no more than moderate assistance from one person. Individuals with the following exclusion criteria were not considered for participation: current orthopedic problems preventing participation, history of cardiac irregularity, or progressive or potentially progressive spinal lesions.

2.2. Study Design

This study was a supplemental investigation of TSS, incorporated into a single-blind, randomized clinical trial comparing the dose–response effects of WBV on spasticity in individuals with chronic (≥ 6 months) motor-incomplete SCI. Participants received a single session of TSS within the schema of the WBV dose/frequency randomization. Complete details of the methods of the larger study are described in a prior publication [16]. In this analysis, we compare the effects of TSS with the effects of the WBV dose that had the largest effect on spasticity. Sessions were scheduled at least 1 week apart to minimize the possibility of carryover effects. Within a session, testing was performed prior to the intervention, immediately post-intervention, and 15 and 45 min post-intervention.

2.3. Intervention

2.3.1. TSS

To administer TSS, one 5 cm round self-adhesive electrode (cathode) was placed on the lower back over the T11-T12 spinous interspace. One large (10 × 15 cm) self-adhesive butterfly electrode (anode) was placed on the abdomen, over the umbilicus. Tonic stimulation (EMPI Continuum, EMPI, Inc., Clear Lake, SD, USA) was applied using a charge-balanced, biphasic waveform with a pulse width of 400 µs at 50 Hz for 15 min with participants supine. Stimulation intensity was adjusted to the level that evoked paresthesia in the legs, without visible muscle contraction. Stimulation intensity was increased slowly to allow for comfortable adjustment.

2.3.2. WBV

Participants began the WBV session seated on the edge of an adjustable height mat, with feet resting on the WBV platform (Power Plate Pro5, Performance Health Systems, LLC, Northbrook, IL, USA). The mat height was adjusted to allow the participant to rise to standing and return to sitting with minimal effort. The participant rose to stand on the vibration platform with knees slightly flexed (~30° from full extension). Eight cycles of WBV were delivered in 45-s bouts, with 1 min of seated rest between bouts, as previously described [16].

2.4. Spasticity Measurement

To evaluate the timecourse of the effects of TSS and WBV on spasticity, spasticity was assessed four times during each session: prior to the start of the intervention (baseline), immediately after the conclusion of the intervention (immediate), 15 min after the intervention (15-min post-intervention), and 45 min after the conclusion of the intervention (45-min post-intervention). We tested the leg the participant reported to be most spastic at the time of study enrollment, and the same leg was tested in each session.

The pendulum test was used to assess stretch-induced quadriceps spasticity. Participants were positioned semi-reclined with the test leg flexed at the knee, lower leg pendant over the edge of the mat, and shoe removed. An electrogoniometer (SG150, Biometrics Ltd., Newport, UK) was affixed to the test leg with the arms aligned with the thigh and shank, and the axis aligned with the knee joint center, to record knee angle during the pendulum test. The non-test leg was supported on a padded chair with the knee extended. Grasping the heel of the test leg, the examiner extended the knee and held the leg in this position for 30 s to allow movement-related excitability to dissipate. The examiner then released the heel, allowing the test leg to drop. The first swing excursion (FSE), the angle at which the swinging knee first reversed from flexion to extension in response to the reflexive contraction of the quadriceps, was the primary measure of spasticity, wherein a larger angle indicates less spasticity [24]. Comparisons have shown FSE to be the best measure of quadriceps spasticity relative to other outcomes of the pendulum test, including the relaxation index and number of oscillations [25]. Acquisition and analysis of FSE was conducted using Spike software (Cambridge Electronic Design Limited, Cambridge, England). The average FSE of 3 trials of each test session was used for analysis.

2.5. Data Analysis

All data analyses were performed in SPSS version 27 (IBM, London, UK). Data are presented as the mean ± SD. All analyses were completed for the entire sample, and for high spasticity and low spasticity subgroups, to determine the effect of spasticity severity on responsiveness to intervention. Participants were grouped into high- and low-spasticity subgroups, based on the previously published median baseline FSE (46.6°) of this study sample, which showed differential effects of WBV based on baseline spasticity [16]. The subgroups were: high spasticity = baseline FSE < 46.6° and low spasticity = baseline

FSE > 46.6°. Subgrouping was determined by the FSE of the baseline test for the TSS and WBV sessions individually, as some participants who met the criteria for high spasticity during one session did not meet that criteria for the other session.

Effect sizes were calculated using Cohen's *d* based on the pooled variance of the compared values. Effect sizes were categorized as small (0.08), moderate (0.31), or large (0.55) based on the recommendations of a recent meta-analysis of rehabilitation research outcomes [26]. For significance testing $\alpha = 0.05$ was considered significant. Paired *t*-tests were used to test for differences between baseline FSE of the two interventions, and to identify between-intervention differences in change scores in FSE at each timepoint. Independent *t*-tests were used to determine differences in the change from baseline between interventions. Paired *t*-tests were used to identify within-condition effects of TSS, comparing baseline FSE to each post-intervention measurement (immediate, 15-min post-intervention, and 45-min post-intervention). Pearson correlations were calculated to determine the relationship of change scores between interventions. Within each intervention, Pearson correlations were calculated to determine the relationship between spasticity severity, as measured by baseline FSE, and responsiveness to each intervention.

3. Results

Thirty-two participants completed both the TSS intervention and the high-frequency/long-duration WBV intervention. Participants were aged from 23 to 65 years old, and included 26 men and 6 women. Of the 32 participants, 9 were classified as American Spinal Injury Association Impairment Scale (AIS) C and 23 were classified as AIS D. Among the participants, 15 used no antispasmodic medications, 11 used baclofen only, 1 used gabapentin only, 3 used two medications (baclofen plus either gabapentin or tizanidine), and 2 used three medications (baclofen, gabapentin, and dantrolene). For the TSS intervention session, 18 participants met the high-spasticity subgroup criteria. For the WBV intervention, 13 participants met the high-spasticity subgroup criteria. Detailed demographic information, including pharmacological use by participants, is available elsewhere [16].

3.1. Effect of TSS on Quadriceps Spasticity

Mean baseline FSE for all participants during the TSS session was $44.8° \pm 16.7°$. Analysis of the full sample showed no overall effect of TSS. Baseline FSE was not different from any of the post-intervention assessments, including baseline vs. immediate post-intervention ($p = 0.42$), baseline vs. 15-min post-intervention ($p = 0.58$), and baseline vs. 45-min post-intervention ($p = 0.50$) (Table 1).

Table 1. Mean first swing excursion (FSE) by group.

	Whole Group FSE			
	Baseline	**Immediate**	**15-min Post-Intervention**	**45-min Post-Intervention**
TSS (*n* = 32)	44.78 ± 16.70	46.64 ± 15.91 (0.11)	45.97 ± 16.65 (0.07)	46.46 ± 15.72 (0.10)
WBV (*n* = 32)	50.34 ± 20.75	51.96 ± 20.85 (0.05)	49.69 ± 17.48 (−0.03)	49.86 ± 17.10 (−0.03)
	High Spasticity Group FSE			
	Baseline	**Immediate**	**15-min Post-Intervention**	**45-min Post-Intervention**
TSS (*n* = 18)	32.69 ± 9.50	37.33 ± 9.80 * (0.48)	35.85 ± 10.31 (0.32)	38.24 ± 10.60 * (0.55)
WBV (*n* = 13)	29.36 ± 8.03	32.10 ± 8.51 (0.33)	37.91 ± 10.81 * (0.90)	37.13 ± 12.60 (0.74)
	Low Spasticity Group FSE			
	Baseline	**Immediate**	**15-min Post-Intervention**	**45-min Post-Intervention**
TSS (*n* = 14)	60.33 ± 9.17	58.60 ± 14.28 (−0.14)	58.97 ± 14.09 (−0.11)	57.04 ± 15.11 (−0.26)
WBV (*n* = 19)	64.62 ± 13.10	64.51 ± 15.85 (−0.01)	57.76 ± 16.71 * (−0.46)	58.58 ± 14.15 * (-0.44)

Results represent means ± SD. Asterisks (*) denote significant ($p < 0.05$) difference from baseline mean FSE. Effect sizes for within-condition pre- and posttest comparisons listed in parentheses. TSS, Transcutaneous spinal stimulation; WBV, whole-body vibration.

Mean baseline FSE for the high-spasticity subgroup was 32.7° ± 9.50°, and for the low-spasticity subgroup, mean baseline was 60.3° ± 9.17°. Upon stratification into high-spasticity and low-spasticity subgroups, the differences between baseline FSE and post-intervention assessments were identified only in the high-spasticity subgroup. In the high-spasticity subgroup, the effect size for difference between FSE at baseline and immediately post-intervention was moderate (d = 0.48) and statistically significant (p = 0.036). The difference between FSE at baseline and 15-min post-intervention has a moderate effect size (d = 0.32), and this difference approached statistical significance (p = 0.100). The effect size for difference between FSE at baseline and 45-min post-intervention was large (d = 0.55) and statistically significant (p = 0.035).

3.2. Effect of WBV on Quadriceps Spasticity

As noted earlier, outcomes of the high-frequency/long-duration WBV session have been previously reported, and are included here to allow for comparison with TSS outcomes [16]. Mean baseline FSE for all participants during the high-frequency/long-duration WBV session was 50.34° ± 20.8°. Analysis of the full sample showed no overall effect of WBV. Baseline FSE was no different from any of the post-intervention assessments (p > 0.05 for each pairwise comparison; Table 1).

Mean baseline FSE for the high spasticity subgroup was 29.4° ± 8.0° and for the low -spasticity subgroup, mean baseline was 64.6° ± 13.1°. Upon stratification into high-spasticity and low-spasticity subgroups, differences between baseline FSE and post-intervention assessments were identified only in the high-spasticity subgroup. In the high-spasticity subgroup, the effect size for difference between FSE at baseline and imme-diately post-intervention was moderate (d = 0.33) but not statistically significant (p > 0.05). The effect size for difference between FSE at baseline and 15-min post-intervention was large (d = 0.90) and statistically significant (p = 0.014). The difference between FSE at baseline and 45-min post-intervention showed a large effect size (d = 0.73), but was not statistically significant (p > 0.05). In the low-spasticity subgroup, effect sizes were moderate and significant for difference between FSE at baseline and 15- and 45-min post-intervention (d = −0.46, d = −0.44).

3.3. Differences between TSS and WBV

Analyzing all participants, no significant differences between the TSS and WBV were observed in change in FSE at any timepoint, including baseline vs. immediate (p = 0.63), baseline vs. 15-min post-intervention (p = 0.50), or baseline vs. 45-min post-intervention (p = 0.48) (Figure 1). Significant differences were, however, observed in baseline FSE between the TSS and WBV conditions (p = 0.034). After subgrouping individuals in ac-cordance with baseline FSE, participants who were in the same subgroup at baseline testing for both TSS and WBV were included in between-intervention subgroup analy-ses. Baseline FSE was not significantly different between conditions after subgrouping (high: p = 0.64, low: p = 0.12). When analyzing subgroup change in FSE, no significant differences between TSS vs. WBV were observed at any timepoint, including baseline vs. immediate post-intervention (p = 0.61, p = 0.69), 15-min post-intervention (p = 0.34, p = 0.45), or 45-min post-intervention (p = 0.80, p = 0.98), for the high- and low-spasticity subgroups, respectively.

3.4. Influence of Baseline Spasticity on Change in Quadriceps Spasticity

Correlations between baseline FSE and change in FSE were significant at all timepoints for the TSS intervention (immediate: r = −0.44, p = 0.012, 15-min post-intervention: r = −0.36, p = 0.04) (Figure 2). Correlations were significant for the WBV intervention only at 15-min and 45-min post-intervention timepoints (15-min post-intervention: r = −0.55, p = 0.001). The strongest correlations for both intervention conditions were observed 45-min post-intervention (TSS: r = −0.49, p = 0.005, WBV: r = −0.57, p = <0.001), with the negative correlations indicating that those with high spasticity (smaller FSE) had a greater change.

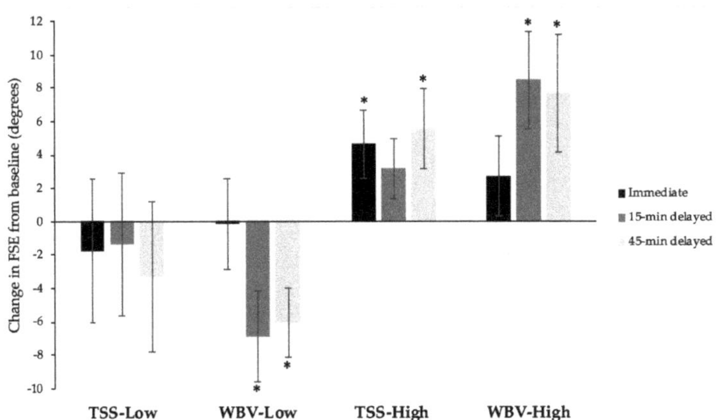

Figure 1. Mean change in FSE from baseline by subgroup. TSS, Transcutaneous spinal stimulation; WBV, whole-body vibration. Immediate (black bars), 15-min post-intervention (dark grey bars), and 45-min post-intervention (light grey bars) results are presented for participants with high (baseline FSE < 46.6°) and low (baseline FSE > 46.6°) spasticity. Asterisks (*) denote a significant ($p < 0.05$) change from baseline FSE.

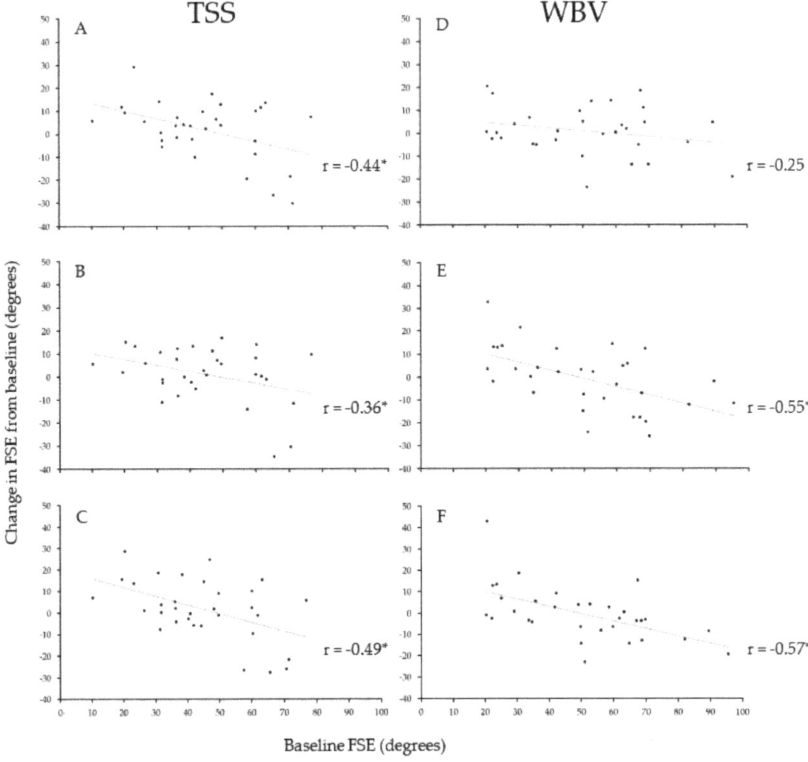

Figure 2. Correlation of baseline spasticity and change. Change in quadriceps spasticity as a function of individuals' baseline spasticity. Correlations of baseline FSE to immediate (**A,D**), 15-min post-intervention (**B,E**), and 45-min post-intervention (**C,F**) timepoints are presented for the whole group. * $p < 0.05$.

4. Discussion

This study compared the single-session effects on quadriceps spasticity of TSS and WBV, two robust forms of afferent stimulation. When all participants were considered together, neither the TSS nor the WBV intervention demonstrated significant differences between baseline and any post-intervention timepoint. Subgrouping participants based on baseline FSE demonstrated the influence of baseline spasticity on responsiveness to TSS, as was reported with WBV [16]. Participants with low spasticity (>46.6° baseline FSE) did not demonstrate significant differences between baseline FSE and any post-intervention timepoint. However, in the high-spasticity subgroup (<46.6° baseline FSE), TSS was associated with an increase in FSE at all post-intervention timepoints, indicating a decrease in quadriceps spasticity persisting for at least 45 min post-intervention. The greatest differences in FSE were found immediately and at 45-min post-intervention. Moreover, the change in FSE immediately post-intervention was associated with a moderate effect size, with a large effect size at 45-min post-intervention.

No differences between TSS and WBV regarding their effect on spasticity were identified at any post-intervention timepoint. The lack of difference between TSS and WBV was true for both the high-spasticity and low-spasticity subgroups, indicating that the two interventions have equivalent effects. Importantly, baseline FSE was significantly correlated with changes in FSE at all timepoints, for both the TSS and WBV interventions. This relationship indicates that participants with higher spasticity demonstrated a greater reduction in spasticity with each of the interventions, compared to those with lower spasticity.

As previously reported, WBV at 50 Hz for eight 45-s bouts resulted in a reduction in quadriceps spasticity in participants with high spasticity at 15-min and 45-min post-intervention assessment timepoints [16]. The differences between baseline FSE and at the 15-min and 45-min post-intervention assessments were associated with large effect sizes. This delayed antispasmodic effect of WBV is consistent with prior reports [15,17]. While no differences were observed between the TSS and WBV interventions, in the TSS condition, the high-spasticity subgroup demonstrated a significant and immediate reduction in spasticity, persisting at the 15- and 45-min post-intervention assessments. Conversely, in the WBV condition, the high-spasticity subgroup did not exhibit a significant reduction in spasticity at the immediate post-intervention assessment. However, a more robust change, as measured by effect size, was observed after WBV as compared to TSS at the 45-min post-intervention assessment. For both interventions, in contrast to the improvement in the high-spasticity subgroups, the low-spasticity subgroup demonstrated increased spasticity after intervention.

Although both TSS and WBV preferentially activate large-diameter afferent fibers [18–20], mechanical and electrical forms of stimulation may be associated with different neuromodulatory effects that account for the different effect sizes observed between interventions. While the same presynaptic inhibitory effects attributed to reductions in spasticity after vibration have been suggested in studies of TSS based on reflex testing [7,22,27], vibration also has excitatory influences on spinal circuits [28]. Vibration at 50 Hz has been shown to increase muscle activation in healthy subjects [29]. In addition, in a study of the multi-session effects of WBV on quadriceps spasticity, within-session testing identified an increase in spasticity immediately post-intervention, followed by reduced spasticity 15-min post-intervention [17]. The balance between the excitatory and inhibitory mechanisms, or the "vibration paradox", was postulated to account for the delayed spasticity reduction [17]. Prior studies have suggested that the activation of spinal circuits by TSS parallels that of epidural stimulation, with frequency-dependent effects [30,31]. While the frequency of WBV (50 Hz) was the same as that of TSS, there is likely some damping that occurs with WBV, which may result in an effective activation frequency on neural structures that is less than 50 Hz.

Unlike WBV, electrical stimulation can activate smaller diameter afferents including Aδ and C fibers in addition to Ia fibers (dependent upon stimulation intensity) [32]. Moreover, an increase in gain of the muscle spindle responses via augmented gamma motoneuron drive has been proposed as a potential mechanism for the hyper-responsiveness to stretch after spinal cord injury [33]. In preclinical studies, nociceptor activation has been shown to increase muscle spindle firing rates associated with increased gamma motoneuron activity [34–36]. In humans, however, the induced nociception during a relaxed state and during muscle contraction did not affect muscle spindle discharge rates, confuting the excitatory gamma motoneuron response to nociceptive stimuli observed in preclinical models [35]. The same absence of excitation to the gamma motoneuron exists in individuals with chronic spinal cord injury; therefore, it is unlikely that the activation of nociceptor fibers is responsible for the differences seen between TSS and WBV [37].

TSS and WBV target different spinal segments during stimulation. The T11-T12 interspace placement of the cathode during TSS targets the rostral portion of the lumbar enlargement, highly favoring motoneuron pools of the quadriceps [38,39]. Although the quadriceps are targeted when standing on the WBV platform with a slight bend at the knee, the muscle afferents of all lower extremity musculature are active during vibration, as evidenced by soleus reflex modulation in non-injured individuals and those with SCI [20,28,40]. Neuromodulatory effects, as elucidated by paired-pulse stimulation studies, have demonstrated the potential for inhibition by heteronymous circuits [38,41–43], providing evidence of intersegmental modulation. Moreover, the voluntary contraction of non-tested musculature during spinal reflex testing has also demonstrated the influence of intersegmental circuits on reflex modulation [44]. The amount of excitation or inhibition from other muscles onto the quadriceps may influence its responsiveness to stretch post-vibration, accounting for the differential effects of TSS and WBV.

The change in responsiveness to stretch in persons with higher levels of spasticity observed in the present study adds to the growing body of literature of the beneficial effects of physical therapeutic interventions on spasticity in persons with SCI. In a comparative study of physical therapeutic interventions, TSS, stretching, and continuous passive movement all demonstrated immediate and persistent reductions in spasticity in persons with chronic SCI [15], congruent with the results found in this study. A reduction in spasticity in persons with SCI has been observed after other forms of electrical stimulation, including transcutaneous electrical nerve stimulation [14] and neuromuscular electrical stimulation/functional electrical stimulation [45].

Physical therapeutic interventions such as TSS and WBV are clinically accessible and avoid the negative side effects associated with pharmacological interventions for spasticity management, such as fatigue, sleepiness [46] and muscle weakness [47]. It is worth noting that antispasmodic medications are prescribed to be taken two or three times daily, as the persistence of their effects is limited. Likewise, it is possible that it is necessary to administer physical therapeutic interventions several times a day to achieve optimal spasticity management. In this case, the greater portability of TSS compared to WBV, and its potential to be "wearable", may make it the more advantageous approach.

Limitations

One limitation of this work relates to the within-participant variability in the quadriceps spasticity measurement. When comparing TSS and WBV based on spasticity at baseline, it was necessary to exclude individuals who did not demonstrate consistently high or low spasticity at baseline for both interventions from the analysis. As demonstrated in our previous work, an individual's quadriceps spasticity may vary day-to-day. By excluding the nine individuals whose baseline spasticity did not result in the same high versus low spasticity subgroup for both conditions, we were able to compare responses in the same individuals across interventions. However, this resulted in a reduced sample size that may have decreased the power to detect differences between interventions. Another limitation of this study is the use of a single set of parameters to achieve antispasmodic effects. While studies of epidural stimulation have been used as guides for parameter selection, systematic studies are needed to explore the optimal parameters, frequency and intensity, for spasticity reduction. Lastly, comparisons between and within interventions can only be drawn up to 45-min post-intervention. The persistence of effects beyond 45 min should be explored and may prove different between interventions.

5. Conclusions

In persons with chronic motor-incomplete SCI with higher levels of quadriceps spasticity, transcutaneous spinal stimulation and whole-body vibration are associated with reduction in spasticity. Both interventions provide persistent effects for at least 45 min after cessation of the intervention. While this is a single-session comparison study, it is necessary to determine the potential of an intervention when treating this participant population. With the limited financial resources allocated for therapeutic intervention throughout the recovery period, future research should focus on identifying whether an individual with SCI is a potential responder prior to performing an intervention.

Author Contributions: Conceptualization, E.S., K.C. and E.C.F.-F.; methodology, E.S., K.C. and E.C.F.-F.; formal analysis, E.S.; writing—original draft preparation, E.S., K.C. and E.C.F.-F.; funding acquisition, E.C.F.-F. All authors reviewed the manuscript and contributed to the editing. All authors have read and agreed to the published version of the manuscript.

Funding: This research was funded by NIH National Institute of Child Health and Human Development (NICHD) R01 HD079009-02 (E.C.F.-F.).

Institutional Review Board Statement: The study was conducted according to the guidelines of the Declaration of Helsinki, and approved by the Shepherd Center Research Review Committee (protocol code 629 and date of approval 7 October 2014).

Informed Consent Statement: Informed consent was obtained from all subjects involved in the study.

Data Availability Statement: The data presented in this study are available upon request from the corresponding author.

Acknowledgments: The authors thank research participants who volunteered their time to participate in this study. We also thank Temple Moore, OTR/L, Elizabeth Sasso-Lance, PT, DPT, and Rachel Beltzer for their recruitment efforts along with Nicholas Evans, MHS, CEP, and Adam Holzwarth, PT, DPT for their assistance with conducting interventions.

Conflicts of Interest: The authors declare no conflict of interest. The funders had no role in the design of the study; in the collection, analyses, or interpretation of data; in the writing of the manuscript, or in the decision to publish the results.

References

1. Holtz, K.A.; Lipson, R.; Noonan, V.K.; Kwon, B.K.; Mills, P.B. Prevalence and Effect of Problematic Spasticity After Traumatic Spinal Cord Injury. *Arch. Phys. Med. Rehabil.* **2017**, *98*, 1132–1138. [CrossRef] [PubMed]
2. Pandyan, A.D.; Gregoric, M.; Barnes, M.P.; Wood, D.; Van Wijck, F.; Burridge, J.; Hermens, H.; Johnson, G.R. Spasticity: Clinical perceptions, neurological realities and meaningful measurement. *Disabil. Rehabil.* **2005**, *27*, 2–6. [CrossRef]
3. Sköld, C. Spasticity in spinal cord injury: Self- and clinically rated intrinsic fluctuations and intervention-induced changes. *Arch. Phys. Med. Rehabil.* **2000**, *81*, 144–149. [CrossRef]
4. Taricco, M.; Adone, R.; Pagliacci, C.; Telaro, E. Pharmacological interventions for spasticity following spinal cord injury. *Cochrane Database Syst. Rev.* **2000**, CD001131. [CrossRef]
5. Field-Fote, E.C.; Furbish, C.L.; Tripp, N.E.; Zanca, J.; Dyson-Hudson, T.; Kirshblum, S.; Heinemann, A.; Chen, D.; Felix, E.; Worobey, L.; et al. Characterizing the Experience of Spasticity after Spinal Cord Injury: A National Survey Project of the Spinal Cord Injury Model Systems Centers. *Arch. Phys. Med. Rehabil.* **2021**. [CrossRef] [PubMed]
6. Minassian, K.; Persy, I.; Rattay, F.; Pinter, M.M.; Kern, H.; Dimitrijevic, M.R. Human lumbar cord circuitries can be activated by extrinsic tonic input to generate locomotor-like activity. *Hum. Mov. Sci.* **2007**, *26*, 275–295. [CrossRef]
7. Hofstoetter, U.S.; McKay, W.B.; Tansey, K.E.; Mayr, W.; Kern, H.; Minassian, K. Modification of spasticity by transcutaneous spinal cord stimulation in individuals with incomplete spinal cord injury. *J. Spinal Cord Med.* **2014**, *37*, 202–211. [CrossRef]
8. Pierrot-Deseilligny, E. Assessing changes in presynaptic inhibition of Ia afferents during movement in humans. *J. Neurosci. Methods* **1997**, *74*, 189–199. [CrossRef]
9. Butler, J.E.; Godfrey, S.; Thomas, C.K. Depression of involuntary activity in muscle paralyzed by spinal cord injury. *Muscle Nerve* **2006**, *33*, 637–644. [CrossRef]
10. Jankowska, E. Interneuronal relay in spinal pathways from proprioceptors. *Prog. Neurobiol.* **1992**, *38*, 335–378. [CrossRef]

11. Bajd, T.; Gregoric, M.; Vodovnik, L.; Benko, H. Electrical stimulation in treating spasticity resulting from spinal cord injury. *Arch. Phys. Med. Rehabil.* **1985**, *66*, 515–517. [PubMed]
12. Seib, T.P.; Price, R.; Reyes, M.R.; Lehmann, J.F. The quantitative measurement of spasticity: Effect of cutaneous electrical stimulation. *Arch. Phys. Med. Rehabil.* **1994**, *75*, 746–750. [CrossRef]
13. DeForest, B.A.; Bohorquez, J.; Perez, M.A. Vibration attenuates spasm-like activity in humans with spinal cord injury. *J. Physiol.* **2020**, *598*, 2703–2717. [CrossRef] [PubMed]
14. Gómez-Soriano, J.; Serrano-Muñoz, D.; Bravo-Esteban, E.; Avendaño-Coy, J.; Ávila-Martin, G.; Galán-Arriero, I.; Taylor, J. Afferent stimulation inhibits abnormal cutaneous reflex activity in patients with spinal cord injury spasticity syndrome. *NeuroRehabilitation* **2018**, *43*, 135–145. [CrossRef] [PubMed]
15. Estes, S.P.; Iddings, J.A.; Field-Fote, E.C. Priming Neural Circuits to Modulate Spinal Reflex Excitability. *Front. Neurol.* **2017**, *8*, 17. [CrossRef] [PubMed]
16. Estes, S.; Iddings, J.A.; Ray, S.; Kirk-Sanchez, N.J.; Field-Fote, E.C. Comparison of Single-Session Dose Response Effects of Whole Body Vibration on Spasticity and Walking Speed in Persons with Spinal Cord Injury. *Neurotherapeutics* **2018**, *15*, 684–696. [CrossRef]
17. Ness, L.L.; Field-Fote, E.C. Effect of whole-body vibration on quadriceps spasticity in individuals with spastic hypertonia due to spinal cord injury. *Restor. Neurol. Neurosci.* **2009**, *27*, 621–631. [CrossRef]
18. Danner, S.M.; Hofstoetter, U.S.; Ladenbauer, J.; Rattay, F.; Minassian, K. Can the human lumbar posterior columns be stimulated by transcutaneous spinal cord stimulation? A modeling study. *Artif. Organs* **2011**, *35*, 257–262. [CrossRef]
19. Holsheimer, J. Computer modelling of spinal cord stimulation and its contribution to therapeutic efficacy. *Spinal Cord* **1998**, *36*, 531–540. [CrossRef] [PubMed]
20. Ritzmann, R.; Kramer, A.; Gollhofer, A.; Taube, W. The effect of whole body vibration on the H-reflex, the stretch reflex, and the short-latency response during hopping. *Scand. J. Med. Sci. Sports* **2013**, *23*, 331–339. [CrossRef]
21. Elbasiouny, S.M.; Moroz, D.; Bakr, M.M.; Mushahwar, V.K. Management of spasticity after spinal cord injury: Current techniques and future directions. *Neurorehabil. Neural Repair* **2010**, *24*, 23–33. [CrossRef] [PubMed]
22. Hofstoetter, U.S.; Freundl, B.; Danner, S.M.; Krenn, M.J.; Mayr, W.; Binder, H.; Minassian, K. Transcutaneous Spinal Cord Stimulation Induces Temporary Attenuation of Spasticity in Individuals with Spinal Cord Injury. *J. Neurotrauma* **2020**, *37*, 481–493. [CrossRef]
23. Karacan, I.; Cidem, M.; Cidem, M.; Türker, K.S. Whole-body vibration induces distinct reflex patterns in human soleus muscle. *J. Electromyogr. Kinesiol.* **2017**, *34*, 93–101. [CrossRef] [PubMed]
24. Stillman, B.; McMeeken, J. A video-based version of the pendulum test: Technique and normal response. *Arch. Phys. Med. Rehabil.* **1995**, *76*, 166–176. [CrossRef]
25. Fowler, E.G.; Nwigwe, A.I.; Ho, T.W. Sensitivity of the pendulum test for assessing spasticity in persons with cerebral palsy. *Dev. Med. Child Neurol.* **2000**, *42*, 182–189. [CrossRef] [PubMed]
26. Kinney, A.R.; Eakman, A.M.; Graham, J.E. Novel Effect Size Interpretation Guidelines and an Evaluation of Statistical Power in Rehabilitation Research. *Arch. Phys. Med. Rehabil.* **2020**, *101*, 2219–2226. [CrossRef]
27. Danner, S.M.; Krenn, M.; Hofstoetter, U.S.; Toth, A.; Mayr, W.; Minassian, K. Body Position Influences Which Neural Structures Are Recruited by Lumbar Transcutaneous Spinal Cord Stimulation. *PLoS ONE* **2016**, *11*, e0147479. [CrossRef]
28. Pollock, R.D.; Woledge, R.C.; Martin, F.C.; Newham, D.J. Effects of whole body vibration on motor unit recruitment and threshold. *J. Appl. Physiol.* **2012**, *112*, 388–395. [CrossRef] [PubMed]
29. Cardinale, M.; Lim, J. Electromyography activity of vastus lateralis muscle during whole-body vibrations of different frequencies. *J. Strength Cond. Res.* **2003**, *17*, 621–624. [CrossRef]
30. Hofstoetter, U.S.; Freundl, B.; Binder, H.; Minassian, K. Common neural structures activated by epidural and transcutaneous lumbar spinal cord stimulation: Elicitation of posterior root-muscle reflexes. *PLoS ONE* **2018**, *13*, e0192013. [CrossRef]
31. Pinter, M.M.; Gerstenbrand, F.; Dimitrijevic, M.R. Epidural electrical stimulation of posterior structures of the human lumbosacral cord: 3. Control of spasticity. *Spinal Cord* **2000**, *38*, 524–531. [CrossRef]
32. Holsheimer, J. Which Neuronal Elements are Activated Directly by Spinal Cord Stimulation. *Neuromodulation* **2002**, *5*, 25–31. [CrossRef] [PubMed]
33. Dietz, V.; Sinkjaer, T. Spastic movement disorder: Impaired reflex function and altered muscle mechanics. *Lancet. Neurol.* **2007**, *6*, 725–733. [CrossRef]
34. Mense, S. Nociception from skeletal muscle in relation to clinical muscle pain. *Pain* **1993**, *54*, 241–289. [CrossRef]
35. Lima, C.R.; Sahu, P.K.; Martins, D.F.; Reed, W.R. The Neurophysiological Impact of Experimentally-Induced Pain on Direct Muscle Spindle Afferent Response: A Scoping Review. *Front. Cell. Neurosci.* **2021**, *15*, 649529. [CrossRef]
36. ALNAES, E.; JANSEN, J.K.; RUDJORD, T. FUSIMOTOR ACTIVITY IN THE SPINAL CAT. *Acta Physiol. Scand.* **1965**, *63*, 197–212. [CrossRef]
37. Macefield, V.G. Discharge rates and discharge variability of muscle spindle afferents in human chronic spinal cord injury. *Clin. Neurophysiol.* **2013**, *124*, 114–119. [CrossRef]
38. Sayenko, D.G.; Atkinson, D.A.; Floyd, T.C.; Gorodnichev, R.M.; Moshonkina, T.R.; Harkema, S.J.; Edgerton, V.R.; Gerasimenko, Y.P. Effects of paired transcutaneous electrical stimulation delivered at single and dual sites over lumbosacral spinal cord. *Neurosci. Lett.* **2015**, *609*, 229–234. [CrossRef] [PubMed]

39. Sayenko, D.G.; Atkinson, D.A.; Dy, C.J.; Gurley, K.M.; Smith, V.L.; Angeli, C.; Harkema, S.J.; Edgerton, V.R.; Gerasimenko, Y.P. Spinal segment-specific transcutaneous stimulation differentially shapes activation pattern among motor pools in humans. *J. Appl. Physiol.* **2015**, *118*, 1364–1374. [CrossRef] [PubMed]
40. Sayenko, D.G.; Masani, K.; Alizadeh-Meghrazi, M.; Popovic, M.R.; Craven, B.C. Acute effects of whole body vibration during passive standing on soleus H-reflex in subjects with and without spinal cord injury. *Neurosci. Lett.* **2010**, *482*, 66–70. [CrossRef] [PubMed]
41. Bergmans, J.; Delwaide, P.J.; Gadea-Ciria, M. Short-latency effects of low-threshold muscular afferent fibers on different motoneuronal pools of the lower limb in man. *Exp. Neurol.* **1978**, *60*, 380–385. [CrossRef]
42. Hultborn, H.; Meunier, S.; Pierrot-Deseilligny, E.; Shindo, M. Changes in presynaptic inhibition of Ia fibres at the onset of voluntary contraction in man. *J. Physiol.* **1987**, *389*, 757–772. [CrossRef] [PubMed]
43. Crone, C.; Hultborn, H.; Mazières, L.; Morin, C.; Nielsen, J.; Pierrot-Deseilligny, E. Sensitivity of monosynaptic test reflexes to facilitation and inhibition as a function of the test reflex size: A study in man and the cat. *Exp. Brain Res.* **1990**, *81*, 35–45. [CrossRef] [PubMed]
44. Masugi, Y.; Sasaki, A.; Kaneko, N.; Nakazawa, K. Remote muscle contraction enhances spinal reflexes in multiple lower-limb muscles elicited by transcutaneous spinal cord stimulation. *Exp. Brain Res.* **2019**. [CrossRef] [PubMed]
45. Bekhet, A.H.; Bochkezanian, V.; Saab, I.M.; Gorgey, A.S. The Effects of Electrical Stimulation Parameters in Managing Spasticity After Spinal Cord Injury: A Systematic Review. *Am. J. Phys. Med. Rehabil.* **2019**, *98*, 484–499. [CrossRef] [PubMed]
46. McCormick, Z.L.; Chu, S.K.; Binler, D.; Neudorf, D.; Mathur, S.N.; Lee, J.; Marciniak, C. Intrathecal Versus Oral Baclofen: A Matched Cohort Study of Spasticity, Pain, Sleep, Fatigue, and Quality of Life. *PM R* **2016**, *8*, 553–562. [CrossRef]
47. Thomas, C.K.; Häger-Ross, C.K.; Klein, C.S. Effects of baclofen on motor units paralysed by chronic cervical spinal cord injury. *Brain* **2010**, *133*, 117–125. [CrossRef]

Journal of
Clinical Medicine

Article

Combined Transcutaneous Spinal Stimulation and Locomotor Training to Improve Walking Function and Reduce Spasticity in Subacute Spinal Cord Injury: A Randomized Study of Clinical Feasibility and Efficacy

Stephen Estes [1,†], Anastasia Zarkou [1,†], Jasmine M. Hope [1,2], Cazmon Suri [1] and Edelle C. Field-Fote [1,3,4,*]

[1] Shepherd Center, Crawford Research Institute, Atlanta, GA 30309, USA; stephenpestes@gmail.com (S.E.); anastasia.zarkou@shepherd.org (A.Z.); jasmine.hope@shepherd.org (J.M.H.); cazmon.suri@shepherd.org (C.S.)

[2] Graduate Division of Biological and Biomedical Sciences, Laney Graduate School, Emory University, Atlanta, GA 30322, USA

[3] Division of Physical Therapy, Emory University School of Medicine, Atlanta, GA 30322, USA

[4] Program in Biomedical Sciences, Georgia Institute of Technology, Atlanta, GA 30332, USA

* Correspondence: Edelle.Field-Fote@Shepherd.org; Tel.: +1-404-603-4274

† These authors contributed equally to the manuscript.

Citation: Estes, S.; Zarkou, A.; Hope, J.M.; Suri, C.; Field-Fote, E.C. Combined Transcutaneous Spinal Stimulation and Locomotor Training to Improve Walking Function and Reduce Spasticity in Subacute Spinal Cord Injury: A Randomized Study of Clinical Feasibility and Efficacy. *J. Clin. Med.* **2021**, *10*, 1167. https://doi.org/10.3390/jcm10061167

Academic Editors: Ursula S. Hofstoetter and Karen Minassian

Received: 1 February 2021
Accepted: 6 March 2021
Published: 11 March 2021

Publisher's Note: MDPI stays neutral with regard to jurisdictional claims in published maps and institutional affiliations.

Abstract: Locomotor training (LT) is intended to improve walking function and can also reduce spasticity in motor-incomplete spinal cord injury (MISCI). Transcutaneous spinal stimulation (TSS) also influences these outcomes. We assessed feasibility and preliminary efficacy of combined LT + TSS during inpatient rehabilitation in a randomized, sham-controlled, pragmatic study. Eighteen individuals with subacute MISCI (2–6 months post-SCI) were enrolled and randomly assigned to the LT + TSS or the LT + TSS$_{sham}$ intervention group. Participants completed a 4-week program consisting of a 2-week wash-in period (LT only) then a 2-week intervention period (LT + TSS or LT + TSS$_{sham}$). Before and after each 2-week period, walking (10 m walk test, 2-min walk test, step length asymmetry) and spasticity (pendulum test, clonus drop test, modified spinal cord injury—spasticity evaluation tool) were assessed. Sixteen participants completed the study. Both groups improved in walking speed and distance. While there were no significant between-groups differences, the LT + TSS group had significant improvements in walking outcomes following the intervention period; conversely, improvements in the LT + TSS$_{sham}$ group were not significant. Neither group had significant changes in spasticity, and the large amount of variability in spasticity may have obscured ability to observe change in these measures. TSS is a feasible adjunct to LT in the subacute stage of SCI and may have potential to augment training-related improvements in walking outcomes.

Keywords: activity-based therapy; gait; locomotion; neuromodulation; paraplegia; task-specific training; tetraplegia; use-dependent plasticity

1. Introduction

The loss of motor control is a hallmark of spinal cord injury (SCI), impacting the health and quality of life for persons with a SCI. To overcome these impairments, rehabilitation therapies aim to improve motor control and maximize functional recovery by optimally activating spared pathways. For walking function, rehabilitation therapy focuses on locomotor training (LT), which involves repetitive stepping with or without manual or robotic assistance, often augmented by body-weight support in either the treadmill or overground environments [1]. Through repetition, LT is thought to activate use-dependent neural mechanisms within spared neuronal circuits to improve motor control and walking function [2]. While improvements in walking function have been observed following LT, randomized studies have demonstrated that no single approach appears to be superior to others [3]. Overall, improvements have been modest and it has been suggested that combination approaches may offer the greatest promise. In recent years, there has been increased interest in combining traditional LT strategies with novel neurotherapeutic approaches to amplify these mechanisms [4–6].

Spinal stimulation is a neuromodulation strategy that has the potential to amplify the inputs of therapies like LT that are intended to promote use-dependent neuroplasticity and improve walking outcomes. Case reports and series on persons with chronic motor-complete and severe incomplete SCI have shown that intensive LT alone was insufficient for improving volitional motor control and walking function; however, with the addition of surgically implanted epidural spinal stimulators, these same individuals showed improvements in volitional muscle activation and walking function [7–9].

While often referred to as "spinal cord stimulation", evidence has shown that the effects of epidural stimulation are mainly attributable to activation of peripheral nerve roots, primarily the large-diameter afferent fibers. The same spinal circuits that are engaged by epidural stimulation are also engaged via transcutaneous spinal stimulation (TSS), which uses electrodes placed on the skin over the vertebrae [10,11]. Definitive evidence for the mechanism whereby spinal stimulation facilitates volitional movement in persons with SCI is not available; however, a reasonable theory is that summation of the weak, subthreshold descending signals related to volitional effort and the stimulation-induced increase in excitability of spinal circuits together are sufficient to bring the motoneurons to threshold. Therefore, combining LT and TSS may represent a non-invasive and clinically accessible approach to improving walking function. While proof-of-concept studies have suggested that the combination of TSS and LT appears to have value for improving walking speed [12] and step kinematics [5], the lack of a control group in these studies makes it difficult to determine whether these effects were due to the addition of TSS or simply to LT.

For many persons with SCI, the loss of volitional muscle activation accounts for only part of their motor control impairments. Upwards of 78% of persons with a SCI experience spasticity [13], characterized by sustained or intermittent involuntary muscle activity that manifests as spasms and stiffness [14]. While the impact of spasticity can vary from person to person, it has been shown to have an overall negative impact on an individual's quality of life and can interfere with daily activities such as sleep and mobility [15–17]. Spasticity can hinder rehabilitative efforts, and lead to a range of gait abnormalities that impact walking [16]. In those who experience problematic spasticity, its management is a top priority. Oral antispasmodic medications are often used as the first line of treatment for managing spasticity; however, other than for tizanidine, evidence for their effectiveness is lacking, and they carry unwanted side effects such as drowsiness, muscle weakness, and lethargy [18,19]. People with SCI report that movement-related activities are among the strategies that they find most useful for managing their spasticity [20], and studies have shown that LT is associated with reductions in spasticity [21–23]. Likewise, there is evidence that spinal stimulation may have value for the management of spasticity in persons with SCI [24–28].

Since both LT and TSS each have the potential to improve walking function and reduce spasticity, it seems reasonable that using these approaches concurrently could be associated with larger effects. Since one of the advantages of TSS is that it is a clinically accessible approach, the purpose of this study was to assess the feasibility and preliminary efficacy of combining LT with TSS as a component of real-world clinical rehabilitation. Using a randomized, sham-controlled, pragmatic study design, we aimed to assess the effects of TSS when used as an adjuvant to 4 weeks of therapist-directed, usual care LT. While this design deviates from the standard clinical explanatory study design and increases variability based on the variety of LT approaches the clinician can use, this pragmatic design more closely mirrors real-world clinical practice and can therefore improve the translatability of study findings to the clinical setting.

2. Materials and Methods

This study was carried out with the ethics approval from the Shepherd Center Research Review Committee (Project #696). All participants provided written informed consent prior to study enrollment in accordance with the Declaration of Helsinki, and the study was conducted in accordance with the Health Insurance Portability and Accountability Act (HIPAA) guidelines. This study was registered with clinicaltrials.gov (NCT03240601).

2.1. Subjects

We captured data related to both feasibility of applying TSS as part of a program of usual care LT, as well as data related to the impact of LT + TSS on walking and spasticity measures. Participants were recruited from the inpatient clinical services of a specialty rehabilitation hospital. Recruitment was via advertisements, information provided in a monthly research education class, and information conveyed to potential participants by clinical staff. To be eligible for the study, participants had to be in the subacute rehabilitation phase of the SCI (2–6 months post-injury) and had to qualify for participation in a clinical LT program as determined by their physical therapist. Individuals were eligible for participation if they met the following inclusion criteria: 16–65 years of age with SCI, ability to take at least one step with or without an assistive device, and presence of at least mild spasticity affecting the lower extremity muscles (as determined by participant self-report). Individuals with the following exclusion criteria were not considered for participation: neurological level of injury at or below T12, progressive or potentially progressive spinal lesions (including degenerative or progressive vascular disorders of the spine and/or spinal cord), history of cardiovascular irregularities, difficulty following instructions, orthopedic problems that would prevent participation in study interventions (i.e., knee or hip flexion contractures > 10 degrees), women who were pregnant, or had reason to believe may become pregnant, persons who have implanted stimulators/electronic devices of any type, and persons with an active infection of any type.

2.2. Study Design

We used a wash-in, randomized, control study design (Figure 1) consisting of four consecutive weeks of LT directed by physical therapists. For the first two weeks, participants received 6 bouts of LT. For the last 2 weeks, participants were randomized to receive 6 bouts of either 30 min of TSS coupled with LT or a sham-control stimulation coupled with LT. Primary outcome assessments for walking function and spasticity were conducted at the beginning and end of each 2-week intervention block (first 2 weeks (wash-in phase) and second 2 weeks (intervention phase)). The assessments at the end of each 2-week block were performed prior to the training on that day. Pre- and post-training assessments of spasticity and assessments of tolerability (see below) were performed on each training day.

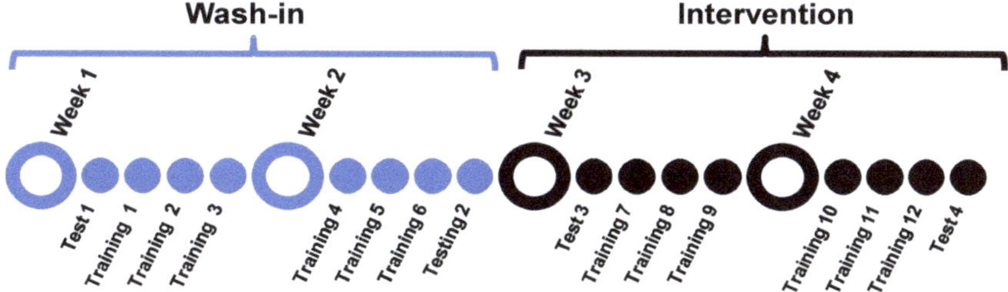

Figure 1. Study design.

2.3. Intervention

Intervention: LT approaches used in the study included treadmill-based training with or without body weight support and with or without manual or robotic assistance (Lokomat, Hocoma, Volketswil, Switzerland), as well as overground locomotor training with or without body weight support and with or without manual assistance. LT approaches for each participant were chosen by the treating therapist in accordance with standard clinical practice. LT duration was approximately 1 h and included both setup and take down time of all equipment.

Transcutaneous spinal cord stimulation (TSS): Electrical stimulation (50 Hz, biphasic pulses) was applied using a portable electrotherapy device (Empi Continuum, DJO Global, Vista, CA, USA), as previously described [28]. Briefly, with the participant seated, the stimulating electrode (5 cm round electrode) was placed over vertebral levels T11/T12, identified via manual palpation, and the reference electrode (10 cm × 15 cm butterfly electrode) was placed over the umbilicus. With the participant standing, the stimulator was turned on and adjusted to a level at which the participant reported sensations of tingling in the legs and feet or to the highest level the participant could tolerate. For those with impaired sensation as well as those with sensation, target stimulation was subthreshold for lower extremity muscle activation and verified by the absence of visible muscle contraction. Stimulation was delivered for 30 min at the target stimulation intensity and occurred concomitantly with LT. Following 30 min of stimulation, the intensity was ramped down and the stimulation unit was turned off and disconnected. If LT finished prior to the 30 min of stimulation, then participants completed their 30 min of stimulation while seated on a high-lo mat where training day assessments would occur. Conversely, if the 30 min of stimulation finished prior to the end of LT, then the LT session was completed prior to the assessments.

Sham-control stimulation—TSS_{sham}: The sham-control was designed to control for placebo effects associated with the perception of intervention. For the TSS_{sham}, a 2-inch round electrode was placed over T11/T12 and a reference electrode was placed over the umbilicus, as performed with TSS. The intensity of electrical stimulation was briefly ramped up to a level at which the participants reported perceiving the stimulation, then ramped down and turned off for the remainder of the intervention. The stimulator and electrodes remained attached to the participant for 30 min of the LT session, similar to the active TSS intervention. After 30 min, the stimulator unit was disconnected from the electrode leads.

2.4. Outcome Measures

2.4.1. Walking Assessments

All walking-related data were captured during overground walking: participants were given the standardized instruction "walk as quickly and safely as possible". Participants wore a gait belt during walking assessments and were guarded by a physical therapist for safety. During testing, participants used the walking assistive and orthotic devices they typically used during therapy, however no body weight support was provided during testing.

10-Meter Walk Test (10MWT): We utilized the 10MWT, which was our primary outcome measure for walking speed to assess changes in walking speed at the beginning and end of each 2-week LT block. A stopwatch was used to capture walking speed. Participants completed 3 walks in each assessment session with a 2 min rest between trials.

Additionally, spatiotemporal gait characteristics were recorded during the 10MWT using a 6 m instrumented walkway (GAITRite, CIR Systems, Franklin, NJ, USA) positioned in the middle of the 10 m walking path. Each walk consisted of at least 4 consecutive footfalls over the mat. Footfall data generated by the GAITRite software were used to compute step asymmetry between the stronger and weaker leg as follows: Step length asymmetry: larger step length/(larger step length + smaller step length). Step length asymmetry was averaged across trials and used for analysis.

2-Minute Walk Test (2MWT) was used to assess changes in walking distance at the beginning and end of each 2-week LT block. Distance was measured with a measuring wheel.

2.4.2. Spasticity Assessments

Pendulum test: The pendulum test, our primary measure of spasticity, was used to assess stretch-induced quadriceps reflex excitability based on first swing excursion (FSE) angle (recorded using motion capture software), wherein a larger angle indicates a greater excursion of the limb before the onset of reflex muscle contraction, and therefore less spasticity [29]. Briefly, participants were positioned reclined at the edge of an adjustable height mat with both legs flexed at the knee, and with the lower leg hanging over the mat with shoes removed. Wireless sensors were strapped to both lower extremities to capture knee joint angles using inertial motion capture software (XSENS MVN, Xsens Technologies BV, Enschede, The Netherlands). Calibrated angles were verified with the leg in full extension prior to starting the test; each leg was extended and dropped separately. Pendulum responses in each leg were tested 3 times. For each leg, the average FSE in the 3 trials of each test session was used for analysis. To confirm stretch-induced quadriceps activation during the pendulum test, electromyographic (EMG) data was concurrently monitored via electrodes (Motion Lab Systems, Baton Rogue, LA) placed over the rectus femoris muscle. EMG data were monitored using Spike software (Cambridge Electronic Design Limited, Cambridge, England). Knee angle data were analyzed off-line using customized MATLAB software (MATLAB, Mathworks, Natick, MA, USA).

Ankle clonus drop test (drop test): The drop test, the plantar flexor analog of the pendulum test, was used to assess stretch-induced reflex excitability in the ankle plantar flexors. The number of clonic oscillations captured using the drop test correlates with both electrophysiologic and clinical measures of spasticity [30]. Participants sat upright with back support on the edge of a mat table, with shoes removed and socks left on. Wireless sensors were strapped to both lower extremities to capture ankle joint angles using inertial motion capture software (XSENS MVN, Xsens Technologies BV, Enschede, The Netherlands). The ball (metatarsal heads) of one foot was positioned on the edge of a platform (10 cm height). The mat height was adjusted to ensure that the hip, knee, and ankle joints were at 90-degree angles. The participants' leg was lifted from beneath the knee until it came into contact with a T-bar positioned 10 cm above the resting position of the knee. The examiner quickly released the leg allowing the forefoot to impact the edge of the platform, eliciting a quick stretch of the plantar flexors. Responses in each ankle were tested 3 times. For each leg, the number of clonic oscillations in each trial was counted off-line, and the average number of oscillations for the 3 trials of each test session was used for analysis. Soleus EMG data were concurrently monitored using Spike software (Cambridge Electronic Design Limited, Cambridge, England), and testing resumed only when the muscle had been quiet for 30 s. Ankle joint oscillations were analyzed off-line using customized MATLAB software (MATLAB, Mathworks, Natick, MA, USA).

Modified spinal cord injury—spasticity evaluation tool (mSCI-SET): The mSCI-SET is a self-report measure of the effect of spasticity on 33 aspects of life. Scores range from −2 to +1, where negative scores indicate problematic effects of spasticity, while positive scores indicate helpful effects of spasticity [31].

2.4.3. Pre/Post-Training Assessments

Spinal Cord Assessment Tool for Spastic Reflexes (SCATS): The SCATS was used as a clinical measure of spasticity that has been shown to be reliable and valid in persons with SCI [32]. The SCATS was performed on each leg immediately before and after every intervention session by masked clinical assessors. Assessors rated clonus, flexor spasms, and extensor spasms on a 4-point scale for each lower extremity. For clonus and extensor spasms, the scale was as follows: 0—no reaction, 1—mild, lasting < 3 s, 2—moderate, lasting 3–10 s, and 3—severe, lasting > 10 s. The scale for flexor spasms was determined by measuring degrees of flexion at the knee and hip on the following scale: 0—no reaction, 1—mild, <10 degrees, 2—moderate, 10–30 degrees, and 3—severe, >30 degrees.

Tolerability of stimulation: During the 2-week intervention period, participants were asked to rate the tolerability of their stimulation (LT + TSS and LT + TSS$_{sham}$). Using a numeric rating scale (NRS) from 0 to 10, participants were asked how tolerable the stimulation was in regard to pain, with 0 being no pain and 10 being the worst pain imaginable. Responses were taken prior to the start of the stimulation to determine baseline tolerability, and then at 1 and 30 min during stimulation.

2.5. Data Analysis

Data were analyzed using SPSS (version 26; SPSS Inc., Chicago, IL, USA). Numerical and ordinal data were presented as the mean (standard error (SE)) and median (interquartile range (IQR)), respectively. For spasticity measures, data in the more- and less-impaired legs (as determined by lower extremity motor score (LEMS)), were analyzed separately. Due to our small sample size, nonparametric tests were performed [33]. Mann–Whitney U tests were conducted to examine differences between the LT + TSS and LT + TSS$_{sham}$ groups in change in walking ability (speed, distance, step length asymmetry) and spasticity (pendulum test, clonus drop test, mSCI-SET) during the wash-in (LT only) period, and during the intervention (LT + TSS, LT + TSS$_{sham}$) period. The Wilcoxon signed-ranks test was used to examine changes within each group during the 2 study periods. Significance was set at $\alpha = 0.05$ for all analyses. We did not adjust the significance level for multiple comparisons as our measures cannot be assumed to be independent [34], and because such adjustments are known to increase the likelihood of false negatives (Type II errors) [34–37], which is problematic in preliminary studies. For within-group comparisons, we computed the effect sizes as these are considered more meaningful than *p*-values for clinical interpretation of results [35,38,39]. We computed the effect size using Hedges' *g*, as this approach is recommended for small samples [38]. Effect sizes were interpreted per recommendations for rehabilitation studies as small effects: 0.08 to 0.15, medium effects: 0.19 to 0.36, and large effects: 0.41 to 0.67 [40].

3. Results

3.1. Feasibility

A total of 107 potential participants were screened for eligibility, and 18 were enrolled, resulting in an enrollment rate of 17% (See Flow Diagram, Figure 2). Of the 18 enrolled participants, 16 completed the study. Two participants were withdrawn after enrollment. One participant was withdrawn before training activities began because they decided to focus their time on therapy rather than study-related assessments. Another participant had developed hip pain from an event unrelated to the study, limiting mobility and thus potentially affecting study outcomes. Demographic information is presented in Table 1. No study-related adverse events were experienced by any of the participants in the study. There were no difficulties encountered with incorporating TSS into the clinical LT program. The stimulator used for the study was selected because of its availability for clinical use. However, over the course of the study, the manufacturer discontinued the production of this device.

Of the 16 participants who completed the study (LT + TSS, $n = 8$; LT + TSS$_{sham}$, $n = 8$), walking data was collected only from 12 participants because 4 individuals met the inclusion criteria of having the ability to take a single step, but were unable to complete the walking tests. Eleven participants completed all 12 training sessions, and 5 completed 11 sessions. Specifically, 2 participants missed a training session during the wash-in phase and 3 participants missed a single session during the intervention phase (LT + TSS group: $n = 1$; LT + TSS$_{sham}$ group: $n = 2$). The length of each session ranged from 11 to 38 min. The mean time spent walking during LT did not significantly differ between wash-in and intervention for either group (LT + TSS group: wash-in mean time walking 27 (6) min, intervention mean time walking 29 (2) min; LT + TSS$_{sham}$ group: wash-in mean time walking 27 (5) min, intervention mean time walking 25 (5) min). In the LT + TSS group, stimulation intensities over the course of the intervention phase were 67.60 ± 7.35 (mean $\pm$ SE, per tphase of the biphasic stimulation pulse), with a range of 39 to 100 mA. Of the 8 participants in the LT + TSS group, 4 were able to tolerate stimulation intensities that induced tingling sensations in the lower extremity dermatomes, and in 4 participants, stimulation intensity at the highest tolerable level did not induce tingling sensations. In terms of stimulation tolerability, median NRS pain scores increased from 0.00 (0.09) at baseline to 3.71 (0.35) after the first minute of stimulation application, and 3.88 (0.77) at the end of stimulation application for the LT + TSS group. For the LT + TSS$_{sham}$ group, the NRS pain scores remained the same throughout the training sessions (baseline: 0.69 (0.50), first min of stimulation: 0.69 (0.41), end of stimulation: 0.69 (0.50)). There were no participants who withdrew from the study due to stimulation-related pain.

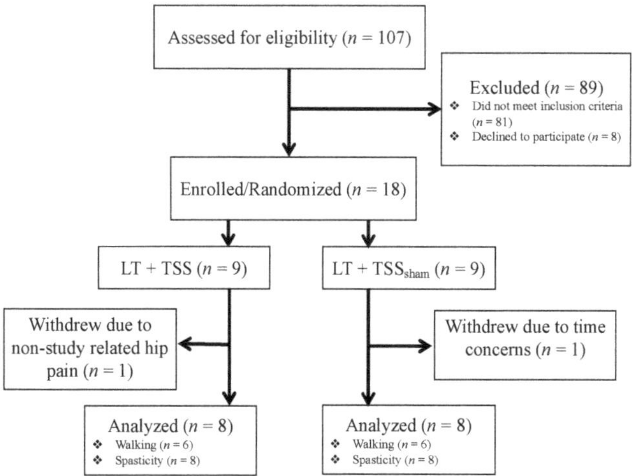

Figure 2. Flow diagram.

Table 1. Demographics.

Subject ID	Sex	Age (Years)	Time Since Injury (Days)	AIS	Neurological Injury Level	LEMS (R)	LEMS (L)	LEMS(Total)
1 *	M	51	56	D	C4	20	17	37
2	M	43	80	C	C4	14	13	27
3 †	M	65	88	D	C5	20	22	42
4	F	53	36	D	C2	22	20	42
5	M	56	84	D	C4	20	24	44
6	M	37	103	C	C3	17	22	39
7 †,‡	F	55	75	C	C3	9	11	20
8	M	47	119	D	C2	21	21	42
9 ‡	M	18	83	D	C7	22	11	33
10	F	24	82	C	C7	17	21	38
11 †,‡	M	32	121	C	C4	24	6	30
12 *	M	54	171	C	C4	14	22	36
13	M	20	68	D	C4	11	25	36
14	M	52	195	B	T11	22	19	41
15 ‡	M	54	141	D	C5	18	14	32
16	M	63	185	D	C1	22	24	46
17 †	F	58	71	D	C6	8	22	30
18	M	18	47	D	C5	25	8	33

Abbreviations: AIS, American Spinal injury Association Impairment Scale (C: Motor function is preserved below the neurological level, and more than half of key muscles below the neurological level have a muscle grade less than 3; D: Motor function is preserved below the neurological level, and at least half of key muscles below the neurological level have a muscle grade of 3 or more); LEMS, Lower Extremity Motor Score. * participants who withdrew from the study; † participants with insufficient walking ability to complete walking tests; ‡ participants in the locomotor training and transcutaneous spinal cord stimulation (LT + TSS) group who reported tingling in the legs during stimulation.

3.2. Effects of TSS on Walking Ability

Outcomes related to walking and spasticity measures are presented in Table 2. Walking speed and step length asymmetry were similar between the LT + TSS and LT + TSS$_{sham}$ groups at baseline. The distance that participants covered during the 2MWT at baseline was significantly longer in the LT + TSS group compared to the LT + TSS$_{sham}$ group ($Z = -2.74$, $p = 0.004$). The change in walking speed ($n = 12$), walking distance ($n = 11$), and step symmetry ($n = 10$) are illustrated in Figures 3–5, respectively. Between-group comparisons indicated that there were no differences in change between the LT + TSS and LT + TSS$_{sham}$ groups following the wash-in phase (walking speed: U = 13, $p = 0.42$; walking distance: U = 14, $p = 0.86$; step length asymmetry: U = 8, $p = 0.39$; pendulum (more impaired): U = 25, $p = 0.46$; pendulum (less impaired): U = 27, $p = 0.60$; ankle clonus (more impaired): U = 27.5, $p = 0.64$; ankle clonus (less impaired): U = 18, $p = 0.14$) or following the intervention phase (walking speed: U = 8, $p = 0.11$; walking distance: U = 10, $p = 0.20$; step length asymmetry: U = 8, $p = 0.39$; pendulum (more impaired): U = 22, $p = 0.29$; pendulum (less impaired): U = 30, $p = 0.83$; ankle clonus (more impaired): U = 31, $p = 0.92$; ankle clonus (less impaired): U = 25, $p = 0.46$).

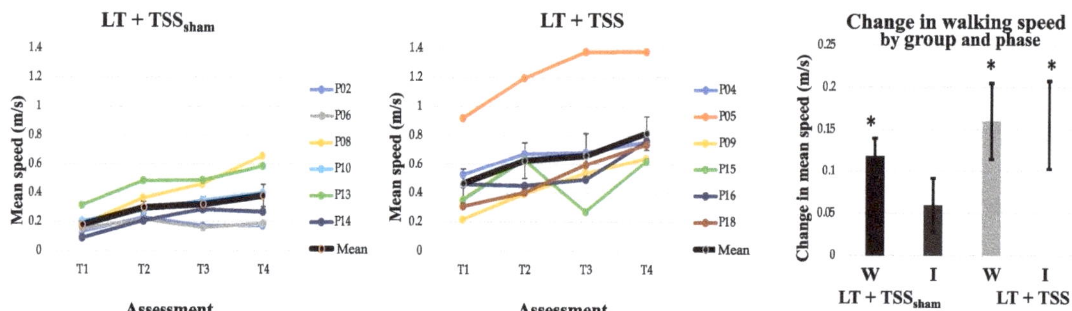

Figure 3. Walking speed. The mean walking speed (bold black line) increased in both the LT + TSS$_{sham}$ (**left**) and LT + TSS (**center**) groups over time. Note that the colored lines represent the mean walking speeds of individual participants. The change in mean walking speed, calculated from the start and end of each 2-week LT phase (T1 to T2, T3 to T4; **right**) showed improvements in both groups; however, only the LT + TSS group showed continued significant improvements during the 2-week TSS intervention phase. W: Wash-in; I: Intervention. Error bars represent standard errors; * $p < 0.05$.

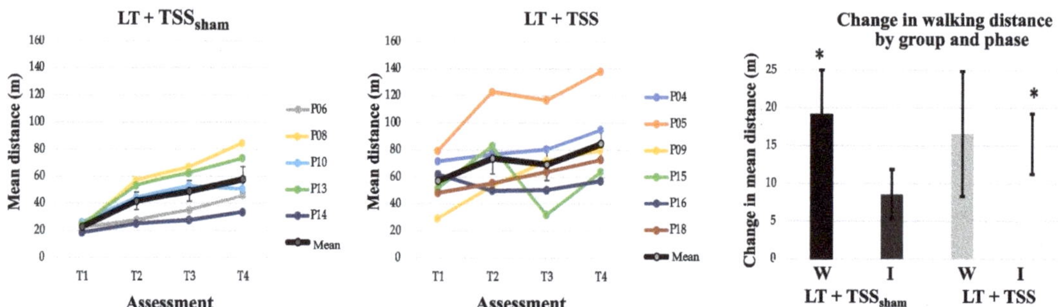

Figure 4. Walking distance. The average walking distance (bold black line) increased in both the LT + TSS$_{sham}$ (**left**) and LT + TSS (**center**) groups over time. Note that the colored lines represent the average walking distance of individual participants. The change in mean walking distance calculated from the start and end of each 2-week LT bout (T1 to T2, T3 to T4; **right**) showed significant improvements in the LT + TSS$_{sham}$ group during the wash-in phase; whereas the LT + TSS group showed significant improvements during the intervention phase. W: Wash-in; I: Intervention. Error bars represent standard errors; * $p < 0.05$.

Table 2. Walking and spasticity outcomes.

	Wash-in Phase					Intervention Phase				
	T1	T2	Difference	*p*-Value	Effect Size	T3	T4	Difference	*p*-Value	Effect Size
Walking Assessments										
Walking Speed (m/s)										
LT + TSS	0.47 (0.25)	0.63 (0.30)	0.16 (0.11)	0.05	0.53	0.66 (0.38)	0.82 (0.28)	0.16 (0.13)	0.03	0.43
LT + TSS$_{sham}$	0.18 (0.08)	0.30 (0.11)	0.12 (0.05)	0.03	1.15	0.32 (0.14)	0.38 (0.21)	0.06 (0.08)	0.12	0.31
Walking Distance (m)										
LT + TSS	57.07 (7.88)	73.65 (27.64)	16.58 (20.21)	0.12	0.66	69.27 (28.84)	84.52 (29.30)	15.25 (9.81)	0.03	0.48
LT + TSS$_{sham}$	22.70 (2.81)	39.13 (14.69)	19.08 (13.17)	0.04	1.35	45.05 (18.43)	51.90 (23.59)	6.85 (7.82)	0.12	0.30
Step Length Asymmetry (unitless)										
LT + TSS	0.54 (0.02)	0.54 (0.02)	0.00 (0.03)	0.92	0.09	0.53 (0.01)	0.53 (0.02)	0.00 (0.02)	0.75	0.13
LT + TSS$_{sham}$	0.57 (0.07)	0.54 (0.05)	−0.03 (0.06)	0.27	0.49	0.54 (0.03)	0.52 (0.02)	−0.02 (0.03)	0.47	0.68
Spasticity Assessments										
Pendulum: More Impaired LE (degrees)										
LT + TSS	60.00 (11.84)	54.61 (21.95)	−5.39 (22.32)	0.40	0.29	54.18 (20.67)	54.90 (15.22)	0.72 (8.88)	0.89	0.04
LT + TSS$_{sham}$	60.43 (23.60)	63.24 (18.83)	2.81 (19.00)	0.48	0.12	57.42 (19.74)	63.98 (20.04)	6.56 (18.00)	0.40	0.31
Pendulum: Less Impaired LE (degrees)										
LT + TSS	67.42 (20.06)	69.47 (24.98)	2.05 (16.09)	0.48	0.09	64.97 (34.24)	71.55 (27.90)	6.58 (13.19)	0.26	0.20
LT + TSS$_{sham}$	60.52 (22.06)	66.62 (19.01)	6.10 (15.88)	0.26	0.28	56.72 (16.86)	63.61 (22.19)	6.88 (18.72)	0.16	0.33
Ankle clonus: More Impaired LE (number of oscillations)										
LT + TSS	8.21 (7.10)	6.54 (7.58)	−1.67 (9.62)	0.50	0.21	9.88 (10.03)	10.42 (11.57)	0.54 (3.69)	1.00	0.05
LT + TSS$_{sham}$	14.75 (16.36)	13.96 (16.63)	−0.79 (7.93)	1.00	0.05	14.13 (15.83)	13.19 (18.51)	−0.94 (6.57)	0.74	0.05
Ankle clonus: Less Impaired LE (number of oscillations)										
LT + TSS	4.67 (2.87)	4.83 (4.86)	0.17 (5.47)	0.89	0.04	6.29 (6.62)	5.88 (7.52)	−0.42 (3.08)	0.92	0.06
LT + TSS$_{sham}$	17.83 (19.04)	18.25 (22.48)	0.42 (12.62)	0.23	0.02	12.38 (18.47)	16.00 (21.97)	3.63 (10.96)	0.35	0.17

Abbreviations: T1, testing session 1; T2, testing session 2; T3, testing session 3; T4, testing session 4; LT + TSS, locomotor training and transcutaneous spinal cord stimulation group; LT + TSS$_{sham}$, locomotor training and sham stimulation group; LE, lower extremity.

Within-group analyses of walking speed (Table 2) indicated that during the wash-in phase, there were significant changes in walking speed for each of the groups (LT + TSS group: $Z = -1.99$, $p = 0.05$; LT + TSS$_{sham}$ group: $Z = -2.20$, $p = 0.03$), with both groups exhibiting large effect sizes ($g = 0.53$ and 1.15, for the LT + TSS and LT + TSS$_{sham}$ groups, respectively). However, during the intervention phase, only the LT + TSS group showed a significant change in walking speed ($Z = -2.20$, $p = 0.03$), with a large effect size ($g = 0.43$). Within-group analyses of walking distance showed that during the wash-in phase, walking distance significantly improved only for the LT + TSS$_{sham}$ group ($Z = -2.02$, $p = 0.04$), however, both groups exhibited large effect sizes ($g = 0.66$ and 1.35 for the LT + TSS and LT + TSS$_{sham}$ groups, respectively). During the intervention phase, only the LT + TSS group showed a significant change in walking distance ($Z = -2.20$, $p = 0.03$), with a large effect size ($g = 0.48$). Within-group analyses of step symmetry revealed that this measure did not significantly change for either of the groups during the wash-in phase, or the intervention phase, however in the LT + TSS$_{sham}$ group, large effect sizes were observed during both the wash-in and intervention phase (g = 0.49 and 0.68, respectively).

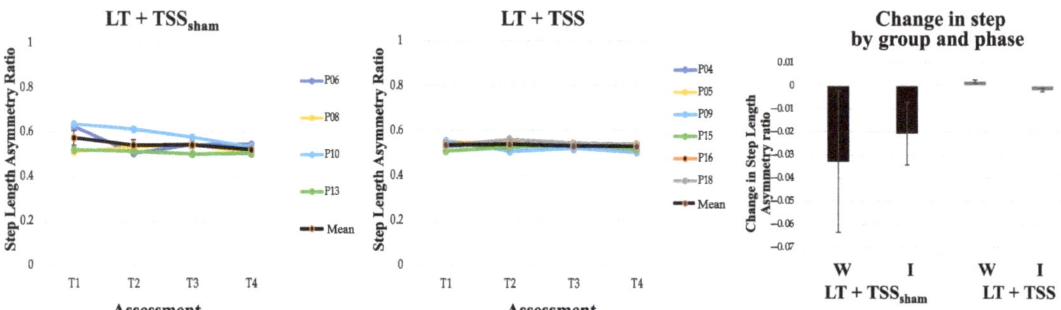

Figure 5. Step length asymmetry. Mean step length asymmetry (bold black line) remained the same in both the LT + TSS$_{sham}$ (**left**) and LT + TSS (**center**) groups over time. Colored lines represent the mean step length asymmetry of individual participants. Change in mean step length asymmetry calculated from the start and end of each 2-week LT bout (T1 to T2, T3 to T4; **right**) showed only a small improvement in the LT + TSS$_{sham}$ group. W: Wash-in; I: Intervention. Error bars represent standard errors.

3.3. Effects of TSS on Spasticity

Spasticity outcome measures were not different at baseline between groups (Table 2). Our findings did not demonstrate significant changes over the course of the study for either quadriceps spasticity (Figure 6) or ankle clonus (Figure 7). Within-group comparisons yielded similar findings before and after training for both the LT + TSS and LT + TSS$_{sham}$ groups for all the measures of interest (Table 2). Further, no significant differences between or within groups were found when SCATS was used to determine the immediate effects of TSS on spasticity.

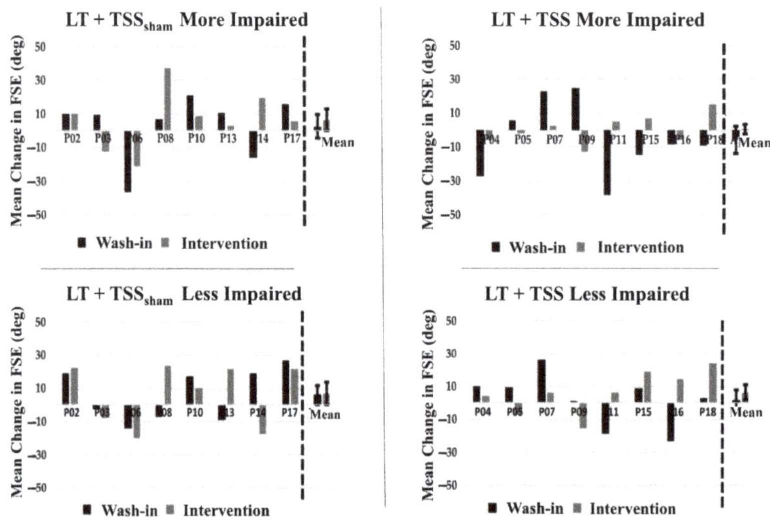

Figure 6. Pendulum test outcomes. No significant change in mean first swing excursion (FSE) was observed in participants in the LT + TSS$_{sham}$ group in either the more impaired (**upper left**) or less impaired (**lower left**) limb. Likewise, no significant change in mean first swing excursion (FSE) was observed in participants in the LT + TSS group in either the more impaired (**upper right**) or less impaired (**lower right**) limb.

Finally, at the beginning of the study, participants reported their spasticity to be mildly problematic, as determined by mSCI-SET (LT + TSS group: −0.22 (0.55); LT + TSS$_{sham}$ group: −0.58 (0.31)). The mSCI-SET scores did not significantly change throughout the study for either of the two groups (wash-in phase: LT + TSS group: −0.31 (0.55); LT + TSS$_{sham}$ group: −0.43 (0.18)/intervention phase: LT + TSS group: −0.24 (0.50); LT + TSS$_{sham}$ group: −0.44 (0.13)). Effect sizes for all spasticity measures were small to medium.

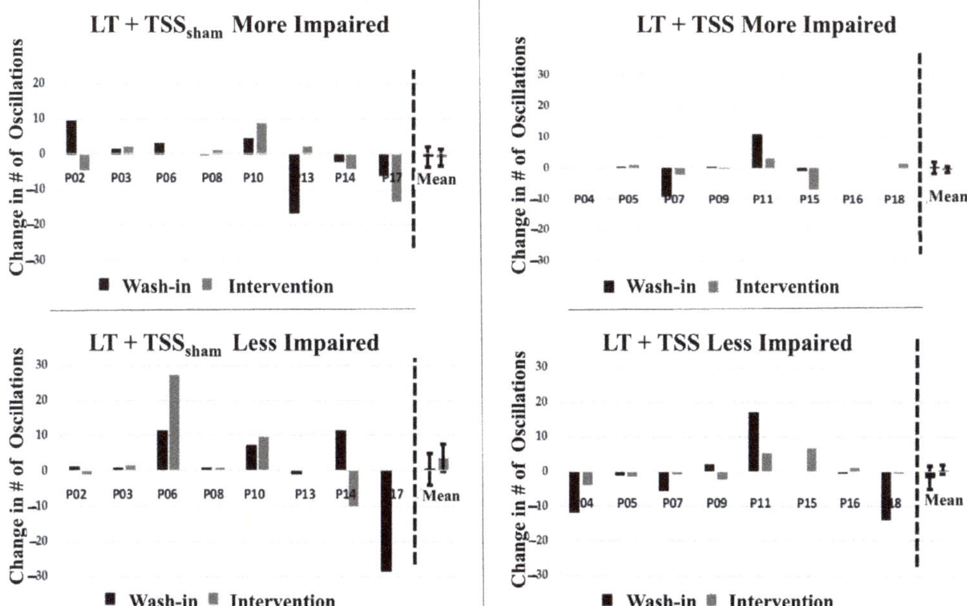

Figure 7. Drop test outcomes. No significant change in mean number of oscillations was observed in participants in the LT + TSS$_{sham}$ group in either the more impaired (**upper left**) or less impaired (**lower left**) limb. Likewise, no significant change in mean number of oscillations was observed in participants in the LT + TSS group in either the more impaired (**upper right**) or less impaired (**lower right**) limb.

4. Discussion

4.1. Feasibility

Integration into Clinical Practice

This study utilized a pragmatic clinical trial design to assess whether it is feasible to incorporate TSS into a clinical LT program and to acquire a preliminary evaluation of the efficacy of this combined training. The interest and willingness of participants to enroll in a study provides insight into the feasibility of a novel neuromodulation intervention. With a 17% enrollment rate, there was moderate interest in using TSS in conjunction with LT among those persons with subacute SCI. While this enrollment rate is relatively low compared with the number of potential participants (n = 107), it is higher than reported for other randomized LT studies enrolling participants during the subacute rehabilitation phase post-SCI [41] and in an exoskeleton LT study [42], wherein the enrollment rates were 9% and 12%, respectively. Moreover, a similar 17% enrollment rate was found in another exoskeleton study [43], indicating that there is a comparable interest among those with subacute SCI to try novel approaches to LT that have minimal risk.

The interest in using TSS as an adjunct to LT may be higher than reported by the enrollment rate. At the time of this study, additional clinical trials were enrolling individuals with SCI from both in-patient and out-patient programs. While individuals often expressed interest in participating in all of the available studies, unknown carryover effects prevented individuals from simultaneously participating in multiple studies. Interestingly, when persons with SCI met criteria for both studies and had to choose between the current study of improving walking function and lower extremity spasticity versus a study focused on neuromodulation for improving hand function, 8 individuals declined to participate in the current study and chose the study focused on hand function. This observation is similar to findings in the literature, wherein when asked to rank functional priorities that would most enhance their quality of life, individuals with tetraplegia ranked improving hand function higher than improving walking function [44].

In assessing study feasibility, we identified early study design obstacles that slowed participant recruitment and that may have also impacted the overall enrollment rate. The design did not fully anticipate the concerns of participants with subacute SCI enrolled in a daily, intensive rehabilitation program with the time required for multiple weekly research assessments. To address this concern, assessment time points were reduced from twice per week (i.e., at the start and end of each week) to once per week (i.e., at the start and end of the 2-week wash-in period, and at the start and end of the 2-week intervention period). It should be noted that these concerns were due to the study assessments and not to the application of TSS as an adjunct intervention. For a typical 1 h LT session (i.e., usual care) with setup time, patients usually walk for approximately 30 min. No difference was found in LT walking time between wash-in and intervention weeks, indicating that TSS can be applied as an adjunct without significant interference with LT. An additional design issue that may have impacted the enrollment rate was recruiting only from patients enrolled in the Shepherd Center's day program. The duration of services for the day program is based on therapy goals and progress, and therefore is individualized for each patient. For some patients, this time may be less than 4 weeks, making them ineligible for the study and ultimately slowing recruitment. To facilitate enrollment, we revised the study protocol to include individuals participating in the in-patient locomotor training program, which utilized an identical clinician-driven LT program, transitioned into the day program LT, and allowed for 4 weeks of LT. Collectively, these design issues may have negatively impacted the enrollment rate, skewing the perception of interest in using TSS as an adjunct LT intervention.

4.2. Safety and Tolerability

The application of TSS was found to be safe and tolerable, with no study-related adverse events being reported. Of the 18 participants enrolled in the study, 16 (89%) completed the study. The high completion rate of this study is comparable to those reported in other LT feasibility studies in persons with subacute [45] and chronic SCI [46], and falls within the studies anticipated 15% attrition rate. TSS was found to be generally tolerable with some reported discomfort and a pain rating of 4 out of 10. Participants often described feeling intense tightness in the abdomen and lower back around the site of electrodes.

While the stimulation intensity was set at a level below motor threshold for the legs and at a level that produced paresthesia in the legs, the stimulation intensity would activate paraspinal muscles and abdominal muscles, as observed from visual inspection, and which may have contributed to the experienced discomfort. The contraction of abdominal and paraspinal muscles is commonly reported in other TSS studies [5,25]; however, the tolerability of the stimulation appears to vary based on how TSS is applied. In short bursts, TSS was not found to cause discomfort [5,25]. In a case series of persons with chronic SCI undergoing LT combined with TSS, TSS was found to be minimally painful with a mean numeric rating scale for pain of 0.12 and a range in responses from 0 to 4 out of 10 [12]. It is possible that the larger rectangular electrode used in the case series helped to disperse the sensation of stimulation on the lower back, providing a different mean pain rating than observed in this current study. Nonetheless, while TSS may be uncomfortable for some individuals, the sensation of TSS is within a tolerable pain range that did not interfere with LT. These findings indicate that TSS can be a feasible adjunct treatment for LT.

In recent years, there has been growing interest in the use of neuromodulation approaches to promote use-dependent plasticity and augment the effects of LT on walking function. Epidural stimulation studies focus on improving walking function in persons with severe SCI, but they often consist of extensive LT schedules with over 250 training sessions [47] that are unrealistic to incorporate into usual clinical care. TSS studies have shown promise in improving walking function but also lack a control group to accurately assess efficacy [5,12]. We specifically designed this study to be integrated into standard clinical care consisting of 1 h LT sessions, 3 days a week, with an appropriate control group to better assess the preliminary efficacy of TSS on three aspects of walking function: walking speed, distance (endurance), and symmetry.

4.3. Preliminary Efficacy—Walking Function

During the 2-week wash-in phase, wherein all participants received LT alone, the change in walking speed was significant for both groups. The LT + TSS group (0.16 m/s) exceeded the threshold of 0.13 m/s for clinically relevant change [48], while the change for the LT + TSS$_{sham}$ group (0.12 m/s) was within 0.01 m/s of the threshold (see Table 2). Further, both groups exhibited large effect sizes for walking speed during the wash-in phase. The finding that there was meaningful improvement in speed during the first 2 weeks suggests that there is a rapid improvement in walking speed early after the onset of LT. As has been noted in the literature, when individuals are exposed to new activities, it is not unusual to observe early improvements regardless of the intervention, and this phenomenon is the rationale for recommending use of a wash-in (phase-in) study design [49] such as that we used in our study.

Following the wash-in phase, during the subsequent 2-week intervention phase, walking speed in the LT + TSS group also improved by 0.16 m/s, continuing to meet the threshold for clinically relevant change, and this change was statistically significant with a large effect size. Conversely, in the LT + TSS$_{sham}$ group, performance plateaued during the wash-in phase, such that the change of 0.06 m/s during the intervention phase failed to meet the threshold for clinically relevant change and was not significant. The inability to detect a significant difference between the LT + TSS and LT + TSS$_{sham}$ groups during the intervention phase may have been due to the failure to meet the recruitment target of 24 participants on which the study was powered.

As with walking speed, the changes in 2MWT distance were similar during the LT-only wash-in phase for the LT + TSS and the LT + TSS$_{sham}$ groups, at 16.58 and 19.08 m, respectively. This change was not significant for the LT + TSS group but was significant for the LT + TSS$_{sham}$ group. However, the effect size was large in both groups. During the intervention phase, the LT + TSS group continued to improve with a change of 15.25 m, which was statistically significant with a large effect size. The change in the LT + TSS$_{sham}$ group of 6.85 m was more modest, and not significant. Although threshold values for clinically relevant change in the 2MWT have not been established for persons with SCI, extrapolation from the 6MWT threshold of 45.8 m [48] would suggest that a change of 16.3 m is a clinically relevant change. The change in the 2MWT distance in the LT + TSS group during the intervention phase fell short of this threshold by 1 m.

4.4. Preliminary Efficacy—Spasticity

Prior studies of locomotor training in persons with SCI have demonstrated that walking is associated with reductions in spasticity [21–23,28], and growing evidence suggests that TSS can reduce spasticity as well [24,25,28]. For this reason, the finding of no significant between-group or within-group effects in our clinical measure of immediate post-treatment effects (SCATS) or our measures of persistent (one day after intervention) effects on quadriceps spasticity (pendulum test) or ankle plantar flexor clonus was unexpected. The literature indicates that an FSE of 96 degrees represents the 95th percentile for minimum FSE in individuals without neurologic conditions [29]. Based on our FSE data, there are some individuals who approached this value, indicating minimal excitability of the quadriceps stretch reflex. Conversely, in other participants, the small FSE angle indicates greater excitability of this reflex. Likewise, prior work has indicated that individuals exhibiting 4 or more oscillations of clonic activity following biomechanical stretch of the plantar flexors also exhibit electrophysiologic evidence of impaired reflex modulation [50]. In the TSS group, the number of clonic oscillations fell within the range of neurologically intact individuals in both the more- and less-impaired legs. However, the TSS$_{sham}$ group demonstrated greater levels of clonic activity in both legs. It is possible that the reason we failed to identify an effect of TSS on spasticity is that the pendulum test and drop test assessments were performed the day following the last intervention. In one prior study of TSS, the antispasmodic effects persisted for up to 7 days, although inspection of the published graphical data indicates that the pendulum test values were highly variable over the course of that study [24]. It may be that SCATS, while a fast, low-tech clinical tool to assess severity of multiple manifestations of spasticity, is not sufficiently sensitive for determining small changes in spasticity, as an ordinal measure. Nonetheless, while the lack of effects on spasticity in the present study were unexpected and may be due to study design limitations, the effects of TSS and LT on spasticity are unclear and warrant further study to understand its efficacy as an approach for management of spasticity.

4.5. Limitations

All participants were in the subacute stage of injury, a time that is thought to have the greatest opportunity for neural plasticity [51]. The fact that this early post-injury period is the time when large changes in function are most likely means that it is also a phase of injury wherein there is a large amount of inherent variability in the course of recovery. The effects of inter-individual differences in patterns of recovery may have been compounded by variability in the rehabilitation program among the participants. For most participants, the LT sessions consisted of both overground and treadmill-based training sessions, with the amount of each dependent on clinical judgement of the treating physical therapist. While prior work has shown that outcomes are equivalent for these two LT approaches when used in the subacute stage of SCI [41], differences in the type of LT received, as well as variations in the types and amounts of other forms of physical therapy, could have added to the variability in outcomes, making it more difficult to discern treatment effects.

The pragmatic nature of our study design required that set-up time be maintained at a minimum with an emphasis on approaches that can be readily incorporated into clinical practice. A limitation of this approach is that we used observation of visible muscle contraction to verify that the stimulation was subthreshold for a motor response rather than measuring electromyographic output. For this reason, it is possible that the stimulation intensity was not optimal for achieving activation of the target neural structures.

The slow initial enrollment rate and withdrawal of two subjects left the study underpowered to detect between-group differences. Likewise, the lack of significant within-group changes in the LT + TSS$_{sham}$ group may also reflect the small sample size and highlights that efficacy findings should be interpreted cautiously. Future studies sufficiently powered to assess these differences will provide better insight into the efficacy of TSS for augmenting the effects of LT.

5. Conclusions

Combination approaches for improving walking function in persons with SCI have been cited most likely to result in meaningful change [52]. Here, we demonstrated that applying TSS as an adjunct to LT is both feasible and tolerable when applied as part of the usual care in a clinical setting. Moreover, the improvements in walking outcomes provides insight into the potential benefits of this combined approach for improving walking function and warrant further research aimed at understanding the optimal stimulation parameters for improving functional outcomes in persons with subacute as well as chronic SCI.

J. Clin. Med. **2021**, 10, 1167

Author Contributions: Conceptualization, S.E. and E.C.F.-F.; methodology, S.E., A.Z., and E.C.F.-F.; software, C.S.; formal analysis, A.Z.; investigation, S.E., A.Z., and J.M.H.; writing—original draft preparation, S.E., A.Z., J.M.H., and E.C.F.-F.; visualization, S.E., A.Z., and C.S.; project administration, S.E.; funding acquisition, S.E. All authors reviewed the manuscript and contributed to the editing. All authors have read and agreed to the published version of the manuscript.

Funding: This research was funded by Wings for Life: WFL-US-2/17 (S.E.) and by the Hulse Spinal Cord Injury Research Fund.

Institutional Review Board Statement: The study was conducted according to the guidelines of the Declaration of Helsinki, and approved by the Shepherd Center Research Review Committee (protocol code 696 and date of approval 09/08/2016).

Informed Consent Statement: Informed consent was obtained from all subjects involved in the study.

Data Availability Statement: The data presented in this study are available upon request from the corresponding author.

Acknowledgments: The authors thank the participants for being a part of this study. We also thank Allison McIntyre, Ryan Koter, Elizabeth Sasso-Lance, Casey Kandilakis, Kyle Condon, Evan Sandler, Nicholas Evans, Cathy Furbish, and all the physical therapists and physical therapist assistants at Shepherd SCI Day Program who contributed to this study.

Conflicts of Interest: The authors declare no conflict of interest. The funders had no role in the design of the study; in the collection, analyses, or interpretation of data; in the writing of the manuscript, or in the decision to publish the results.

References

1. Field-Fote, E.C. Therapeutic Interventions to Improve Mobility with Spinal Cord Injury Related Upper Motor Neuron Syndromes. *Phys. Med. Rehabil. Clin. N. Am.* **2020**, *31*, 437–453. [CrossRef]
2. Smith, A.C.; Knikou, M. A Review on Locomotor Training after Spinal Cord Injury: Reorganization of Spinal Neuronal Circuits and Recovery of Motor Function. *Neural. Plast.* **2016**, *2016*, 1216258. [CrossRef]
3. Mehrholz, J.; Harvey, L.A.; Thomas, S.; Elsner, B. Is body-weight-supported treadmill training or robotic-assisted gait training superior to overground gait training and other forms of physiotherapy in people with spinal cord injury? A systematic review. *Spinal Cord* **2017**, *55*, 722–729. [CrossRef]
4. Powell, E.S.; Carrico, C.; Raithatha, R.; Salyers, E.; Ward, A.; Sawaki, L. Transvertebral direct current stimulation paired with locomotor training in chronic spinal cord injury: A case study. *NeuroRehabilitation* **2016**, *38*, 27–35. [CrossRef]
5. Hofstoetter, U.S.; Krenn, M.; Danner, S.M.; Hofer, C.; Kern, H.; Mckay, W.B.; Mayr, W.; Minassian, K. Augmentation of voluntary locomotor activity by transcutaneous spinal cord stimulation in motor-incomplete spinal cord-injured individuals. *Artif. Organs* **2015**, *39*, E176–E186. [CrossRef] [PubMed]
6. Shapkova, E.Y.; Pismennaya, E.V.; Emelyannikov, D.V.; Ivanenko, Y. Exoskeleton Walk Training in Paralyzed Individuals Benefits From Transcutaneous Lumbar Cord Tonic Electrical Stimulation. *Front. Neurosci.* **2020**, *14*, 416. [CrossRef]
7. Carhart, M.R.; He, J.; Herman, R.; D'Luzansky, S.; Willis, W.T. Epidural spinal-cord stimulation facilitates recovery of functional walking following incomplete spinal-cord injury. *IEEE Trans. Neural. Syst. Rehabil. Eng.* **2004**, *12*, 32–42. [CrossRef] [PubMed]
8. Wagner, F.B.; Mignardot, J.-B.; Le Goff-Mignardot, C.G.; Demesmaeker, R.; Komi, S.; Capogrosso, M.; Rowald, A.; Seáñez, I.; Caban, M.; Pirondini, E.; et al. Targeted neurotechnology restores walking in humans with spinal cord injury. *Nature* **2018**, *563*, 65–71. [CrossRef] [PubMed]
9. Angeli, C.A.; Edgerton, V.R.; Gerasimenko, Y.P.; Harkema, S.J. Altering spinal cord excitability enables voluntary movements after chronic complete paralysis in humans. *Brain* **2014**, *137*, 1394–1409. [CrossRef] [PubMed]
10. Danner, S.M.; Hofstoetter, U.S.; Ladenbauer, J.; Rattay, F.; Minassian, K. Can the human lumbar posterior columns be stimulated by transcutaneous spinal cord stimulation? A modeling study. *Artif. Organs* **2011**, *35*, 257–262. [CrossRef]
11. Hofstoetter, U.S.; Freundl, B.; Binder, H.; Minassian, K. Common neural structures activated by epidural and transcutaneous lumbar spinal cord stimulation: Elicitation of posterior root-muscle reflexes. *PLoS ONE* **2018**, *13*, e0192013. [CrossRef]
12. McHugh, L.V.; Miller, A.A.; Leech, K.A.; Salorio, C.; Martin, R.H. Feasibility and utility of transcutaneous spinal cord stimulation combined with walking-based therapy for people with motor incomplete spinal cord injury. *Spinal Cord Ser. Cases* **2020**, *6*, 104. [CrossRef]
13. Maynard, F.M.; Karunas, R.S.; Waring, W.P. Epidemiology of spasticity following traumatic spinal cord injury. *Arch. Phys. Med. Rehabil.* **1990**, *71*, 566–569. [PubMed]
14. Pandyan, A.D.; Gregoric, M.; Barnes, M.P.; Wood, D.; Van Wijck, F.; Burridge, J.; Hermens, H.; Johnson, G.R. Spasticity: Clinical perceptions, neurological realities and meaningful measurement. *Disabil. Rehabil.* **2005**, *27*, 2–6. [CrossRef]
15. Jorgensen, S.; Iwarsson, S.; Lexell, J. Secondary Health Conditions, Activity Limitations, and Life Satisfaction in Older Adults With Long-Term Spinal Cord Injury. *PM R* **2017**, *9*, 356–366. [CrossRef]

16. Krawetz, P.; Nance, P. Gait analysis of spinal cord injured subjects: Effects of injury level and spasticity. *Arch. Phys. Med. Rehabil.* **1996**, *77*, 635–638. [CrossRef]
17. Tibbett, J.A.; Field-Fote, E.C.; Thomas, C.K.; Widerström-Noga, E.G. Spasticity and Pain after Spinal Cord Injury: Impact on Daily Life and the Influence of Psychological Factors. *PM R* **2020**. [CrossRef] [PubMed]
18. Taricco, M.; Pagliacci, M.C.; Telaro, E.; Adone, R. Pharmacological interventions for spasticity following spinal cord injury: Results of a Cochrane systematic review. *Eur. Medicophys.* **2006**, *42*, 5–15.
19. Montane, E.; Vallano, A.; Laporte, J.R. Oral antispastic drugs in nonprogressive neurologic diseases: A systematic review. *Neurology* **2004**, *63*, 1357–1363. [CrossRef] [PubMed]
20. Field-Fote, E.C. Characterizing the Experience of Spasticity after Spinal Cord Injury: A National Survey Project of the Spinal Cord Injury Model Systems Centers. *Arch. Phys. Med. Rehabil.* **2020**.
21. Manella, K.J.; Field-Fote, E.C. Modulatory effects of locomotor training on extensor spasticity in individuals with motor-incomplete spinal cord injury. *Restor. Neurol. Neurosci.* **2013**, *31*, 633–646. [CrossRef]
22. Mirbagheri, M.M. Comparison between the therapeutic effects of robotic-assisted locomotor training and an anti-spastic medication on spasticity. *Annu. Int. Conf. IEEE Eng. Med. Biol. Soc.* **2015**, *2015*, 4675–4678. [CrossRef]
23. Adams, M.M.; Hicks, A.L. Comparison of the effects of body-weight-supported treadmill training and tilt-table standing on spasticity in individuals with chronic spinal cord injury. *J. Spinal Cord Med.* **2011**, *34*, 488–494. [CrossRef]
24. Hofstoetter, U.S.; Freundl, B.; Danner, S.M.; Krenn, M.J.; Mayr, W.; Binder, H.; Minassian, K. Transcutaneous spinal cord stimulation induces temporary attenuation of spasticity in individuals with spinal cord injury. *J. Neurotrauma* **2020**, *37*, 481–493. [CrossRef]
25. Hofstoetter, U.S.; McKay, W.B.; Tansey, K.E.; Mayr, W.; Kern, H.; Minassian, K. Modification of spasticity by transcutaneous spinal cord stimulation in individuals with incomplete spinal cord injury. *J. Spinal Cord Med.* **2015**, *37*, 202–211. [CrossRef] [PubMed]
26. Dimitrijevic, M.M.; Dimitrijevic, M.R.; Illis, L.S.; Nakajima, K.; Sharkey, P.C.; Sherwood, A.M. Spinal cord stimulation for the control of spasticity in patients with chronic spinal cord injury: I. Clinical observations. *Cent. Nerv. Syst. Trauma.* **1986**, *3*, 129–144. [CrossRef] [PubMed]
27. Dimitrijevic, M.M.R.; Dimitrijevic, M.M.R.; Illis, L.S.; Nakajima, K.; Sharkey, P.C.; Sherwood, A.M. Spinal cord stimulation for the control of spasticity in patients with chronic spinal cord injury: II. Neurophysiologic observations. *Cent. Nerv. Syst. Trauma* **1986**, *3*, 145–152. [CrossRef]
28. Estes, S.P.; Iddings, J.A.; Field-Fote, E.C. Priming neural circuits to modulate spinal reflex excitability. *Front. Neurol.* **2017**, *8*, 17. [CrossRef] [PubMed]
29. Stillman, B.; McMeeken, J. A video-based version of the pendulum test: Technique and normal response. *Arch. Phys. Med. Rehabil.* **1995**, *76*, 166–176. [CrossRef]
30. Manella, K.J.; Roach, K.E.; Field-Fote, E.C. Temporal indices of ankle clonus and relationship to electrophysiologic and clinical measures in persons with spinal cord injury. *J. Neurol. Phys. Ther.* **2017**, *41*, 229–238. [CrossRef] [PubMed]
31. Sweatman, W.M.; Heinemann, A.W.; Furbish, C.L.; Field-Fote, E.C. Modified PRISM and SCI-SET Self-Reported Spasticity Measures for Persons with Traumatic Spinal Cord Injury: Outcomes of a Rasch Analysis. *Arch. Phys. Med. Rehabil.* **2020**. in review. [CrossRef]
32. Benz, E.N.; Hornby, T.G.; Bode, R.K.; Scheidt, R.A.; Schmit, B.D. A physiologically based clinical measure for spastic reflexes in spinal cord injury. *Arch. Phys. Med. Rehabil.* **2005**, *86*, 52–59. [CrossRef]
33. Portney, L.G.; Watkins, M.P. *Foundations of Clinical Research: Applications to Practice*; Pearson/Prentice Hall: Upper Saddle River, NJ, USA, 2009; Volume 892.
34. Hart, T.; Bagiella, E. Design and implementation of clinical trials in rehabilitation research. *Arch. Phys. Med. Rehabil.* **2012**, *93*, S117–S126. [CrossRef] [PubMed]
35. Nakagawa, S. A farewell to Bonferroni: The problems of low statistical power and publication bias. *Behav. Ecol.* **2004**, *15*, 1044–1045. [CrossRef]
36. Perneger, T.V. What's wrong with Bonferroni adjustments. *BMJ* **1998**, *316*, 1236–1238. [CrossRef] [PubMed]
37. KJ, R. No adjustments are needed for multiple comparisons. *Epidemiology* **1990**, *1*, 43–46.
38. Lakens, D. Calculating and reporting effect sizes to facilitate cumulative science: A practical primer for t-tests and ANOVAs. *Front. Psychol.* **2013**, *4*, 863. [CrossRef] [PubMed]
39. Borenstein, M. Hypothesis Testing and Effect Size Estimation in Clinical Trials. *Ann. Allergy Asthma Immunol.* **1997**, *78*, 5–16. [CrossRef]
40. Kinney, A.R.; Eakman, A.M.; Graham, J.E. Novel effect size interpretation guidelines and an evaluation of statistical power in rehabilitation research. *Arch. Phys. Med. Rehabil.* **2020**, *101*, 2219–2226. [CrossRef]
41. Dobkin, B.; Apple, D.; Barbeau, H.; Basso, M.; Behrman, A.; Deforge, D.; Ditunno, J.; Dudley, G.; Elashoff, R.; Fugate, L.; et al. Weight-supported treadmill vs over-ground training for walking after acute incomplete SCI. *Neurology* **2006**, *66*, 484–493. [CrossRef]
42. Platz, T.; Gillner, A.; Borgwaldt, N.; Kroll, S.; Roschka, S. Device-Training for Individuals with Thoracic and Lumbar Spinal Cord Injury Using a Powered Exoskeleton for Technically Assisted Mobility: Achievements and User Satisfaction. *Biomed. Res. Int.* **2016**, *2016*, 8459018. [CrossRef]
43. Benson, I.; Hart, K.; Tussler, D.; van Middendorp, J.J. Lower-limb exoskeletons for individuals with chronic spinal cord injury: Findings from a feasibility study. *Clin. Rehabil.* **2016**, *30*, 73–84. [CrossRef] [PubMed]
44. Anderson, K.D. Targeting recovery: Priorities of the spinal cord-injured population. *J. Neurotrauma* **2004**, *21*, 1371–1383. [CrossRef]

45. Delgado, A.D.; Escalon, M.X.; Bryce, T.N.; Weinrauch, W.; Suarez, S.J.; Kozlowski, A.J. Safety and feasibility of exoskeleton-assisted walking during acute/sub-acute SCI in an inpatient rehabilitation facility: A single-group preliminary study. *J. Spinal Cord Med.* **2020**, *43*, 657–666. [CrossRef]

46. Gagnon, D.H.; Escalona, M.J.; Vermette, M.; Carvalho, L.P.; Karelis, A.D.; Duclos, C.; Aubertin-Leheudre, M. Locomotor training using an overground robotic exoskeleton in long-term manual wheelchair users with a chronic spinal cord injury living in the community: Lessons learned from a feasibility study in terms of recruitment, attendance, learnability, performance and safety. *J. Neuroeng. Rehabil.* **2018**, *15*, 12. [CrossRef]

47. Angeli, C.A.; Boakye, M.; Morton, R.A.; Vogt, J.; Benton, K.; Chen, Y.; Ferreira, C.K.; Harkema, S.J. Recovery of over-ground walking after chronic motor complete spinal cord injury. *N. Engl. J. Med.* **2018**, *379*, 1244–1250. [CrossRef] [PubMed]

48. Lam, T.; Noonan, V.K.; Eng, J.J.; Team, S.R. A systematic review of functional ambulation outcome measures in spinal cord injury. *Spinal Cord* **2008**, *46*, 246–254. [CrossRef]

49. Dobkin, B.H. Progressive staging of pilot studies to improve phase III trials for motor interventions. *Neurorehabil. Neural Repair* **2009**, *23*, 197–206. [CrossRef]

50. Koelman, J.H.; Bour, L.J.; Hilgevoord, A.A.; van Bruggen, G.J.; Ongerboer, V.D. Soleus H-reflex tests and clinical signs of the upper motor neuron syndrome. *J. Neurol. Neurosurg. Psychiatry* **1993**, *56*, 776–781. [CrossRef] [PubMed]

51. Fouad, K.; Krajacic, A.; Tetzlaff, W. Spinal cord injury and plasticity: Opportunities and challenges. *Brain Res. Bull.* **2011**, *84*, 337–342. [CrossRef]

52. Dobkin, B.H.; Duncan, P.W. Should body weight-supported treadmill training and robotic-assistive steppers for locomotor training trot back to the starting gate? *Neurorehabil. Neural Repair* **2012**, *26*, 308–317. [CrossRef] [PubMed]

Journal of
Clinical Medicine

Article

Immediate Effects of Transcutaneous Spinal Cord Stimulation on Motor Function in Chronic, Sensorimotor Incomplete Spinal Cord Injury

Christian Meyer [1,†], Ursula S. Hofstoetter [2,*,†], Michèle Hubli [1], Roushanak H. Hassani [1], Carmen Rinaldo [1], Armin Curt [1] and Marc Bolliger [1]

[1]   Spinal Cord Injury Center, Balgrist University Hospital, Forchstrasse 340, 8008 Zurich, Switzerland; christian.meyer@balgrist.ch (C.M.); Michele.Hubli@balgrist.ch (M.H.); Roushanak.Hassani@balgrist.ch (R.H.H.); carmen_rinaldo@hotmail.com (C.R.); Armin.Curt@balgrist.ch (A.C.); marc.bolliger@balgrist.ch (M.B.)

[2]   Center for Medical Physics and Biomedical Engineering, Medical University of Vienna, 1090 Vienna, Austria

*   Correspondence: ursula.hofstoetter@meduniwien.ac.at; Tel.: +43-1-40400-19720

†   These authors contributed equally to this manuscript.

Received: 12 October 2020; Accepted: 30 October 2020; Published: 2 November 2020

Abstract: Deficient ankle control after incomplete spinal cord injury (iSCI) often accentuates walking impairments. Transcutaneous electrical spinal cord stimulation (tSCS) has been shown to augment locomotor activity after iSCI, presumably due to modulation of spinal excitability. However, the effects of possible excitability modulations induced by tSCS on ankle control have not yet been assessed. This study investigated the immediate (i.e., without training) effects during single-sessions of tonic tSCS on ankle control, spinal excitability, and locomotion in ten individuals with chronic, sensorimotor iSCI (American Spinal Injury Association Impairment Scale D). Participants performed rhythmic ankle movements (dorsi- and plantar flexion) at a given rate, and irregular ankle movements following a predetermined trajectory with and without tonic tSCS at 15 Hz, 30 Hz, and 50 Hz. In a subgroup of eight participants, the effects of tSCS on assisted over-ground walking were studied. Furthermore, the activity of a polysynaptic spinal reflex, associated with spinal locomotor networks, was investigated to study the effect of the stimulation on the dedicated spinal circuitry associated with locomotor function. Tonic tSCS at 30 Hz immediately improved maximum dorsiflexion by +4.6° ± 0.9° in the more affected lower limb during the rhythmic ankle movement task, resulting in an increase of +2.9° ± 0.9° in active range of motion. Coordination of ankle movements, assessed by the ability to perform rhythmic ankle movements at a given target rate and to perform irregular movements according to a trajectory, was unchanged during stimulation. tSCS at 30 Hz modulated spinal reflex activity, reflected by a significant suppression of pathological activity specific to SCI in the assessed polysynaptic spinal reflex. During walking, there was no statistical group effect of tSCS. In the subgroup of eight assessed participants, the three with the lowest as well as the one with the highest walking function scores showed positive stimulation effects, including increased maximum walking speed, or more continuous and faster stepping at a self-selected speed. Future studies need to investigate if multiple applications and individual optimization of the stimulation parameters can increase the effects of tSCS, and if the technique can improve the outcome of locomotor rehabilitation after iSCI.

Keywords: human; non-invasive; spinal cord injury; spinal cord stimulation; spinal reflexes; voluntary ankle control; walking

1. Introduction

Supraspinal input to spinal circuits is critical for the initiation and control of human locomotion [1,2]. In particular, the activation of ankle dorsiflexors in the swing phase is strongly related to corticospinal transmission [3,4]. After an incomplete spinal cord injury (iSCI), reduced corticospinal input causes muscle weakness and spasticity, which impair voluntary ankle control [5–7]. Deficient ankle control after iSCI often accentuates walking impairments and poses a challenge to regaining walking function [8,9]. Assisting locomotor recovery after iSCI and improving its outcome is of high relevance for affected individuals, and would considerably increase their quality of life [10,11].

Intensive training, in combination with electrical stimulation of the lumbar spinal cord using epidurally implanted electrodes, has been shown to improve, or induce regains of, voluntary control over lower limb movements and over-ground walking capabilities in individuals with chronic, severe SCI [12–16]. Epidural spinal cord stimulation (eSCS) activates afferent fibers in the posterior roots, and generates multisegmental input to spinal networks [17–21]. Previous studies have demonstrated a variety of immediate (i.e., without training), frequency-dependent effects of eSCS on lower limb motor control after SCI. Specifically, eSCS in the range of 5–15 Hz was shown to induce bilateral lower limb extension and upright standing within single application sessions [15,22]. At frequencies around 30 Hz, eSCS immediately generated rhythmic flexion-extension movements of the lower limbs of individuals with motor complete SCI lying supine [21,23,24], allowed enhanced rhythmic lower limb muscle activities induced by proprioceptive input generated during assisted treadmill stepping [15,25], and enabled voluntary hip and knee flexion and extension [15], as well as ankle dorsiflexion [16]. Stimulation at frequencies ≥50 Hz was associated with the attenuation of lower-limb spasticity [26]. However, eSCS is an invasive procedure, with the corresponding risks for treated patients. The same neural input structures to the spinal cord can be less specifically, but non-invasively, stimulated with surface electrodes [27–29]. Similarly to eSCS, transcutaneous spinal cord stimulation (tSCS) has been shown to immediately induce lower limb extension at 15 Hz [30,31] and to attenuate spasticity at 50 Hz [32–34]. Furthermore, initiation of flexion–extension movements and an immediate augmentation of locomotor activity was also reported for 30-Hz tSCS [35–37]. No study so far has compared the impact of different stimulation frequencies on the augmentation of residual voluntary motor control. Specifically, the effects of tSCS on more finely controlled ankle movements, which could be an origin of improvements in locomotor function, has not yet been assessed. Several mechanisms were suggested to be responsible for the modulatory effects of electrical lumbar SCS, including an increase in excitability of spinal locomotor networks [16,38,39] and the regulation of activity in multiple segmental circuits [34,40]. However, the recruitment of specific spinal circuits, also dependent on the different stimulation frequencies applied, has not yet been shown.

In this study, we investigated the immediate effects of a single application of tSCS at three different frequencies (15 Hz, 30 Hz, and 50 Hz) on voluntary ankle control, studied in the supine position in ten individuals with chronic, sensorimotor iSCI. We hypothesized that tSCS would immediately facilitate ankle control similar to eSCS, and that this facilitation would be strongest at 30 Hz [16,21], i.e., at a frequency previously shown for eSCS to promote rhythmic flexion-extension movements Moreover, we investigated whether the effects observed in the ankle control task would transfer to over-ground walking with a body weight support (BWS) system and if other stimulation-induced effects were present during locomotion.

Besides effects on motor control, this is the first study to explore the effect of tSCS on dedicated spinal networks by the electromyographic (EMG) representation of a polysynaptic spinal reflex elicited by electrical stimulation of the distal tibial nerve, hereinafter termed spinal reflex. Previous studies have suggested an association of this reflex with spinal locomotor circuits and their functional state after SCI [41–43]. In neurologically intact individuals, the spinal reflex is characterized by a dominant early EMG component occurring at latencies of 60–120 ms, while in individuals with severe SCI, an additional late EMG component at latencies of 120–450 ms gradually develops in the weeks post-injury [41,42,44,45]. Dominant late spinal reflex components were associated with more impaired locomotor function [42]. We hypothesized that afferent input generated by 30-Hz tSCS would interact with the spinal locomotor circuits, resulting in facilitated early, and diminished late reflex components under tonic stimulation.

2. Materials and Methods

2.1. Participants

Data were derived from ten individuals with chronic, sensorimotor iSCI, classified as grade D on the American Spinal Injury Association Impairment Scale [46]. Details on the neurological status of the participants and the assistance required for walking are provided in Table 1. Lower extremity motor scores and sensory scores were taken from previous clinical assessments, and the walking index for spinal cord injury (WISCI II) [47] was assessed in this study. Participant 3 had additional lesions affecting the efferent system associated with his left lower limb. None of the participants had previous experience with tSCS. Among the exclusion criteria were neurological lesion levels caudal to T12, and active or passive implants at vertebral level T9 or caudally, such as osteosynthesis material. The study was approved by the cantonal ethics committee of Zurich, Switzerland (KEK-ZH 2017-00053), and conducted in accordance with the Declaration of Helsinki. All participants signed written informed consent prior to their enrollment into the study.

Table 1. Neurological status of the participants according to the International Standards for Neurological Classification of Spinal Cord Injury.

Nr.	Sex	Age (y)	Neurol. Level of SCI	Time Post-SCI (y)	LEMS Total (L/R)	PP Sensory Score Total (L/R)	LT Sensory Score Total (L/R)	WISCI II Score	FLOAT-BWS (%)
1	M	28	C5	8	5/25	32/32	32/32	13	7
2	M	53	C3	38	25/17	29/30	30/30	20	7
3	M	65	T10	15	16/20	47/49	45/45	20	7
4	M	40	C6	4	21/20	20/18	26/25	16	10
5	M	45	C7	15	14/15	29/26	28/27	20	6
6	M	48	C5	8	20/22	31/31	31/31	20	6
7	M	31	T4	13	19/19	40/36	40/40	13	45
8	M	40	C7	5	17/21	22/22	39/39	13	6
9	F	40	C4	4	23/21	31/31	31/31	9	NT
10	M	64	T3	6	25/23	39/37	44/45	13	NT

BWS, body weight support; FLOAT, free levitation for over-ground active training [48]; LEMS, lower extremity motor score (max. 25 per leg); LT, light touch (max. 56 per side); PP, pin prick (max. 56 per side); Neurol., neurological; Part., participant; WISCI, walking index for spinal cord injury [47]; y, years; SCI, spinal cord injury.

2.2. Study Protocol

Two independent study sessions were conducted on separate days (time between sessions: 1–21 days, same order of sessions for all participants). In session 1, the participants' ability to perform controlled ankle movements was investigated. In session 2, spinal reflex activity and walking performance were assessed.

For the ankle control assessment, participants were lying supine with their thighs and shanks stabilized by a vacuum pillow without restricting the full range of motion (ROM) of the ankle joints. Participants were able to see their feet during the tests. The assessment comprised two protocols. First, participants performed 25 cycles of unilateral rhythmic ankle movements, from maximum dorsi- to maximum plantar flexion angles, at a target rate given by an auditory cue. Three different movement rates were tested, in this sequence, 0.8 Hz, 1.6 Hz, and 2.4 Hz [49]. Second, participants performed a unilateral irregular precision task. It consisted of unilateral ankle movements following a predefined, irregular, sinusoidal trajectory projected on a screen. A motion capture system (Vicon Motion Systems Ltd., Oxford, UK) provided visual feedback of the executed ankle movements in relation to the targeted trajectory. The maxima of the trajectory were adjusted to 90% of the individual maximum ankle ROM. Each protocol was repeated five times (first right side and then left side per repetition), first without tSCS (tSCS-off), followed by three repetitions under tonic tSCS at 15 Hz, 30 Hz, and 50 Hz (applied in a randomized order), and finally without tSCS again (tSCS-off$_{rep}$). Between repetitions, participants rested as needed.

Spinal reflexes were elicited by applying monopolar electrical stimulation to the distal tibial nerve at the dorsal aspect of the medial malleolus (anode and cathode over the nerve with minimal distance but without contact) through self-adhesive hydrogel surface electrodes (Neuroline 700, Ambu, Ballerup, Denmark) with participants in the supine position (knees supported with a cushion). A constant current stimulator (Dantec Keypoint Focus Workstation, Natus Medical Incorporated, Pleasanton, CA, USA) was set to deliver trains of five (five participants) or eight monophasic rectangular pulses, each with a duration of 1.0 ms and an interpulse interval of 4 ms (250 Hz). The total time of a stimulus was 17 ms (five pulses) or 29 ms (eight pulses). Stimulation amplitude was gradually increased in 1-mA increments from 1 mA up to the reflex threshold (spinal reflexes with amplitudes > 20 μV in three out of five stimuli). Subsequently, five stimuli with an amplitude of 1.5 times the reflex threshold were applied. The protocol was conducted first without and then under tonic 30-Hz tSCS on the right limb, and subsequently in the same order on the left limb. For the five above-threshold stimuli, the same stimulation amplitudes were used in the tSCS-off and tSCS-on conditions (1.5 times the reflex threshold of tSCS-off). To minimize habituation, stimulation trains were delivered at randomized intervals of 30–40 s [50], and participants were given a backward counting task [51,52]. The positioning of the participants and the limb was kept constant between the two conditions.

For the gait assessments, participants were fitted with a harness to the BWS system (The FLOAT, Reha-Stim Medtec AG, Schlieren, Switzerland) [48,53,54], set to the individual minimum that allowed unrestricted over-ground walking over a distance of 7 m (Table 1; same BWS provided for tSCS-off and tSCS-on conditions), but always ≥ 4 kg (fall detection limit). Participants 1, 7, and 8 additionally required a walker, and participant 4 used two crutches. Walk tests comprised the assessment (i) of the minimum time required to walk 7 m without tSCS, as well as under tonic 30-Hz tSCS, used to calculate the maximum walking speed (m/s); and (ii) of gait kinematics and lower limb muscle activity without (first condition) tSCS, and during tSCS at 15 Hz, 30 Hz, and 50 Hz (applied in a randomized sequence) while walking 7 m at a self-selected speed. To avoid confounding effects, trials under tonic tSCS were repeated if self-selected walking speed deviated by more than ± 10% from that of the tSCS-off condition. Two trials of each assessment were conducted per condition. Participants 1–6 completed both assessment parts and fulfilled the criterion of constant walking speed in the self-selected walking condition (four trials repeated). Participants 7 and 8 showed signs of fatigue during walking, completed the 7 m with major fluctuation of walking speed across all tSCS-off and tSCS-on trials, and were unable to fulfill the criterion of constant walking speed across repetitions. In both participants, the maximum walking speed was not assessed, and self-selected walking speed under different tSCS conditions are reported instead. Walk tests were not conducted in participants 9 and 10 since no posterior root-muscle (PRM) reflexes [27,28] were elicitable in the BWS- and walker-assisted standing position (see below).

2.3. EMG and Kinematic Data

EMG activity was recorded from bilateral rectus femoris (RF), vastus medialis (VM), semitendinosus (ST), tibialis anterior (TA), and medial gastrocnemius (MG) with surface electrodes (Kendall H124SG, Covidien, Medtronic, Dublin, Ireland), placed in accordance with the European recommendations for surface electromyography (SENIAM [55]), and using the Aktos system (myon AG, Schwarzenberg, Switzerland). Data were amplified with a gain of 1000 over a bandwidth of 10–500 Hz and digitized at 2000 samples per second and channel. For the spinal reflex protocol, the rectified EMG activity of bilateral TA was recorded using surface electrodes (BlueSensor NF, Ambu, Ballerup, Denmark), and a Dantec Keypoint Focus Workstation over a bandwidth of 20 Hz to 3000 Hz and digitized at 6000 samples per second. A copper strip covered in a wet absorbent felt placed around the ankle joint was used as ground electrode.

Kinematic data were acquired using the Vicon motion capture system (Vicon Motion Systems Ltd., Oxford, UK). For the ankle control assessment, three reflective markers were placed bilaterally on the fibula head, the lateral malleolus, and the fifth metatarsal head. For the walk tests, a total of 42 markers were placed according to the plug-in gait model (Vicon Motion Systems Ltd., Oxford, UK), with additional locations for gap filling purposes. Kinematic data were sampled at 100 Hz (ankle control assessment) or 200 Hz (walk tests), labelled and gap filled using Vicon Nexus 2.6, and synchronized to the EMG data (without upscaling). Trajectories of the walk tests were additionally filtered (Woltring filter, volume specific mean-square-error value of 15) and modelled with the plug-in gait model [56]. Gait events were manually detected based on marker position data. Angular excursions of the lower limb joints were calculated based on vectors using toe, ankle, knee, and sacrum markers, as well as the modelled hip joint rotation center [57,58].

2.4. Transcutaneous Spinal Cord Stimulation

Lumbar tSCS was applied through a self-adhesive surface electrode (5 × 9 cm; RehaTrode, Hasomed GmbH, Magdeburg, Germany) positioned longitudinally over the spine, covering the T11 and T12 spinous processes, and a coupled pair of electrodes (7.5 × 13 cm each) placed over the lower abdomen, in symmetry to the umbilicus [27,28,59]. A current-controlled stimulator (RehaMove 3.0, Hasomed GmbH) was set to deliver charge-balanced, symmetric, biphasic rectangular pulses of 1 ms width per phase. With reference to the abdominal electrodes, the paraspinal electrode was the anode for the first, and the cathode for the second, pulse phase [28]. Paravertebral electrode position over the lumbar and upper sacral spinal cord was verified by the elicitation of PRM reflexes in the L2–S2 innervated RF, VM, ST, TA, and MG bilaterally [27,28]. Stimulation of afferent fibers was tested using a paired-pulse paradigm with interstimulus intervals of 30 ms, 50 ms, and 120 ms [27,28,60–64]. In session 1, the mean PRM-reflex threshold ± SD across participants was 31.0 ± 18.8 mA (per phase of the biphasic stimulation pulse), ranging from 15 mA to 70 mA. Tonic tSCS was applied with a target amplitude of 0.8–1.0 times the individual PRM-reflex threshold [16,36], yet always below a level causing discomfort, and which amounted to 26.4 ± 17.3 mA across participants (same for all stimulation frequencies, 13–65 mA; 0.84 ± 0.10 × PRM-reflex threshold). In session 2, the mean PRM-reflex threshold was 32.0 ± 14.4 mA (14–65 mA) in the supine position (spinal reflex protocol), and 40.6 ± 13.5 mA (20–60 mA) in the BWS-supported standing position with walking aids, as used for stepping. The mean stimulation amplitude of tonic tSCS was 28.8 ± 14.7 mA (13–65 mA, 0.89 ± 0.08 × PRM reflex threshold) for the spinal reflex protocol, and 34.8 ± 13.1 mA (same for all stimulation frequencies, 16–60 mA, 0.85 ± 0.10 × PRM reflex threshold) for the walk tests. The maximum durations of tSCS application with the target amplitude were approximately 5 min for the ankle control and walk tests, and 15 min for the spinal reflex assessment. Participants 4, 5, 8, and 10 reported paresthesia in lower limb dermatomes during the tonic stimulation [34,65].

2.5. Data Analysis and Statistics

Data were analyzed offline using Matlab 2017b (The Mathworks Inc., Natick, MA, USA), IBM SPSS Statistics 26.0 for Windows (IBM Corporation, Armonk, NY, USA), and R 3.6.1 (The R project for Statistical Computing, R Core Team) with R Studio 1.2 (R Studio, Inc., Boston, MA, USA). α-errors of $p < 0.05$ were considered significant. Prior to the data analysis, the two legs of each participant were grouped into more and less affected sides, based on the total lower extremity motor scores (LEMS) per leg [66–68]. In participant 7, total light touch sensory scores were additionally considered, as LEMS were equivalent in both legs. This separation was applied to account for asymmetrical damages in SCI, to make legs within a group more comparable, and to allow inferences on impairment level.

For the unilateral rhythmic ankle dorsi- and plantar flexion movement, ankle ROMs, root-mean-square (RMS) values of the EMG signals of TA during dorsiflexion, and MG during plantar flexion, as well as deviations from the target movement rate, were determined for 20 movement cycles of each repetition of the task (the first five of the 25 cycles per repetition were omitted) and averaged. Separate linear mixed models (due to several factors and missing values), with movement rate (0.8 Hz, 1.6 Hz, and 2.4 Hz) and tSCS condition (tSCS-off, 15-Hz tSCS, 30 Hz tSCS, 50-Hz tSCS, tSCS-off$_{rep}$) as fixed effects and between-subject differences as random effect, were fit to the data of the more and the less affected lower limbs. Residuals of linear mixed models were visually screened for normal distribution (QQ-plots), and input datasets were transformed if necessary (log or square root transformations). Post-hoc tests were based on estimated marginal means and Bonferroni-corrected to adjust for multiple comparisons. Effect sizes are reported by the partial eta-squared (η_p^2). Participant 4 did not perform the tSCS-off$_{rep}$ condition (fatigue) and participant 2 was unable to perform the highest movement rate with his more affected lower limb. For the irregular ankle movements, deviations from the predefined target trajectory were calculated as the RMS-error normalized to the total range of the trajectory (90% of the individual maximum ankle ROM) and compared using linear mixed models. In participants 4 and 8, the irregular ankle movement test was not conducted due to fatigue. Across ankle control assessments, data derived from the less affected lower limb of participant 1 were considered only (additional lesions affecting efferent system in more affected limb), and the more affected leg of participant 1 was excluded from analysis because of 0° ankle ROM in all conditions.

In the recordings of session 2, stimulation artifacts produced by tSCS were superimposed on the EMG activities, generated in TA during the spinal reflex protocol, and on those of the thigh muscles during the walk tests. Prior to further analysis, the artefacts were removed offline by adjustable blanking intervals of 3–6 ms, beginning with the leading edge of the stimulus and covering any die-away effects of the falling stimulus edge [36] (cf. Figure 1 in [36] for illustration). Within the blanking intervals, EMG traces were set to not-a-number. In the spinal reflex protocol, the same procedure was applied to the EMG recordings in the tSCS-off condition for an assumed 30-Hz stimulation train to avoid confounding effects on the evaluation of the early and late reflex components (see below). For the walk tests, where only increased EMG activity was expected during stimulation, the procedure was applied to the tSCS-on conditions only, thus rather under- than overestimating the EMG activity of the thigh muscles generated in the trials with tonic stimulation.

Spinal reflexes obtained with above-threshold stimulation were analyzed for the occurrence of early and late components [41,42]. Time windows were set at 60–120 ms for the detection of the early component, and at 120–450 ms for the late component [41]. Maximum EMG amplitudes within these windows were determined, and noise-corrected RMS values (a period of noise was manually selected for each recording) of the EMG within ±25 ms of the maxima were calculated. Values of 0 were assigned if no response was detectable within a given time window (i.e., RMS noise ≥ RMS reflex). EMG-RMS values of the early and the late response components, as well as spinal reflex thresholds, were compared by running separate linear mixed models (because more than one factor) with stimulation condition (tSCS-off, 30-Hz tSCS) and lower limb (more affected, less affected) as fixed effects and between-subject differences as random effect (see above for further methodological details of applied linear mixed models). Data derived from the less affected lower limb of participant 3 were excluded from analysis due to the regular occurrence of non-stimulation related muscle twitches in TA that were superimposed on the EMG and impeded reflex threshold estimation. No spinal reflexes were recorded from the less affected lower limb of participant 4 because of the perception of discomfort at the stimulation amplitude required for their elicitation.

Maximum walking speed was compared between tSCS-off and tSCS-on at 30 Hz using a Wilcoxon signed-rank test (due to the limited sample size of 6 participants, correlation coefficient as effect size). For the analysis of gait kinematics at a self-selected speed, mean hip, knee, and ankle ROMs, as well as relative durations of double limb support phases, step lengths, and stride times were derived for the more and less affected lower limbs, separately for the different tSCS-off and tSCS-on conditions. Lower limb muscle activity was characterized by mean EMG-RMS values per muscle, and separately for stance and swing phases. Each outcome was statistically compared between conditions in participants 1–6 (constant self-selected speed) using a Friedman test (due to limited sample size, Bonferroni-corrected for more and less affected leg). Additionally, the results of the walk tests were reported descriptively. For this purpose, data obtained under tonic stimulation were normalized to the respective values in the tSCS-off condition using z-scores (changes in standard deviations from baseline). Given the high kinematic fluctuations in participants 7 and 8, the variability of double limb support phases, step lengths, and stride times was additionally calculated as the coefficient of variation (SD divided by the mean).

For the ankle control tasks, as well as the walk tests, participants were asked to rate their comfort and performance perception in the tSCS-off and tSCS-on conditions (all frequencies combined) based on the visual analogue scale (VAS; 0–100, 0 being the lowest subjective score and 100 being the highest subjective score). Data were not normally distributed (Shapiro–Wilk test) and on-off conditions were consequently compared using Wilcoxon signed-rank tests (correlation coefficient as effect size).

3. Results

3.1. Voluntary Ankle Control

Tonic tSCS applied for a few minutes had several immediate effects on voluntary alternating ankle flexion and extension at three different movement rates (Figure 1A). In the more affected lower limb, tSCS condition (tSCS-off, 15-Hz tSCS, 30 Hz tSCS, 50-Hz tSCS, tSCS-off$_{rep}$) had a significant main effect on ankle ROM ($F_{4;89.016} = 4.368$, $p = 0.003$, $\eta_p^2 = 0.164$), maximum dorsiflexion ($F_{4;89.011} = 6.779$, $p < 0.001$, $\eta_p^2 = 0.234$) as well as plantar flexion angles ($F_{4;89.005} = 3.557$, $p = 0.010$, $\eta_p^2 = 0.138$), EMG activity of MG during the plantar flexion phases ($F_{4;88.037} = 3.290$, $p = 0.015$, $\eta_p^2 = 0.130$), and on the deviation from the target movement rate ($F_{4;89.028} = 2.872$, $p = 0.027$, $\eta_p^2 = 0.114$). Post-hoc pairwise comparisons revealed significantly larger ankle ROM during 30-Hz tSCS than tSCS-off (absolute mean difference: $2.9° \pm 0.9°$, $p = 0.013$) and tSCS-off$_{rep}$ ($3.4° \pm 0.9°$, $p = 0.003$). Additionally, dorsiflexion angles during 30-Hz tSCS were significantly increased by $4.6° \pm 0.9°$ compared to tSCS-off ($p < 0.001$), by $2.7° \pm 0.9°$ compared to 15-Hz tSCS ($p = 0.041$), and by $3.8° \pm 1.0°$ compared to tSCS-off$_{rep}$ ($p = 0.002$). At the same time, absolute plantar flexion angles were smaller by $1.7° \pm 0.5°$ during 30-Hz tSCS than tSCS-off ($p = 0.008$). MG-RMS values during plantar flexion were reduced in tSCS-off$_{rep}$ compared to tSCS-off (5.7 ± 1.6 µV, $p = 0.007$). In the less affected lower limb, tSCS condition had a significant main effect on ankle ROM ($F_{4;123.011} = 2.522$, $p = 0.044$, $\eta_p^2 = 0.076$), plantar flexion angles ($F_{4;123.003} = 8.125$, $p < 0.001$, $\eta_p^2 = 0.290$), as well as on TA activity during dorsiflexion ($F_{4;123.005} = 2.673$, $p = 0.035$, $\eta_p^2 = 0.080$) and MG activity during plantar flexion ($F_{4;123.017} = 8.127$, $p < 0.001$, $\eta_p^2 = 0.209$). Post-hoc pairwise comparisons demonstrated significantly larger absolute plantar flexion angles during tSCS-off than during 15-Hz tSCS ($1.9° \pm 0.4°$, $p < 0.001$), 30-Hz tSCS ($1.5° \pm 0.4°$, $p = 0.001$), and tSCS-off$_{rep}$ ($1.8° \pm 0.4°$, $p < 0.001$). MG-RMS values during plantar flexion were larger during tSCS-off than during 15-Hz tSCS (8.5 ± 2.4 µV, $p = 0.005$) and tSCS-off$_{rep}$ (12.9 ± 2.5 µV, $p < 0.001$) as well as larger during 50-Hz tSCS than during tSCS-off$_{rep}$ (9.8 ± 2.5 µV, $p = 0.001$). No significant interaction between tSCS condition and movement rate existed for any outcome measure derived from the more and less affected lower limbs.

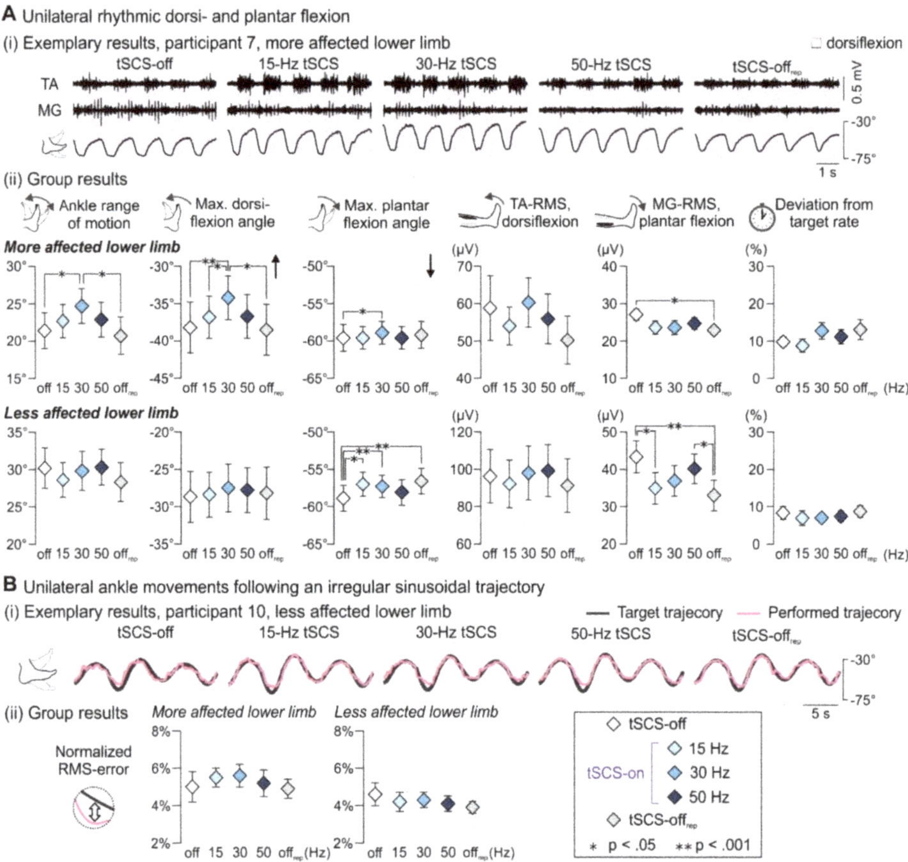

Figure 1. Effect of tSCS on voluntary unilateral ankle control. (**A**) Unilateral rhythmic dorsi- and plantar flexion movements. (**i**) EMG activities of tibialis anterior (TA) and medial gastrocnemius (MG) shown for five movement cycles at a rate of 1.6 Hz during tSCS-off, tSCS at stimulation frequencies as indicated, and tSCS-off$_{rep}$ for the more affected lower limb of participant 7. Shaded backgrounds mark dorsiflexion phases, identified based on kinematic data. Under tonic tSCS, maximum ankle angles and TA activity during dorsiflexion were increased, while clonus-like activity in MG that was present in the tSCS-off condition was visibly reduced. (**ii**) Group results (mean ± SE) across movement rates of ankle range of movement; maximum dorsiflexion angles (black arrow indicates direction of dorsiflexion, zero reference represents an absolute joint angle of 90°, i.e., as during an upright standing position); maximum plantar flexion angles (black arrow indicates direction of plantar flexion); TA EMG-RMS values during dorsiflexion; MG EMG-RMS values during plantar flexion; and deviation from target movement rate during tSCS conditions as indicated, separately shown for the more and the less affected lower limbs. (**B**) Unilateral ankle movements following an irregular sinusoidal target trajectory. (**i**) Exemplary recordings of performed ankle movements (purple line) relative to the predefined irregular target trajectory (grey line); less affected lower limb, participant 10. (**ii**) Group results (mean ± SE) of the normalized RMS-error, providing information on the deviation of the performed movement from the target trajectory, across tSCS conditions, shown for the more and the less affected lower limb. Asterisks mark significant results of post-hoc pairwise comparisons. deg., degree; EMG, electromyographic; MG, medial gastrocnemius; RMS, root mean square; TA, tibialis anterior; * $p < 0.05$; ** $p < 0.001$.

tSCS did not modify the ability to follow a predefined, irregular sinusoidal trajectory (Figure 1B) as quantified by the normalized RMS-error, neither in the more ($F_{4;17.259} = 0.541$, $p = 0.708$, $\eta_p^2 = 0.111$) nor the less affected lower limb ($F_{4;26.903} = 1.172$, $p = 0.345$, $\eta_p^2 = 0.148$).

The measures derived from the more and the less affected lower limbs were separately analyzed, since the following measures were, in the tSCS-off condition, significantly lower in the more affected lower limb: (i) the median LEMS (more affected, 18.0 (interquartile range (IQR): 15.5–20); less affected, 21.5 (19.75–25.0), $z = -2.673$, $p = 0.008$, $r = 0.845$), (ii) the ankle ROM during active rhythmic movements (F_1; $38.548 = 22.978$, $p < 0.001$, $\eta_p^2 = 0.373$), and (iii) the EMG activities of TA during dorsiflexion ($F_{1;37.076} = 7.846$, $p = 0.008$, $\eta_p^2 = 0.175$), as well as of MG during plantar flexion ($F_{1;37.181} = 10.043$, $p = 0.003$, $\eta_p^2 = 0.213$).

All group results are detailed in Table S1.

3.2. Spinal Reflex Activity

Tonic 30-Hz tSCS did not alter the spinal reflex threshold ($F_{1;21.142} = 0.493$, $p = 0.490$, $\eta_p^2 = 0.023$) nor the EMG-RMS of the early reflex component ($F_{1;22.668} = 0.775$, $p = 0.388$, $\eta_p^2 = 0.033$; Figure 2, Table S2). On the other hand, it significantly reduced the EMG-RMS of the late reflex component ($F_{1;21.952} = 6.337$, $p = 0.020$, $\eta_p^2 = 0.224$). Lower limb (levels: more and less affected) was not a significant factor for any of the three measures (threshold, $F_{1;21.500} = 0.262$, $p = 0.614$, $\eta_p^2 = 0.012$; early component, $F_{1;23.935} < 0.001$, $p = 0.988$, $\eta_p^2 < 0.001$; late component, $F_{1;22.330} = 0.037$, $p = 0.848$, $\eta_p^2 = 0.002$). Further, no interaction between tSCS condition (tSCS-off, 30-Hz tSCS) and lower limb (levels: more and less affected) existed (threshold, $F_{1;21.142} = 0.251$, $p = 0.621$, $\eta_p^2 = 0.012$; early component, $F_{1;22.668} = 0.009$, $p = 0.926$, $\eta_p^2 < 0.001$; late component, $F_{1;21.952} = 1.272$, $p = 0.272$, $\eta_p^2 = 0.002$). All data are reported in Table S2.

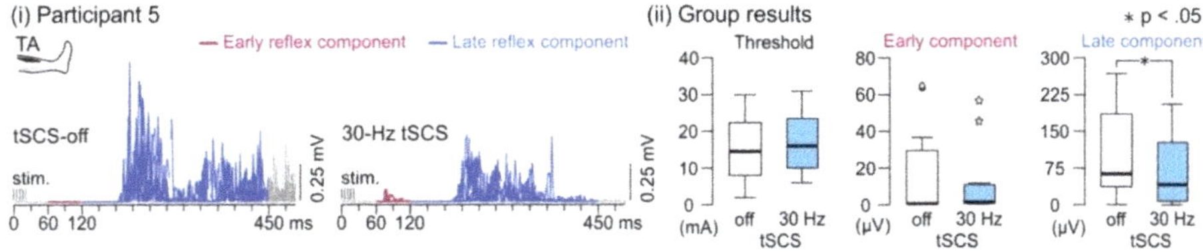

Figure 2. Effect of tSCS on spinal reflex activity. (**i**) Rectified electromyographic (EMG) responses of tibialis anterior (TA) to distal tibial nerve stimulation, applied to evoke a spinal reflex in the tSCS-off and the 30-Hz tSCS conditions in one participant. Early reflex components (purple lines) were identified within time windows of 60–120 ms post-stimulation train onset (stim.), and late reflex components (blue lines) within 120–450 ms. Shown are four (tSCS-off; one reflex excluded due to pre-activation) or five repetitions superimposed; participant 5, less affected lower limb. Note the shorter duration of the late reflex component under tonic 30-Hz tSCS. (**ii**) Group results of spinal reflex thresholds and TA EMG-root mean square (RMS) values associated with the early and late reflex component, respectively, during tSCS-off and 30-Hz tSCS conditions. The EMG-RMS of the late component was significantly reduced under tonic 30-Hz tSCS (asterisk; $p < 0.05$). Box plots represent medians by bold horizontal lines within boxes that span the interquartile range (IQR). Whiskers extend to the smallest and largest values that are not outliers (here, values between 1.5 and 3 times the IQR of the upper quartile; shown as separately plotted points) or extreme values (>3 times the IQR of the upper quartile; white asterisks).

3.3. Walking Performance

Walking performance, assessed using the FLOAT-BWS system and assistive devices as needed, showed considerable inter-individual differences, both in the baseline recordings without tSCS, as well as with respect to the effects of tonic tSCS at the three stimulation frequencies tested (individual values, see Table S3).

Maximum walking speed without tSCS, as well as under tonic 30-Hz tSCS, was assessed in participants 1–6 (Figure 3A(i); not tested in participants 7 and 8, see Methods) and no significant group-effect between conditions was found (z = 0.734, *p* = 0.463, r = 0.212). Individual maximum walking speed was increased by 12.7% (+0.05 m/s) and 9.6% (+0.11 m/s) in participants 1 and 2, respectively, and reduced during stimulation by 14.4% (−0.05 m/s) in participant 6. Participants 3–5 showed no differences during tSCS-on compared to tSCS-off (+0.9%, +0.01 m/s; +0.4%, 0.00 m/s; −0.2%, 0.00 m/s). In the assessment of gait kinematics and lower limb muscle EMG activity, the six participants were able to maintain a constant (±10%) self-selected walking speed across tSCS conditions (tSCS-off, 15-Hz, 30-Hz, and 50-Hz tSCS; Table S3), while participants 7 and 8 completed the 7 m at variable, individually possible speed in the different conditions. The self-selected speed was increased during tonic tSCS at all three stimulation frequencies in participant 8 and reduced during 15-Hz and 30-Hz tSCS, and increased during 50-Hz tSCS, in participant 7 (Figure 3A(ii)).

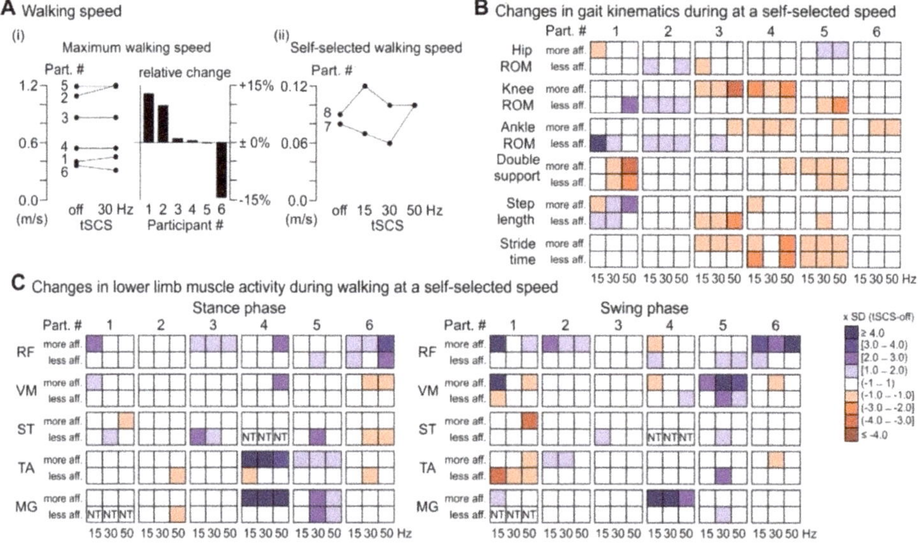

Figure 3. Effect of tSCS on over-ground walking performance. (**A**) Walking speed assessed on a 7 m track using the FLOAT-BWS system. (**i**) Absolute maximum walking speed of the six participants (see Section 2.2 Study protocol) completing the assessment during tSCS-off and 30-Hz tSCS, along with relative changes. (**ii**) Self-selected walking speed across tSCS conditions as indicated of the two participants (see Section 2.2 Study protocol) unable to maintain a constant speed over 7 m (both did not perform maximum walking speed assessment). (**B**) Changes in gait kinematics during 15-Hz, 30-Hz, and 50-Hz tSCS, given as multiples of the SD of the respective values in the tSCS-off condition (z-scores), and illustrated by the opacity of blue (indicating an increase) and red (decrease) boxes. Shown are data derived from the more and less affected lower limbs of participants 1–6 who walked at a constant self-selected speed across tSCS conditions. (**C**) Changes in lower limb muscle activation (RMS values) during stance (left) and swing phases (right). MG values of the less affected lower limb of participant 1, and ST values of the less affected lower limb of participant 4 are missing data. BWS, body-weight support; FLOAT, free levitation for over-ground active training; MG, medial gastrocnemius; RF, rectus femoris; ST, semitendinosus; TA, tibialis anterior; VM, vastus medialis.

For participants 1–6, with constant self-selected walking speed, there was no group effect of tSCS on kinematic and EMG outcome measures (for detailed statistics see Table S3), except for the stride time in the more affected leg, which was shorter during tonic stimulation (tSCS-off, median (IQR): 1.84 (1.36–3.22); 15 Hz: 1.75 (1.28–2.99); 30Hz: 1.78 (1.28–3.10); 50 Hz: 1.84 (1.27–2.85); $\chi^2(3)$ = 9.8,

$p = 0.041$). An overview of kinematics and muscle activity is provided for participants 1–6 in Figure 3B,C, and this revealed individual changes during tSCS, whereof some are briefly highlighted in the following. Participants 1 and 2 generally showed increased ROMs in one or more lower limb joints under tonic stimulation, and participant 1 additionally demonstrated decreased double limb support and increased step length. Participants 3–5 exhibited decreases in ROM, step length, and stride time during tSCS. Double limb support was additionally reduced in participant 5. Muscle activity of RF was enhanced during tSCS in the swing phase in participant 2 and in the stance phase in participant 3. Participants 1, 5, and 6 also exhibited augmented activity in RF or the synergistic VM, however, closer inspection revealed that the increase was attributable to stimulus triggered responses superimposed on the naturally generated EMG activities. TA and MG activity was increased under tSCS during stance in participant 5, whereas participant 1 exhibited decreased TA activity during swing. Participant 4 showed clonus-like activity in the ankle joint of the more affected limb during stimulation, resulting in increased muscle activity of the TA and MG.

In participants 7 and 8, with fluctuations of self-selected speed, mean hip and knee ROMs, as well as phases of double limb support were reduced in both lower limbs under tonic tSCS across the stimulation frequencies (Table S3), and in participant 8, the mean step length was additionally augmented. Both participants walked the 7 m with less variation in double limb support, stride time, and step length across tSCS frequencies, except for the step length of the more affected lower limb of participant 7 at 15 Hz, and of his less affected lower limb at 30 Hz (Table S3).

3.4. Subjective Reports

tSCS at 15 Hz, 30 Hz, and 50 Hz with targeted stimulation amplitudes was generally well tolerated by all participants with the exception of participant 7 in the first study session (stimulation amplitude hence set to 0.65 times the PRM-reflex threshold, a level clear below the threshold of discomfort). In the second study session, participant 7 reported no discomfort when being stimulated with the target amplitude. Across participants, no differences existed between comfort as well as performance perception between ankle control tests without and with tSCS (median VAS score (IQR), comfort: tSCS-off, 68 (55.8–84.5); tSCS-on, 73.5 (54.3–89.3); $z = 0.059$, $p = 0.953$, $r = 0.019$; performance: tSCS-off, (70 (51.5–78.3); tonic tSCS, 77 (57.0–85.0); $z = 1.487$, $p = 0.137$, $r = 0.470$). During walking, comfort perception did not differ between tSCS-off and tSCS-on conditions (62 (48.0–72.0) and 63 (53.0–82.5); $z = 1.461$, $p = 0.144$, $r = 0.653$). However, the performance perception during stimulation was significantly increased under tonic tSCS (65 (47.0–71.0) and 71 (61.0–79.5), $z = 2.023$, $p = 0.043$, $r = 0.905$).

4. Discussion

Single sessions of tonic tSCS produced modulatory effects on lower limb motor activity within minutes of application on group-level in individuals with chronic, sensorimotor iSCI. During rhythmic voluntary ankle movements, maximum dorsiflexion was significantly enhanced in the more affected lower limb during 30-Hz stimulation compared to tSCS-off. Despite a concomitant reduction of maximum plantar flexion, this resulted in an augmented active ROM. No effects on rhythmic ankle movements were observed for 15-Hz and 50-Hz tSCS. Furthermore, coordination of ankle joint movements was unchanged during stimulation across the frequencies. Electrophysiological recordings of a polysynaptic spinal reflex demonstrated, for the first time, an immediate interaction of tSCS and the spinal interneuronal networks associated with locomotion. There was no group effect of tSCS on locomotor function in a subset of eight participants, with high levels of inter-individual variability. Among the eight participants, the three with the lowest (score 13) and one with the maximum WISCI II scores showed increased maximum walking speed (participants 1 and 2) or more continuous (participants 7 and 8) and faster stepping (participant 8) at a self-selected speed. Some participants exhibited reduced ROM, step lengths, and strides times during tSCS.

The target neural structures of tSCS, applied over the lumbar spinal cord, are the large-to-medium diameter afferent fibers in posterior roots originating in distant dermatomes and myotomes, and are comparable to target structures of eSCS [17,19,20,28,29,69]. Here, stimulation of these target structures was verified by the elicitation of PRM reflexes with single stimulation pulses and their suppression and gradual recovery when using a paired-pulse paradigm [36,70]. Stimulation at 15 Hz, 30 Hz, and 50 Hz was applied, with an amplitude below the threshold for the elicitation of PRM reflexes at rest, thus not reflexively depolarizing alpha-motoneurons, while providing repetitive, multisegmental afferent input to the spinal cord [16,36]. The stimulation of afferent structures within the L2–S2 posterior roots during tonic tSCS was further supported by the occurrence of tingling sensations in the respective lower extremity dermatomes reported by some of the participants [34]. Repetitive afferent input to spinal networks generated by SCS is hypothesized to modulate spinal excitability, regulate activity in segmental spinal circuitries [34,40], and enhance their responsiveness to remaining descending input, ultimately resulting in improved motor control after SCI [16,38,39].

In this study, we demonstrated that tSCS at 30 Hz improved voluntary control of the ankle joint in the more affected lower limb of participants with chronic iSCI. Active ankle ROM and voluntary dorsiflexion were immediately increased, by 2.9° ± 0.9° and 4.6° ± 0.9°, respectively. A previous study investigated the effects of one-month Lokomat training on voluntary ankle movements in an iSCI cohort comparable to this study (11.80 ± 2.54 years post-injury, WISCI II 9–20) [71]. Post-training, dorsiflexion movements were improved, reflected by an increase of the active ROM by 5.1° ± 1.6, which is in a similar range as the tSCS-induced improvements in this study. Voluntary ankle control has been shown to be substantially improved with eSCS at 25–30 Hz, which enabled ankle movements in paralyzed muscles after motor complete SCI, including relatively fine controlled dorsiflexion [16]. The lack of facilitation in the less affected lower limb in the participants of the present study was likely related to the significantly higher LEMS on this side, with three participants scoring the maximum of 25, and another five participants having scores ≥20. Additionally, in the baseline recording without tSCS, ankle ROM and EMG activity of TA during dorsiflexion were greater than on the more affected side. Together, this leads to the assumption that the participants had maintained a critical level of control over voluntary ankle movements in their less affected lower limbs, holding less potential for further improvement. Absolute plantar flexion angles were unchanged, or even decreased, for both legs across stimulation conditions in this study. The plantar flexion movement was in the direction of gravity, and EMG activities recorded from MG were at a relatively low level across tSCS-off and tSCS-on conditions, suggesting that the participants did not perform this movement phase at full voluntary capacity. This may also explain the lack of the hypothesized facilitation of the plantar flexion phase by 15-Hz tSCS. A decrease of maximal plantarflexion angle from the off-condition before, to the off-condition after, the intervention might additionally indicate that participants were less engaged in plantarflexion over time. Since there was no change over time for the other variables, fatigue can be excluded as an origin of this reduction. No effects on rhythmic ankle movements were also observed during 50-Hz tSCS, where previous studies demonstrated an antispastic effect [32–34]. While we also observed signs of an antispastic effect (i.e., reduced clonus-like activity when stimulation was applied at 50 Hz in participant 7 (cf. Figure 1A)), stimulation at 50 Hz did not translate into functional improvements of ankle motor control over the whole group in the first application. However, as clonus-like muscle activity was not present in the majority of patients during initial tSCS-off, spasticity might not have been a major problem for ankle control in the assessed cohort. Coordination of the ankle joint was unchanged during tSCS across all frequencies. However, previous studies employing similar tasks demonstrated that ankle dexterity is largely retained after iSCI, while the strength of muscles acting on the ankle joint is constrained resulting in a reduced ROM [49,72,73].

This study showed for the first time a modulatory effect of electrical SCS on spinal locomotor networks by means of electrophysiological measures of spinal reflex activity. The activity of this reflex is substantially altered after SCI, and was suggested to reflect the functional state of spinal locomotor circuits [41,42,45]. While the physiological early reflex component is generally reduced after severe SCI, an additional late component gradually occurs after SCI, and dominance of this late component is associated with higher degrees of walking impairment [41,42,45]. However, if appropriate proprioceptive input is provided in individuals with chronic motor complete SCI (e.g., in a physiological movement task) this effect can be temporarily reversed in a single session, and the late reflex component can be decreased, while the early component reappears [74]. Furthermore, locomotor training after SCI was shown to increase the early, and decrease the late, reflex component during walking [75] or at rest [42]. In this study, the late reflex component was decreased under tonic 30-Hz stimulation for the more and less affected leg, suggesting an interaction of the afferent input provided by tSCS with spinal locomotor networks, and a resulting modulation of their excitability. No significant group effect on the early reflex component existed, yet, in some of the participants, it was increased or reoccurred (the latter in participant 5 only) when the stimulation was applied (cf. Figure 2A). This probably reflects inter-individual differences in initial spinal excitability levels, and may indicate that the afferent input provided by tSCS did not modulate the spinal excitability level in a similar manner to movement-related afferent feedback during stepping. Another mechanism involved in the observed effect could potentially be the spasticity reducing impact of tSCS. A previous study reported that the late component in the TA EMG activity after cutaneous reflex elicitation was reduced during active plantarflexion in controls, and that this modulation was absent after iSCI, however, only if spasticity was present [76]. This may indicate that the inhibitory effects of tSCS, involved in the known antispastic impact of the stimulation (possibly mechanisms of presynaptic inhibition, post activation depression, reciprocal inhibition [26,32–34]), contributed to the suppression of the late reflex component. Since prolonged reflex activity was observed in individuals with spasticity [76], observations of shortened late reflex components further indicate that such mechanisms were involved in the reflex modulation (cf. Figure 2A). An antispastic effect for tSCS and eSCS has been reported for stimulation frequencies around 50 Hz [26,32–34], but also for lower frequencies in eSCS [77]. As both tSCS, and the electrical stimulation to elicit the spinal reflex, share common stimulated neuronal structures [34,41], suppression of the spinal reflex caused by antidromic volleys following the posterior root stimulation may have led to decreases in amplitudes of the reflex components. Yet, likely only to a minor extent, considering the suppression of the late reflex component only across all participants, as well as the facilitation or recurrence of the early component in some participants.

There was no group effect of tSCS on walking ability, and high variability between participants was observed. The only statistical significance across participants 1–6 involved a reduction of the stride time in the more affected leg, which does probably not represent a clinically relevant change, as the highest difference between two conditions was 0.09 s. Whereas some participants showed enhanced maximum walking speed, increased joint ROMs, as well as decreased double limb support during tSCS, and can hence be considered as treatment responders, others showed only minor changes, or even reduced ROMs. Specifically, the increase in ankle ROM under tonic stimulation when participants were in the supine position did not generally translate to a similar increase during walking; nevertheless, this was present in responder participants 1 and 2 who increased their maximum walking speed during 30-Hz tSCS by, or beyond, the minimally clinically important difference of 0.05 m/s for individuals with SCI [78]. Participants 7 and 8, who walked with high kinematic variability and were unable to maintain a constant walking speed across repetitions, exhibited more continuous stepping when tSCS was applied. Notably, participants 1, 7, and 8 had WISCI II scores of 13, i.e., the lowest scores among the eight participants completing the walk tests, indicating that individuals with higher walking impairments are more likely treatment responders. Several other factors may have contributed to the variable group results. Walking is a much more complex motor task than rhythmic single joint movements, locomotor deficits may be highly variable between participants [79], and possible deficit-specific improvements dilute group effects. Additionally, the stimulation may not hold the potential to target each deficit in an equally efficient manner. Furthermore, all participants were in the chronic stage of recovery, and likely had difficulties to acutely adopt their walking strategies during a single session, and within the relatively short track of 7 m. Four of the eight participants had maximum WISCI II scores of 20, with probably even less space for further immediate improvement. It needs to be determined if individually optimized stimulation parameters and multiple applications can enhance the effects of tSCS on ambulation, and increase the number of treatment responders [36,80], as shown for eSCS [12–14]. The higher subjective performance ratings during walking under tonic stimulation may have been related to the co-activation of the paraspinal and abdominal muscles, leading to a perceived increased stabilization of the trunk. Yet, in some participants, this co-activation may have resulted in a more rigid gait, perhaps partially explaining the findings of reduced ROM, and highlighting the need of familiarization to the stimulation. Most participants were additionally unexperienced with the BWS setup, and familiarization to the system was possibly not completed in the first session.

In participants 9 and 10, PRM reflexes could not be elicited in the BWS standing position, with additional assistance of a walker required to stabilize equilibrium. Both participants exhibited a slightly forward bent posture during standing. It was previously shown that PRM reflex thresholds are generally higher in the standing than the supine position, and that changes in the volume conductor in between the paraspinal and abdominal electrodes, as in alterations of posture (e.g., a forward bend), can substantially influence the effects of tSCS [81].

Despite applying stimulation amplitudes below the PRM reflexes threshold in the standing position, stimulus-triggered responses were superimposed on the EMG signals of the thigh muscles of participants 1, 5, and 6 during walking. Additional afferent input and posture alterations during the dynamic task possibly changed response thresholds, as well as activation sites of tSCS to the intervertebral foramina, where posterior and anterior roots approach each other to form the spinal nerve [81,82]. The clonus-like activity in TA and MG of participant 4 during walking under tSCS (not observed during ankle assessments) was probably caused by the combination of afferent input from tSCS and the proprioceptive feedback related to the motor task.

A limitation of the current study, and a general challenge for clinical studies using tSCS, is the lack of a sham intervention with confirmed ineffectiveness [33,34,83]. Yet, in the assessment of rhythmic ankle movements, effects were only observed in the more affected lower limb with significantly lower LEMS, indicating that placebo effects likely played a minor role. The findings of the study are further limited due to the low sample size, especially in the walking condition. A limitation of tSCS itself is the constrained target cohort among individuals with SCI, given by the exclusion criteria comprising osteosynthesis material at the site of stimulation, as well as caudal injury sites that are regularly accompanied by secondary peripheral lesions. The order of the assessments (ankle control tasks, spinal reflex, walking tests) might have influenced the results, because participants gained experience with tSCS. However, this was probably only present to a minor extent, since the biggest effects were observed in the first assessment.

5. Conclusions

This study demonstrated that tonic tSCS can acutely facilitate residual voluntary ankle control after chronic iSCI, and modulates spinal locomotor networks, i.e., polysynaptic spinal reflex behavior. Effects on locomotion were variable across participants, yet, the ones with lower ambulatory function showed increased maximum walking speed or more continuous stepping in the presence of tSCS. This indicates that effects on walking performance likely also depend on the degree of impairment of the baseline walking ability. Individually tailored stimulation parameter settings, as well as multiple applications of the intervention together with task-specific training paradigms, may lead to enhanced clinical outcomes.

Supplementary Materials: Supplementary materials can be found at http://www.mdpi.com/2077-0383/9/11/3541/s1. Table S1: Ankle control assessments, Table S2: Spinal reflex activity, Table S3: Walk tests.

Author Contributions: Conceptualization, U.S.H., A.C. and M.B.; methodology, C.M., M.H., U.S.H., A.C., and M.B.; software, R.H.H.; formal analysis, C.M. and U.S.H.; investigation, C.M., U.S.H., M.H., and C.R.; resources, M.H., M.B. and A.C.; data curation, C.M. and C.R.; writing—original draft preparation, C.M. and U.S.H.; writing—review and editing, C.M., U.S.H., M.H., R.H.H., C.R., A.C., M.B.; visualization, C.M. and U.S.H.; supervision, A.C. and M.B.; project administration, C.M. and M.B.; funding acquisition, A.C. and M.B. All authors have read and agreed to the published version of the manuscript.

Funding: This study was supported by the Clinical Research Priority Program (CRPP) for NeuroRehab of the University of Zurich. Movement analysis was supported by the Swiss Center for Clinical Movement Analysis, SCMA, Balgrist Campus AG, Zürich.

Acknowledgments: The authors would like to thank all volunteers for their study participants. We thank Adrian Cathomen and Romina Willi for their support of the assessments.

Conflicts of Interest: The authors declare no conflict of interest.

References

1. Grillner, S. Control of Locomotion in Bipeds, Tetrapods, and Fish. In *Comprehensive Physiology*; Wiley: Hoboken, NJ, USA, 2011; ISBN 9780470650714.
2. Drew, T.; Marigold, D.S. Taking the next step: Cortical contributions to the control of locomotion. *Curr. Opin. Neurobiol.* **2015**, *33*, 25–33. [CrossRef] [PubMed]
3. Schubert, M.; Curt, A.; Jensen, L.; Dietz, V. Corticospinal input in human gait: Modulation of magnetically evoked motor responses. *Exp. Brain Res.* **1997**, *115*, 234–246. [CrossRef] [PubMed]
4. Capaday, C.; Lavoie, B.A.; Barbeau, H.; Schneider, C.; Bonnard, M. Studies on the corticospinal control of human walking. I. Responses to focal transcranial magnetic stimulation of the motor cortex. *J. Neurophysiol.* **1999**, *81*, 129–139. [CrossRef]
5. Jayaraman, A.; Gregory, C.M.; Bowden, M.; Stevens, J.E.; Shah, P.; Behrman, A.L.; Vandenborne, K. Lower extremity skeletal muscle function in persons with incomplete spinal cord injury. *Spinal Cord* **2006**, *44*, 680–687. [CrossRef]
6. Hansen, N.L.; Conway, B.A.; Halliday, D.M.; Hansen, S.; Pyndt, H.S.; Biering-Sørensen, F.; Nielsen, J.B. Reduction of common synaptic drive to ankle dorsiflexor motoneurons during walking in patients with spinal cord lesion. *J. Neurophysiol.* **2005**, *94*, 934–942. [CrossRef] [PubMed]
7. Adams, M.M.; Hicks, A.L. Spasticity after spinal cord injury. *Spinal Cord* **2005**, *43*, 577–586. [CrossRef]
8. Barbeau, H.; Pépin, A.; Norman, K.E.; Ladouceur, M.; Leroux, A. Walking after spinal cord injury: Control and recovery. *Neuroscientist* **1998**, *4*, 14–24. [CrossRef]
9. Wirth, B.; Van Hedel, H.J.A.; Curt, A. Ankle paresis in incomplete spinal cord injury: Relation to corticospinal conductivity and ambulatory capacity. *J. Clin. Neurophysiol.* **2008**, *25*, 210–217. [CrossRef]
10. Anderson, K.D. Targeting recovery: Priorities of the spinal cord-injured population. *J. Neurotrauma* **2004**, *21*, 1371–1383. [CrossRef]
11. Ditunno, P.L.; Patrick, M.; Stineman, M.; Ditunno, J.F. Who wants to walk? Preferences for recovery after SCI: A longitudinal and cross-sectional study. *Spinal Cord* **2008**, *46*, 500–506. [CrossRef]
12. Angeli, C.A.; Boakye, M.; Morton, R.A.; Vogt, J.; Benton, K.; Chen, Y.; Ferreira, C.K.; Harkema, S.J. Recovery of over-ground walking after chronic motor complete spinal cord injury. *N. Engl. J. Med.* **2018**, *379*, 1244–1250. [CrossRef]
13. Gill, M.L.; Grahn, P.J.; Calvert, J.S.; Linde, M.B.; Lavrov, I.A.; Strommen, J.A.; Beck, L.A.; Sayenko, D.G.; Van Straaten, M.G.; Drubach, D.I.; et al. Neuromodulation of lumbosacral spinal networks enables independent stepping after complete paraplegia. *Nat. Med.* **2018**, *24*, 1677–1682. [CrossRef]
14. Wagner, F.B.; Mignardot, J.B.; Le Goff-Mignardot, C.G.; Demesmaeker, R.; Komi, S.; Capogrosso, M.; Rowald, A.; Seáñez, I.; Caban, M.; Pirondini, E.; et al. Targeted neurotechnology restores walking in humans with spinal cord injury. *Nature* **2018**, *563*, 65–93. [CrossRef]
15. Harkema, S.; Gerasimenko, Y.; Hodes, J.; Burdick, J.; Angeli, C.; Chen, Y.; Ferreira, C.; Willhite, A.; Rejc, E.; Grossman, R.G.; et al. Effect of epidural stimulation of the lumbosacral spinal cord on voluntary movement, standing, and assisted stepping after motor complete paraplegia: A case study. *Lancet* **2011**, *377*, 1938–1947. [CrossRef]
16. Angeli, C.A.; Edgerton, V.R.; Gerasimenko, Y.P.; Harkema, S.J. Altering spinal cord excitability enables voluntary movements after chronic complete paralysis in humans. *Brain* **2014**, *137*, 1394–1409. [CrossRef]
17. Minassian, K.; Persy, I.; Rattay, F.; Pinter, M.M.; Kern, H.; Dimitrijevic, M.R. Human lumbar cord circuitries can be activated by extrinsic tonic input to generate locomotor-like activity. *Hum. Mov. Sci.* **2007**, *26*, 275–295. [CrossRef]

18. Murg, M.; Binder, H.; Dimitrijevic, M.R. Epidural electric stimulation of posterior structures of the human lumbar spinal cord: 1. Muscle twitches—A functional method to define the site of stimulation. *Spinal Cord* **2000**, *38*, 394–402. [CrossRef]

19. Rattay, F.; Minassian, K.; Dimitrijevic, M.R. Epidural electrical stimulation of posterior structures of the human lumbosacral cord: 2. Quantitative analysis by computer modeling. *Spinal Cord* **2000**, *38*, 473–489. [CrossRef] [PubMed]

20. Ladenbauer, J.; Minassian, K.; Hofstoetter, U.S.; Dimitrijevic, M.R.; Rattay, F. Stimulation of the human lumbar spinal cord with implanted and surface electrodes: A computer simulation study. *IEEE Trans. Neural Syst. Rehabil. Eng.* **2010**, *18*, 637–645. [CrossRef]

21. Minassian, K.; Jilge, B.; Rattay, F.; Pinter, M.M.; Binder, H.; Gerstenbrand, F.; Dimitrijevic, M.R. Stepping-like movements in humans with complete spinal cord injury induced by epidural stimulation of the lumbar cord: Electromyographic study of compound muscle action potentials. *Spinal Cord* **2004**, *42*, 401–416. [CrossRef] [PubMed]

22. Jilge, B.; Minassian, K.; Rattay, F.; Pinter, M.M.; Gerstenbrand, F.; Binder, H.; Dimitrijevic, M.R. Initiating extension of the lower limbs in subjects with complete spinal cord injury by epidural lumbar cord stimulation. *Exp. Brain Res.* **2004**, *154*, 308–326. [CrossRef]

23. Dimitrijevic, M.R.; Gerasimenko, Y.; Pinter, M.M. Evidence for a spinal central pattern generator in humans. *Ann. N. Y. Acad. Sci.* **1998**, *860*, 360–376. [CrossRef]

24. Danner, S.M.; Hofstoetter, U.S.; Freundl, B.; Binder, H.; Mayr, W.; Rattay, F.; Minassian, K. Human spinal locomotor control is based on flexibly organized burst generators. *Brain* **2015**, *138*, 577–588. [CrossRef] [PubMed]

25. Minassian, K.; Persy, I.; Rattay, F.; Dimitrijevic, M. Effect of peripheral afferent and central afferent input to the human lumbar spinal cord isolated from brain control. *Biocybern. Biomed. Eng.* **2005**, *25*, 11–29.

26. Pinter, M.M.; Gerstenbrand, F.; Dimitrijevic, M.R. Epidural electrical stimulation of posterior structures of the human lumbosacral cord: 3. Control of spasticity. *Spinal Cord* **2000**, *38*, 524–531. [CrossRef] [PubMed]

27. Minassian, K.; Persy, I.; Rattay, F.; Dimitrijevic, M.R.; Hofer, C.; Kern, H. Posterior root-muscle preflexes elicited by transcutaneous stimulation of the human lumbosacral cord. *Muscle Nerve* **2007**, *35*, 327–336. [CrossRef]

28. Hofstoetter, U.S.; Freundl, B.; Binder, H.; Minassian, K. Common neural structures activated by epidural and transcutaneous lumbar spinal cord stimulation: Elicitation of posterior root-muscle reflexes. *PLoS ONE* **2018**, *13*, e0192013. [CrossRef]

29. Danner, S.M.; Hofstoetter, U.S.; Ladenbauer, J.; Rattay, F.; Minassian, K. Can the Human Lumbar Posterior Columns Be Stimulated by Transcutaneous Spinal Cord Stimulation? A Modeling Study. *Artif. Organs* **2011**, *35*, 257–262. [CrossRef]

30. Sayenko, D.G.; Rath, M.; Ferguson, A.R.; Burdick, J.W.; Havton, L.A.; Edgerton, V.R.; Gerasimenko, Y.P. Self-assisted standing enabled by non-invasive spinal stimulation after spinal cord injury. *J. Neurotrauma* **2019**, *36*, 1435–1450. [CrossRef]

31. Hofstoetter, U.S.; Freundl, B.; Binder, H.; Minassian, K. Spinal cord stimulation as a neuromodulatory intervention for altered motor control following spinal cord injury. In *Biosystems and Biorobotics*; Sandrini, G., Homberg, V., Saltuari, L., Smania, N., Pedrocchi, A., Eds.; Springer Verlag GmbH: Berlin/Heidelberg, Germany, 2018; Volume 19, pp. 501–521.

32. Hofstoetter, U.S.; McKay, W.B.; Tansey, K.E.; Mayr, W.; Kern, H.; Minassian, K. Modification of spasticity by transcutaneous spinal cord stimulation in individuals with incomplete spinal cord injury. *J. Spinal Cord Med.* **2014**, *37*, 202–211. [CrossRef]

33. Estes, S.P.; Iddings, J.A.; Field-Fote, E.C. Priming neural circuits to modulate spinal reflex excitability. *Front. Neurol.* **2017**, *8*, 17. [CrossRef]

34. Hofstoetter, U.S.; Freundl, B.; Danner, S.M.; Krenn, M.J.; Mayr, W.; Binder, H.; Minassian, K. Transcutaneous Spinal Cord Stimulation Induces Temporary Attenuation of Spasticity in Individuals with Spinal Cord Injury. *J. Neurotrauma* **2020**, *37*, 481–493. [CrossRef]

35. Gad, P.; Gerasimenko, Y.; Zdunowski, S.; Turner, A.; Sayenko, D.; Lu, D.C.; Edgerton, V.R. Weight bearing over-ground stepping in an exoskeleton with non-invasive spinal cord neuromodulation after motor complete paraplegia. *Front. Neurosci.* **2017**, *11*, 333. [CrossRef]

36. Hofstoetter, U.S.; Krenn, M.; Danner, S.M.; Hofer, C.; Kern, H.; Mckay, W.B.; Mayr, W.; Minassian, K. Augmentation of Voluntary Locomotor Activity by Transcutaneous Spinal Cord Stimulation in Motor-Incomplete Spinal Cord-Injured Individuals. *Artif. Organs* **2015**, *39*, E176–E186. [CrossRef]

37. Gerasimenko, Y.P.; Lu, D.C.; Modaber, M.; Zdunowski, S.; Gad, P.; Sayenko, D.G.; Morikawa, E.; Haakana, P.; Ferguson, A.R.; Roy, R.R.; et al. Noninvasive reactivation of motor descending control after paralysis. *J. Neurotrauma* **2015**, *32*, 1968–1980. [CrossRef]

38. Van den Brand, R.; Mignardot, J.B.; von Zitzewitz, J.; Le Goff, C.; Fumeaux, N.; Wagner, F.; Capogrosso, M.; Martin Moraud, E.; Micera, S.; Schurch, B.; et al. Neuroprosthetic technologies to augment the impact of neurorehabilitation after spinal cord injury. *Ann. Phys. Rehabil. Med.* **2015**, *58*, 232–237. [CrossRef] [PubMed]

39. Minassian, K.; Hofstoetter, U.S. Spinal Cord Stimulation and Augmentative Control Strategies for Leg Movement after Spinal Paralysis in Humans. *CNS Neurosci. Ther.* **2016**, *22*, 262–270. [CrossRef]

40. Dimitrijevic, M.R.; Illis, L.S.; Nakajima, K.; Sharkey, P.C.; Sherwood, A.M. Spinal Cord Stimulation for the Control of Spasticity in Patients with Chronic Spinal Cord Injury: II. Neurophysiologic Observations. *Cent. Nerv. Syst. Trauma* **1986**, *3*, 145–152. [CrossRef]

41. Dietz, V.; Grillner, S.; Trepp, A.; Hubli, M.; Bolliger, M. Changes in spinal reflex and locomotor activity after a complete spinal cord injury: A common mechanism. *Brain* **2009**, *132*, 2196–2205. [CrossRef]

42. Hubli, M.; Dietz, V.; Bolliger, M. Spinal reflex activity: A marker for neuronal functionality after spinal cord injury. *Neurorehabil. Neural Repair* **2012**, *26*, 188–196. [CrossRef]

43. Hubli, M.; Dietz, V.; Bolliger, M. Influence of spinal reflexes on the locomotor pattern after spinal cord injury. *Gait Posture* **2011**, *34*, 409–414. [CrossRef] [PubMed]

44. Dietz, V. Behavior of spinal neurons deprived of supraspinal input. *Nat. Rev. Neurol.* **2010**, *6*, 167–174. [CrossRef]

45. Hubli, M.; Bolliger, M.; Dietz, V. Neuronal dysfunction in chronic spinal cord injury. *Spinal Cord* **2011**, *49*, 582–587. [CrossRef]

46. Kirshblum, S.; Waring, W. Updates for the international standards for neurological classification of Spinal Cord Injury. *Phys. Med. Rehabil. Clin. N. Am.* **2014**, *25*, 505–517. [CrossRef]

47. Morganti, B.; Scivoletto, G.; Ditunno, P.; Ditunno, J.F.; Molinari, M. Walking index for spinal cord injury (WISCI): Criterion validation. *Spinal Cord* **2005**, *43*, 27–33. [CrossRef]

48. Vallery, H.; Lutz, P.; Von Zitzewitz, J.; Rauter, G.; Fritschi, M.; Everarts, C.; Ronsse, R.; Curt, A.; Bolliger, M. Multidirectional transparent support for overground gait training. In Proceedings of the IEEE International Conference on Rehabilitation Robotics, Seattle, WA, USA, 24–26 June 2013.

49. Wirth, B.; Van Hedel, H.; Curt, A. Foot control in incomplete SCI: Distinction between paresis and dexterity. *Neurol. Res.* **2008**, *30*, 52–60. [CrossRef]

50. Marcus, J. Effects of stimulus intensity on the habituation of flexor withdrawal activity mediated by the functionally transected human spinal cord. *Physiol. Psychol.* **1977**, *5*, 321–326.

51. Willer, J.C. Anticipation of pain-produced stress: Electrophysiological study in man. *Physiol. Behav.* **1980**, *25*, 49–51. [CrossRef]

52. Willer, J.C.; Albe-Fessard, D. Electrophysiological evidence for a release of endogenous opiates in stress-induced 'Analgesia' in man. *Brain Res.* **1980**, *198*, 419–426. [CrossRef]

53. Bannwart, M.; Bolliger, M.; Lutz, P.; Gantner, M.; Rauter, G. Systematic analysis of transparency in the gait rehabilitation device the FLOAT. In Proceedings of the 2016 14th International Conference on Control, Automation, Robotics and Vision, ICARCV 2016, Phuket, Thailand, 13–15 November 2016.

54. Easthope, C.S.; Traini, L.R.; Awai, L.; Franz, M.; Rauter, G.; Curt, A.; Bolliger, M. Overground walking patterns after chronic incomplete spinal cord injury show distinct response patterns to unloading 11 Medical and Health Sciences 1103 Clinical Sciences. *J. Neuroeng. Rehabil.* **2018**, *15*, 102. [CrossRef]

55. Hermens, H.J.; Freriks, B.; Disselhorst-Klug, C.; Rau, G. Development of recommendations for SEMG sensors and sensor placement procedures. *J. Electromyogr. Kinesiol.* **2000**, *10*, 361–374. [CrossRef]

56. Vicon. Plug-in-Gait modelling instructions. In *Plug-in-Gait Manual*; Oxford Metrics plc: Oxford, UK, 2002; p. 612.

57. Meyer, C.; Killeen, T.; Easthope, C.S.; Curt, A.; Bolliger, M.; Linnebank, M.; Zörner, B.; Filli, L. Familiarization with treadmill walking: How much is enough? *Sci. Rep.* **2019**, *9*, 5232. [CrossRef]

58. Schmidt, K.; Duarte, J.E.; Grimmer, M.; Sancho-Puchades, A.; Wei, H.; Easthope, C.S.; Riener, R. The myosuit: Bi-articular anti-gravity exosuit that reduces hip extensor activity in sitting transfers. *Front. Neurorobot.* **2017**, *11*, 57. [CrossRef]

59. Minassian, K.; Hofstoetter, U.; Rattay, F. Transcutaneous Lumbar Posterior Root Stimulation for Motor Control Studies and Modification of Motor Activity after Spinal Cord Injury. In *Restorative Neurology of Spinal Cord Injury*; Dimitrijevic, M., Kakulas, B., McKay, W., Vrbova, G., Eds.; Oxford University Press: New York, NY, USA, 2012; pp. 226–255, ISBN 9780199918768.

60. Minassian, K.; McKay, W.B.; Binder, H.; Hofstoetter, U.S. Targeting Lumbar Spinal Neural Circuitry by Epidural Stimulation to Restore Motor Function After Spinal Cord Injury. *Neurotherapeutics* **2016**, *13*, 284–294. [CrossRef]

61. Andrews, J.C.; Stein, R.B.; Roy, F.D. Reduced postactivation depression of soleus h reflex and root evoked potential after transcranial magnetic stimulation. *J. Neurophysiol.* **2015**, *114*, 485–492. [CrossRef]

62. Hultborn, H.; Illert, M.; Nielsen, J.; Paul, A.; Ballegaard, M.; Wiese, H. On the mechanism of the post-activation depression of the H-reflex in human subjects. *Exp. Brain Res.* **1996**, *108*, 450–462. [CrossRef]

63. Pierrot-Deseilligny, E.; Burke, D. *The Circuitry of the Human Spinal Cord: SPINAL and Corticospinal Mechanisms of Movement*; Cambridge University Press: Cambridge, UK, 2012; ISBN 9781139026727.

64. Hofstoetter, U.S.; Freundl, B.; Binder, H.; Minassian, K. Recovery cycles of posterior root-muscle reflexes evoked by transcutaneous spinal cord stimulation and of the H reflex in individuals with intact and injured spinal cord. *PLoS ONE* **2019**, *14*, e0227057. [CrossRef]

65. Barolat, G.; Myklebust, J.B.; Wenninger, W. Effects of spinal cord stimulation on spasticity and spasms secondary to myelopathy. *Appl. Neurophysiol.* **1988**, *51*, 29–44. [CrossRef]

66. Zörner, B.; Filli, L.; Reuter, K.; Kapitza, S.; Lörincz, L.; Sutter, T.; Weller, D.; Farkas, M.; Easthope, C.S.; Czaplinski, A.; et al. Prolonged-release fampridine in multiple sclerosis: Improved ambulation effected by changes in walking pattern. *Mult. Scler.* **2016**, *22*, 1463–1475. [CrossRef]

67. Meyer, C.; Filli, L.; Stalder, S.A.; Awai Easthope, C.; Killeen, T.; von Tscharner, V.; Curt, A.; Zörner, B.; Bolliger, M. Targeted Walking in Incomplete Spinal Cord Injury: Role of Corticospinal Control. *J. Neurotrauma* **2020**. [CrossRef] [PubMed]

68. Filli, L.; Sutter, T.; Easthope, C.S.; Killeen, T.; Meyer, C.; Reuter, K.; Lörincz, L.; Bolliger, M.; Weller, M.; Curt, A.; et al. Profiling walking dysfunction in multiple sclerosis: Characterisation, classification and progression over time. *Sci. Rep.* **2018**, *8*, 4984. [CrossRef]

69. Capogrosso, M.; Wenger, N.; Raspopovic, S.; Musienko, P.; Beauparlant, J.; Luciani, L.B.; Courtine, G.; Micera, S. A computational model for epidural electrical stimulation of spinal sensorimotor circuits. *J. Neurosci.* **2013**, *33*, 19326–19340. [CrossRef]

70. Formento, E.; Minassian, K.; Wagner, F.; Mignardot, J.B.; Le Goff-Mignardot, C.G.; Rowald, A.; Bloch, J.; Micera, S.; Capogrosso, M.; Courtine, G. Electrical spinal cord stimulation must preserve proprioception to enable locomotion in humans with spinal cord injury. *Nat. Neurosci.* **2018**, *21*, 1728–1741. [CrossRef] [PubMed]

71. Varoqui, D.; Niu, X.; Mirbagheri, M.M. Ankle voluntary movement enhancement following robotic-assisted locomotor training in spinal cord injury. *J. Neuroeng. Rehabil.* **2014**, *11*, 46. [CrossRef] [PubMed]

72. Wirth, B.; Van Hedel, H.J.A.; Curt, A. Ankle dexterity is less impaired than muscle strength in incomplete spinal cord lesion. *J. Neurol.* **2008**, *255*, 273–279. [CrossRef]

73. Van Hedel, H.J.A.; Wirth, B.; Curt, A. Ankle motor skill is intact in spinal cord injury, unlike stroke: Implications for rehabilitation. *Neurology* **2010**, *74*, 1271–1278. [CrossRef]

74. Bolliger, M.; Trepp, A.; Zörner, B.; Dietz, V. Modulation of spinal reflex by assisted locomotion in humans with chronic complete spinal cord injury. *Clin. Neurophysiol.* **2010**, *121*, 2152–2158. [CrossRef]

75. Smith, A.C.; Mummidisetty, C.K.; Rymer, W.Z.; Knikou, M. Locomotor training alters the behavior of flexor reflexes during walking in human spinal cord injury. *J. Neurophysiol.* **2014**, *112*, 2164–2175. [CrossRef]

76. Gómez-Soriano, J.; Bravo-Esteban, E.; Pérez-Rizo, E.; Ávila-Martín, G.; Galán-Arriero, I.; Simón-Martinez, C.; Taylor, J. Abnormal cutaneous flexor reflex activity during controlled isometric plantarflexion in human spinal cord injury spasticity syndrome. *Spinal Cord* **2016**, *54*, 687–694. [CrossRef] [PubMed]

77. Campos, R.J.; Dimitrijevic, M.R.; Sharkey, P.C.; Sherwood, A.M. Epidural spinal cord stimulation in spastic spinal cord injury patients. *Stereotact. Funct. Neurosurg.* **1987**, *50*, 453–454. [CrossRef]

78. Musselman, K.E. Clinical significance testing in rehabilitation research: What, why, and how? *Phys. Ther. Rev.* **2007**, *12*, 287–296. [CrossRef]
79. Field-Fote, E.C.; Tepavac, D. Improved intralimb coordination in people with incomplete spinal cord injury following training with body weight support and electrical stimulation. *Phys. Ther.* **2002**, *82*, 707–715. [CrossRef] [PubMed]
80. Minassian, K.; Hofstoetter, U.S.; Danner, S.M.; Mayr, W.; Bruce, J.A.; McKay, W.B.; Tansey, K.E. Spinal rhythm generation by step-induced feedback and transcutaneous posterior root stimulation in complete spinal cord-injured individuals. *Neurorehabil. Neural Repair* **2016**, *30*, 233–243. [CrossRef]
81. Danner, S.M.; Krenn, M.; Hofstoetter, U.S.; Toth, A.; Mayr, W.; Minassian, K. Body position influences which neural structures are recruited by lumbar transcutaneous spinal cord stimulation. *PLoS ONE* **2016**, *11*, e0147479. [CrossRef]
82. Hofstoetter, U.S.; Danner, S.M.; Minassian, K. Paraspinal Magnetic and Transcutaneous Electrical Stimulation. In *Encyclopedia of Computational Neuroscience*; Jaeger, D., Jung, R., Eds.; Springer New York: New York, NY, USA, 2014; pp. 1–21, ISBN 978-1-4614-6676-5.
83. Estes, S.; Iddings, J.A.; Ray, S.; Kirk-Sanchez, N.J.; Field-Fote, E.C. Comparison of Single-Session Dose Response Effects of Whole Body Vibration on Spasticity and Walking Speed in Persons with Spinal Cord Injury. *Neurotherapeutics* **2018**, *15*, 684–696. [CrossRef]

Publisher's Note: MDPI stays neutral with regard to jurisdictional claims in published maps and institutional affiliations.

Journal of
Clinical Medicine

Article

The Effects of Adding Transcutaneous Spinal Cord Stimulation (tSCS) to Sit-To-Stand Training in People with Spinal Cord Injury: A Pilot Study

Yazi Al'joboori [1,2,*], Sarah J. Massey [2], Sarah L. Knight [3], Nick de N. Donaldson [1] and Lynsey D. Duffell [1,2]

[1] Department of Medical Physics & Biomedical Engineering, UCL, London WC1E 6BT, UK; n.donaldson@ucl.ac.uk (N.d.N.D.); l.duffell@ucl.ac.uk (L.D.D.)
[2] Aspire CREATe, UCL, Stanmore HA7 4LP, UK; sarah.massey.13@ucl.ac.uk
[3] London Spinal Cord Injury Centre, Royal National Orthopaedic Hospital, Stanmore HA7 4LP, UK; sarah.knight23@nhs.net
* Correspondence: y.aljoboori@ucl.ac.uk; Tel.: +44-020-3108-4083

Received: 10 July 2020; Accepted: 21 August 2020; Published: 26 August 2020

Abstract: Spinal cord stimulation may enable recovery of volitional motor control in people with chronic Spinal Cord Injury (SCI). In this study we explored the effects of adding SCS, applied transcutaneously (tSCS) at vertebral levels T10/11, to a sit-to-stand training intervention in people with motor complete and incomplete SCI. Nine people with chronic SCI (six motor complete; three motor incomplete) participated in an 8-week intervention, incorporating three training sessions per week. Participants received either tSCS combined with sit-to-stand training (STIM) or sit-to-stand training alone (NON-STIM). Outcome measures were carried out before and after the intervention. Seven participants completed the intervention (STIM N = 5; NON-STIM N = 2). Post training, improvements in International Standards for Neurological Classification of Spinal Cord Injury (ISNCSCI) motor scores were noted in three STIM participants (range 1.0–7.0), with no change in NON-STIM participants. Recovery of volitional lower limb muscle activity and/or movement (with tSCS off) was noted in three STIM participants. Unassisted standing was not achieved in any participant, although standing with minimal assistance was achieved in one STIM participant. This pilot study has shown that the recruitment of participants, intervention and outcome measures were all feasible in this study design. However, some modifications are recommended for a larger trial.

Keywords: human; neuromodulation; neurorehabilitation; non-invasive; spinal cord injury; transcutaneous spinal cord stimulation

---

1. Introduction

Spinal cord injury (SCI) is a life-long condition, which can substantially impact the health and well-being of affected individuals. Appropriate management is essential to maximise health-related quality of life. Standing remains one of the only clinical interventions used both acutely and following discharge from hospital, in people with chronic SCI, because it is affordable and relatively simple to do at home. Regular standing has many known health benefits, including reduced muscle tone, improved blood flow in the lower limbs, beneficial effects on bladder and bowel function and improvements in quality of life [1]. Active stand training, with additional electrical stimulation, may facilitate independent or minimally-assisted standing in people with chronic SCI [2].

Non-patterned spinal cord stimulation (SCS), delivered by electrodes implanted in the epidural space of the spinal cord has been shown to elicit lower limb extensor movements in rats [3,4] and humans [5–7] with motor complete SCI, via activation of large-to-medium diameter sensory fibres within the posterior roots (below the level of the injury). Transcutaneous SCS (tSCS), which uses non-invasive electrodes placed over the T12—L1 vertebrae and abdomen, has been shown, by neurophysiological [8] and computer simulation [9] studies, to recruit similar neural structures as epidural SCS. Single pulses of tSCS elicited compound muscle action potentials in lower limb muscles, due to transynaptic activation of α-motoneurons, termed Posterior Root Reflexes (PRRs), which are susceptible to homosynaptic depression when paired pulses are applied [10–12]. Similar to epidural SCS, supra-threshold tSCS has been shown to augment or generate lower limb extension in people with motor complete and incomplete SCI [13,14]. This delivery method removes the requirement for surgery and is simple to apply, so it may be readily transferable for clinical use; however, it can cause some discomfort and unwanted muscle contractions as the current passes through skin and trunk musculature. Transcutaneous SCS also has less specificity than epidural SCS and may be more affected by changes in body position [15].

When delivered at an intensity below that required to produce motor response (i.e., sub-threshold), epidural SCS has been shown to enable individuals with motor complete SCI to produce volitional movement in their otherwise paralysed muscles [16–18], and to augment volitional standing and stepping [17,19–21], offering an important therapeutic pathway for people living with chronic SCI [22]. Transcutaneous SCS, delivered at an intensity that is sub-threshold for generating lower limb activity, either muscle contractions or whole-limb responses, but high enough to produce paraesthesia in lower limb dermatomes, has also been shown to augment volitional stepping on a treadmill [23,24], and to suppress lower limb spasticity [25–27] in people with incomplete SCI.

After several months of rehabilitative training combined with epidural SCS, progressive improvements in volitional movements, standing and stepping (in the presence of SCS) have been reported in pre-clinical [28–30] and clinical [16,17,19,31,32] studies. Furthermore, it has recently been reported that this recovery of volitional motor control after SCS training was possible even when the SCS was switched off, indicative of neuroplastic recovery. This was achieved in people with incomplete SCI after several months of epidural SCS combined with intensive locomotor training [33], and in people with motor complete injuries after one month of daily epidural SCS, used to achieve volitional movement and specific autonomic functions (such as blood pressure regulation) [34].

The only studies to report progressive improvements in the generation of lower limb activity in the presence of non-invasive tSCS [14,35] used a specific stimulation waveform for tSCS (each pulse filled by a 10 kHz carrier frequency [36]); this is thought to reduce the discomfort associated with tSCS. With this waveform, supra-threshold tSCS has been used to achieve minimally-assisted standing in people with motor complete and incomplete injuries, and progressive improvements in standing were reported after several weeks of training [14]. In the current study, we explored the effects of sub-threshold tSCS combined with sit-to-stand training on recovery of motor control in people with complete and incomplete SCI, using a conventional stimulation waveform. This was compared to sit-to-stand training alone, in order to specifically explore the importance of additional tSCS to achieve motor recovery. We also measured the effects on health-related quality of life and functional independence.

The main aim of this pilot study was to explore the feasibility of comparing a sit-to-stand training intervention, with and without additional sub-threshold tSCS, in people with motor complete and incomplete SCI. We found that people with both complete and incomplete injuries were willing to take part in the trial, however participants were more willing to participate in the tSCS group than sit-to-stand training alone. In this small group of participants, we found increases in volitional activation of specific lower limb muscles after training, only in the participants that received tSCS.

2. Experimental Section

This trial had ethical approval from the London–Stanmore Research Ethics Committee (REC reference: 18/LO/0784) and all participants provided written informed consent to take part. The trial was registered with ClinicalTrials.gov (NCT03536338).

2.1. Participants

Participants were recruited from the London Spinal Cord Injury Centre (Royal National Orthopaedic Hospital) and Neurokinex (specialised neurological activity-based rehabilitation facilities). Inclusion criteria were (1) spinal cord injury for >1 year, (2) SCI level C5-T12, (3) aged >18 years, (4) AIS A-D, (5) unable to stand from a chair unaided. Exclusion criteria were (1) cardiac pacemaker (2) any other musculoskeletal diagnosis affecting the lower limbs, (3) pregnancy, (4) complex regional pain syndrome, (5) implanted metal or active device at electrodes caudal to T9 (e.g., screws, contraceptive coil), (6) spinal malignancy, (7) uncontrolled autonomic dysreflexia, (8) neurological degenerative diseases, (9) peripheral nerve damage affecting the lower limbs, (10) currently on any form of anti-spasticity treatment (e.g., Botox, but not including bladder Botox), (11) osteoporotic-bone density T-score less than −2.5. The demographics and injury characteristics of the 9 participants assessed at baseline are provided in Table 1.

Table 1. Participant demographic information.

	Age (Years)	Height (Meters)	Weight (Kg)	Sex (M/F)	Injury Level	Cause of Injury		AIS Grade	Time Since Injury	Group
P1	37	1.83	100	M	T3	Traumatic	Motor vehicle	A	1 yr 6 m	STIM
P2	38	1.92	83	M	T5	Traumatic	Motor vehicle	A	2 yr 2 m	STIM
P3	38	1.72	59	F	T6	Traumatic	Motor vehicle	A	8 yr 4 m	NON-STIM
P4	41	1.71	58	F	T5	Non-Traumatic	Spinal tumor	C	1 yr 5 m	STIM
P5	28	1.98	80	M	C6/7	Traumatic	Sports injury	D	1 yr 1 m	STIM
P6	40	1.59	61	F	T10	Traumatic	Fall	A *	9 yr 0 m	STIM
P7 [W]	49	1.98	121	M	T10	Non-Traumatic	Vascular	B	2 yr 11 m	NON-STIM
P8	31	1.81	80	M	C5	Traumatic	Fall	C	7 yr 11 m	NON-STIM
P9 [W]	69	1.75	70	M	T6	Non-Traumatic	Vascular	A	1 yr 4 m	NON-STIM

* Partial preservation; [W] withdrawn; yr, Years; m, Months.

2.2. Study Design

This was a purposefully sampled cohort study, designed to balance Training + tSCS (STIM) and Training Only (NON-STIM) groups by AIS grade. Participants were initially assessed against the inclusion/exclusion criteria, and then assigned to either the STIM or NON-STIM group: we aimed to recruit at least two participants in each group with motor complete injuries (AIS A/B), and two participants in each group with incomplete injuries (AIS C/D). Thirty-five participants were assessed for eligibility; of these, 26 did not participate in the trial. Of these, applicants unable to meet criteria ($n = 7$) was primarily due to existing implanted metal work being too low and applicants being less than 1-year post-injury. Others expressed that the training center locations were too far to travel 3 times a week for 8 weeks ($n = 7$), that they were unwilling to participate in the study after being assigned to the control arm ($n = 4$), or did not disclose the reason ($n = 7$). One applicant was willing to participate, but unable to start the study due to early termination (due to the COVID-19 pandemic). Therefore, nine participants were assessed at baseline; of these, 7 completed the intervention and 2 were withdrawn due to a lower limb injury ($n = 1$) or early termination ($n = 1$). See CONSORT flow diagram (Figure 1).

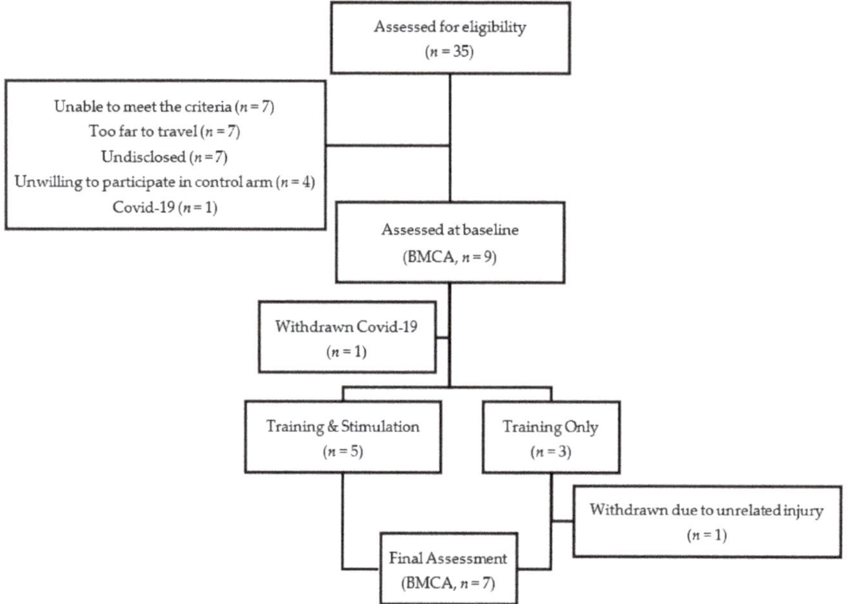

Figure 1. CONSORT flow diagram.

2.3. Intervention

All training was carried out at Neurokinex and consisted of 24 sessions (3 sessions per week for 8 weeks). During each session, participants transferred to a physiotherapy couch and were fitted with a body weight support (BWS) harness. The harness was attached to a Keiser Power Rack (Keiser UK Ltd., Gloucestershire, UK), which incorporates a pneumatic resistance system to assist the participant in the sit-to-stand manoeuvre by partially supporting their bodyweight. During each training session, participants stood up 5 times, taking approximately one-hour. Each sit-to-stand manoeuvre was initiated by increasing the BWS to assist the participant into stand; two therapists were available to stabilise the participant if required. Standing was then maintained for 4–5 min, during which postural exercises such as deep and shallow squats, lateral and anterior/posterior weight shifts, squat holds, single leg bends, hip thrusts, kettle bell arm presses, trunk strengthening (arms off bar and straighten posture), hip rotations and squat rotations were performed. The participants were then returned to a seated position and rested for 2–3 min before the cycle was repeated. In the STIM group only, continuous tSCS was applied during active standing (sit-to-stand, standing and sit-from-stand), no tSCS was provided during seated rest.

2.4. Stimulation Parameters

The self-adhesive electrodes for tSCS were (5 × 5 cm Axelgard, Fallbrook, CA, United States), placed on the midline, the cathode at T10/11 and the anode at T12/L1. Electrode placement was confirmed using single pulses (monophasic, 1 ms pulse width) applied using a Digitimer DS8R Constant Current Stimulator (Digitimer, Welwyn Garden City, Hertfordshire, UK), driven by Signal software (Cambridge Electronic Design, Cambridge, UK), to test for PRR threshold in lower limb muscles. Stimulation current (mA) was increased until PRRs were observed in all measured lower limb muscles, which was defined as motor threshold. Paired pulses at an interstimulus interval (ISI) of 30 ms were then used to test the presence of post-activation depression in order to confirm activation of afferent roots. Previous studies have reported post-activation depression with tSCS paired pulses applied at ISIs between 25 and 50 ms in people with SCI [25] and in able-bodied subjects [37].

PRRs due to single and paired tSCS pulses are shown in Figure 2a. Optimal electrode location was recorded for each participant and replicated during training for participants in the STIM group. During training, tSCS was fixed at 30 Hz (biphasic, 1 ms pulse width) and applied below motor threshold (did not elicit any visible muscle contractions), at a level that induced paraesthesia in lower limb dermatomes, or the maximum level tolerated by the participant (whichever was lower), using the Chattanooga Intelect Mobile stimulator (Chattanooga Group International, Chattanooga, TN, USA). Stimulation intensity was initially established at the start of each session, and modified throughout each session as required, due to habituation to the stimulation. Transcutaneous SCS was applied tonically during each sit-to-stand manoeuvre. Stimulation was present during transitions from sit-to-stand and sit-from-stand and maintained throughout standing. No stimulation was applied during rest periods between stands.

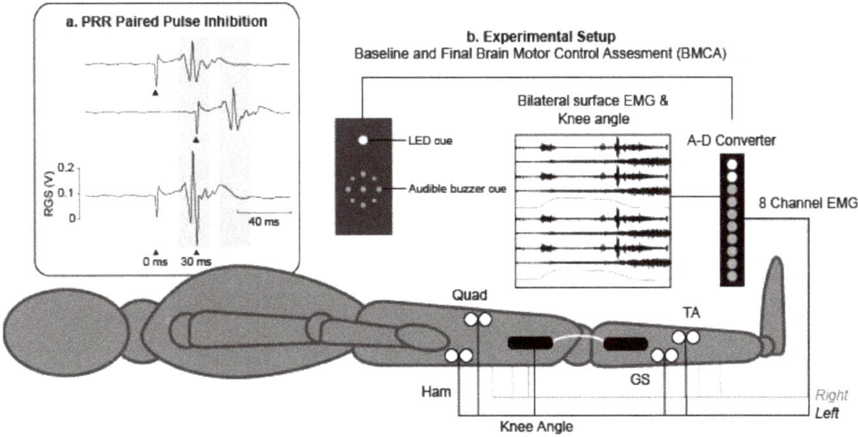

Figure 2. (**a**) Example traces during posterior root reflex (PRR) testing; single pulses were delivered at 0 and 30 ms (upper traces) and as a pair with an interstimulus interval (ISI) of 30 ms (lower trace) to demonstrate paired pulse inhibition (arrows denote the time at which the stimulus was applied). (**b**) Experimental setup for baseline and final Brain Motor Control Assessments. Participants were placed in a supine position with bilateral electromyography (EMG) electrodes placed over the Quadriceps (Quad), Hamstring (Ham), Tibialis Anterior (TA) and Gastrocnemius (GS) muscles to record EMG and electro-goniometers were placed laterally across the knee joints to synchronously record knee joint range of motion.

2.5. Outcome Measures

2.5.1. Baseline and Follow-Up Measures

Baseline and follow-up outcome measures were completed before and after the 8-week training/stimulation intervention. The International Standards for Neurological Classification of Spinal Cord Injury (ISNCSCI) assessment (without the anorectal exam) was used to observe changes in 10 myotomes on both sides of the body at baseline and final assessments. The Brain Motor Control Assessment (BMCA) was used to characterise previously undetected motor responses, from multi-muscle lower limb electromyography (EMG). A neurophysiological protocol was developed from the original BMCA procedure [38]. Recordings were performed at baseline and final assessments (experimental set-up is shown in Figure 2b). Pairs of surface 35 × 52 mm Ag-AgCl electromyography (EMG) electrodes (Covidien, Watford, UK) were placed bilaterally over the Quadriceps (Qu), Hamstring (Ham), Tibialis Anterior (TA) and Gastrocnemius (GS) muscles of the participant's lower limbs, according to SENIAM guidelines [38]. EMG data were amplified (×1000) using a Digitimer Isolated Patient preamplifier/amplifier system (D360 8-channel Patient Amplifier System, Digitimer, Welwyn

Garden City, Hertfordshire, UK) digitised at 2 kHz (Power 1401, Cambridge Electronic Design, Cambridge, UK). The data were then filtered with a pass-band of 10–200 Hz and stored on a personal computer for analysis. Additional synchronisation of bilateral electro-goniometers (DLK800, Biometrics Ltd., Wales, UK) recorded the range of movement (ROM) of knee joints. Participants lying supine, were asked to perform two bilateral voluntary manoeuvres: hip-knee flexion/extension and ankle dorsiflexion/plantar-flexion. All manoeuvres were repeated three times, cued by a light and audible tone. Health-related quality of life was assessed using the SF-36 Health Survey [39], which has been shown to be discriminative in the SCI population [40]. The SF-36 assesses 8 health-related domains: physical functioning; physical role limitations; emotional role functioning; vitality; mental health; social functioning; bodily pain; general health perception. Functional independence was assessed using the Spinal Cord Independence Measure (SCIM III) [41,42], which assesses functional status in 3 sub-categories: self-care, respiration and sphincter management and mobility.

2.5.2. Weekly Measures

At 0, 4 and 8 weeks of training, weekly outcome measures of BWS and upper- and lower-limb loading were recorded. During sit-to-stand, the participant's feet were positioned on a force platform (Wii Balance Board, Nintendo Co Ltd., Kyoto, Japan) to record leg loading, with their hands grasping handles (each fixed to a customised S-type load cell which measured the vertical force) to record arm loading, used for additional balance, and to lift their body weight (Figure 3). Body weight support was manually recorded from an output screen of the pneumatic resistance system.

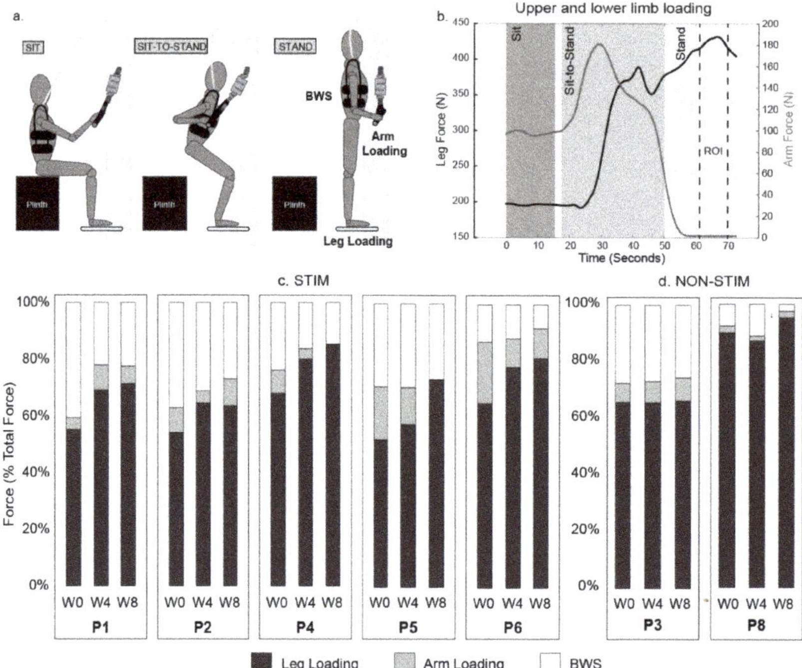

Figure 3. Maximum loading through the lower and upper limbs during standing. Experimental set up (**a**) and example plot of measured leg force (black) and arm force (grey) from one participant with dashed lines indicating an isolated region of interest (ROI) used to quantify force as described in 2.6 (**b**). The weekly changes in distribution of forces (measured from the ROI) are shown in (**c,d**). Loading though lower limbs (black), upper limbs (grey) and manually recorded bodyweight support (white) at week 0, 4 and 8 of the intervention, for all participants.

2.6. Data Analysis

Vertical ground reaction force was calculated for weeks 0, 4 and 8 from the sum of the forces measured by the load cells at the four corners of the Wii Boards. Upper limb arm loading was measured for left and right handle load cells, previously calibrated with known weights. Analogue signals were processed using a moving average (2.5 s sliding window) and the maximum load on the feet during standing at weeks 0, 4, and 8 was extracted from an isolated region of interest (ROI). This ROI, a time window of <1 min, was selected based on review of video footage to isolate a period during the training session where static standing is present and not during any exercises outlined in Section 2.3. This ROI produces a less variable period for biomechanical measures to be examined.

To assess changes in muscle activation patterns between baseline and final assessments compared to published patterns of neurologically intact subjects [38], EMG envelopes (sEMG) were produced using a root mean square algorithm (0.4 s sliding window) for both flexion and extension phases for each manoeuvre (6 total). Integrated EMG (IEMG) was then calculated to express the cumulative area under the curve over time for each phase (flexion/extension) of each manoeuvre, and this was averaged across the 3 repeat trials. Simultaneous knee joint range of motion (ROM) for the flexion phase was quantified by subtracting the angle (°) at rest from the maximum angle during the flexion. The subsequent ROM during extension was quantified by subtracting the maximum flexion angle from the minimum angle in the extension phase.

SF-36 and SCIM scores for each participant were considered significant if a change greater than the Minimal Detectable Change (MDC) for each sub-category was found. For SF-36, MDCs were calculated from standard deviations reported in each sub-category from a cohort of 187 people with SCI [43]. These were Physical Functioning = 59.4, Role Physical = 40.8, Role Emotional = 11.3, Vitality = 12.7, Mental Health = 20.6, Social Functioning = 16.3, Pain = 20.6 and General Health = 21.8 For SCIM, reported MDCs were used [44]; these were Self-care = 2.6, Respiration and Sphincter Management = 6.1 and Mobility = 3.6.

3. Results

3.1. Intervention

Of the nine participants who underwent baseline measures, seven completed the 8-week intervention, and attended all training sessions. One participant experienced symptoms of autonomic dysreflexia in the evening following a training session and was later found to have an injury to the left calf. The participant was withdrawn from participating in any further training, and the incident was reported to the trial sponsor. No further adverse events were reported. Stimulation thresholds found to elicit PRRs at baseline are shown in Table 2 with the subsequent current amplitude used during training ranging between 40 and 110 mA in all participants. Paraesthesia was experienced by all participants in the STIM group during training, and tSCS at this intensity was tolerated in all participants, however some reported discomfort due to the tSCS current. In particular, when tSCS was provided at higher intensities (>80 mA), the activation of trunk muscles caused an anterior pelvic tilt in some participants, which placed strain on the spine during standing, their stimulation intensities were reduced.

Table 2. Participant stimulation thresholds.

	PRR Threshold (mA)	Average Weekly Stimulation Thresholds during Training (mA)								Paraesthesia during Training (Y/N)
	BL	W1	W2	W3	W4	W5	W6	W7	W8	
P1	50	68.8	75.2	66.3	68.3	65.7	59.1	67.5	75.4	Y
P2	150	70.8	64.9	63.7	61.1	69.6	60.0	71.7	71.8	Y
P3	120	-	-	-	-	-	-	-	-	-
P4	145	74.9	73.0	95.6	94.5	105.8	64.3	51.1	68.7	Y
P5	53	78.2	86.3	102.5	105.6	97.4	95.4	75.7	80.0	Y
P6	52	21.7	20.0	19.5	19.6	22.3	40.6	41.3	49.4	Y
P8	90	-	-	-	-	-	-	-	-	-

BL, Baseline.

During training, BWS was provided at the minimum level required to achieve standing for each participant, and was adjusted throughout the 8-week intervention, with the aim of progressively reducing the level of BWS. The set-up and an example plot of the upper and lower limb forces, recorded during a single sit-to-stand, are shown in Figure 3a,b. For each participant, peak upper and lower limb forces and BWS during standing at the first, fourth and eighth week of training is shown in Figure 3c,d.

For all participants in the STIM group, loading through the lower limbs increased progressively throughout the intervention, which was due to reduced BWS and/or reduced lifting through the upper limbs. In the NON-STIM group, neither participant showed any change in BWS with training; however, P8 had high initial leg loading, with little scope for improvement. Participants in the STIM group also reported enhanced voluntary control (ability to "actively engage" in the standing) and proprioceptive (sensory) feedback during tSCS standing activities; these effects developed over several weeks and were evident in all participants in the STIM group by Week 5. P2 and P4 regained the ability to voluntarily activate knee flexor and extensor muscles on-command in the presence of tSCS combined with standing (see P2 (AIS A) in Video S1); this occurred after 5 weeks of training. After 8 weeks of training, P4 (AIS C) regained the ability to stand with a standing frame and minimal assistance, but only in the presence of tSCS.

3.2. ISNCSCI Motor Scores

For each participant, ISNCSCI motor scores before and after the intervention are shown in Table 3. Lower limb motor scores increased in three of the five participants in the STIM group (+1 (AIS A), +5 (AIS C) and +7 (AIS D)) and were unchanged in the other two (both AIS A). P1 presented palpable trace function in his Achilles tendon during ankle plantarflexion in the gravity-eliminated position. P4 showed palpable trace function in her hip flexors, ankle dorsiflexors, long toe extensors and ankle plantar flexors which were absent at baseline. P5 was able to perform full ROM of his ankle dorsiflexors and knee extensors in addition to full ROM of his ankle plantar flexors in the gravity-eliminated position and was able to resist moderate pressure whilst sustaining full ROM of the long toe extensors. Lower limb motor scores were unchanged in both participants in the NON-STIM group (AIS A and AIS C). Upper limb motor scores were unchanged in all participants except for a reduction of one point recorded in one participant in the NON-STIM group.

Table 3. International Standard for Neurological Classification of Spinal Cord Injury (ISNCSCI) before (pre) and after (post) the intervention for participants in transcutaneous spinal cord stimulation (tSCS) combined with sit-to-stand training (STIM (S)) and sit-to-stand training alone (NON-STIM (NS)) groups.

	Group	Injury Level	AIS Grade	Upper Motor Max = 50			Lower Motor Max = 50		
				Pre	Post	Diff	Pre	Post	Diff
P1	S	T3	A	50	50	0	0	1	+1
P2	S	T5	A	50	50	0	0	0	0
P3	NS	T6	A	50	50	0	0	0	0
P4	S	T5	C	50	50	0	1	6	+5
P5	S	C6/7	D	49	49	0	24	31	+7
P6	S	T10	A	50	50	0	1	1	0
P8	NS	C5	C	20	19	−1	0	0	0

3.3. Brain Motor Control Assessment (BMCA)

Before and after the intervention, participants were requested to perform voluntary hip/knee and ankle flexion and extension movements to audible cues, whilst lying supine on a couch; no tSCS was provided. Normative data in non-injured subjects demonstrates rapid and sustained recruitment of motor units in the prime mover (agonist) muscles and smaller residual amplitude response in antagonist muscles during voluntary motor tasks [38]. In SCI individuals, patterns of muscle activity range from no spinal motor output (paralysis) to appropriately sequenced reciprocal activation during controlled joint movements [38,45,46].

3.3.1. Hip/Knee Flexion and Extension

During the voluntary hip/knee manoeuvre, Range of Motion (ROM) at the knee joint was unchanged at the onset of the cue, in all but one participant (P5), indicating that these participants were unable to perform the movements both before and after the intervention (all measured ROMs were <2°). P5 (AIS D, STIM group) was able to reliably perform hip/knee flexion but not extension before the intervention; uncontrolled coactivation of agonist and antagonist muscles was observed during both phases. Although coactivation was still present after training, a more appropriate activation pattern was observed, permitting the ability to extend his left lower limb from flexed (P5 left ROM extension increased from 34.4° to 98.3°). This was supported by the left quadriceps EMG activity increasing considerably during hip flexion (see P5 LQuad in Figures 4 and 5) followed by an increased and sustained contraction in the left hamstrings during extension (see P5 LHam in Figures 4 and 5). Although no change in knee ROM was detected, increases in EMG activity were also noted in P4 (AIS C, STIM group) and P6 (AIS A, STIM group) post training compared with pre (see P4 RQuad during extension in Figures 4 and 5; P6 LHam and RHam during flexion in Figure 5). Despite this activity being present, the activation pattern was inappropriate for the phase of the task.

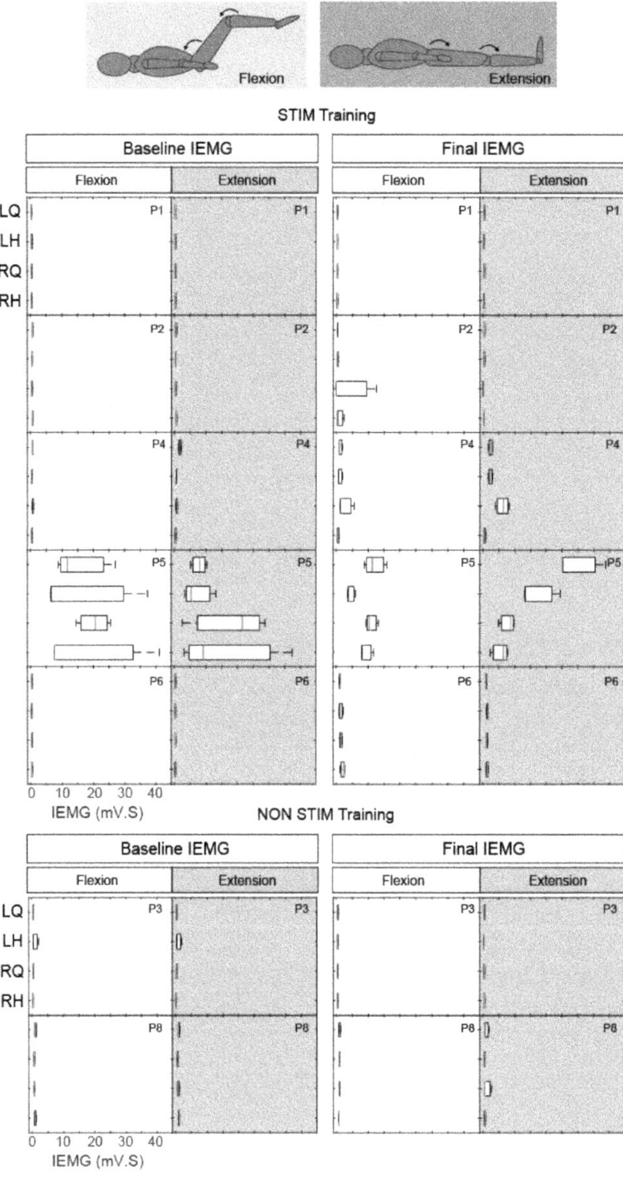

Figure 4. Box plots of integrated EMG activity recorded from the left (L) and right (R) quadriceps (Q) and hamstrings (H) during hip/knee flexion (light grey) and extension (dark grey) movements without tSCS, before (Baseline) and after (Final) the intervention, for all participants. Patterns of integrated EMG (IEMG) (averaged over 3 trials) are displayed for each participant during both flexion and extension phases. Each plot indicates the 25–75th percentiles (box), minimum and maximum values (whiskers), and median value (central red line).

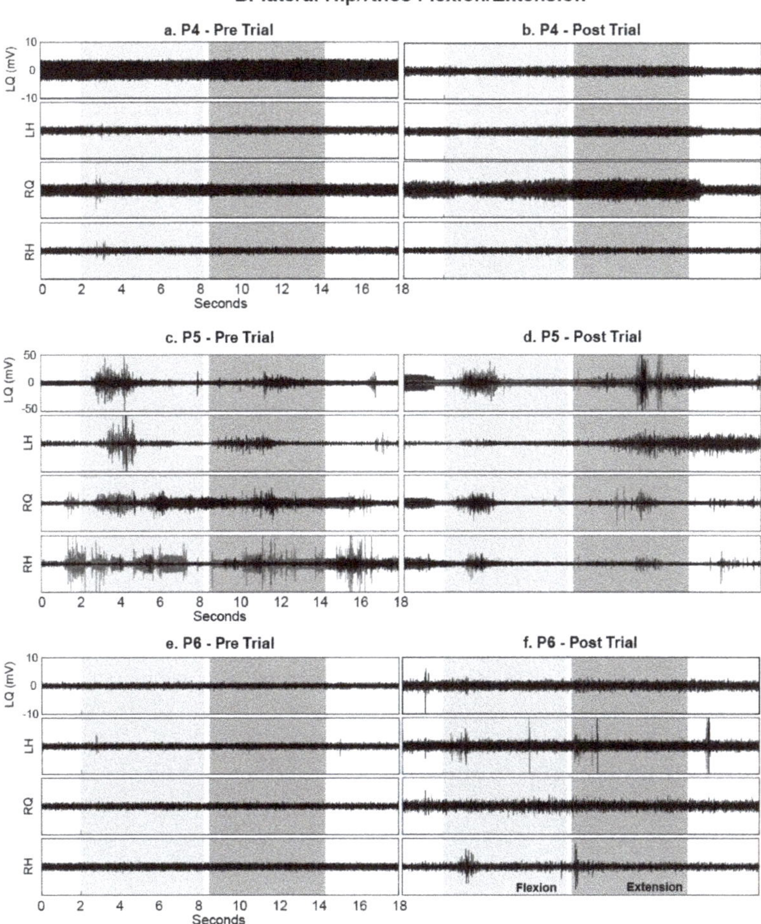

Figure 5. EMG activity recorded from the left (L) and right (R) quadriceps (Q) and hamstrings (H) during hip/knee flexion (light grey) and extension (dark grey) movements without tSCS, before and after the intervention. Data are shown for (**a,b**) P4, (**c,d**) P5 and (**e,f**) P6 (all in the STIM group); each movement was repeated three times, and EMG data from all three movements are overlaid.

3.3.2. Ankle Flexion and Extension

During the voluntary ankle manoeuvre, participants were asked to bring their toes up and point their toes down in response to the audio/visual cue (Figure 6). P5 (AIS D, STIM group) was unable to perform both flexion and extension of the right ankle before the intervention (Figure 7c) but after training, he regained the ability (with a delayed recruitment) to flex and extend the right foot (Figure 7c,d). The EMG quantification revealed that although this functional movement was achieved, the pattern of activity was seemingly uncoordinated (Figure 6). P5 displayed an additional increase in volitional muscle activity with control over the left ankle movement in comparison to baseline recordings (Figure 7c,d). An increase in volitional EMG activity was also present in P4 (AIS C, STIM group) post training compared with pre (see P4 RTA and RGS during extension in Figures 6 and 7a,b) showing coactivation of both agonist and antagonist muscles during the extension phase. Both P6

(AIS A, STIM group) and P8 (AIS C NON-STIM group) displayed increased tonic firing (see P6 LTA Figure 7e,f; P8 LGS and RGS Figure 7g,h) post training compared with pre.

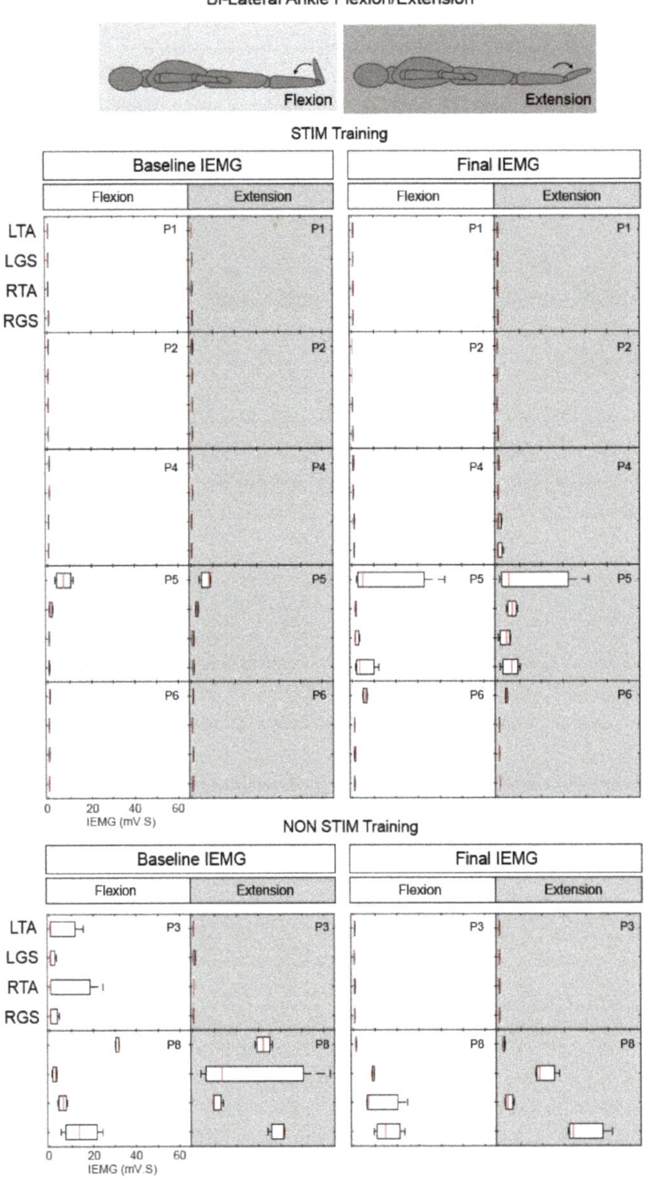

Figure 6. Box plots of integrated EMG activity recorded from the left (L) and right (R) Tibialis Anterior (TA) and Gastrocnemius (GS) during ankle flexion (light grey) and extension (dark grey) movements without tSCS, before (Baseline) and after (Final) the intervention for all participants. Patterns of IEMG (averaged over 3 trials) are displayed for each participant during both flexion and extension phases. Each plot indicates the 25–75th percentiles (box), minimum and maximum values (whiskers), and median value (central red line).

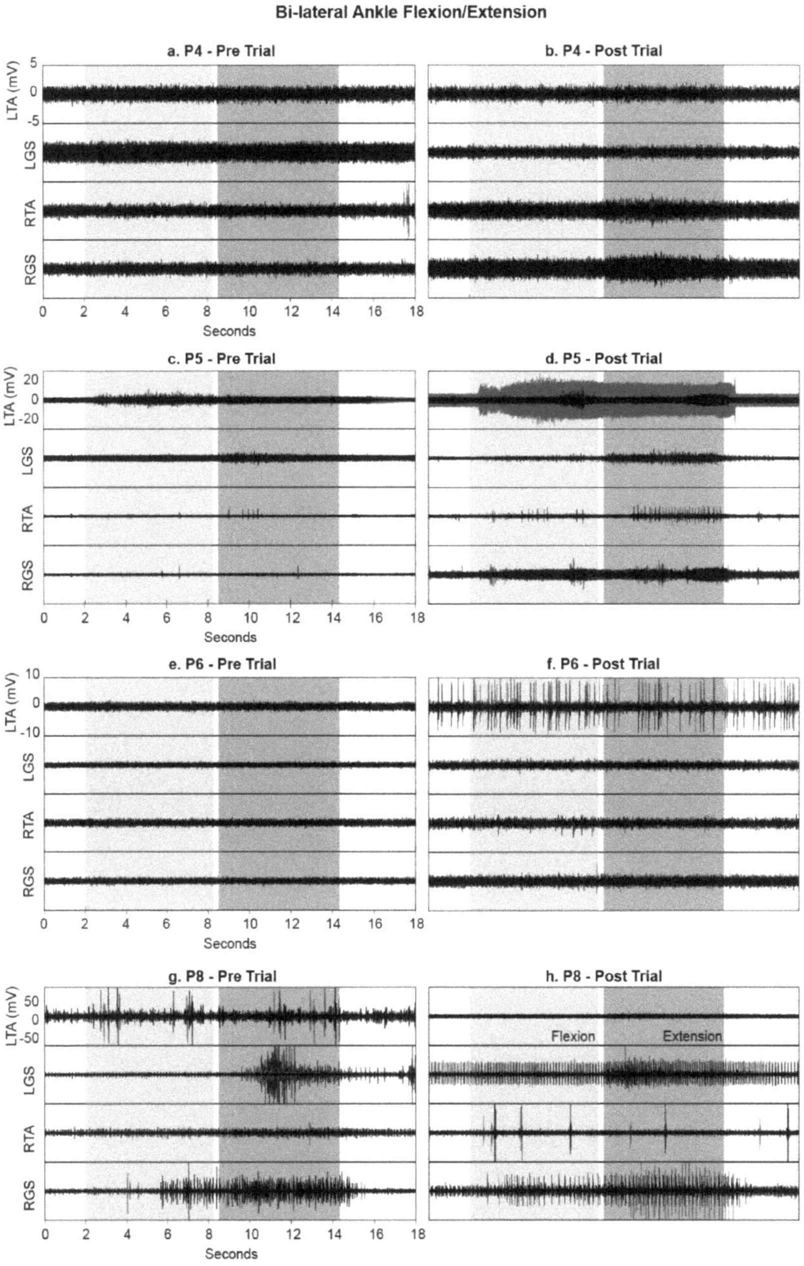

Figure 7. EMG activity recorded from the left (L) and right (R) Tibialis Anterior (TA) and Gastrocnemius (GS) during ankle flexion (light grey) and extension (dark grey) movements without tSCS, before and after the intervention. Data are shown for (**a**,**b**) P4, (**c**,**d**) P5, (**e**,**f**) P6 (STIM group) and (**g**,**h**) P8 (NON-STIM group); each movement was repeated three times, and EMG data from all three movements are overlaid.

3.4. Health-Related Quality of Life Questionnaires

All participants in the STIM group, who had increased lower limb motor scores after training, showed improvements >MDC [43], in at least one sub-category (Figure 8). P5 showed increases in three sub-categories: Role limitations due to physical health (+50), Role limitations due to emotional problems (+67) and Social functioning (+25). P4 increased in one sub-category: Vitality (+33), and P6 increased in two sub-categories: Vitality (+15) and Pain (+23). In contrast, one participant in the STIM group (P1) showed a reduction in one sub-category: Pain (−23; i.e., the participant reported increased pain after training). In the NON-STIM participants, one participant (P3) showed a change >MDC in one sub-category: Role limitations due to physical health (+50). No changes in SCIM III scores, greater than the MDC for each sub-category [44], were noted in any participant (Figure 9).

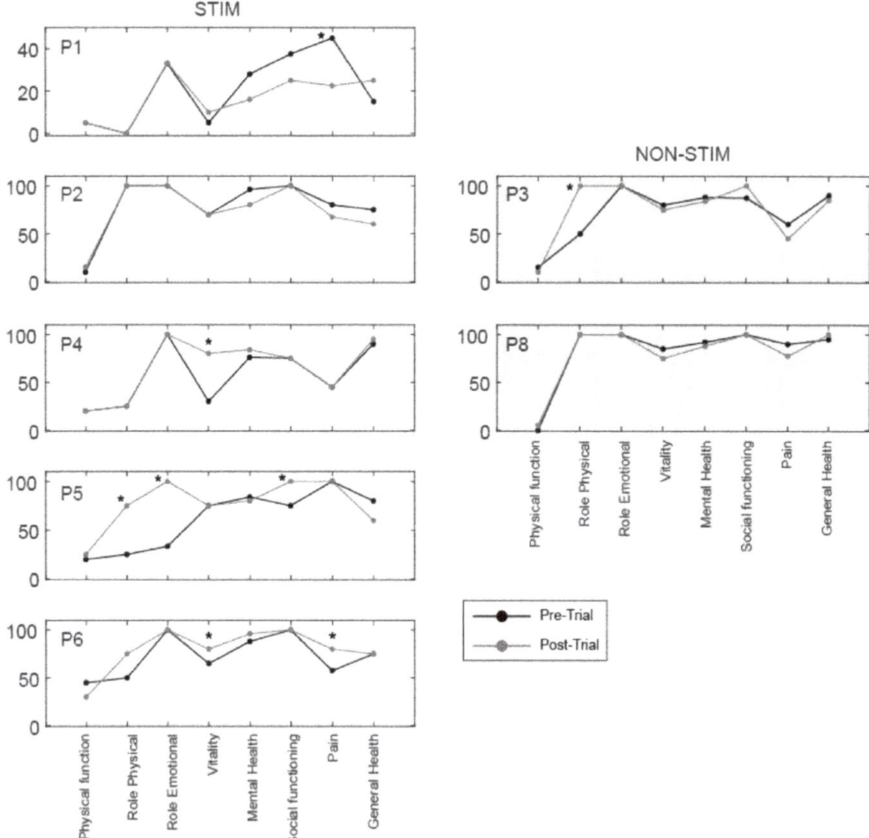

Figure 8. SF-36 scores across 8 sub-categories before (black) and after (grey) the intervention for participants in STIM and NON-STIM groups. * denotes a pre-post intervention change of >MDC for each participant within each sub-category [43].

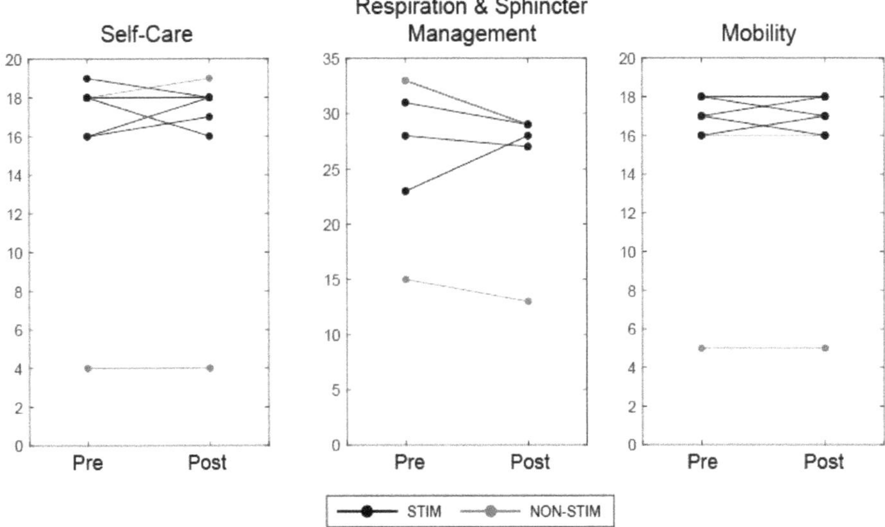

Figure 9. SCIM scores across three sub-categories before (pre) and after (post) the intervention for participants in STIM (black) and NON-STIM (grey) groups.

4. Discussion

The aim of this pilot study was to assess the feasibility of adding sub-threshold transcutaneous SCS to an 8-week sit-to-stand training intervention in individuals with chronic motor complete and incomplete SCI. Of the 35 participants assessed for eligibility, 9 participated, and 7 completed the trial. For all participants in the STIM (tSCS + sit-to-stand training) group, loading through the lower limbs increased progressively throughout the intervention. Both participants with motor incomplete injuries in the STIM group considerably improved their ISNCSCI motor scores (by 5–7 points) and voluntary muscle activation (during the BMCA), when tSCS was switched off. Two participants in the STIM group with motor complete injuries showed small increases in volitional muscle activation in the BMCA, with tSCS switched off, and one of these participants also improved their ISNCSCI motor score by one point. No changes were found in either participant in the NON-STIM (sit-to-stand training alone) group (AIS A and C). Improvements in health-related quality of life were detected in SF-36 subcategories including physical, emotional, vitality, social functioning and pain in STIM group participants.

4.1. Feasibility of Study Design

There was considerable interest in the trial from people with SCI worldwide. In total, 35 potential participants were assessed for eligibility. The majority ($n = 28$) met the study criteria. The known reasons for not participating once deemed eligible were distance to travel, and assignment to the NON-STIM group. To mitigate these drop-outs, a larger trial should be carried out across multiple sites, and consider adding "sham" tSCS to the sit-to-stand alone intervention. The sensation caused by tSCS often precludes a suitable sham intervention (i.e., participants would be aware of their group assignment); however, it may be possible to inform participants that they will receive a tSCS intervention that is delivered either above or below sensory threshold. Of the nine participants that carried out the intervention, none voluntarily withdrew (two participants were withdrawn for other reasons), indicating that they found the intervention acceptable. Some participants (in the STIM group) also reported that they would be willing to continue beyond 8 weeks, indicating that a longer intervention period may also be feasible if a suitable sham intervention can be incorporated. The outcome measures were feasible overall, however stimulation artefacts caused by the tSCS precluded use of the EMG data collected during training: this issue should be further explored prior to completing a larger trial. In addition, increased pain after training was reported in one participant in the STIM group, therefore more detailed reporting of these changes during the intervention is important to distinguish the source (i.e., do they relate to the physical activity or the tSCS *per se*). Up-to 1.5 years post-trial, participants (continuing active training regimes) anecdotally reported the development of further functional changes. This highlights the importance of including long-term follow-up assessments within the study design.

4.2. Recovery of Motor Control in the Presence of tSCS in the STIM Group

None of our participants were able to achieve immediate unassisted or minimally assisted standing during the first session with tSCS, as has been reported by other groups [14]. This was most likely due to the tSCS being delivered at an intensity below motor threshold, so the stimulation did not directly elicit lower limb extension movements or EMG activity. However, all the participants in the STIM group showed some evidence of motor recovery in the presence of tSCS during training, which occurred after at least 4 weeks of training. This included progressive reductions in the required BWS to stand up (Figure 3c), and restored ability to voluntarily activate the quadriceps to generate lower limb extension on-command during training exercises (see Video S1). One participant (P4, AIS C) also regained the ability to stand with a standing frame with minimal assistance after 8 weeks, which was only possible in the presence of tSCS. The combination of tSCS with proprioceptive feedback, due to task-specific stand training and progressive increases in lower limb loading, may have enhanced appropriate muscle activity to enable minimally assisted standing [47].

Similar progressive recovery has previously been reported in the presence of either epidural [16,17,19,31] or transcutaneous [14,35] SCS, but recovery is dependent on preserved functional neurons passing the lesion site, even in people diagnosed with motor complete injuries [45,48]. Of the participants in the STIM group with a clinically complete diagnosis (P1, P2 and P6), none showed evidence of volitional EMG activity at baseline, however P6 was known to have retained partial innervation below the level of injury [49,50] and the other two presumably had "discomplete" injuries. It has been hypothesised that tonic SCS shifts the baseline level of spinal network excitability closer to motor threshold enabling people with incomplete and discomplete SCI to voluntarily generate movements by descending input via these preserved functional neurons [35,51].

4.3. Recovery of Volitional Motor Control (in the Absence of tSCS)

Improvements in voluntary control of movement or an increase in volitional muscle activity after training (when tSCS was off) were found in four of the five participants in the STIM group (P1-AIS A, P4-AIS C, P5-AIS D and P6-AIS A with partial preservation). Only one other study has reported functional recovery (with tSCS switched off) following an intervention of non-invasive SCS [35]; that study used gravity-neutral step training combined with sub-motor threshold tSCS, delivered with a 10 kHz carrier frequency. No changes were observed in either participant in our NON-STIM group (P3-AIS A and P8-AIS C), indicating the importance of the additional tSCS for these changes to occur.

The participant in the STIM group with the least severe injury (P5-AIS D), and the only participant in our study who had any evidence of voluntary motor control at baseline, recovered considerable motor function after training. He was able to fully control flexion and extension in his left leg, which was absent at baseline, and regained the ability to freely move his right foot; his ISNCSCI motor scores had improved by a total of seven points after training. Similar functional recovery (with SCS off) has previously been reported in four people with incomplete injuries (ISNCSCI sores improved by between 4 and 11 points) following intensive activity-based training combined with epidural SCS, applied at an intensity that enabled voluntarily-driven movements [33]. The other two participants with incomplete neurological diagnosis in our study had more extensive motor deficits (both showed little/no voluntary EMG activity at baseline); one was in the STIM group (P4-AIS C) and the other was in the NON-STIM group (P8-AIS C). After training, P4 (STIM) showed increases in appropriate muscle activity during the BMCA, and her ISNCSCI score improved by five points; no changes were observed in P8 (NON-STIM), indicating the importance of additional tSCS.

Among the three participants with motor complete injuries in the STIM group (P1, P2 and P6), some recovery of motor control (with tSCS off) was observed in two of them (P1 and P6). P6 had increased voluntarily driven muscle activity (during the BMCA) after training, and P1 improved his ISNCSCI motor score by one point (palpable contraction in Achilles tendon). No recovery (with tSCS off) was observed in P2, despite evidence of voluntary motor performance in the presence of tSCS during training sessions (see Video S1). Previous studies using epidural SCS in people with motor complete SCI have also reported considerable improvements in motor control (in the presence of SCS) in all participants (n = 12) [19,32,34], but recovery of volitional motor control with SCS off was only observed in a subset of these (5/12). Pre-clinical trials, using epidural SCS combined with step training, have also reported that animals with severe complete transections only show detectable improvements in stepping when stimulation is on, whereas animals with less severe injuries and greater lesion sparing were able to recover voluntary motor control when stimulation was absent [28–30,52]. Overall, in our STIM participants, the amount of recovery that occurred was related to the severity of the injury; as the progressive reduction in BWS had not reached a plateau after 8 weeks, a longer intervention may bring about further recovery.

One factor thought to be important in the effectiveness of SCS is the baseline level of excitatory support from supraspinal and peripheral systems on lumbar spinal circuitry [53]: SCI disrupts this process, causing an imbalance in descending and ascending transmission and dysregulated spinal activity [4,54]. SCS had been shown to regulate this with better prognosis for those with underlying excitability at baseline, for example, in spastic conditions [28,55]. Indeed, one recent study reported that the recovery of volitional control (with SCS off), observed after one-month of epidural SCS, was correlated with spasticity scores at baseline [34]; those authors proposed that baseline spasticity might be a marker for preserved corticospinal tract axons [56]. Given that tonic sub-threshold SCS has been reported to attenuate neural hyper-excitability and recover spinal inhibitory control in people with SCI [25,26,57], and has been successfully used in the treatment of spasticity [58,59], it is also possible that acute reductions in spasticity (due to SCS) enabled the participants access to these retained pathways, which were otherwise masked by hyper-excitability in the central nervous system. Repeated activation of these pathways by carrying out sit-to-stand training in combination with SCS, over several weeks, may have contributed to the observed improvements in voluntary function without tSCS [4].

4.4. Stimulation Parameters

Previous studies have reported optimal parameters to elicit lower limb extension movements or standing as being at 5–15 Hz [5–7]. We chose a relatively higher stimulation frequency (30 Hz), which is closer to the frequency shown to suppress lower limb spasticity when applied at sub-threshold intensities [25–27]. Relatively higher frequencies (25–30 Hz) have also been found to be as effective for standing as 15 Hz tonic stimulation [17,60]. Another study reported that higher stimulation frequencies (80–100 Hz) augmented the activity of lower limb flexor muscles specifically, whereas lower frequencies (20–30 Hz) augmented extensor muscle activity [33]. Guidelines reported by Rejc and co-workers [60] recommend an initial stimulation frequency of 25 Hz at near-motor threshold (that does not directly elicit lower limb movements) to enable standing without BWS, however further investigation into the optimal frequency for standing is warranted, including patient-specific customisation approaches.

In this study, we used sub-threshold SCS in order to alter baseline excitability of the spinal cord, enabling movements triggered by descending inputs that remain intact after SCI [54]. While lower limb extension and standing have been directly elicited by both epidural [6] and transcutaneous [13,14] supra-threshold SCS, such high stimulation intensities have been found to cause, in some people, pulsatile contractions or rhythmic bursting that interfered with standing [60,61], and these effects may be augmented during activities in which body position is altered such as sit-to-stand, causing inconsistency in the structures being activated during movement [15,62,63]. Some of our participants also reported discomfort with higher stimulation intensities and, in some cases, co-activation of the trunk musculature caused discomfort or poor postural alignment. Therefore, to permit adequate free range of movement during stand exercises and prevent discomfort, lower stimulation intensities were selected in this trial.

4.5. Health-Related Quality of Life and Functional Independence

The SF-36 and SCIM III questionnaires were completed before and after training in order to explore whether any recovery of motor control was associated with improvements in healthy-related quality of life and functional independence. We used MDCs as a threshold to determine whether or not changes took place in each participant. All participants that showed evidence of functional changes after training (P4, P5 and P6) also had increased SF-36 in at least one sub-category, indicating that the motor recovery may have improved their health-related quality of life. In the pain sub-category, there were contradictory findings in the STIM group: one participant reported an improvement (P6), and one reported more pain after training (P1). This requires further investigation, including the source of any increases in pain. One participant in the NON-STIM group reported an improvement in one category (role limitations due to physical health), suggesting that the changes may have been associated with the sit-to-stand training alone, or may have occurred with any intervention provided in addition to their usual activities.

4.6. Limitations

The main limitation in the present study was the imbalance of participants between groups. Our intention was to recruit the same number of participants with a motor complete and motor incomplete injury in each group; which we did still achieve, however three of the five participants in the NON-STIM group were unable to complete the trial. In the STIM group participants, tSCS intensity may not have always been optimal during training, due to discomfort from the stimulation. While this cannot be avoided when using traditional tSCS waveforms, PRRs should be elicited during training to enable stimulation intensity to be defined relative to PRR threshold. Another limitation was stimulation artefacts in EMG data when tSCS was switched on. This meant that we were unable to quantify changes in EMG activity due to volitional drive in the presence of tSCS in out STIM group participants. Evidence of the motor recovery in the presence of tSCS can however be viewed in one participant in Video S1.

4.7. Clinical Implications and Future Work

This study has shown that the addition of non-invasive sub-threshold tSCS to sit-to-stand training is an acceptable intervention for people with motor complete and incomplete SCI. In this pilot work, we found that this intervention caused some recovery of volitional drive and control in people with clinically diagnosed complete and incomplete SCI; these findings should now be verified in a larger trial. This intervention is simple and could be achieved by people living with chronic SCI in their own homes i.e., using a standing frame and a commercially available stimulator. If larger trials support our early observations, the accessibility of this intervention could enable many people living with SCI to achieve progressive improvements in motor recovery in the presence of tSCS and, in some cases, neuroplastic change, which may also benefit health-related quality of life.

Future work should further explore the effects of sub-threshold SCS on corticospinal excitability, focusing on the effects of different stimulation intensities, in order to determine the optimal intensity for neuroplastic change, and to improve our understanding of the underlying mechanisms. Future work should also consider optimal rehabilitative interventions to combine with SCS. While there is ample pre-clinical evidence to suggest that SCS is more effective when combined with afferent feedback due to locomotor training [28–30], our study, and the recent clinical studies using epidural SCS [18,34], suggest that recovery of volitional motor control can occur when SCS is combined with simple exercises (incorporating descending volitional drive), which are cheaper and more accessible to the SCI population.

Supplementary Materials: The following are available online at http://www.mdpi.com/2077-0383/9/9/2765/s1, Video S1: Volitional motor recovery with tSCS.

Author Contributions: Conceptualization, L.D.D., N.d.N.D. and S.L.K.; methodology, L.D.D.; software, Y.A. and S.J.M.; validation, Y.A., L.D.D. and N.d.N.D.; formal analysis, Y.A.; investigation, Y.A. and S.J.M.; resources, S.L.K. and N.d.N.D.; data curation, Y.A.; writing—original draft preparation, Y.A.; writing—review and editing, L.D.D., N.d.N.D., S.L.K. and S.J.M.; visualization, Y.A.; supervision, L.D.D., N.d.N.D. and S.L.K.; project administration, L.D.D. and Y.A.; funding acquisition, L.D.D., N.d.N.D. and S.L.K. All authors have read and agreed to the published version of the manuscript.

Funding: This research was funded by the INSPIRE Foundation.

Acknowledgments: The authors are grateful to all of the trainers at Neurokinex (Hemel Hempstead and Gatwick), particularly Jane Symonds, Taylor Omran, Morgan Price-King, Claire Ryan, Stephen Sims and Ben Smith, for their assistance with sit-to-stand training sessions, and to interns from University of Bath and Queen Mary University. The authors also wish to express their sincere thanks to all of the participants for their time, effort and dedication to the trial.

Conflicts of Interest: The authors declare no conflict of interest. The funders had no role in the design of the study; in the collection, analyses, or interpretation of data; in the writing of the manuscript, or in the decision to publish the results.

References

1. Spinal Cord Injury Centre Physiotherapy Lead Clinicians United Kingdom and Ireland. *Clinical Guideline for Standing Adults Following Spinal Cord Injury*; MASCIP: Aylesbury, UK, 2013.
2. Hofer, A.S.; Schwab, M.E. Enhancing rehabilitation and functional recovery after brain and spinal cord trauma with electrical neuromodulation. *Curr. Opin. Neurol.* **2019**, *32*, 828–835. [CrossRef] [PubMed]
3. Ichiyama, R.M.; Gerasimenko, Y.P.; Zhong, H.; Roy, R.R.; Edgerton, V.R. Hindlimb stepping movements in complete spinal rats induced by epidural spinal cord stimulation. *Neurosci. Lett.* **2005**, *383*, 339–344. [CrossRef] [PubMed]
4. Lavrov, I.; Dy, C.J.; Fong, A.J.; Gerasimenko, Y.; Courtine, G.; Zhong, H.; Roy, R.R.; Edgerton, V.R. Epidural stimulation induced modulation of spinal locomotor networks in adult spinal rats. *J. Neurosci. Off. J. Soc. Neurosci.* **2008**, *28*, 6022–6029. [CrossRef] [PubMed]
5. Dimitrijevic, M.R.; Gerasimenko, Y.; Pinter, M.M. Evidence for a spinal central pattern generator in humans. *Ann. N. Y. Acad. Sci.* **1998**, *860*, 360–376. [CrossRef] [PubMed]
6. Minassian, K.; Jilge, B.; Rattay, F.; Pinter, M.M.; Binder, H.; Gerstenbrand, F.; Dimitrijevic, M.R. Stepping-like movements in humans with complete spinal cord injury induced by epidural stimulation of the lumbar cord: Electromyographic study of compound muscle action potentials. *Spinal Cord* **2004**, *42*, 401–416. [CrossRef]
7. Jilge, B.; Minassian, K.; Rattay, F.; Pinter, M.M.; Gerstenbrand, F.; Binder, H.; Dimitrijevic, M.R. Initiating extension of the lower limbs in subjects with complete spinal cord injury by epidural lumbar cord stimulation. *Exp. Brain Res.* **2004**, *154*, 308–326. [CrossRef]
8. Hofstoetter, U.S.; Freundl, B.; Binder, H.; Minassian, K. Common neural structures activated by epidural and transcutaneous lumbar spinal cord stimulation: Elicitation of posterior root-muscle reflexes. *PLoS ONE* **2018**, *13*, e0192013. [CrossRef]
9. Ladenbauer, J.; Minassian, K.; Hofstoetter, U.S.; Dimitrijevic, M.R.; Rattay, F. Stimulation of the human lumbar spinal cord with implanted and surface electrodes: A computer simulation study. *IEEE Trans. Neural Syst. Rehabil. Eng. A Publ. IEEE Eng. Med. Biol. Soc.* **2010**, *18*, 637–645. [CrossRef]
10. Knikou, M. Neurophysiological characterization of transpinal evoked potentials in human leg muscles. *Bioelectromagnetics* **2013**, *34*, 630–640. [CrossRef]
11. Hofstoetter, U.S.; Freundl, B.; Binder, H.; Minassian, K. Recovery cycles of posterior root-muscle reflexes evoked by transcutaneous spinal cord stimulation and of the H reflex in individuals with intact and injured spinal cord. *PLoS ONE* **2019**, *14*, e0227057. [CrossRef]
12. Minassian, K.; Persy, I.; Rattay, F.; Dimitrijevic, M.R.; Hofer, C.; Kern, H. Posterior root-muscle reflexes elicited by transcutaneous stimulation of the human lumbosacral cord. *Muscle Nerve* **2007**, *35*, 327–336. [CrossRef] [PubMed]

13. Minassian, K.; Persy, I.; Rattay, F.; Pinter, M.M.; Kern, H.; Dimitrijevic, M.R. Human lumbar cord circuitries can be activated by extrinsic tonic input to generate locomotor-like activity. *Hum. Mov. Sci.* **2007**, *26*, 275–295. [CrossRef] [PubMed]

14. Sayenko, D.G.; Rath, M.; Ferguson, A.R.; Burdick, J.W.; Havton, L.A.; Edgerton, V.R.; Gerasimenko, Y.P. Self-Assisted Standing Enabled by Non-Invasive Spinal Stimulation after Spinal Cord Injury. *J. Neurotrauma* **2019**, *36*, 1435–1450. [CrossRef] [PubMed]

15. Danner, S.M.; Krenn, M.; Hofstoetter, U.S.; Toth, A.; Mayr, W.; Minassian, K. Body Position Influences Which Neural Structures Are Recruited by Lumbar Transcutaneous Spinal Cord Stimulation. *PLoS ONE* **2016**, *11*, e0147479. [CrossRef] [PubMed]

16. Angeli, C.A.; Edgerton, V.R.; Gerasimenko, Y.P.; Harkema, S.J. Altering spinal cord excitability enables voluntary movements after chronic complete paralysis in humans. *Brain J. Neurol.* **2014**, *137*, 1394–1409. [CrossRef] [PubMed]

17. Harkema, S.; Gerasimenko, Y.; Hodes, J.; Burdick, J.; Angeli, C.; Chen, Y.; Ferreira, C.; Willhite, A.; Rejc, E.; Grossman, R.G.; et al. Effect of epidural stimulation of the lumbosacral spinal cord on voluntary movement, standing, and assisted stepping after motor complete paraplegia: A case study. *Lancet* **2011**, *377*, 1938–1947. [CrossRef]

18. Darrow, D.; Balser, D.; Netoff, T.I.; Krassioukov, A.; Phillips, A.; Parr, A.; Samadani, U. Epidural Spinal Cord Stimulation Facilitates Immediate Restoration of Dormant Motor and Autonomic Supraspinal Pathways after Chronic Neurologically Complete Spinal Cord Injury. *J. Neurotrauma* **2019**, *36*, 2325–2336. [CrossRef]

19. Angeli, C.A.; Boakye, M.; Morton, R.A.; Vogt, J.; Benton, K.; Chen, Y.; Ferreira, C.K.; Harkema, S.J. Recovery of Over-Ground Walking after Chronic Motor Complete Spinal Cord Injury. *N. Engl. J. Med.* **2018**, *379*, 1244–1250. [CrossRef]

20. Grahn, P.J.; Lavrov, I.A.; Sayenko, D.G.; Van Straaten, M.G.; Gill, M.L.; Strommen, J.A.; Calvert, J.S.; Drubach, D.I.; Beck, L.A.; Linde, M.B.; et al. Enabling Task-Specific Volitional Motor Functions via Spinal Cord Neuromodulation in a Human With Paraplegia. *Mayo Clin. Proc.* **2017**, *92*, 544–554. [CrossRef]

21. Huang, H.; He, J.; Herman, R.; Carhart, M.R. Modulation effects of epidural spinal cord stimulation on muscle activities during walking. *IEEE Trans. Neural Syst. Rehabil. Eng. A Publ. IEEE Eng. Med. Biol. Soc.* **2006**, *14*, 14–23. [CrossRef]

22. Dimitrijevic, M.R.; Dimitrijevic, M.M.; Faganel, J.; Sherwood, A.M. Suprasegmentally induced motor unit activity in paralyzed muscles of patients with established spinal cord injury. *Ann. Neurol.* **1984**, *16*, 216–221. [CrossRef] [PubMed]

23. Hofstoetter, U.S.; Hofer, C.; Kern, H.; Danner, S.M.; Mayr, W.; Dimitrijevic, M.R.; Minassian, K. Effects of transcutaneous spinal cord stimulation on voluntary locomotor activity in an incomplete spinal cord injured individual. *Biomed. Tech. Biomed. Eng.* **2013**. [CrossRef] [PubMed]

24. Hofstoetter, U.S.; Krenn, M.; Danner, S.M.; Hofer, C.; Kern, H.; McKay, W.B.; Mayr, W.; Minassian, K. Augmentation of Voluntary Locomotor Activity by Transcutaneous Spinal Cord Stimulation in Motor-Incomplete Spinal Cord-Injured Individuals. *Artif. Organs* **2015**, *39*, E176–E186. [CrossRef]

25. Hofstoetter, U.S.; McKay, W.B.; Tansey, K.E.; Mayr, W.; Kern, H.; Minassian, K. Modification of spasticity by transcutaneous spinal cord stimulation in individuals with incomplete spinal cord injury. *J. Spinal Cord Med.* **2014**, *37*, 202–211. [CrossRef] [PubMed]

26. Hofstoetter, U.S.; Freundl, B.; Danner, S.M.; Krenn, M.J.; Mayr, W.; Binder, H.; Minassian, K. Transcutaneous Spinal Cord Stimulation Induces Temporary Attenuation of Spasticity in Individuals with Spinal Cord Injury. *J. Neurotrauma* **2019**. [CrossRef] [PubMed]

27. Estes, S.P.; Iddings, J.A.; Field-Fote, E.C. Priming Neural Circuits to Modulate Spinal Reflex Excitability. *Front. Neurol.* **2017**, *8*, 17. [CrossRef] [PubMed]

28. Courtine, G.; Gerasimenko, Y.; van den Brand, R.; Yew, A.; Musienko, P.; Zhong, H.; Song, B.; Ao, Y.; Ichiyama, R.M.; Lavrov, I.; et al. Transformation of nonfunctional spinal circuits into functional states after the loss of brain input. *Nat. Neurosci.* **2009**, *12*, 1333–1342. [CrossRef]

29. Ichiyama, R.M.; Courtine, G.; Gerasimenko, Y.P.; Yang, G.J.; van den Brand, R.; Lavrov, I.A.; Zhong, H.; Roy, R.R.; Edgerton, V.R. Step training reinforces specific spinal locomotor circuitry in adult spinal rats. *J. Neurosci.* **2008**, *28*, 7370–7375. [CrossRef]

30. Van den Brand, R.; Heutschi, J.; Barraud, Q.; DiGiovanna, J.; Bartholdi, K.; Huerlimann, M.; Friedli, L.; Vollenweider, I.; Moraud, E.M.; Duis, S.; et al. Restoring voluntary control of locomotion after paralyzing spinal cord injury. *Science* **2012**, *336*, 1182–1185. [CrossRef]

31. Carhart, M.R.; He, J.; Herman, R.; D'Luzansky, S.; Willis, W.T. Epidural spinal-cord stimulation facilitates recovery of functional walking following incomplete spinal-cord injury. *IEEE Trans. Neural Syst. Rehabil. Eng. A Publ. IEEE Eng. Med. Biol. Soc.* **2004**, *12*, 32–42. [CrossRef]

32. Gill, M.L.; Grahn, P.J.; Calvert, J.S.; Linde, M.B.; Lavrov, I.A.; Strommen, J.A.; Beck, L.A.; Sayenko, D.G.; Van Straaten, M.G.; Drubach, D.I.; et al. Neuromodulation of lumbosacral spinal networks enables independent stepping after complete paraplegia. *Nat. Med.* **2018**, *24*, 1677–1682. [CrossRef] [PubMed]

33. Wagner, F.B.; Mignardot, J.B.; Le Goff-Mignardot, C.G.; Demesmaeker, R.; Komi, S.; Capogrosso, M.; Rowald, A.; Seanez, I.; Caban, M.; Pirondini, E.; et al. Targeted neurotechnology restores walking in humans with spinal cord injury. *Nature* **2018**, *563*, 65–71. [CrossRef] [PubMed]

34. Peña Pino, I.; Hoover, C.; Venkatesh, S.; Ahmadi, A.; Sturtevant, D.; Patrick, N.; Freeman, D.; Parr, A.; Samadani, U.; Balser, D.; et al. Long-Term Spinal Cord Stimulation After Chronic Complete Spinal Cord Injury Enables Volitional Movement in the Absence of Stimulation. *Front. Syst. Neurosci.* **2020**, *14*. [CrossRef]

35. Gerasimenko, Y.P.; Lu, D.C.; Modaber, M.; Zdunowski, S.; Gad, P.; Sayenko, D.G.; Morikawa, E.; Haakana, P.; Ferguson, A.R.; Roy, R.R.; et al. Noninvasive Reactivation of Motor Descending Control after Paralysis. *J. Neurotrauma* **2015**, *32*, 1968–1980. [CrossRef] [PubMed]

36. Gerasimenko, Y.; Gorodnichev, R.; Puhov, A.; Moshonkina, T.; Savochin, A.; Selionov, V.; Roy, R.R.; Lu, D.C.; Edgerton, V.R. Initiation and modulation of locomotor circuitry output with multisite transcutaneous electrical stimulation of the spinal cord in noninjured humans. *J. Neurophysiol.* **2015**, *113*, 834–842. [CrossRef]

37. Andrews, J.C.; Stein, R.B.; Roy, F.D. Post-activation depression in the human soleus muscle using peripheral nerve and transcutaneous spinal stimulation. *Neurosci. Lett.* **2015**, *589*, 144–149. [CrossRef]

38. McKay, W.B.; Sherwood, A.M.; Tang, S.F. Appendix 1. A Manual for the Neurophysiological Assessment of Motor Control: The Brain Motor Control Assessment (BMCA) Protoco. In *Restorative Neurology of the Injured Spinal Cord*; Dimitrijevic, M.R., Kakulas, B., McKay, W.B., Vrbova, G., Eds.; Oxford University Press: New York, NY, USA, 2011; pp. 262–283.

39. Ware, J.E., Jr. SF-36 Health Survey. In *The Use of Psychological Testing for Treatment Planning and Outcomes Assessment*, 2nd ed.; Lawrence Erlbaum Associates Publishers: Mahwah, NJ, USA, 1999; pp. 1227–1246.

40. Haran, M.J.; Lee, B.B.; King, M.T.; Marial, O.; Stockler, M.R. Health status rated with the Medical Outcomes Study 36-Item Short-Form Health Survey after spinal cord injury. *Arch. Phys. Med. Rehabil.* **2005**, *86*, 2290–2295. [CrossRef]

41. Catz, A.; Itzkovich, M.; Agranov, E.; Ring, H.; Tamir, A. SCIM—Spinal cord independence measure: A new disability scale for patients with spinal cord lesions. *Spinal Cord* **1997**, *35*, 850–856. [CrossRef]

42. Catz, A.; Itzkovich, M.; Agranov, E.; Ring, H.; Tamir, A. The spinal cord independence measure (SCIM): Sensitivity to functional changes in subgroups of spinal cord lesion patients. *Spinal Cord* **2001**, *39*, 97–100. [CrossRef]

43. Lin, M.R.; Hwang, H.F.; Chen, C.Y.; Chiu, W.T. Comparisons of the brief form of the World Health Organization Quality of Life and Short Form-36 for persons with spinal cord injuries. *Am. J. Phys. Med. Rehabil. Assoc. Acad. Physiatr.* **2007**, *86*, 104–113. [CrossRef]

44. Scivoletto, G.; Tamburella, F.; Laurenza, L.; Molinari, M. The spinal cord independence measure: How much change is clinically significant for spinal cord injury subjects. *Disabil. Rehabil.* **2013**, *35*, 1808–1813. [CrossRef]

45. McKay, W.B.; Lim, H.K.; Priebe, M.M.; Stokic, D.S.; Sherwood, A.M. Clinical neurophysiological assessment of residual motor control in post-spinal cord injury paralysis. *Neurorehabilit. Neural Repair* **2004**, *18*, 144–153. [CrossRef]

46. McKay, W.B.; Ovechkin, A.V.; Vitaz, T.W.; Terson de Paleville, D.G.; Harkema, S.J. Neurophysiological characterization of motor recovery in acute spinal cord injury. *Spinal Cord* **2011**, *49*, 421–429. [CrossRef] [PubMed]

47. Lavrov, I.; Gerasimenko, Y.; Burdick, J.; Zhong, H.; Roy, R.R.; Edgerton, V.R. Integrating multiple sensory systems to modulate neural networks controlling posture. *J. Neurophysiol.* **2015**, *114*, 3306–3314. [CrossRef]

48. Sherwood, A.M.; Dimitrijevic, M.R.; McKay, W.B. Evidence of subclinical brain influence in clinically complete spinal cord injury: Discomplete SCI. *J. Neurol. Sci.* **1992**, *110*, 90–98. [CrossRef]

49. Fawcett, J.W.; Curt, A.; Steeves, J.D.; Coleman, W.P.; Tuszynski, M.H.; Lammertse, D.; Bartlett, P.F.; Blight, A.R.; Dietz, V.; Ditunno, J.; et al. Guidelines for the conduct of clinical trials for spinal cord injury as developed by the ICCP panel: Spontaneous recovery after spinal cord injury and statistical power needed for therapeutic clinical trials. *Spinal Cord* **2007**, *45*, 190–205. [CrossRef] [PubMed]

50. Dietz, V.; Fouad, K. Restoration of sensorimotor functions after spinal cord injury. *Brain* **2014**, *137*, 654–667. [CrossRef] [PubMed]

51. Minassian, K.; Hofstoetter, U.S. Spinal Cord Stimulation and Augmentative Control Strategies for Leg Movement after Spinal Paralysis in Humans. *CNS Neurosci. Ther.* **2016**, *22*, 262–270. [CrossRef]

52. Calvert, J.S.; Grahn, P.J.; Zhao, K.D.; Lee, K.H. Emergence of Epidural Electrical Stimulation to Facilitate Sensorimotor Network Functionality After Spinal Cord Injury. *Neuromodul. J. Int. Neuromodul. Soc.* **2019**, *22*, 244–252. [CrossRef]

53. Minassian, K.; McKay, W.B.; Binder, H.; Hofstoetter, U.S. Targeting Lumbar Spinal Neural Circuitry by Epidural Stimulation to Restore Motor Function After Spinal Cord Injury. *Neurotherapeutics* **2016**, *13*, 284–294. [CrossRef]

54. Edgerton, V.R.; Courtine, G.; Gerasimenko, Y.P.; Lavrov, I.; Ichiyama, R.M.; Fong, A.J.; Cai, L.L.; Otoshi, C.K.; Tillakaratne, N.J.; Burdick, J.W.; et al. Training locomotor networks. *Brain Res. Rev.* **2008**, *57*, 241–254. [CrossRef] [PubMed]

55. Pinter, M.M.; Gerstenbrand, F.; Dimitrijevic, M.R. Epidural electrical stimulation of posterior structures of the human lumbosacral cord: 3. Control Of spasticity. *Spinal Cord* **2000**, *38*, 524–531. [CrossRef] [PubMed]

56. Sangari, S.; Lundell, H.; Kirshblum, S.; Perez, M.A. Residual descending motor pathways influence spasticity after spinal cord injury. *Ann. Neurol.* **2019**, *86*, 28–41. [CrossRef] [PubMed]

57. Knikou, M.; Murray, L.M. Repeated transspinal stimulation decreases soleus H-reflex excitability and restores spinal inhibition in human spinal cord injury. *PLoS ONE* **2019**, *14*, e0223135. [CrossRef] [PubMed]

58. Barolat, G.; Singh-Sahni, K.; Staas, W.E., Jr.; Shatin, D.; Ketcik, B.; Allen, K. Epidural spinal cord stimulation in the management of spasms in spinal cord injury: A prospective study. *Stereotact. Funct. Neurosurg.* **1995**, *64*, 153–164. [CrossRef]

59. Dimitrijevic, M.M.; Dimitrijevic, M.R.; Illis, L.S.; Nakajima, K.; Sharkey, P.C.; Sherwood, A.M. Spinal cord stimulation for the control of spasticity in patients with chronic spinal cord injury: I. Clinical observations. *Cent. Nerv. Syst. Trauma J. Am. Paralys. Assoc.* **1986**, *3*, 129–144. [CrossRef]

60. Rejc, E.; Angeli, C.; Harkema, S. Effects of Lumbosacral Spinal Cord Epidural Stimulation for Standing after Chronic Complete Paralysis in Humans. *PLoS ONE* **2015**, *10*, e0133998. [CrossRef]

61. Gerasimenko, Y.; Roy, R.R.; Edgerton, V.R. Epidural stimulation: Comparison of the spinal circuits that generate and control locomotion in rats, cats and humans. *Exp. Neurol.* **2008**, *209*, 417–425. [CrossRef]

62. Sayenko, D.G.; Angeli, C.; Harkema, S.J.; Edgerton, V.R.; Gerasimenko, Y.P. Neuromodulation of evoked muscle potentials induced by epidural spinal-cord stimulation in paralyzed individuals. *J. Neurophysiol.* **2014**, *111*, 1088–1099. [CrossRef]

63. Hofstoetter, U.S.; Minassian, K.; Hofer, C.; Mayr, W.; Rattay, F.; Dimitrijevic, M.R. Modification of reflex responses to lumbar posterior root stimulation by motor tasks in healthy subjects. *Artif. Organs* **2008**, *32*, 644–648. [CrossRef]

Journal of
Clinical Medicine

Article

Cervical Electrical Neuromodulation Effectively Enhances Hand Motor Output in Healthy Subjects by Engaging a Use-Dependent Intervention

Hatice Kumru [1,2,3,*], África Flores [4], María Rodríguez-Cañón [4], Victor R. Edgerton [1,5], Loreto García [1,2,3], Jesús Benito-Penalva [1,2,3], Xavier Navarro [1,4], Yury Gerasimenko [6,7], Guillermo García-Alías [1,4] and Joan Vidal [1,2,3]

1 Fundación Institut Guttmann, Institut Universitari de Neurorehabilitació Adscrit a la Universitat Autònoma de Barcelona, 08916 Badalona, Spain; vre@ucla.edu (V.R.E.); loretogarcia@guttmann.com (L.G.); jbenito@guttmann.com (J.B.-P.); xavier.navarro@uab.cat (X.N.); guillermo284@gmail.com (G.G.-A.); jvidal@guttmann.com (J.V.)
2 Universitat Autònoma de Barcelona, Bellaterra, 08193 Barcelona, Spain
3 Fundació Institut d'Investigació en Ciències de la Salut Germans Trias i Pujol, 08916 Badalona, Spain
4 Departament de Biologia Cel·lular, Fisiologia i Immunologia & Insititute of Neuroscience, Universitat Autònoma de Barcelona, and CIBERNED, Bellaterra, 08193 Barcelona, Spain; africa.flores@uab.cat (Á.F.); rodriguezcanonmaria@gmail.com (M.R.-C.)
5 Department of Integrative Biology and Physiology, University of California Los Angeles, Los Angeles, CA 90095, USA
6 Pavlov Institute of Physiology, 199034 St. Petersburg, Russia; yuryg@ucla.edu
7 Department of Physiology and Biophysics, University of Louisville, Louisville, KY 40292, USA
* Correspondence: hkumru@guttmann.com; Tel.: +34-934-977-700

Citation: Kumru, H.; Flores, Á.; Rodríguez-Cañón, M.; Edgerton, V.R.; García, L.; Benito-Penalva, J.; Navarro, X.; Gerasimenko, Y.; García-Alías, G.; Vidal, J.et al. Cervical Electrical Neuromodulation Effectively Enhances Hand Motor Output in Healthy Subjects by Engaging a Use-Dependent Intervention. *J. Clin. Med.* **2021**, *10*, 195. https://doi.org/10.3390/jcm10020195

Received: 12 December 2020
Accepted: 5 January 2021
Published: 7 January 2021

Publisher's Note: MDPI stays neutral with regard to jurisdictional clai-ms in published maps and institutio-nal affiliations.

Abstract: Electrical enabling motor control (eEmc) through transcutaneous spinal cord stimulation is a non-invasive method that can modify the functional state of the sensory-motor system. We hypothesize that eEmc delivery, together with hand training, improves hand function in healthy subjects more than either intervention alone by inducing plastic changes at spinal and cortical levels. Ten voluntary participants were included in the following three interventions: (i) hand grip training, (ii) eEmc, and (iii) eEmc with hand training. Functional evaluation included the box and blocks test (BBT) and hand grip maximum voluntary contraction (MVC), spinal and cortical motor evoked potential (sMEP and cMEP), and resting motor thresholds (RMT), short interval intracortical inhibition (SICI), and F wave in the abductor pollicis brevis muscle. eEmc combined with hand training retained MVC and increased F wave amplitude and persistency, reduced cortical RMT and facilitated cMEP amplitude. In contrast, eEmc alone only increased F wave amplitude, whereas hand training alone reduced MVC and increased cortical RMT and SICI. In conclusion, eEmc combined with hand grip training enhanced hand motor output and induced plastic changes at spinal and cortical level in healthy subjects when compared to either intervention alone. These data suggest that electrical neuromodulation changes spinal and, perhaps, supraspinal networks to a more malleable state, while a concomitant use-dependent mechanism drives these networks to a higher functional state.

Keywords: transcutaneous spinal cord stimulation; hand training; combined intervention; neuromodulation; cervical spinal cord

1. Introduction

Electrical stimulation of the spinal cord is an emerging valuable tool in clinical practice for facilitating the recovery of sensory and motor function in subjects with spinal cord injury (SCI) [1]. Outstanding clinical achievements have been recently reported showing that lumbar epidural stimulation facilitates the recovery of posture, stepping, and voluntary control of the lower limbs in subjects with chronic SCI [2–6]. Cervical epidural stimulation increased hand grip force and hand volitional control in tetraplegic patients [7].

On the other hand, recently developed non-invasive transcutaneous electrical spinal cord stimulation (electrical enabling motor control, eEmc) has demonstrated its efficacy to improve lower limb motor function after paralysis [8–10], as well as hand grip strength and voluntary control in tetraplegic patients [11–14]. In both stimulation conditions, the underlying hypothesis states that the spinal sensory–motor networks above, within, and below the lesion are neuromodulated and raised into an elevated functional state that enables and amplifies voluntary motor control [11–14]. In the context of SCI rehabilitation, the term eEmc [8–10] emphasizes and distinguishes this technique from other transcutaneous electrical stimulation approaches that directly induce muscular contraction, instead of facilitating and enabling voluntary control.

Since transcutaneous spinal cord stimulation (tSCS) is a non-invasive and safe method, it can be easily applied in healthy subjects and can also be employed to unravel the neurophysiological mechanisms acting on sensory and motor recovery in individuals suffering from SCI [12]. By applying high-frequency trains of stimuli (i.e., eEmc) to the thoracolumbar segments, locomotor-related neuronal networks can be neuromodulated to physiological states that enable the generation of bilateral rhythmic step-like movements with the legs placed in a gravity-neutral position in non-injured subjects [15] and in spinally injured subjects categorized as a complete SCI (ASIA A) [16]. One of the mechanisms considered to be important in that movement is mediated by activation of the spinal central pattern generators of locomotion [17]. The degree to which this specific mechanism contributes to upper limb movements, however, remains unclear, since it may act together with other adaptations to allow the recovery of a wide range of movements that are not predominantly repetitive, such as stepping. Importantly, continuous eEmc delivery also improves hand motor performance after one single exposure from 20 min to 2 h in SCI patients [11–14]. A recent work reveals that high-frequency eEmc induces plastic changes on neuronal circuits controlling upper limb function also in intact subjects [14].

Activity-dependent plasticity is the key phenomenon thought to underlie functional recovery observed after both physical activity and stimulation-based rehabilitative approaches in SCI [18–20]. However, the mechanisms recruited by physical training and by electrical stimulation, although partially overlapping, may involve different and perhaps synergistic processes leading to more effective neural circuit reorganization. However, reported studies do not address whether eEmc applied alone or together with hand training showed better clinical or neurophysiological effects in SCI patients. Indeed, the majority of the studies only applied eEmc combined with hand training [9,10,12,13]. More studies are needed to directly assess the role of use-dependent mechanisms in an attempt to optimize therapeutic efficacy of stimulation approaches and training. Here we hypothesized that the application of eEmc together with hand grip training can increase hand motor output more than either of the two interventions alone. We further hypothesized that the improved function would be the result of a synergistic reorganization of both spinal and supraspinal networks that could enable the performance of the neuromuscular unit being engaged. To answer these questions, each participant was subjected to a hand motor functional and neurophysiological assessment before and after each of three interventions (training, eEmc, and training + eEmc). To detect hand muscle strength and a manual dexterity, we used the hand grip maximum voluntary contraction (MVC) and the box and blocks test (BBT). Spinal cord excitability was assessed by F wave and by recruitment curves of spinal motor evoked potentials (sMEPs) in hand and arm muscles induced by single-pulse tSCS. Corticospinal excitability was measured by resting motor threshold (RMT) and cortical motor evoked potentials (cMEPs) in hand muscles induced by transcranial magnetic stimulation (TMS), and changes in intracortical inhibitory circuits were tested by short interval intracortical inhibition (SICI).

2. Experimental Section

2.1. Participants

Thirty healthy subjects were selected for participating in this study (Consolidated Standards of Reporting Trials (CONSORT) flow diagram on Figure 1). Recruitment of participants consisted of self-selection within the current users of our installations as well as asking the participants for referrals. Eleven healthy volunteers accepted informed consent and completed inclusion criteria with no known history of neurological disorders and accepted to participate in the study, and ten of them (7 men, 3 women; mean age = 38 ± 11 years, age range: 24–60 years; Table 1) completed all three interventions, and were included in the data analysis (CONSORT flow diagram on Figure 1). One subject rejected to continue at the beginning of the study because of unpleasant sensation through cervical electrical stimulation. Inclusion criteria were: male or female between 18 and 65 years old without any neurological disorder and other disorder which could limit the experiment (uncontrolled cancer, arthritis, etc.), and had written informed consent. Exclusion criteria were any metal implants, implanted electrical devices, medications that could raise seizure threshold, cardiac conditions, and history of syncope or concussion with loss of consciousness, tinnitus, or pregnancy. The study protocol was approved by the Research Ethics Committee of the Institute Guttmann and was conducted in accordance with the Declaration of Helsinki [21].

Flow Diagram

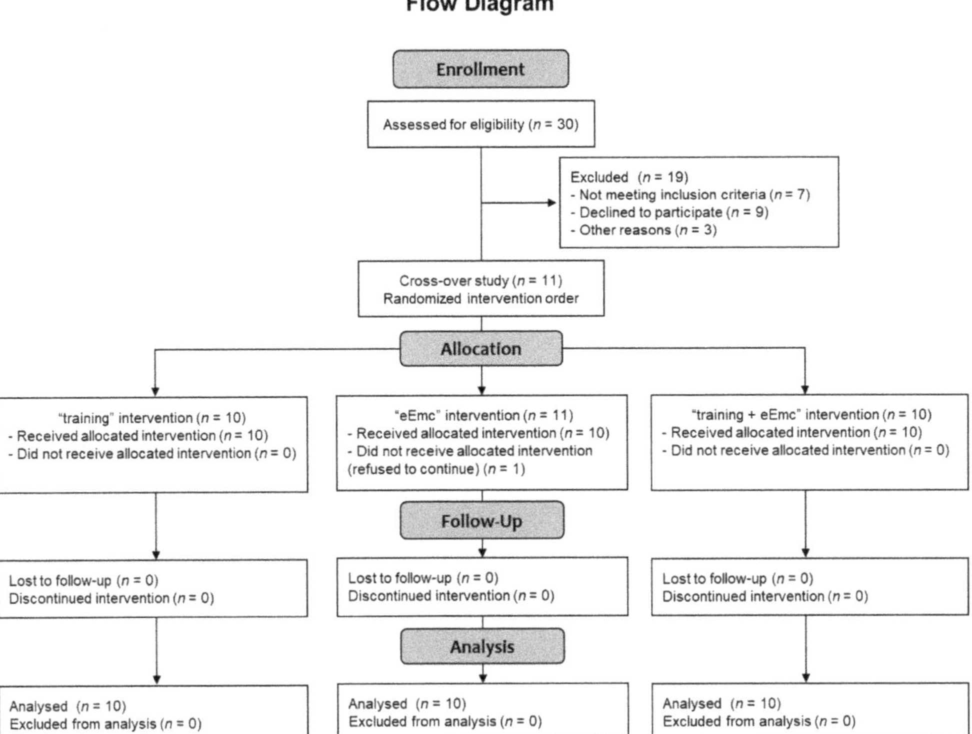

Figure 1. CONSORT 2010 flow diagram showing the number of subjects involved in each phase of the study.

Table 1. Demographic data from the ten subjects participating in the present study. Stimulation intensities employed during high-frequency electrical enabling motor control (eEmc) are also shown.

Subject	Sex	Age	Hand Preference	eEmc Intensity (mA)	
				C3–C4	C6–C7
1	male	44	right	29	32–38
2	male	60	right	59–67	69–77
3	female	25	right	28–30	30–32
4	female	27	left	34	36
5	female	33	right	52–54	59–63
6	male	41	right	34–38	45
7	male	24	right	36–50	56–59
8	male	51	right	52–63	63–77
9	male	39	right	59	72–74
10	male	38	both	54–67	63–77

2.2. Experimental Design

It was a randomized crossover study, which included three different interventions. Interventions consisted of: (i) hand grip training, (ii) eEmc, or (iii) eEmc combined with hand grip training. All participants received each intervention once with at least 1 week between each intervention.

For hand functional outcomes, we assessed MVC during hand grip force and BBT. For neurophysiological assessment at spinal cord level, we evaluated F wave of abductor pollicis brevis (APB) muscle and recruitment curves of sMEPs in hand and arm muscles induced by single-pulse tSCS, applied to C3–C4 and then to C6–C7. At cortical level: RMT and cMEPs induced by TMS in hand muscles at 120% of RMT and SICI were recorded. All neurophysiological recordings were done in the dominant upper extremity muscles. The functional and neurophysiological assessments were realized at baseline, just after intervention (starting at $0'$ post) and one hour after the end of intervention (starting at $60'$ post, follow-up) to evaluate the possible short- and long-lasting aftereffects of each intervention. The duration of each experiment was around 4 h including the set up. Each complete functional and neurophysiological assessment took 30–45 min for each timepoint and was carried out in the following order: BBT, MVC, F wave, cortical RMT, cMEPs, SICI, cMEP recruitment curve, spinal RMT, and sMEP recruitment curve.

2.3. Functional Assessment

2.3.1. Maximum Voluntary Contraction during Hand Grip Strength

The hydraulic hand grip dynamometer (Jamar Model 5030J1, Sammons Preston, NJ, USA) was used for MVC. Participants were asked to perform MVC for 4 s, as soon as they noticed a triggering stimulus, which was a mild electrical pulse at an intensity of 3 mA with 0.5 ms of duration delivered with ring electrodes to the fifth finger of the dominant hand. We recorded the maximal and maintained force during these 4 s in three consecutive trials, with at least one minute of rest between each trial. We registered the electromyographic (EMG) activity from the arm and hand muscles.

2.3.2. Box and Blocks Test

This is a test of manual dexterity consisting of a box with a center partition [22]. The participant must pick up the maximum number of small cube-shaped blocks, one at a time, and drop them at the other side of the partition in 60 s. The score of BBT is represented by the number of blocks transported.

2.4. Neurophysiological Assessment

2.4.1. Electromyographical Recording

EMG activity was recorded with a conventional EMG machine (Medelec Synergy, Oxford Instruments; Surrey, UK). After standard skin preparation, disposable adhesive surface electrodes (outer diameter of 20 mm; Technomed) were placed over the muscle belly of the APB, abductor digiti minimi (ADM), flexor carpi radialis (FCR), extensor digitorum (ED), and biceps brachii (BB) muscles of the dominant arm, with the cathode proximal and the anode approximately 2 cm distally. The EMG signal was amplified and then filtered using a band-pass of 30 Hz–10 kHz, amplitude sensitivity of 0.1–0.5 mV and epochs of 100 ms sweep duration and recorded at a sampling rate of 50 kHz. During assessments, we ensured that the baseline EMG activity of all recorded muscles was lower than 50 μV of amplitude before delivering each single stimulus: EMG activity was checked out online and, if necessary, the subject was reminded to be relaxed. If any background EMG activity was observed after stimulus delivery, this recording was eliminated in situ and stimulation procedure was repeated. EMG signals were stored in a Synergy computer and analyzed offline with MATLAB.

2.4.2. F Wave

To measure the spinal motoneuron pool excitability, F wave was recorded in APB muscle using suprathreshold electrical median nerve stimulation at the wrist level [23,24]. Both stimulus delivery and EMG signal recording were processed by a conventional EMG machine (Medelec Synergy, Oxford Instruments; Surrey, UK). Electrical stimuli of 0.5 ms of duration were delivered at 1 Hz and the electrical intensity was set at supramaximal level to induce the maximal amplitude in M wave (Mmax) (range: 17–30 mA) [24]. A minimum of 10 stimuli were delivered to calculate F wave persistency [25].

2.4.3. Recruitment Curve of sMEPs

To assess spinal cord excitability, we used monophasic rectangular 1-ms single-pulse tSCS delivered through 2 cm diameter hydrogel adhesive electrodes (axion GmbH, Hamburg, Germany) as cathodes at C3–C4 and then at C6–C7 and two 5 × 12 cm rectangular electrodes placed symmetrically over the iliac crests as anodes. The electric intensity for spinal RMT in APB muscle of the dominant hand was determined at baseline condition in each intervention. RMT was defined as the lowest intensity that elicited sMEPs of ≥50 μV peak-to-peak amplitude at APB muscle in at least 5 out of 10 consecutive trials. Recruitment curves of sMEPs were established at gradual increasing intensities from 90% to 150% of RMT (at 10% increments, three recordings at each intensity) in all recorded muscles of the dominant hand [14,26].

2.4.4. Transcranial Magnetic Stimulation

Changes in corticospinal excitability and SICI were evaluated using a Magstim[®] BiStim[2] TMS (Magstim Company, Whitland, Wales, UK). Subjects were seated in a chair, resting their pronated forearms on a desk in front of them and were asked to stay relaxed but awake throughout the assessments. A figure-of-eight coil was held tangentially to the scalp over the motor area of the dominant hand in the optimal position for activating the APB in a posterior-anterior current direction. The hot point for evoking the largest cMEP in APB was marked over scalp. The following parameters were measured before, after, and during follow-up of each intervention: 1. RMT, defined as the lowest intensity of TMS that evoked a cMEP of ≥50 μV peak-to-peak amplitude in the APB muscle in at least 5 of 10 consecutive trials. 2. Mean amplitude of cMEPs using single-pulse TMS at 120% of RMT of ABP in five recordings. 3. SICI using paired-pulse TMS with a subthreshold conditioning stimulus (80% of RMT) and a suprathreshold test stimulus (120% of RMT) at interstimulus interval of 2 ms in 5 recordings without background activity (between 1 to 5 recordings for each subject were rejected because of background activity) [27]. 4. Recruitment curves were obtained at increasing intensities from 90% to 150% of RMT of APB, at 10% increments (three recordings at each intensity). The absence of baseline activation was verified before carrying out each of the abovementioned recordings.

2.5. Interventions

The study consisted of three interventions: (1) hand training, (2) eEmc, (3) eEmc combined with hand training. There was at least 1 week between each experiment.

Hand training protocol was adapted from previous studies [12] and consisted in grasping a hand grip dynamometer at maximum maintainable contraction for 20 s, followed by an 80-s resting period. This was repeated alternating right and left hands for 30 min (9 times per hand). Thus, the total duration of contractions for each treatment was 3 min for each hand and a total of 9 min of maximum effort for all three treatments for each muscle group. Mean sustained force was registered for each contraction. In all cases, the subjects were instructed and closely monitored to assure that a neutral wrist position and a 90° angle of the elbow was maintained while performing MVC with hand grip (Figure 2a).

A second condition consisted of delivering eEmc in the same time pattern (20 s of stimulation followed by an 80-s resting period for 30 min) in absence of hand training. eEmc was carried out with the transcutaneous electrical stimulator BioStim-5 (Cosyma Inc., Moscow, Russia). Previous reports indicate that applying stimulation simultaneously at two sites within the cervical area is consistently more effective than a single stimulation site [12]. Thus, we delivered eEmc simultaneously at two sites along the midline between spinous processes C3–C4 and C6–C7 during the corresponding intervention period. Regarding the possible mechanisms involved, it is critical to recognize that the intensity of stimulation at each spinal level was set at 90% of RMT induced by single-pulse tSCS at APB muscle of the dominant hand (range: ~30–80 mA, Table 1). Stimulation was continuously delivered using 2 cm diameter hydrogel adhesive electrodes (axion GmbH, Hamburg, Germany) as cathodes and two 5×12 cm rectangular electrodes placed symmetrically over the iliac crests as anodes. eEmc employed biphasic rectangular 1-ms pulses, each one filled with a carrier frequency of 10 kHz (i.e., each 1-ms pulse was composed of ten 0.1-ms biphasic rectangular pulses), that were delivered at a frequency of 30 Hz [13,17,28]. During stimulation, the subjects reported a non-painful but uncomfortable tingling sensation down the arms and at the site of stimulation, with some associated tonic contraction of paraspinal and posterior neck muscles.

The last condition was eEmc combined with hand training. The subjects performed maximum sustainable contractions and simultaneously received eEmc, alternating right and left hands for 30 min. The abovementioned tonic contraction of neck muscles during stimulation did not interfere with performance of repeated grip contractions, since force levels achieved in the dynamometer during this intervention were similar to those achieved during hand training without stimulation and participants reported no impediments to perform the training.

2.6. Data Analysis and Statistics

For each assessment, we calculated the mean $\pm$ standard error measurement (SEM) at baseline, right after and sixty minutes after intervention (0′ post and 60′ post intervention respectively) in each condition. We calculated the % changes according to baseline for each timepoint evaluation when appropriate. BBT, cMEP, SICI, and cMEP slopes could be easily compared between interventions without this normalization, so absolute values were preserved. The analysis of these outcome measures revealed the same effects if normalized.

For analyzing EMG activity during MVC, the area under the curve (AUC) in EMG of each muscle was calculated from the beginning of EMG signal during 4 s of MVC.

For the F wave analysis, we measured the peak-to-peak amplitude and for Mmax we measured the maximum amplitude of M wave. F wave was considered if peak-to-peak amplitude was at least 1% of M wave (Mmax) amplitude and F persistence was calculated by dividing the number of F responses by the number of stimuli from 10 recordings. The maximum amplitude of F was normalized to Mmax amplitude to obtain Fmax/Mmax ratio.

For recruitment curve of sMEPs and cMEPs, we measured the peak-to-peak MEP amplitude (μV) induced in each recording of all muscles and then we calculated the mean amplitude of MEPs from three recordings for each intensity for each subject and for each experimental condition. To detect the possible influence of each intervention and counteract the masking effect of interindividual variability, recruitment curves obtained at $0'$ post and $60'$ post intervention were normalized to baseline recruitment curves (which were considered as 0% change for all intensities) registered for each session and each subject. As a measure of corticospinal excitability, in addition, slope of non-normalized recruitment curves was calculated through linear regression from 90% to 150% RMT intensities, an interval that fitted linear distribution in all conditions and muscles [29].

For calculating SICI percentage, averaged peak-to-peak amplitude of the conditioned cMEP (obtained after the conditioning stimulus of 80% RMT with 2 ms separation from conditioned stimulus of 120% RMT) was expressed as a percentage of the averaged amplitude of the test cMEP (obtained at supramaximal 120% RMT stimulus): % = (conditioned cMEP/test cMEP) $\times$ 100.

Data were expressed as mean $\pm$ SEM. Shapiro–Wilk test was used to assess whether data were normally distributed or not. All sets of data fulfilled normal distribution requirement except F wave persistence, cortical RMT, SICI, and cMEP slope. For normally distributed data, repeated measures two-way analysis of variance (ANOVA) was used. For post-hoc analysis, we used Tukey's test. For not normally distributed data, Friedman's analysis was used followed by post-hoc Dunn's test. Least squares regression was applied as linear regression method to calculate slope of sMEP and cMEP recruitment curves. Significance level was set at $p < 0.05$ in all cases.

3. Results

3.1. Hand Motor Output

Figure 2a shows a general setting for MVC assessment. MVC at baseline was 30.6 ± 0.9 kg. This value was consistent between sessions (within-subject variation coefficient was $10.9 \pm 1.2\%$), although inter-subject variability was considerable (between-subjects variation coefficient was $28.7 \pm 3.0\%$). Thus, all values were normalized to baseline for further analysis. Repeated-measures ANOVA showed that intervention ($F_{(2, 18)} = 11.2, p < 0.001$) and timepoint ($F_{(2, 18)} = 3.67$, $p < 0.05$) factors, as well as their interaction ($F_{(4, 36)} = 8.77, p < 0.001$) had a significant effect in MVC. MVC tended to increase following eEmc combined with training and did not change after eEmc alone, whereas training alone reduced MVC significantly after intervention during follow-up ($p < 0.001$ according to Tukey's post hoc). The percentage of MVC vs. baseline levels was bigger after eEmc with training compared to training alone at $0'$ and $60'$ post intervention ($p < 0.001$ for each comparison) and also compared to eEmc intervention alone ($p < 0.01$ for each comparison) (Figure 2b).

The EMG activity of APB and ADM changed significantly according to two-way ANOVA ($F_{(1, 52)} = 16.43, p < 0.001$ and $F_{(1, 52)} = 6.77, p < 0.01$, respectively) (Figure 2c,d). There was a significant interaction between intervention and timepoint factors for APB and ADM muscles ($F_{(4, 36)} = 3.36, p < 0.05$ and $F_{(4, 36)} = 4.34, p < 0.01$, respectively). Tukey's test showed that the training condition significantly reduced EMG activity of APB (at $0'$ and $60'$ post training) and ADM (at $60'$ post intervention) ($p < 0.05$). Moreover, EMG activity in APB and ADM was significantly lower following training or eEmc alone than following eEmc with hand training at both $0'$ and $60'$ post-intervention ($p < 0.05$ at least for each comparison) (Figure 2c,d). The EMG of FCU, ED, or BB muscles did not change significantly by intervention or timepoint according to two-way repeated measures ANOVA. Figure 2e shows a representative EMG of APB muscle (subject #5) corresponding to MVC trials during "training" and "eEmc with training" interventions. This figure depicts similar EMG activity levels at basal conditions before these two interventions, but higher EMG activity at both $0'$ post and $60'$ after the combined intervention, an effect that becomes obvious when calculating the difference in EMG activity between "eEmc with training" and "training" condition.

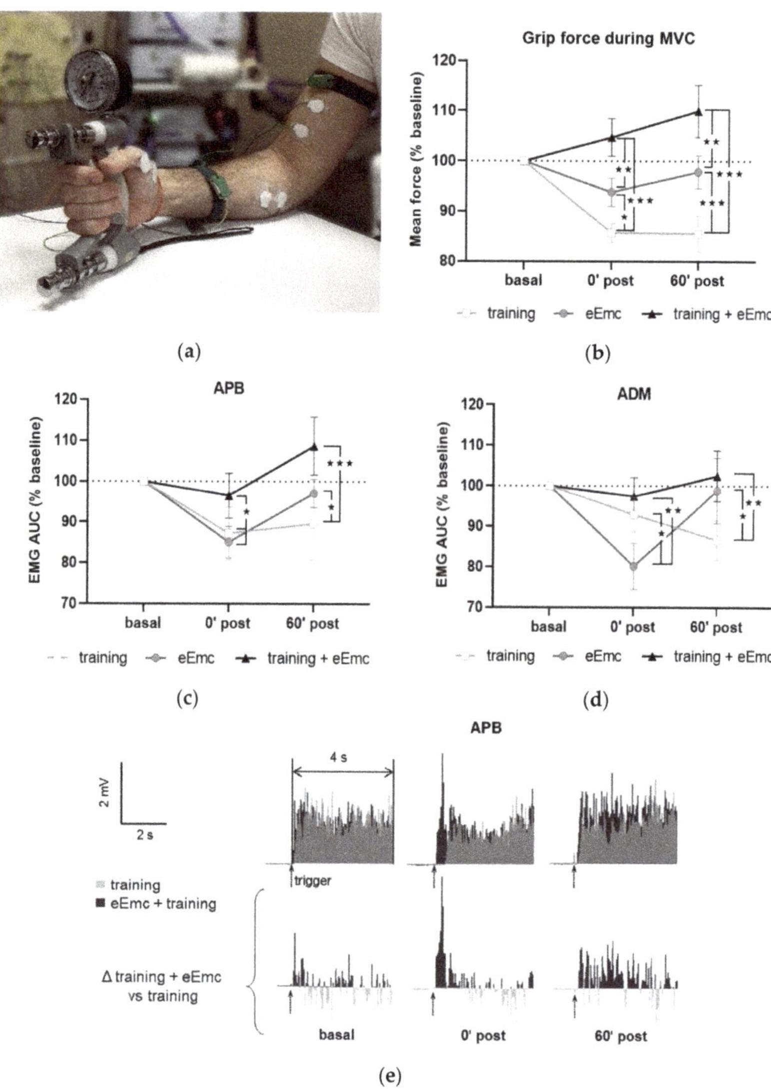

(a)

(b)

(c)

(d)

(e)

Figure 2. Maximal voluntary contraction (MVC) during hand grip strength. (**a**) Picture showing general setting and position for hand grip MVC assessment. (**b**) Percentage of MVC vs. baseline levels at 0 min (0′ post) and 60 min (60′ post) after each intervention. Significant differences between interventions: ★ $p < 0.05$, ★★ $p < 0.01$, ★★★ $p < 0.001$. (**c**,**d**) Percentage of area under the curve (AUC) of electromyographic (EMG) activity vs. baseline levels from APB (**c**) and from ADM (**d**). Significant differences between interventions: ★ $p < 0.05$, ★★ $p < 0.01$, ★★★ $p < 0.001$. (**e**) Representative APB muscle EMG registry (obtained from subject #5) corresponding to maximum voluntary contraction (MVC) trials during training and eEmc with training interventions at basal, 0′ and 60′ after intervention. Upper panels correspond to rectified values, and lower panels show difference between registers of both interventions for better difference appreciation. APB, abductor pollicis brevis; ADM, abductor digiti minimi.

Manual dexterity was assessed with the BBT in order to detect possible alterations induced by power grip training and/or eEmc-based interventions. Two-way ANOVA did not show an effect of intervention ($F_{(2, 18)}$ = 2.62, p = 0.10) neither timepoint × intervention interaction ($F_{(4, 36)}$ = 0.206, p = 0.93). However, there was a significant timepoint effect ($F_{(2, 18)}$ = 43.5, p < 0.001) with increasing number of boxes, being significantly higher just after intervention (0′) and during follow-up (60′) in comparison to basal evaluation when equally considering all interventions (p < 0.001) (Figure 3).

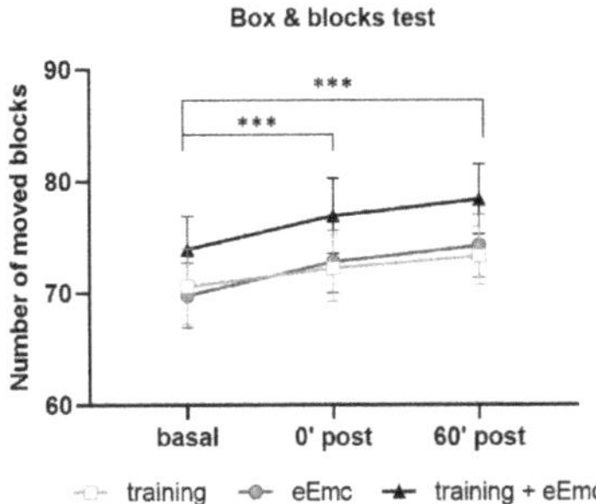

Figure 3. Number of moved blocks in the box and blocks test (BBT) for hand training, eEmc, or combination of both interventions right after intervention (0′ post) and during follow-up (60′ post) in comparison to basal evaluation. Significant differences between timepoints according to Tukey's multiple comparisons (considering all intervention groups as a whole): *** p < 0.001.

3.2. Spinal Cord Excitability

F wave was used to detect excitability of anterior horn motoneurons in the spinal cord. There was a significant effect of intervention over F wave persistence (Friedman's test = 5.84, df 3.30, p = 0.05) (Figure 4a). Dunn's post-hoc comparisons revealed an overall increase in F persistence after eEmc combined with training compared to training alone that almost reached statistical significance (p = 0.053). The intervention factor had a nearly significant impact on F wave amplitude ($F_{(2, 18)}$ = 3.10, p = 0.07). Post hoc comparisons showed that eEmc delivery, with or without hand training, increased F wave amplitude when compared to training alone (p < 0.05), particularly at 0′ post intervention (Figure 4b,c).

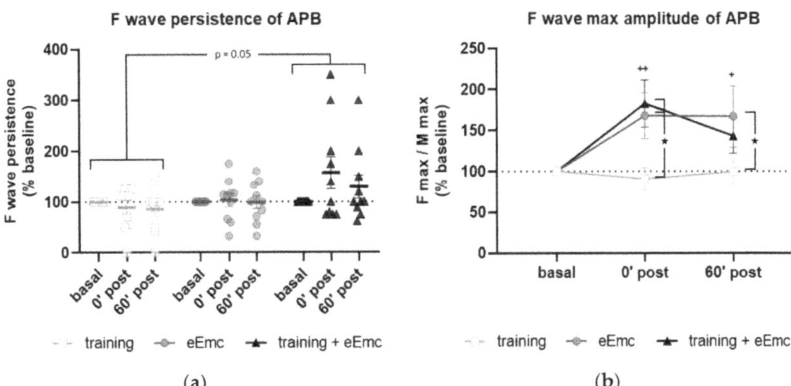

Figure 4. *Cont.*

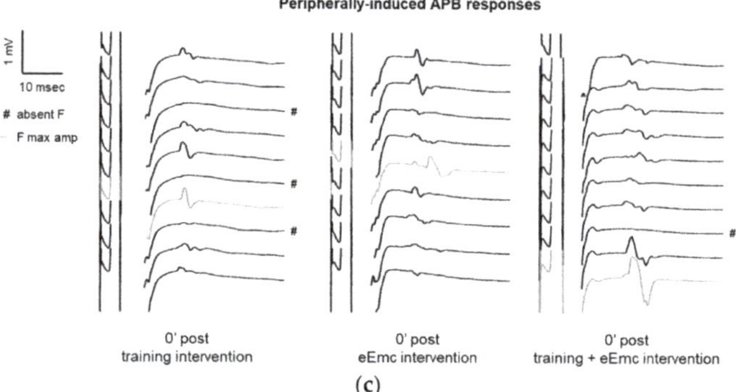

Figure 4. Variables of F waves in APB by median nerve stimulation. (**a**) F wave persistency. Significant difference between eEmc with training and training alone (regardless of the timepoint): $p = 0.05$. Data expressed as median $\pm$ interquartile range. (**b**) Averaged maximum amplitude of F wave for each condition and timepoint. Significant differences vs. baseline: $^{+}$ $p < 0.05$ (for training + eEmc), $^{++}$ $p < 0.01$ (for eEmc alone), and between interventions: ★ $p < 0.05$. (**c**) Representative APB muscle registers (obtained from subject #7) corresponding to ten consecutive median nerve stimulations showing M and F wave at 0' after in each intervention. Light grey registers highlight F waves with maximum amplitude. # Symbol indicates registers where no F wave presence was considered. APB, abductor pollicis brevis.

Excitability of spinal networks was tested by single-pulse tSCS at cervical level. The RMT in APB at baseline was of 51.0 ± 2.9 mA at C3–C4 and 60.8 ± 3.5 mA at C6–C7. The recruitment curve of sMEP from upper extremity muscles did not show any significant changes in any muscle at any intensity and in any intervention ($p > 0.05$) (Table S1).

3.3. Corticospinal Excitability and Intracortical Inhibition

The RMT in APB at basal conditions was $37.9 \pm 1.2\%$ of maximum TMS intensity. RMT was significantly affected by timepoint in the training intervention (Friedman's test = 7.93, df 3.10, $p < 0.05$) and by intervention at 60' post (Friedman's test = 8.06, df 3.10, $p < 0.05$) (Figure 5a). Dunn's post-hoc comparisons confirmed that higher RMT was found 60' after training ($p < 0.05$). eEmc with training significantly reduced RMT when compared to training alone at this timepoint ($p < 0.05$). Corticospinal excitability measured by cMEP amplitude obtained at 120% RMT stimulus did not differ between interventions ($F_{(2, 18)} = 0.25$, $p = 0.77$) or between timepoints ($F_{(2, 18)} = 1.55$, $p = 0.23$) (Figure 5b).

SICI was significantly affected by timepoint in the training intervention (Friedman's test = 7.40, df 3.10, $p < 0.05$). Dunn's post hoc analysis revealed that hand training induced higher inhibition (i.e., smaller conditioned cMEP amplitude) at 0' post-intervention ($p < 0.05$) compared to baseline, with no other significant comparisons (Figure 5c).

The recruitment curves of cMEP at baseline did not significantly differ between interventions in any muscle and appear represented as a single pool in Figure 6a, which shows APB data. Analysis of APB muscle recruitment curves normalized to baseline showed significant intervention $\times$ timepoint interaction after intervention ($F_{(12, 108)} = 3.59$, $p < 0.001$) and during follow-up ($F_{(12, 108)} = 1.58$, $p < 0.05$). Specifically, eEmc with training enhanced cMEP amplitude significantly in comparison to training and eEmc alone at high TMS-stimulation intensities ($1.4\times$ and $1.5\times$ RMT at 0' post-intervention and $1.3\times$ and $1.5\times$ RMT at 60' post-intervention; Figure 6b,c). Corticospinal excitability was also quantified by calculating the slope of non-normalized absolute recruitment curves, which also resulted in a significant effect of intervention (Friedman's test = 8.08, df 3.30, $p < 0.05$), with higher slopes for APB after eEmc with training compared to training or eEmc alone at 0' and 60' post-intervention, according to Dunn's multiple comparisons (Figure 6d, Table S2). There were no significant differences for slopes analyzed for other muscles (Table S2).

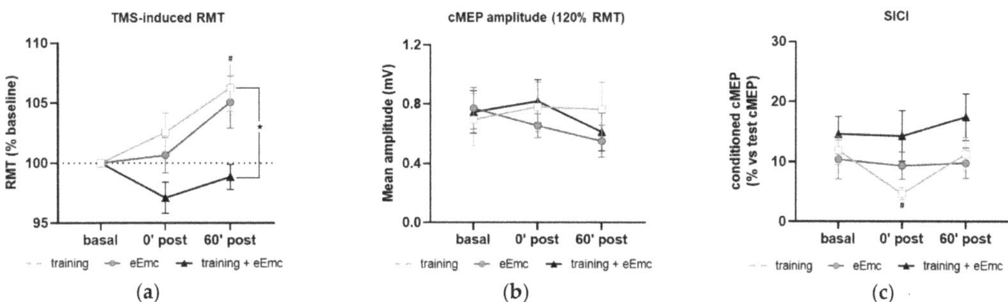

Figure 5. Effect of hand training, eEmc and eEmc + training on the corticospinal excitability and SICI. (**a**) Percentage of resting motor threshold (RMT) vs. baseline levels in APB at 0 min (0′ post) and 60′ min (60′ post) after intervention. Significant differences between interventions: ★ $p < 0.05$, (training vs. eEmc + training), and in comparison to baseline: # $p < 0.05$ (for training). (**b**) Effect of training, eEmc and eEmc + training on the supramaximal cMEP amplitude (120% transcranial magnetic stimulation (TMS)) according to RMT. (**c**) SICI: (average of conditioned cMEP) × 100/ (average of test cMEP) recorded at APB muscle and obtained at 0′ and 60′ after each intervention. # $p < 0.05$ for training at 0′ post in comparison to baseline. APB, abductor pollicis brevis; RMT, resting motor threshold; cMEP, cortically-induced motor evoked potential; SICI, short-interval intracortical inhibition.

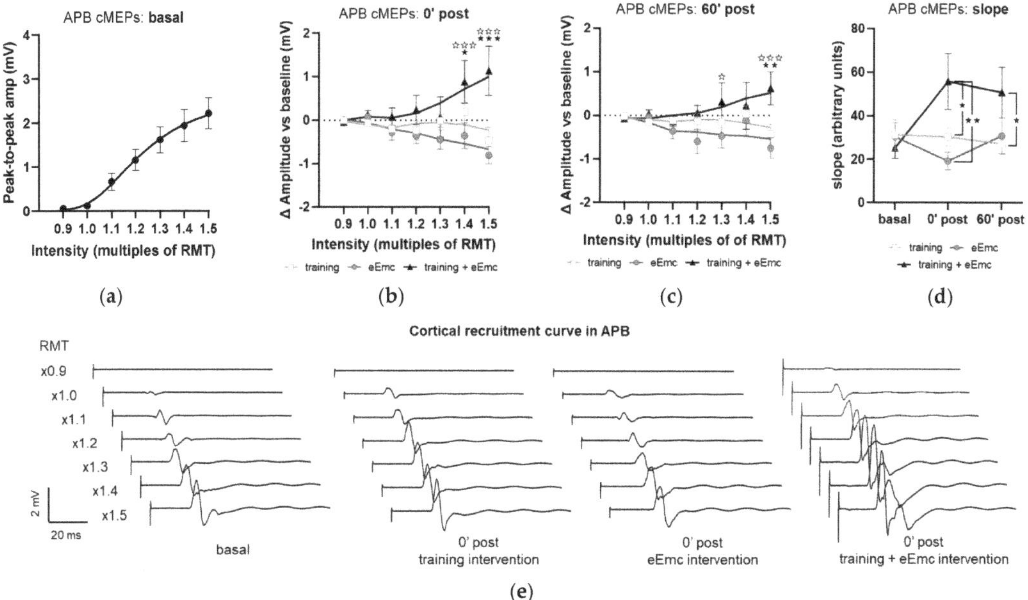

Figure 6. Effect of interventions on cortical recruitment curve. (**a**) Pooled APB recruitment curves obtained at baseline for each subject and intervention. (**b,c**) Increase or decrease (in millivolt (mV); indicated as Δ) in APB recruitment curves with respect to baseline at 0′ (**b**) and 60′ (**c**) post intervention. ★ $p < 0.05$, ★★ $p < 0.01$, ★★★ $p < 0.001$ training vs. training + eEmc; ☆ $p < 0.05$, ☆☆☆ $p < 0.001$ eEmc vs. training + eEmc. (**d**) Slope of APB cMEP recruitment curves. Indicated slopes correspond to the slope of the linear regression line through stimulation intensities (in %RMT) and cMEP values (in mV), multiplied ×1000 to facilitate data visualization. ★ $p < 0.05$, ★★ $p < 0.01$, between interventions at 0′ post and 60′ post. (**e**) Representative APB muscle recording (obtained from subject #1) corresponding to recruitment curve assessment at basal (three conditions averaged) and 0′ post each intervention. APB, abductor pollicis brevis; RMT, resting motor threshold; cMEPs, cortically-induced motor evoked potentials.

4. Discussion

The present study shows that one single session of cervical eEmc modifies the excitability of neuronal networks controlling upper limb function in healthy subjects. These changes strongly depend on eEmc combined with hand training, improving hand grip force and increasing spinal and corticospinal excitability in comparison to each intervention tested alone. Indeed, eEmc alone increased cervical spinal cord excitability measured by anterior horn motoneuron excitability, but had no effect on corticospinal excitability or hand grip force, while hand training alone reduced hand grip force, corticospinal excitability, and intracortical inhibition. The effects of eEmc combined with hand training appeared immediately after the intervention and were observed in hand grip force measured by MVC, in spinal cord excitability measured by F wave, and in corticospinal excitability measured by RMT induced by TMS and cMEP recruitment curve, and these effects lasted at least one hour in most clinical and neurophysiological assessments.

Previous studies in SCI patients have reported improved hand motor performance following eEmc, even after one single session [11–14,17]. However, it is not clear whether eEmc benefits or not from simultaneous physical training to deploy its full potential to facilitate motor improvement. Indeed, although eEmc has been mostly delivered together with training, improvements were reported in hand function immediately after eEmc in absence of any hand training and persisted for at least 75 min following 20 min of 30 Hz stimulation [14].

Diverse modalities of spinal cord stimulation, traditionally employed to treat chronic pain, interfere with proprioceptive and sensitive processing [30]. The lack of differences observed between interventions in the BBT indicates that eEmc does not interfere with manual dexterity as might be observed if stimulation altered sensorial/proprioceptive processing or precise motor control. Of note, progressive improvement shown after repeated BBT testing in all groups suggests some form of motor or cognitive learning that allows improved task performance, and importantly, this process is not compromised by eEmc exposure either. As expected, however, we did not observe any improvement after eEmc exposure since little space for dexterity refinement is left in uninjured participants at this particular task.

Unlike eEmc or the combined treatment, hand training alone resulted in a significant decrease in hand grip force as shown by reduced MVC, an effect that lasted at least one hour. The decline in hand grip force was partially correlated with changes in EMG activity of the hand muscles APB and ADM but not with more proximal arm muscles (Figure 3b,c). This was probably due to fatigue at peripheral and/or central level, since the voluntary effort performed by all subjects was maximal although it did not reach the force level achieved in basal conditions, accordingly to previous observations in similar settings (reviewed in [31]). This was unsurprising due to the nature of the intervention (i.e., repetitive maximal contractions for 30 min). Interestingly, whereas MVC strength remained relatively unaltered after eEmc delivery alone as no exercise-induced fatigue could be present, an increase in grip force was observed after eEmc together with training when compared to both hand training and eEmc alone, even 1 h after the intervention. This suggests that eEmc delivery, together with hand training, is probably counteracting the fatigue effect induced by concomitant exercise, but other mechanisms are required to explain the differences compared to eEmc alone. Moreover, the maintenance of EMG activity levels after the combined intervention suggests that eEmc requires concomitant hand training to more successfully sustain the recruitment of muscle fibers and muscle contraction levels. The fatiguing and the compensatory effects are likely to be a combination of neural and muscular elements [32].

F wave has been considered an indirect measure of motoneuron pool excitability [33] and reflects backfiring of a small number of motoneurons which are reactivated by antidromic impulses following supramaximal stimulation of a peripheral nerve. Increased Fmax/Mmax amplitude ratios (by means of increased Fmax and constant Mmax) observed after eEmc alone, with or without hand training, indicated that a higher portion of the motoneuron pool can be activated by an antidromic stimulus after eEmc intervention. We also observed higher F persistence when eEmc was applied simultaneously to hand training during the intervention. Whereas F max amplitude may reflect de maximal number of motoneurons backfiring simultaneously, F persistence may reflect the probability of these motoneurons to backfire at the same time. Since the probability to successfully produce an orthodromic action potential after antidromic depolarization depends on the refractory period of the axonic cone [34], it is possible that eEmc alone is sufficient to increase motoneuron excitability but somehow does not favor optimal refractory period conditions to increase F wave occurrence. Moreover, it has been shown that pre-activation increases the occurrence of F wave in neurons with a low number of F waves at rest, but not in those neurons with high basal number of F responses [35]. Considering this, a possibility might be that hand grip training focuses eEmc-induced excitability to a discrete spinal neuronal population, perhaps the one contributing to grip performance.

Previous reports show that tSCS at the cervical level generally reaches anterior horn motoneurons through activation of dorsal root afferents, as shown by elicited bilateral monosynaptic reflexes [36,37], although tSCS can however preferentially activate sensory or motor roots depending on the stimulation parameters and electrode location [38,39]. A recent work shows that eEmc applied at the cervical spinal cord is able to increase the excitability of spinal networks in uninjured and SCI participants [14], which goes in line with increased motoneuron excitability observed in our study. In the work of Benavides et al., sMEPs were elicited by cervicomedullary magnetic stimulation and showed increased amplitude in proximal and distal arm muscles for 75 min following tSCS, but not sham-tSCS. In contrast, we could not observe any difference between interventions on sMEPs elicited by single-pulse tSCS. A reason for this discrepancy could rely on the different current distribution within the spinal cord elicited by magnetic and electrical stimulation, or perhaps our single-pulse tSCS-based method to recruit upper limb muscles resulted not sensitive enough to detect such changes at the level of spinal network excitability. However, since eEmc delivers electrical currents at intensities below the RMT, an indirect (more than a direct) effect on cervical spinal networks may have increased motoneuron excitability.

Corticospinal excitability is known to be altered due to physical training and also to diverse modalities of CNS stimulation [14,24,31,40]. Both voluntary muscular activity (e.g., sustained maximal contractions) and "imposed" muscular activity (induced by sustained electrical stimulation at motor point) reaching fatiguing levels result in transiently decreased corticospinal excitability [40]. In agreement, we found increased TMS-evoked RMT after hand grip training and after eEmc without training, an effect that persisted at least 1 h after each intervention. TMS-evoked RMT, however, remained unaltered after eEmc with hand training. In addition, recruitment curves at APB muscle showed higher cMEP amplitude after eEmc with hand training than each intervention separately. Recruitment curves of cMEP from muscles other than APB did not show this enhancement, possibly because TMS was optimized to recruit APB. Increased cMEP amplitude in the APB recruitment curve entails that the number of spinal motoneurons recruited at particular TMS stimulation intensities is higher after eEmc with training than after each intervention alone. This could reflect changes at any level within the corticospinal motor pathway controlling the APB muscle: intracortical circuits favoring cortical primary motoneuron to depolarize, excitability of cortical motoneurons itself, axonal corticospinal conductivity, spinal premotor networks favoring spinal motoneuron depolarization, excitability of spinal motoneurons itself, peripheral axonal conductivity, or efficiency at the neuromuscular junction level [41]. Thus, increased spinal excitability itself reflected as F wave facilitation after eEmc, with or without hand training, could contribute to increased corticospinal excitability. This finding might also be explained by the different mechanisms underlying cMEP depression after exercise or after electrical stimulation. As discussed by Pitcher et al., cMEP depression may emerge as an adaptation to tonic afferent input to the cortex from the exercising muscle (presumably Golgi tendon organs or group III and non-spindle group II afferents [42]), altering cortical excitability without altering spinal excitability. On the other hand, peripheral electrical stimulation initially induces cMEP facilitation [43] until prolonged stimulation leads to cMEP depression reflecting cortical plasticity [40]. In our study, since electrical stimulation is applied at spinal level, and enhances spinal excitability, several concurrent (but opposing) effects might be taking place, and corticospinal output may depend on their final balance. Thus, a possible explanation for increased corticospinal output after eEmc with training includes that spinal cord stimulation could interfere with afferent inputs to the cortex from the exercising muscle, counteracting the effect of muscular activity.

Given that intracortical networks influence corticospinal output, and that exercise-induced decrease in corticospinal excitability has been ascribed to intracortical rather than spinal mechanisms [40], we assessed SICI to reveal the possible modulation of inhibitory circuits in the primary motor cortex [44]. We observed increased intracortical inhibition right after power grip training, which may be surprising given that fatiguing exercise usually leads to reduced inhibition [45,46]. However, this event has been reported *during* maximal contractions, being already absent at 7–10 min after ceasing the exercise. In addition, increased excitability of intracortical inhibitory circuits (i.e., increased SICI) occurs *after* relaxation from voluntary contractions [47], a phenomenon that has been interpreted as a rebound effect after relaxation. This may explain our observations and further supports our vision supporting that eEmc combined with training prevents its fatigue-induced effects. On the other hand, we could not find any effect of eEmc, with or without hand training, on intracortical inhibition. These results differ from a recent study reporting increased SICI after eEmc alone that lasted for more than 1 h [14]. Methodological differences might explain these divergences, such as duration of stimulation, frequency stimulation, and simultaneous evaluation of other parameters together with SICI in the present study. Notably, eEmc delivered simultaneously to hand training prevented the alterations observed in SICI after hand training alone. This might suggest, again, the possible prevention of fatigue induced by simultaneous eEmc, indicating that the modulation of intracortical inhibitory motor circuits by eEmc delivered with hand training contributes to increased corticospinal excitability observed after this intervention. However, further studies are needed to confirm this hypothesis by accurately controlling the time-course of SICI during and after each intervention.

Mechanisms of plasticity observed in unimpaired subjects will, to some extent, be extrapolable to individuals with complete and incomplete spinal cord injury (SCI), since a portion of neural substrates remains preserved after SCI. Indeed, even in very severe injuries with completely abolished motor and sensory function, some spinal axons survive at the epicenter of the injury [48,49]. Thus, SCI population may benefit not only from plastic changes taking place in spinal neurons but also from the possible enhancement of the descending motor drive through residual corticospinal projections (or other descending systems), and via the interneurons that form the propriospinal system forming intersegmental connections ipsilaterally and bilaterally. However, this benefit might be limited, particularly below the injury, by the amount of spared brain-to-spine connections.

Activity-based neural plasticity mechanisms involve both physiological (functional modification of existing synapses and neurones) and structural changes that alter the anatomical connectivity of neurones (circuit reorganization by means of formation, removal, and morphological remodeling of synapses, dendritic spines, and even neurites) (reviewed in [50]). In the present study, signs of neural plasticity were observed after a single training and/or stimulation session, suggesting that short-term plastic changes were taking place. These may primarily affect intrinsic properties of neurons to depolarize and generate an action potential, including, for instance, modulation or relocation of ion channels and surface receptors. Sustained and/or repeated exposure to this activity-dependent plasticity (based either on physical training, electrical stimulation, or both) may lead these short-term changes to promote occurrence of long-term plasticity involving mechanisms, such as long-term potentiation and depression, morphological changes of dendrites, synaptogenesis, and axonal branching/regeneration [20]. Based on our findings, rehabilitative strategies involving the simultaneous application of physical training and spinal electrical neuromodulation may have more chances to induce long-term plastic adaptations than either intervention alone.

Limitations

We acknowledge diverse limitations present in the current study: (i) sample size was relatively small, which might compromise statistical power and subsequent interpretation of the results. Indeed, only a third of all screened candidates finally participated in the study (see Figure 1), and this was in great part due to candidate refusal to participate, probably because of concerns about discomfort associated to cervical stimulation, possible side effects of the intervention and limited availability of participants (each testing session required around 4 h to be completed). (ii) The duration of the testing sessions also resulted exhausting for participants, limiting the number of replicates that could be recorded for some neurophysiological parameters (recruitment curve, cMEP, SICI). This might increase variability in the observed data as well. (iii) Since many parameters were measured, it is possible that testing itself influenced those measured in the last place. Moreover, care should be taken when interpreting temporal patterns observed at $0'$ and $60'$ post-intervention, which may correspond to longer timepoints in fact since complete functional and neurophysiological assessment lasted 30–45 min. (iv) We did not assess afferent processing. For this reason, we cannot conclude about the possible roles of altered afferent pathways on the observed results.

5. Conclusions

As a summary, eEmc combined with hand training enhanced hand motor output in healthy subjects as observed in power grip performance when compared to hand training or eEmc applied separately. This effect was associated to increased corticospinal excitability that was due, at least in part, by plastic changes induced at spinal level and possibly at cortical level. These changes may also possibly reflect prevention of fatigue-associated alterations induced by physical training. Therefore, eEmc combined with use-dependent interventions should be considered when designing rehabilitation protocols for restoring motor performance following stroke, SCI, or other neurological affections compromising motor function.

Supplementary Materials: The following are available online at https://www.mdpi.com/2077-0383/10/2/195/s1, Table S1: Slopes of recruitment curves of tSCS-elicited spinal MEPs for each analysed muscle, intervention and timepoint, Table S2: Slopes of recruitment curves of TMS-elicited cortical MEPs for each analysed muscle, intervention and timepoint.

Author Contributions: Conceptualization, H.K., Á.F., G.G.-A., J.V.; methodology, H.K., Á.F., M.R.-C., L.G., J.B.-P., G.G.-A., J.V.; software, M.R.-C.; formal analysis, H.K., Á.F., M.R.-C.; validation, H.K., Á.F., M.R.-C., G.G.-A., J.V.; investigation, H.K., Á.F., M.R.-C., L.G., G.G.-A., J.V.; resources, H.K., M.R.-C., J.B.-P., G.G.-A., J.V.; data curation, H.K., Á.F., M.R.-C., L.G.; writing—original draft preparation, H.K., Á.F.; writing—review and editing, H.K., Á.F., M.R.-C., V.R.E., L.G., J.B.-P., X.N., Y.G., G.G.-A., J.V.; visualization, H.K., Á.F., G.G.-A.; supervision, H.K., V.R.E., J.B.-P., X.N., Y.G., G.G.-A., J.V.; project administration, H.K., J.B.-P., G.G.-A., J.V.; funding acquisition, H.K., Y.G., G.G.-A., J.V. All authors have read and agreed to the published version of the manuscript.

Funding: This research was funded by H2020-ERA-NET Neuron (AC16/00034) to J.V.; AES 2019 ISCIII (PI19/01680) to J.V.; Fundació La Marató de TV3 2017 (201713.31) to G.G.-A.; Premi Beca "Mike Lane" 2019-Castellers de la Vila de Gràcia to H.K.; and National Institutes of Health Grant 1R01 NS102920-01A1 to Y.G. The APC was funded by Fundació La Marató de TV3 2017 (201713.31).

Institutional Review Board Statement: The study was conducted according to the guidelines of the Declaration of Helsinki, and approved by the Ethics Committee of Institut Guttmann (protocol code 2017255 and date of approval: 29/04/2017).

Informed Consent Statement: Informed consent was obtained from all subjects involved in the study.

Data Availability Statement: The data presented in this study are available on request from the corresponding author. The data are not publicly available due to privacy restrictions.

Conflicts of Interest: V.R.E. and Y.G. hold shareholder interest in NeuroRecovery Technologies and hold certain inventorship rights on intellectual property licensed by The Regents of the University of California to NeuroRecovery Technologies and its subsidiaries. Y.G. also holds shareholder interest in Cosyma Inc., St. Petersburg, Russia. The funders had no role in the design of the study; in the collection, analyses, or interpretation of data; in the writing of the manuscript, or in the decision to publish the results.

References

1. Moritz, C. A giant step for spinal cord injury research. *Nat. Neurosci.* **2018**, *21*, 1647–1648. [CrossRef] [PubMed]
2. Harkema, S.; Gerasimenko, Y.; Hodes, J.; Burdick, J.; Angeli, C.; Chen, Y.; Ferreira, C.; Willhite, A.; Rejc, E.; Grossman, R.G.; et al. Effect of epidural stimulation of the lumbosacral spinal cord on voluntary movement, standing, and assisted stepping after motor complete paraplegia: A case study. *Lancet* **2011**, *377*, 1938–1947. [CrossRef]
3. Minassian, K.; Hofstoetter, U.S.; Danner, S.M.; Mayr, W.; Bruce, J.A.; McKay, W.B.; Tansey, K.E. Spinal Rhythm Generation by Step-Induced Feedback and Transcutaneous Posterior Root Stimulation in Complete Spinal Cord–Injured Individuals. *Neurorehabil. Neural Repair* **2016**, *30*, 233–243. [CrossRef] [PubMed]
4. Angeli, C.A.; Boakye, M.; Morton, R.A.; Vogt, J.; Benton, K.; Chen, Y.; Ferreira, C.K.; Harkema, S.J. Recovery of Over-Ground Walking after Chronic Motor Complete Spinal Cord Injury. *N. Engl. J. Med.* **2018**, *379*, 1244–1250. [CrossRef] [PubMed]
5. Gill, M.L.; Grahn, P.J.; Calvert, J.S.; Linde, M.B.; Lavrov, I.A.; Strommen, J.A.; Beck, L.A.; Sayenko, D.G.; Van Straaten, M.G.; Drubach, D.I.; et al. Neuromodulation of lumbosacral spinal networks enables independent stepping after complete paraplegia. *Nat. Med.* **2018**, *24*, 1677–1682. [CrossRef]
6. Wagner, F.B.; Mignardot, J.-B.; Le Goff-Mignardot, C.G.; Demesmaeker, R.; Komi, S.; Capogrosso, M.; Rowald, A.; Seáñez, I.; Caban, M.; Pirondini, E.; et al. Targeted neurotechnology restores walking in humans with spinal cord injury. *Nature* **2018**, *563*, 65–71. [CrossRef]
7. Lu, D.C.; Edgerton, V.R.; Modaber, M.; AuYong, N.; Morikawa, E.; Zdunowski, S.; Sarino, M.E.; Sarrafzadeh, M.; Nuwer, M.R.; Roy, R.R.; et al. Engaging Cervical Spinal Cord Networks to Reenable Volitional Control of Hand Function in Tetraplegic Patients. *Neurorehabil. Neural Repair* **2016**, *30*, 951–962. [CrossRef]
8. Hofstoetter, U.S.; Hofer, C.; Kern, H.; Danner, S.M.; Mayr, W.; Dimitrijevic, M.R.; Minassian, K. Effects of transcutaneous spinal cord stimulation on voluntary locomotor activity in an incomplete spinal cord injured individual. *Biomed. Tech. (Berl.)* **2013**, *58* (Suppl. 1). [CrossRef]
9. Gerasimenko, Y.P.; Lu, D.C.; Modaber, M.; Zdunowski, S.; Gad, P.; Sayenko, D.G.; Morikawa, E.; Haakana, P.; Ferguson, A.R.; Roy, R.R.; et al. Noninvasive Reactivation of Motor Descending Control after Paralysis. *J. Neurotrauma* **2015**, *32*, 1968–1980. [CrossRef]
10. Sayenko, D.G.; Rath, M.; Ferguson, A.R.; Burdick, J.W.; Havton, L.A.; Edgerton, V.R.; Gerasimenko, Y.P. Self-Assisted Standing Enabled by Non-Invasive Spinal Stimulation after Spinal Cord Injury. *J. Neurotrauma* **2019**, *36*, 1435–1450. [CrossRef]
11. Freyvert, Y.; Yong, N.A.; Morikawa, E.; Zdunowski, S.; Sarino, M.E.; Gerasimenko, Y.; Edgerton, V.R.; Lu, D.C. Engaging cervical spinal circuitry with non-invasive spinal stimulation and buspirone to restore hand function in chronic motor complete patients. *Sci. Rep.* **2018**, *8*, 15546. [CrossRef] [PubMed]
12. Gad, P.; Lee, S.; Terrafranca, N.; Zhong, H.; Turner, A.; Gerasimenko, Y.; Edgerton, V.R. Non-Invasive Activation of Cervical Spinal Networks after Severe Paralysis. *J. Neurotrauma* **2018**, *35*, 2145–2158. [CrossRef]
13. Inanici, F.; Samejima, S.; Gad, P.; Edgerton, V.R.; Hofstetter, C.P.; Moritz, C.T. Transcutaneous Electrical Spinal Stimulation Promotes Long-Term Recovery of Upper Extremity Function in Chronic Tetraplegia. *IEEE Trans. Neural Syst. Rehabil. Eng.* **2018**, *26*, 1272–1278. [CrossRef] [PubMed]
14. Benavides, F.D.; Jo, H.J.; Lundell, H.; Edgerton, V.R.; Gerasimenko, Y.; Perez, M.A. Cortical and subcortical effects of transcutaneous spinal cord stimulation in humans with tetraplegia. *J. Neurosci.* **2020**, *40*, 2633–2643. [CrossRef] [PubMed]
15. Gorodnichev, R.M.; Pivovarova, E.A.; Puhov, A.; Moiseev, S.A.; Savochin, A.A.; Moshonkina, T.R.; Chsherbakova, N.A.; Kilimnik, V.A.; Selionov, V.A.; Kozlovskaya, I.B.; et al. Transcutaneous electrical stimulation of the spinal cord: A noninvasive tool for the activation of stepping pattern generators in humans. *Hum. Physiol.* **2012**, *38*, 158–167. [CrossRef]
16. Gerasimenko, Y.; Gorodnichev, R.; Puhov, A.; Moshonkina, T.; Savochin, A.; Selionov, V.; Roy, R.R.; Lu, D.C.; Edgerton, V.R. Initiation and modulation of locomotor circuitry output with multisite transcutaneous electrical stimulation of the spinal cord in noninjured humans. *J. Neurophysiol.* **2015**, *113*, 834–842. [CrossRef]
17. Gerasimenko, Y.; Gorodnichev, R.; Moshonkina, T.; Sayenko, D.; Gad, P.; Edgerton, V.R. Transcutaneous electrical spinal-cord stimulation in humans. *Ann. Phys. Rehabil. Med.* **2015**, *58*, 225–231. [CrossRef]
18. Behrman, A.L.; Ardolino, E.M.; Harkema, S.J. Activity-Based Therapy: From Basic Science to Clinical Application for Recovery after Spinal Cord Injury. In Proceedings of the Journal of Neurologic Physical Therapy; Lippincott Williams and Wilkins: Philadelphia, PA, USA, 2017; Volume 41, pp. S39–S45.

19. Taccola, G.; Sayenko, D.; Gad, P.; Gerasimenko, Y.; Edgerton, V.R. And yet it moves: Recovery of volitional control after spinal cord injury. *Prog. Neurobiol.* **2018**, *160*, 64–81. [CrossRef]

20. Knikou, M. Neural control of locomotion and training-induced plasticity after spinal and cerebral lesions. *Clin. Neurophysiol.* **2010**, *121*, 1655–1668. [CrossRef]

21. World Medical Association. World Medical Association World Medical Association Declaration of Helsinki. *JAMA* **2013**, *310*, 2191–2194. [CrossRef]

22. Mathiowetz, V.; Volland, G.; Kashman, N.; Weber, K. Adult Norms for the Box and Block Test of Manual Dexterity. *Am. J. Occup. Ther.* **1985**, *39*, 386–391. [CrossRef] [PubMed]

23. Thomas, C.K.; Johansson, R.S.; Bigland-Ritchie, B. Incidence of F waves in single human thenar motor units. *Muscle Nerve* **2002**, *25*, 77–82. [CrossRef] [PubMed]

24. Long, J.; Federico, P.; Perez, M.A. A novel cortical target to enhance hand motor output in humans with spinal cord injury. *Brain* **2017**, *140*, 1619–1632. [CrossRef] [PubMed]

25. Weber, F. The diagnostic sensitivity of different F wave parameters. *J. Neurol. Neurosurg. Psychiatry* **1998**, *65*, 535–540. [CrossRef] [PubMed]

26. Sayenko, D.G.; Atkinson, D.A.; Dy, C.J.; Gurley, K.M.; Smith, V.L.; Angeli, C.; Harkema, S.J.; Edgerton, V.R.; Gerasimenko, Y.P. Spinal segment-specific transcutaneous stimulation differentially shapes activation pattern among motor pools in humans. *J. Appl. Physiol.* **2015**, *118*, 1364–1374. [CrossRef]

27. Kujirai, T.; Caramia, M.D.; Rothwell, J.C.; Day, B.L.; Thompson, P.D.; Ferbert, A.; Wroe, S.; Asselman, P.; Marsden, C.D. Corticocortical inhibition in human motor cortex. *J. Physiol.* **1993**, *471*, 501–519. [CrossRef]

28. Gad, P.N.; Kreydin, E.; Zhong, H.; Latack, K.; Edgerton, V.R. Non-invasive Neuromodulation of Spinal Cord Restores Lower Urinary Tract Function after Paralysis. *Front. Neurosci.* **2018**, *12*, 432. [CrossRef]

29. Schättin, A.; Gennaro, F.; Egloff, M.; Vogt, S.; de Bruin, E.D. Physical activity, nutrition, cognition, neurophysiology, and short-time synaptic plasticity in healthy older adults: A cross-sectional study. *Front. Aging Neurosci.* **2018**, *10*. [CrossRef]

30. Chakravarthy, K.; Richter, H.; Christo, P.J.; Williams, K.; Guan, Y. Spinal Cord Stimulation for Treating Chronic Pain: Reviewing Preclinical and Clinical Data on Paresthesia-Free High-Frequency Therapy. *Neuromodulation* **2018**, *21*, 10–18. [CrossRef]

31. Gandevia, S.C. Spinal and Supraspinal Factors in Human Muscle Fatigue. *Physiol. Rev.* **2001**, *81*, 1725–1789. [CrossRef]

32. Burke, R.E.; Edgerton, V.R. Motor unit properties and selective involvement in movement. *Exerc. Sport Sci. Rev.* **1975**, *3*, 31–82. [CrossRef] [PubMed]

33. Espiritu, M.G.; Lin, C.S.-Y.; Burke, D. Motoneuron excitability and the F wave. *Muscle Nerve* **2003**, *27*, 720–727. [CrossRef] [PubMed]

34. Yates, S.K.; Brown, W.F. Characteristics of the F response: A single motor unit study. *J. Neurol. Neurosurg. Psychiatry* **1979**, *42*, 161–170. [CrossRef]

35. Mesrati, F.; Vecchierini, M.F. F-waves: Neurophysiology and clinical value. *Neurophysiol. Clin.* **2004**, *34*, 217–243. [CrossRef]

36. Sabbahi, M.A.; Sengul, Y.S. Cervical multisegmental motor responses in healthy subjects. *Spinal Cord* **2012**, *50*, 432–439. [CrossRef]

37. Milosevic, M.; Masugi, Y.; Sasaki, A.; Sayenko, D.G.; Nakazawa, K. On the reflex mechanisms of cervical transcutaneous spinal cord stimulation in human subjects. *J. Neurophysiol.* **2019**, *121*, 1672–1679. [CrossRef]

38. Roy, F.D.; Gibson, G.; Stein, R.B. Effect of percutaneous stimulation at different spinal levels on the activation of sensory and motor roots. *Exp. Brain Res.* **2012**, *223*, 281–289. [CrossRef]

39. Minassian, K.; Persy, I.; Rattay, F.; Dimitrijevic, M.R.; Hofer, C.; Kern, H. Posterior root–muscle reflexes elicited by transcutaneous stimulation of the human lumbosacral cord. *Muscle Nerve* **2007**, *35*, 327–336. [CrossRef]

40. Pitcher, J.B.; Miles, T.S. Alterations in corticospinal excitability with imposed vs. voluntary fatigue in human hand muscles. *J. Appl. Physiol.* **2002**, *92*, 2131–2138. [CrossRef]

41. Bestmann, S.; Krakauer, J.W. The uses and interpretations of the motor-evoked potential for understanding behaviour. *Exp. Brain Res.* **2015**, *233*, 679–689. [CrossRef]

42. Taylor, J.L.; Butler, J.E.; Gandevia, S.C. Changes in muscle afferents, motoneurons and motor drive during muscle fatigue. *Eur. J. Appl. Physiol.* **2000**, *83*, 106–115. [CrossRef] [PubMed]

43. Hamdy, S.; Rothwell, J.C.; Aziz, Q.; Singh, K.D.; Thompson, D.G. Long-term reorganization of human motor cortex driven by short-term sensory stimulation. *Nat. Neurosci.* **1998**, *1*, 64–68. [CrossRef]

44. Di Lazzaro, V.; Oliviero, A.; Profice, P.; Pennisi, M.A.; Di Giovanni, S.; Zito, G.; Tonali, P.; Rothwell, J.C. Muscarinic receptor blockade has differential effects on the excitability of intracortical circuits in the human motor cortex. *Exp. Brain Res.* **2000**, *135*, 455–461. [CrossRef] [PubMed]

45. Hunter, S.K.; McNeil, C.J.; Butler, J.E.; Gandevia, S.C.; Taylor, J.L. Short-interval cortical inhibition and intracortical facilitation during submaximal voluntary contractions changes with fatigue. *Exp. Brain Res.* **2016**, *234*, 2541–2551. [CrossRef] [PubMed]

46. Maruyama, A.; Matsunaga, K.; Tanaka, N.; Rothwell, J.C. Muscle fatigue decreases short-interval intracortical inhibition after exhaustive intermittent tasks. *Clin. Neurophysiol.* **2006**, *117*, 864–870. [CrossRef] [PubMed]

47. Buccolieri, A.; Abbruzzese, G.; Rothwell, J.C. Relaxation from a voluntary contraction is preceded by increased excitability of motor cortical inhibitory circuits. *J. Physiol.* **2004**, *558*, 685–695. [CrossRef]
48. Kakulas, B.A.; Kaelan, C. The neuropathological foundations for the restorative neurology of spinal cord injury. *Clin. Neurol. Neurosurg.* **2015**, *129* (Suppl. 1), S1–S7. [CrossRef]
49. Yu, K.; Rong, W.; Li, J.; Jia, L.; Yuan, W.; Yie, X.; Shi, Z. Neurophysiological evidence of spared upper motor conduction fibers in clinically complete spinal cord injury: Discomplete SCI in rats. *J. Neurol. Sci.* **2001**, *189*, 23–36. [CrossRef]
50. Smith, A.C.; Knikou, M. A Review on Locomotor Training after Spinal Cord Injury: Reorganization of Spinal Neuronal Circuits and Recovery of Motor Function. *Neural Plast.* **2016**, *2016*. [CrossRef]

Journal of
Clinical Medicine

Review

Neural Substrates of Transcutaneous Spinal Cord Stimulation: Neuromodulation across Multiple Segments of the Spinal Cord

Trevor S. Barss [1,2,3], Behdad Parhizi [1,2,3], Jane Porter [2,3] and Vivian K. Mushahwar [1,2,3,*]

1   Neuroscience and Mental Health Institute, University of Alberta, Edmonton, AB T6G 2R3, Canada; tbarss@ualberta.ca (T.S.B.); parhizi@ualberta.ca (B.P.)
2   Division of Physical Medicine and Rehabilitation, Department of Medicine, University of Alberta, Edmonton, AB T6G 2R3, Canada; jporter@ualberta.ca
3   Sensory Motor Adaptive Rehabilitation Technology (SMART) Network, University of Alberta, Edmonton, AB T6G 2R3, Canada
*   Correspondence: vivian.mushahwar@ualberta.ca

Abstract: Transcutaneous spinal cord stimulation (tSCS) has the potential to promote improved sensorimotor rehabilitation by modulating the circuitry of the spinal cord non-invasively. Little is currently known about how cervical or lumbar tSCS influences the excitability of spinal and corticospinal networks, or whether the synergistic effects of multi-segmental tSCS occur between remote segments of the spinal cord. The aim of this review is to describe the emergence and development of tSCS as a novel method to modulate the spinal cord, while highlighting the effectiveness of tSCS in improving sensorimotor recovery after spinal cord injury. This review underscores the ability of single-site tSCS to alter excitability across multiple segments of the spinal cord, while multiple sites of tSCS converge to facilitate spinal reflex and corticospinal networks. Finally, the potential and current limitations for engaging cervical and lumbar spinal cord networks through tSCS to enhance the effectiveness of rehabilitation interventions are discussed. Further mechanistic work is needed in order to optimize targeted rehabilitation strategies and improve clinical outcomes.

Keywords: neuromodulation; interlimb coordination; rehabilitation; neurophysiology; Hoffmann (H)-reflex; motor-evoked potential; locomotion; spinal cord injury

Citation: Barss, T.S.; Parhizi, B.; Porter, J.; Mushahwar, V.K. Neural Substrates of Transcutaneous Spinal Cord Stimulation: Neuromodulation across Multiple Segments of the Spinal Cord. *J. Clin. Med.* **2022**, *11*, 639. https://doi.org/10.3390/jcm11030639

Academic Editors: Ursula S. Hofstoetter and Karen Minassian

Received: 19 November 2021
Accepted: 18 January 2022
Published: 27 January 2022

Publisher's Note: MDPI stays neutral with regard to jurisdictional claims in published maps and institutional affiliations.

1. Introduction

Neuromodulation of the spinal cord by means of non-invasive transcutaneous (tSCS) and implanted epidural (eSCS) spinal cord stimulation may improve sensorimotor rehabilitation after spinal cord injury (SCI) [1–4]. However, developing an optimal treatment approach requires taking advantage of the intrinsic ability of the spinal circuits by facilitating preserved sensorimotor pathways that could drive spinal plasticity [5]. The influence of spinal cord stimulation (SCS) does not necessarily depend on the nature of the neurological disorder, but on the operational and functional status of residual neural networks [6]. Epidural SCS has been shown to modulate neuronal circuits in persons with motor-complete SCI, including corticospinal, [7–9] propriospinal [10,11], and corticoreticulospinal [12] tracts. The resulting neuroplasticity is thought to improve spinal motor output and volitional movements even in cases of severely reduced supraspinal input, without negatively impacting residual motor function [3,13–17]. Most recently, eSCS applied to the lumbar spinal cord, in conjunction with intensive locomotor training, enabled persons with clinically motor-complete SCI to walk over ground for short distances [4,13,18]. This demonstrates that dormant neurons that survive the injury may be reengaged with spinal neuromodulation, and can produce stepping-like movements [19,20].

While eSCS has important implications for rehabilitation after SCI, its invasive nature, high cost, and limited accessibility are limitations for rapid translation to a broad population. Transcutaneous SCS is a non-invasive, accessible, and cost-effective alternative that is thought to be a safe assistive technology with important implications for both furthering our understanding of the mechanisms controlling locomotion, and for rehabilitating sensorimotor function after SCI [21–23]. It has been suggested that tSCS of the lumbar spinal cord may activate similar spinal circuitry to eSCS [24–26]; if accurate, tSCS is likely to enhance functional recovery in a similar manner to eSCS when paired with rehabilitation strategies. This would also allow for the tSCS to build on the foundation of knowledge of the intrinsic circuitry recruited by eSCS. In case studies and small clinical trials, tSCS improved hand and arm function [2,27–29], produced locomotor-like stepping [1,30], and improved walking function [22,31,32] in participants with neurological deficits including incomplete and complete SCI, stroke, and cerebral palsy. Evidence suggests that tSCS may also be used as a viable alternative to pharmacological anti-spasticity approaches, altering the excitability of spinal pathways and possibly augmenting pre- and post-synaptic inhibitory mechanisms [33,34]. Understanding the impact that tSCS has on spinal cord circuitry is vital to ensuring that the stimulation is applied at therapeutically appropriate sites, and that the parameters of stimulation are chosen so as to optimize the desired rehabilitative effects.

It is critical to realize that not all of the studies using tSCS follow the same pattern of stimulation. Transcutaneous SCS patterns including single pulses, trains of pulses, and waveforms with and without carrier frequencies have been used. The present review focuses on the use of alternating current (AC) tSCS, because most studies aimed at improving functional recovery after SCI have used this type of stimulation. Direct current (DC) tSCS also modulates spinal excitability, and may be another promising and novel tool to pair with activity-based interventions [35,36]; however, this technique is beyond the scope of this review, and requires further research in order to determine the specific mechanisms involved. In this review, the two common patterns of AC tSCS that have been employed to date will be included and discussed in detail. The first pattern, which will be referred to as unmodulated tSCS, does not include a carrier frequency, and is generally composed of rectangular pulses delivered as single individual pulses, or in trains of 1–90 Hz frequency. The second stimulation pattern, which will be referred to as modulated tSCS, includes rectangular pulses with a carrier frequency of 2.5–10 kHz, delivered at a rate of 5–40 Hz [37]. While both patterns have been reported to modulate neural circuitry across the central nervous system and produce functional outcomes, it is unlikely that they share identical mechanisms of action. The fundamental differences between the two patterns will be highlighted throughout this review as the different studies are discussed.

The aims of this review are as follows: first, to identify the parameters and the potential underlying mechanisms that allow tSCS to facilitate ongoing motor output; secondly, to highlight the effects of tSCS on excitability across multiple segments of the spinal cord; thirdly, to address the ability of multiple sites of tSCS to converge and enhance modulation of spinal reflex and corticospinal pathways; and finally, to explore the potential and limitations for engaging cervical and lumbar spinal cord networks through tSCS to enhance the effectiveness of rehabilitation interventions. This review will also underscore the need for further mechanistic work to optimize tSCS parameters that, when paired with targeted rehabilitation strategies, can effectively improve clinical outcomes.

2. Historical Perspective

The use of electricity for neuromodulation has a storied history, ultimately leading to a variety of therapeutic electrical stimulation techniques that target spinal networks, including tSCS, eSCS, and intraspinal microstimulation (ISMS) [38,39]. Epidural SCS initially emerged in the pain literature in 1967 [40], and is currently most commonly used for the treatment of intractable chronic pain; while originally designed to alleviate pain, it was used in 1971 as a method for facilitating motor control in persons with multiple sclerosis [41], and to reduce spasticity after incomplete SCI [42–44]. In 1979, tonic stimulation of dorsal roots of the spinal cord was shown to generate locomotion in low-spinal cats [45,46]. This work then led to initial investigations demonstrating improved stepping in humans, and providing the potential for this technology to be used as a translational tool to facilitate improved function after neural injury [3,15,47,48].

In humans, eSCS involves implanting electrodes over the dura mater encasing the lumbosacral segments of the spinal cord. Dorsal root fibers are the first to be recruited, with the lowest thresholds, while the ventral root fibers are the least accessible [49]. This recruitment leads to the activation of motor neurons through monosynaptic and polysynaptic proprioceptive circuits, and increases the overall excitability of the spinal cord, allowing for greater responsiveness of spinal circuits to descending signals and sensory feedback [14]. Extensive evidence from animal studies has led to the hypothesis that electrically stimulating the human spinal cord through the epidural space can facilitate improvements in motor function.

Transcutaneous SCS was inspired by high-voltage percutaneous electrical stimulation over the lumbosacral spinal column to activate peripheral motor axons [50]. In 1997, the generation of locomotor-like activity with the application of tSCS over the lumbar enlargement was demonstrated in individuals with SCI [51]. It was then suggested that there are low-threshold sites in the posterior structure of the human lumbosacral cord that could be accessed from the surface [49]. In 2007, and encouraged by earlier discoveries, Minassian et al. revealed that posterior root afferents can be accessed by tSCS with single pulses (unmodulated), and they reported monosynaptic reflex responses in multiple muscles of the legs [52]. Later, it was shown that unmodulated tSCS can enhance voluntary locomotor-like electromyographic (EMG) activity [53] and modify spasticity in individuals with incomplete SCI [54]. In 2015, tSCS was used with a novel waveform that included a carrier frequency (i.e., modulated) to activate spinal networks while reducing the perception of pain associated with the necessarily high stimulus amplitudes [22,30]. The tSCS parameters were based on a previous finding that a 10 kHz carrier frequency of transcutaneous stimulation reduces the likelihood of activating pain fibers [55]. Building on these exciting initial investigations, the tSCS literature has incorporated a diverse set of stimulation parameters that are vital to understand, as they may have important implications for improving function in persons experiencing sensorimotor impairments due to neurological conditions.

3. Properties of Transcutaneous Spinal Cord Stimulation (tSCS)

3.1. Parameters of tSCS

Typically, tSCS is applied through circular adhesive electrodes of 2–3 cm diameter that are placed on the skin overlying the lumbar or cervical segments of the spinal cord (Figure 1). Optimal placement of electrodes is dependent on the individual symptoms, desired rehabilitation outcomes, and paired rehabilitation strategies, on a case-by-case basis. When targeting the lower extremities, the most common cathode placement is over the T11–T12 and/or L1–L2 spinous processes, while C6–C7 or C7–T1 is the most common placement for the upper extremities [37]. The anode electrodes are placed either over the iliac crests or the anterior superior iliac spine [37].

In addition to electrode placement, it is important to consider the waveform characteristics of the applied current for maximal therapeutic outcomes [5,17,21]. With unmodulated tSCS, which evolved from the eSCS literature, rectangular mono- or biphasic pulses of 0.4–2 ms duration are typically delivered at a frequency range of 1–90 Hz and stimulation intensity of up to 170 mA [21,37]. On the other hand, the novelty of the modulated stimulation pattern comes from its unique waveform, which includes a carrier frequency of up to 10 kHz within a given pulse. Such high-frequency stimulation approaches were originally used to reduce the perception of pain during transcutaneous nerve stimulation [55]. The waveform in the modulated stimulation pattern generally consists of 0.3–1 ms long rectangular biphasic or monophasic pulses that repeat at a frequency of 5–40 Hz. Each of these pulses encompasses a carrier frequency of 2.5–10 kHz, aimed at suppressing the user's perceived pain and, thus, allowing for greater current amplitudes to be employed. The amplitude of the current for modulated tSCS is similar to that of unmodulated tSCS, and ranges from 30 to 180 mA, depending on the stimulation site and the desired outcome. In neurologically intact participants, the intensity of modulated tSCS (with a 5 kHz carrier frequency) allows for maximal tolerable current amplitudes of 103 mA, while unmodulated tSCS has maximal tolerable amplitudes of 39 mA. However, when considering maximal tolerable stimulation with respect to the stimulation levels needed to evoke motor responses, tSCS with a carrier frequency was no different than unmodulated tSCS in reducing the perception of pain [56].

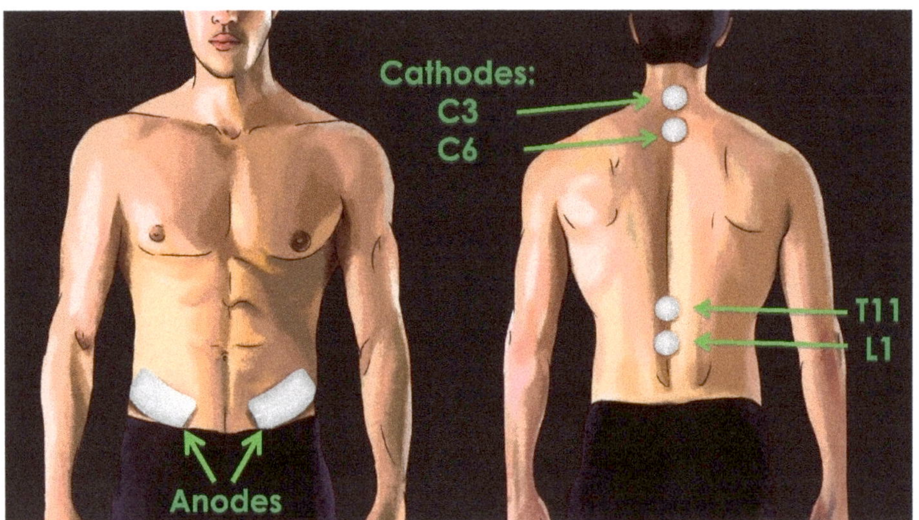

Figure 1. Typical tSCS electrode placement: Transcutaneous SCS is commonly delivered via two 2.5 cm round cathodic electrodes placed over the C3–4 and C6–7 (cervical) or T11 and L1 (lumbar) spinous processes. Two 5 × 10 cm rectangular anodic electrodes are placed bilaterally over the iliac crests.

Interestingly, when using an array of electrodes and adjusting the parameters of stimulation—including intensity and location—different patterns of independent and coordinated upper limb motion at both distal and proximal joints have been elicited, showing the potential of tSCS without a carrier frequency to evoke functional movements [28]. Therefore, the chosen parameters of tSCS can have a meaningful effect on the recruited circuitry and the functional movements that are facilitated or inhibited. Understanding how the applied electrical current is integrated into the spinal circuitry is vital.

3.2. Current Flow Involved in tSCS

The current flow and electrical potential generated by eSCS and tSCS are markedly different [24,25]. With eSCS, 80–90% of the ionic current flows between the active electrodes through the cerebrospinal fluid [57]. In tSCS, the current flow is strongly influenced by the electrical properties of the numerous conductivity boundaries of body tissues (e.g., skin, fat, muscle, and bone), with computer simulations estimating that only ~8% of the overall current flows through the cerebrospinal fluid [25]. With the dramatic difference in current flow and the proximity of neural structures to the electrodes between eSCS and tSCS, both the selectivity of spinal circuitry and the required stimulation intensity are dissimilar. Modelling studies suggest that the superficially located large-diameter posterior column fibers with multiple collaterals have a threshold three times higher than that of posterior root fibers [26]. For both tSCS and eSCS, large-diameter proprioceptive sensory fibers within the posterior rootlets/roots have the lowest thresholds of all neural structures within the vertebral canal [24], making it unlikely that the effects of SCS arise exclusively from dorsal column stimulation [57]. Computer modeling indicates that action potentials generated by tSCS are initiated in the posterior root fibers at their entry into the spinal cord, or along the longitudinal portions of the afferent fiber trajectories, depending on the cathode position [25]. Evidence suggests that the reflex nature of unmodulated tSCS exploits the difference in the strength–duration properties of sensory and motor axons; however, future research should be conducted to explore how modulated tSCS generates action potentials in neural tissue [58]. At stimulation intensities that result in the recruitment of posterior column axons, co-activation of posterior root fibers of large and small diameters is observed, demonstrating the substantial differences in the thresholds of activation of various components of the spinal cord [26]. Moreover, increasing stimulation intensity engages spinal interneurons via synaptic projections which, in turn, activate motor neurons [22,59]. These simulation results provide a biophysical explanation for the electrophysiological findings of lower limb muscle responses that are induced by posterior root stimulation (Figure 2A). However, it should be noted that these computer simulation studies have all applied unmodulated tSCS (i.e., without carrier frequency), and the results may not necessarily be generalizable to other types of pulses. Understanding the potential unique properties associated with modulated tSCS is vital for implementing tSCS in a manner that optimizes functional recovery after neural injury or disease. Similar simulation studies using high carrier frequencies are necessary in order to extend the knowledge regarding current flow in tSCS.

3.3. Transcutaneous SCS Carrier Frequency Is Important for Reducing Discomfort, but Its Role in Restoring Motor Function Remains Unclear

The inclusion of a carrier frequency within a given stimulation pulse is used for its ability to disrupt synchronous firing of the high-threshold C-fibers related to pain perception [60]. Pain management through SCS is based on the gate control theory introduced in 1965 [57], which proposed that the activation of Aβ mechanoreceptor fibers that synapse onto a range of neurons within the dorsal horn that release inhibitory neurotransmitters—including γ-amino butyric acid (GABA) and adenosine [61]—reduces the activity of nociceptive projection neurons in laminae I and V traveling along the spinothalamic tract. It has also been proposed that high-frequency stimulation of the spinal cord blocks discomfort by inactivating paresthesia-inducing large-diameter fibers and activating medium–small-diameter fibers that suppress wide-dynamic-range neurons encoding neuropathic pain [62]. Sub-perception SCS at 1 kHz was more effective for pain relief compared to low-frequency supra-perception stimulation [63]. Moreover, a recent eSCS study suggested that there was no observable difference between 1 kHz and 10 kHz stimulation for the relief of back pain [64]. Charge per pulse is lower in high-frequency eSCS in comparison with low-frequency stimulation, while charge per second is higher [61]. While these studies did not use tSCS, and were only aimed at pain management, they can play an important role in explaining the potential mechanisms that reduce discomfort in modulated tSCS. Manson et al. have recently shown that the maximal tolerable stimulation intensity is significantly greater during modulated tSCS compared to unmodulated tSCS [56]; however, the stimulation intensity required to evoke a muscle response (motor threshold) was correspondingly higher with a carrier frequency, leading to no difference in the relative current amplitude required to evoke a motor response [56]. This study indicated that the addition of a carrier frequency reduces discomfort for a given current amplitude compared to unmodulated tSCS, but does not reduce discomfort when evoking the same motor response.

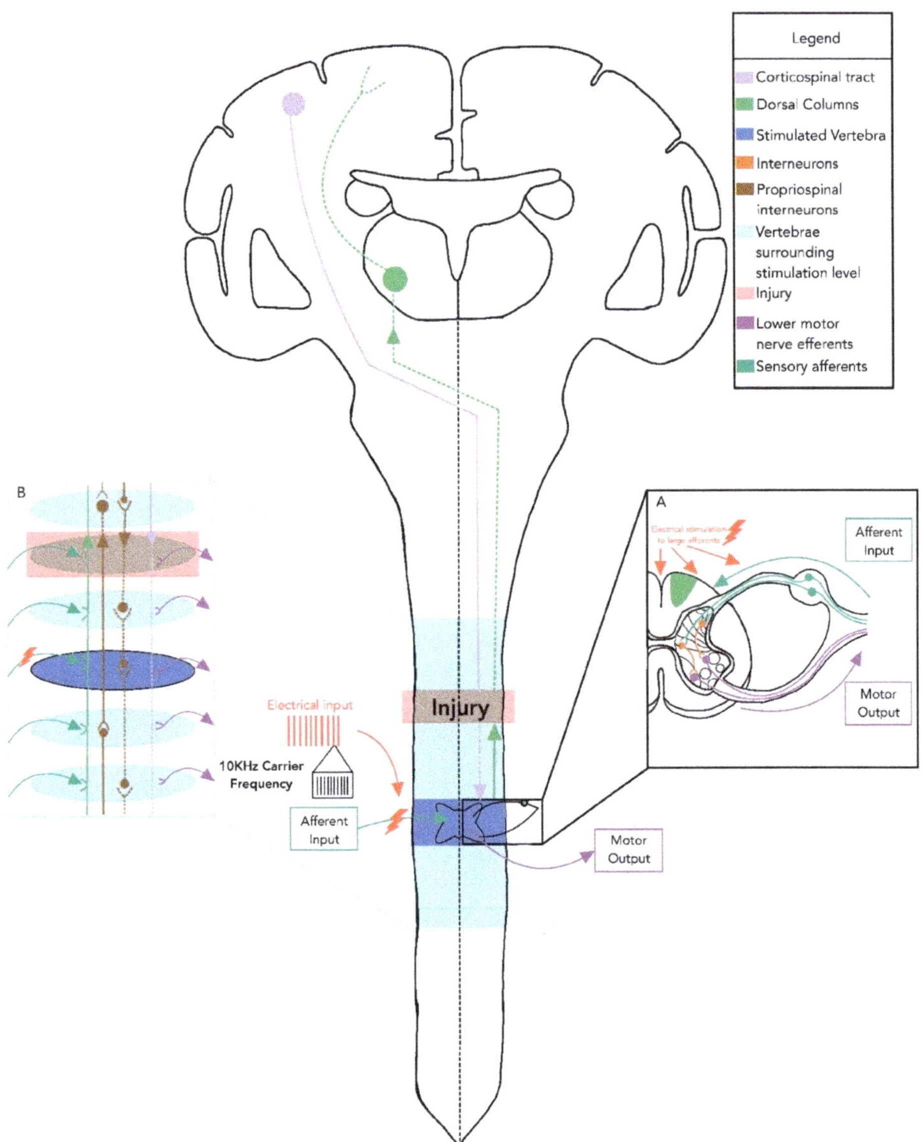

Figure 2. Schematic of networks within the spinal cord that are potentially altered with tSCS: The main figure highlights the ability of tSCS to modulate ongoing motor output through dorsal root afferents that trans-synaptically facilitate motor output by bringing previously inaccessible motor units closer to their threshold, allowing them to contribute to the execution of a desired task. (**A**) Large-diameter afferents are likely activated and synapse on several types of interneurons that facilitate ongoing motor output. (**B**) Among these interneurons are propriospinal interneurons, which transmit this input to multiple segments of the spinal cord in order to alter excitability and impact ongoing motor output throughout the cord. Solid lines indicate that transmission remains intact to the point of injury to the spinal cord, while dashed lines indicate that transmission is impaired, and may be facilitated by tSCS. Typically, tSCS is applied in single unmodulated or modulated monophasic or biphasic pulses or trains of pulses.

What is less clear is the impact that the carrier frequency has on the neural circuitry recruited during tSCS, and the specific role it serves to improve functional recovery when paired with rehabilitation strategies. Recently, hand and arm function improved significantly during a single session of cervical tSCS with a 5 kHz carrier frequency applied in individuals with an SCI compared to when a carrier frequency was not included [65]. However, limited data are available as to the differences in specific neural substrates recruited by tSCS with and without a carrier frequency. Overall, integrating a carrier frequency may be an important feature of tSCS that not only circumvents pain compared to other stimulation profiles, but also promotes effective restoration of function after SCI. Further exploration is required in order to understand whether the carrier frequency is a unique feature necessary for optimizing the use of tSCS for sensorimotor recovery. Incorporating this knowledge into a mechanistic framework for the implementation of tSCS is essential in order to facilitate optimal functional recovery after neurological damage.

3.4. Mechanisms of tSCS Recruitment

The principal mechanism by which tSCS non-invasively activates inaccessible neuronal networks of the spinal cord likely includes the recruitment of afferent fibers (large–medium) in the posterior root in order to elevate spinal network excitability [66,67]. The excitability of spinal interneuronal networks can be readily modulated (changing the networks' physiological state) without directly producing action potentials [22]. The route of stimulation propagation is through the dorsal root afferents, as indicated by the significant inhibition of cervical tSCS responses when using paired stimuli, during passive muscle stretching, and during muscle–tendon vibration [67]. Moreover, it has been suggested that eSCS and modulated tSCS can engage both afferent and efferent pathways, based on observations of early- and medium-response components of evoked potentials that are partially ascribed to posterior roots/group Ia/group II and motor neurons/anterior roots [22,68]. It is proposed that as stimulation intensity is increased, in addition to the Ia afferents, the smaller diameter afferents such as group Ib, larger diameter cutaneous afferents, group II muscle spindle afferents, and even more intraspinal connections and spinal interneurons are recruited through tSCS, similarly to what has been observed in eSCS [22,59]. This, in turn, brings interneurons and motor neurons closer to their firing threshold, making them more likely to respond to limited post-injury descending drive and improving supraspinal control after both modulated and unmodulated tSCS [24,30]. Both electrophysiological and computer modeling studies to date suggest that unmodulated tSCS excites posterior root fibers similarly to eSCS [24,52].

Recently, a few studies have compared the different effects of modulated and unmodulated tSCS on descending input. Benavides et al. reported that single-site tSCS applied with a 5 kHz carrier frequency at the C5–C6 level facilitated the amplitude of cervicomedullary-evoked potentials (CMEPs), but did not increase the amplitude of the motor-evoked potentials (MEPs) [65]; this was accompanied by an increase in the level of short-interval cortical inhibition (SICI). When tSCS was applied without the carrier frequency, both cortically and subcortically driven responses were facilitated. This is similar to our recent investigation, which found that modulated tSCS (33 Hz trains of 1 m long pulses with a 10 kHz carrier frequency) applied over the C3–4 and C6–7 spinous processes in neurologically intact individuals did not alter MEPs assessed in the forearm flexors [69]. Moreover, data from a paired associative stimulation (PAS) paradigm involving single pulses of transcranial magnetic stimulation (TMS) and unmodulated tSCS arriving at the same time at spinal motor neurons revealed increases in corticospinal excitability, but facilitation of MEPs following tSCS was less pronounced when tSCS pulses were filled with a carrier frequency [70]. These studies highlight the fact that in the presence of a carrier frequency, tSCS may be unable to facilitate MEPs. In contrast, it was shown that sub-motor-threshold tSCS without a carrier frequency, applied for a short period of 10 min to the cervical region, did not alter the excitability of the corticospinal and spinal reflex pathways [71]. At first glance, these results seem contradictory; however, the stimulation duration, stimulation amplitude, frequency of stimulation, stimulation waveform (modulated/unmodulated), and target muscles varied across these studies, which may have influenced the neuromodulatory effects of tSCS. By priming neural structures at the level of the spinal cord, unmodulated tSCS modulated spinal reflex excitability and reduced spasticity in a manner similar to that seen with passive cycling movements [34]. This suggests that alterations in spinal circuitry—including presynaptic influences—are likely the primary target of tSCS, and play an important role in the recovery of arm and hand function in persons with SCI.

Importantly, dorsal root stimulation is likely not entirely responsible for the effects of tSCS. Group Ia muscle spindle afferent fibres, which travel in the dorsal roots, have a lower threshold of activation compared to the largest cutaneous fibres [72]. If the effects of tSCS are only due to the activation of dorsal root afferents, then at low stimulation amplitudes the large-diameter group Ia afferents should be activated, leading to muscle contractions and proprioceptive errors via monosynaptic reflexes [57]. However, cutaneous sensation typically occurs over a large range of stimulus amplitudes that are lower than those required to produce motor responses mediated by purely monosynaptic reflex pathways, and proprioceptive errors are not a significant occurrence [73], making it unlikely that tSCS functions entirely by stimulating dorsal root afferents. Epidural SCS at 1–2 Hz has been shown to activate inhibitory interneurons in laminae I–III, albeit with latencies consistent with trans-synaptic (i.e., indirect) activation [74]. Therefore, it is important to consider whether inhibitory neurons in this region are the main or, at least, a contributing mechanism underlying the therapeutic benefit of tSCS; that is, tSCS may restore inhibition by enhancing dorsal horn GABAergic systems. It has been suggested that islet cells in the substantia gelatinosa require further consideration as prime candidates for the inhibitory effects on pain [57].

Moreover, while it is widely believed that tSCS depolarizes sensory afferents in the dorsal roots and dorsal horn that trans-synaptically recruit motor pools, it remains possible that polysynaptic connections from cutaneous mechanoreceptors in the skin act on both sensory processes and motor pools in the spinal cord. This, in turn, alters the excitability at both the level of the spinal cord—where the stimulation is provided—as well as remote levels of the spinal cord, through propriospinal interneuronal connections. Cutaneous inputs are known to have diffuse input that is specific to the task, phase, and amplitude at which stimulation is delivered [75,76]. It is therefore plausible that the recruitment of cutaneous mechanoreceptors surrounding the electrodes may contribute to the neuromodulatory effects of tSCS through these polysynaptic connections. The potential role of cutaneous mechanoreceptors in the skin with tSCS remains an important avenue to explore in future work [77–79].

A potential mechanism by which tSCS improves upon previously developed rehabilitation interventions is potentiation. Guiho et al. observed potentiation of supraspinal evoked responses with both dorsal eSCS and modulated tSCS over the C3–4 and C7–T1 intervertebral spaces in monkeys, but facilitation was stronger with dorsal eSCS [80]. It is vital to identify the capability of unmodulated tSCS to alter supraspinally driven responses compared to eSCS and modulated tSCS, in order to identify whether unique stimulation parameters are required for individual outcomes. Similarly, PAS with tSCS and TMS induced facilitation of corticospinal excitability for at least 30 min after the PAS, which is indicative of long-term potentiation (LTP)-like plasticity in the lower limb region of the primary motor cortex [81]. An important component of tSCS is its neuromodulatory effect on remote segments of the spinal cord, which must be considered during SCI rehabilitation.

4. Transcutaneous SCS Alters Excitability across Multiple Levels of the Spinal Cord

Evidence indicates that tSCS alters the excitability of multiple segments of the spinal cord [70,82]. These multi-segmental effects were specifically investigated in our recent work exploring how stimulation alters excitability across multiple levels of the spinal cord in neurologically intact participants, using the setup described in Figure 3. We first determined that cervical tSCS suppresses the amplitude of the soleus Hoffmann (H)-reflex by 22.9% (Figure 4B), which was similar to the 19.7% reduction produced by rhythmic arm cycling (Figure 4C), demonstrating that cervical tSCS alters lumbar excitability [59]. The suppression of H-reflexes evoked in one limb by rhythmic movements of the remote limbs demonstrates coupling between the arms and legs in humans [83–85]. A bidirectional linkage between the cervical and lumbar segments of the spinal cord exists during rhythmic movements in both quadrupedal mammals and humans [86,87], facilitated primarily by propriospinal connections [83,88]. Therefore, it was hypothesized that a similar reciprocal organization may also be revealed by tSCS applied to the cervical and lumbar networks, suggesting that tonic tSCS activates similar networks to those activated during rhythmic activity of the arms or legs [76,89]. In contrast to our hypothesis, lumbar tSCS significantly facilitated the amplitude of the H-reflex in the flexor carpi radialis (FCR) by 11.1% relative to no stimulation (Figure 4D), as opposed to the expected 13.6% reduction in reflex amplitude during leg cycling (Figure 4E) [69]. This indicates that separate propriospinal networks are likely responsible for the effects of tSCS and rhythmic cycling.

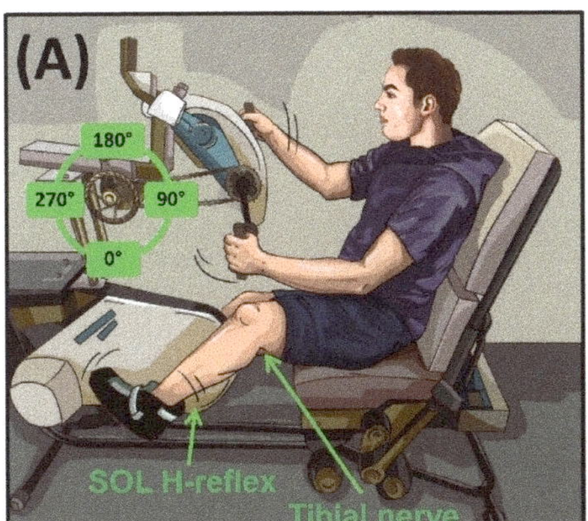

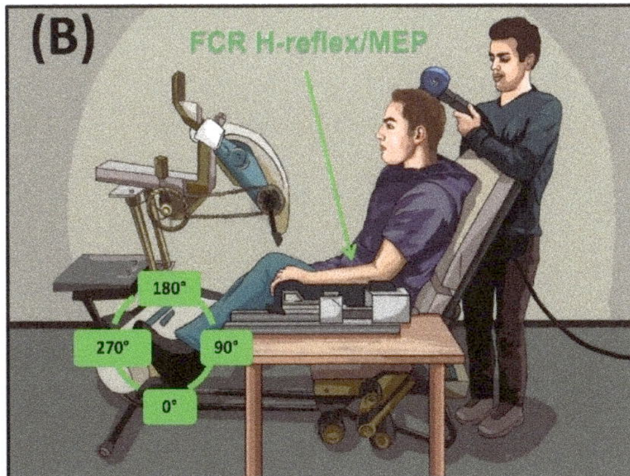

Figure 3. Experimental setup for investigating the effect of modulated tSCS on cervicolumbar connectivity and corticospinal facilitation: (**A**) Hoffmann (H-) reflexes were evoked during tSCS via stimulation of the tibial nerve and recorded in the soleus (SOL) muscle. The left leg was held static in an extended position, and stimulation to evoke the H-reflex was delivered with either the left arm held at 0° or during arm cycling. (**B**) H-reflexes were evoked during tSCS via stimulation of the median nerve and recorded in the flexor carpi radialis (FCR) muscle, while motor evoked potentials (MEPs) were evoked in the contralateral motor cortex and recorded in the FCR muscle, either with the legs held static, or during leg cycling. Responses were evoked during a consistent background contraction of ≈5–10% peak muscle activity at the same position, regardless of condition.

These results are summarized in Figure 4A, as tonic activation of spinal cord networks via tSCS alters excitability over multiple segments of the spinal cord, and is not bidirectional in its effects. The mechanisms responsible for the disinhibition of the H-reflex results between the upper and lower limbs are unknown. Facilitation of the H-reflex pathway through tSCS may be due to reduced Ia presynaptic inhibition, or to facilitation of the motor pool through activation of posterior root afferents and interneuronal projections [24]. It also remains possible that the stimulation of skin itself may be a larger contributing factor in altering the excitability of the spinal cord with tSCS than previously considered [77]. Understanding the integration of tSCS across multiple segments of the spinal cord across the range of stimulation parameters is critical in order to determine whether facilitating or inhibiting the circuitry involved is desirable based on the individual, the available technology, and the primary clinical outcome. While single-site tSCS neuromodulates remote segments of the spinal cord, multiple sites of tSCS appear to converge and facilitate the spinal and corticospinal circuitry.

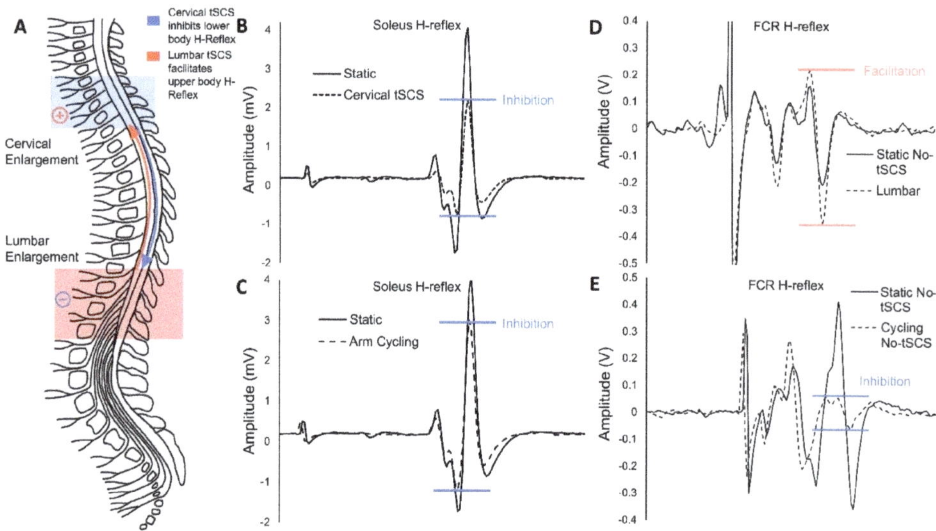

Figure 4. Effects of tSCS on interlimb connectivity are not similar to those of cycling in terms of reciprocal organization: (**A**) The schematic highlights common spinal segments activated by tSCS, including the cervical (blue) and lumbar (pink) enlargements. The blue arrow indicates that tonic cervical tSCS inhibits lumbar excitability, while the red arrow indicates that lumbar tSCS facilitates cervical excitability in neurologically intact individuals. (**B**) Spinal reflex excitability as assessed by the H-reflex in the soleus muscle is significantly inhibited in the presence of cervical tSCS [82]. (**C**) Spinal reflex excitability is similarly reduced in the lower limbs during arm cycling, which is a known condition for altering interlimb connectivity via presynaptic mechanisms [84,90]. (**D**) Conversely, spinal reflex excitability as assessed by the H-reflex in the FCR muscle is significantly facilitated in the presence of lumbar tSCS [70]. (**E**) Leg cycling continues to inhibit spinal reflex excitability in the upper limbs. Panels (**B–D**) adapted from published data in [70,82].

5. Multiple Sites of tSCS Converge to Facilitate Alterations in Excitability

Further improvements to the reengagement of previously inaccessible networks may be possible using multiple stimulation sites of tSCS. Previous investigations have indicated that unmodulated tSCS delivered at the vertebral level T11 can activate the locomotor circuitry in neurologically intact study participants when their legs are placed in a gravity-neutral position [91]. Simultaneous stimulation of cervical, thoracic, and lumbar levels (i.e., C5, T11, and L1, respectively) with a carrier frequency induced coordinated stepping movements with a greater range of motion at multiple joints in five of six neurologically intact participants, compared to stimulation of T11 alone [92]. The addition of stimulation at L1 and/or at C5 to stimulation at T11 immediately resulted in enhancing the kinematics and interlimb coordination as well as the EMG patterns in proximal and distal leg muscles. Moreover, paired tSCS at the L2 and S1 segments of the spinal cord resulted in greater potentiation of the evoked response than from either site alone, indicating synergistic effects of multi-segmental pathways [93]. The interactive and synergistic effects indicate multi-segmental convergence of descending, ascending and, most likely, propriospinal influences on the neuronal circuitry during tSCS [93].

Interestingly, multisite (i.e., combined) modulated tSCS in both the cervical and lumbar segments of the spinal cord led to a convergence in the upper limbs (FCR muscle) that significantly increased H-reflex and MEP amplitude, by 19.6% (Figure 5B) and 19.7% (Figure 5C), respectively. Cervical tSCS alone did not increase H-reflex or MEP amplitude in the FCR, but both were significantly facilitated with the addition of lumbar tSCS. This indicates that tSCS alters excitability across multiple segments of the spinal cord, and converges to facilitate both spinal and corticospinal transmission, as demonstrated in Figure 5A. The facilitation of MEPs in the FCR by combined cervical and lumbar tSCS could be due to reinforced projection of ascending propriospinal and corticospinal axons onto cervical spinal motor neurons [94]. Therefore, the activation of proprioceptive inputs at both the cervical and lumbar spinal cord by tSCS, which synapse on cervical motor neurons, may be a major contributor to the facilitation of H-reflexes and MEPs to the FCR muscle. An important consideration with the potential use of multisite tSCS is the role that spasticity plays in the rehabilitation strategy; facilitating H-reflexes in muscles that have significant spasticity could compound the effect. Further study is required for understanding the effects of multisite tSCS in individuals living with an SCI, as well as its effects on spasticity both within a session and after training.

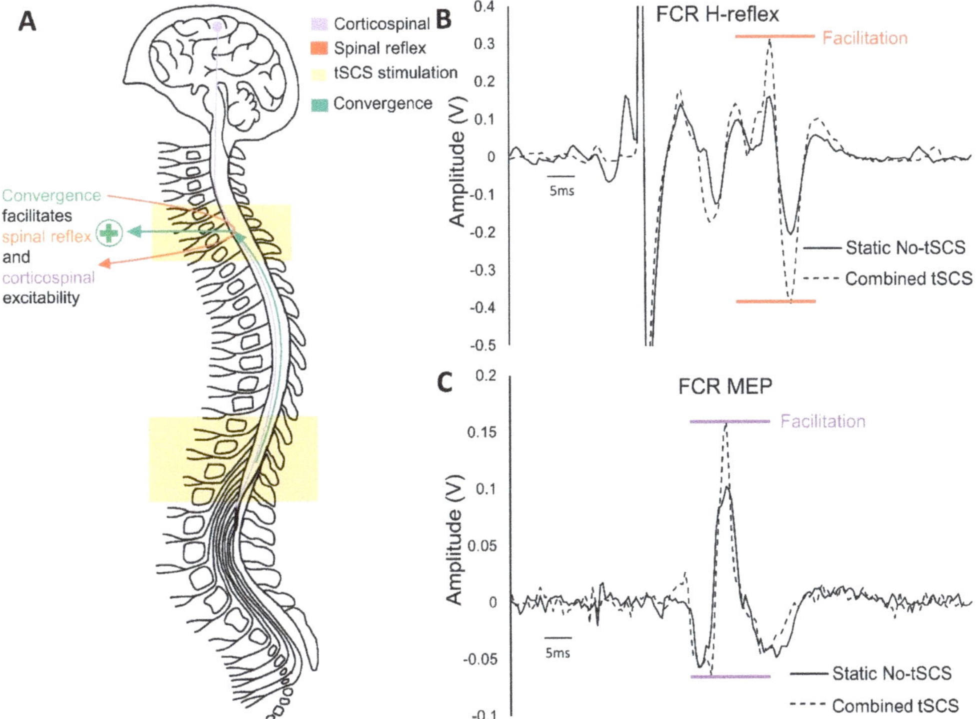

Figure 5. Convergence across multiple spinal segments facilitates spinal and corticospinal excitability: (**A**) The schematic highlights that simultaneous cervical and lumbar tSCS (yellow) significantly facilitates cervical spinal reflex and corticospinal excitability. (**B**) Spinal reflex excitability as assessed by the H-reflex in the flexor carpi radialis (FCR) muscle is significantly facilitated in the presence of combined cervical and lumbar tSCS. (**C**) Similarly, corticospinal excitability as assessed by MEPs in the FCR elicited from the contralateral motor cortex was also significantly facilitated by combined cervical and lumbar tSCS. Panels (**B**,**C**) adapted from published data in [69].

Interestingly, in neurologically intact study participants, modulated tSCS was unable to alter the excitability of either H-reflexes or MEPs when combined with either arm or leg cycling, regardless of whether single-site or multisite tSCS was applied [70,82]. This indicates that in neurologically intact individuals where interlimb coordination and the corticospinal tract are intact, the effects of arm or leg cycling on cervicolumbar coupling and corticospinal drive were not impacted significantly by the tSCS intensity used. Therefore, it will be a vital next step to determine the role that multisite tSCS has on interlimb connectivity after SCI. The potential impact of using multisite tSCS as a strategy to neuromodulate the spinal circuitry has significant implications in furthering our understanding of the mechanisms controlling posture and locomotion, and for regaining significant sensorimotor function even after neural injury.

6. Is There a Role for tSCS to Facilitate Cervicolumbar Coupling to Improve Walking?

Since single-site modulated tSCS alters excitability at remote segments of the spinal cord, and multisite modulated tSCS shows a significant convergence effect, it is possible that tSCS may influence coupling between the arms and legs after SCI. The coordination between the legs and arms is an inherent feature of locomotor neural networks [80], with coupling between the cervical (arms) and lumbar (legs) spinal networks (cervicolumbar coupling) well demonstrated in both animals and humans [95–97]. Oscillatory movements are governed by separate locomotor centers known as central pattern generators (CPGs), which are located in the cervical and lumbar spinal cord segments [76,89]. Similarly to quadrupedal mammals, a bidirectional linkage between the cervical and lumbar segments of the spinal cord during rhythmic movements is present in humans [86,87], facilitated primarily by propriospinal connections [83,88].

Engaging these connections with simultaneous arm and leg (A&L) cycling training improves walking capacity after both chronic incomplete SCI [98] and stroke [99,100]. Highlighting the importance of these interlimb connections, arms-only cycling has also been shown to improve overground walking function after stroke [101]. A&L cycling often capitalizes on the incompleteness of the injury to the spinal cord, even in cases where the injury is clinically classified as complete. The effect of neuromodulation is maximized when accompanied by a residual intact descending/ascending input. While the beneficial effects of rehabilitation strategies such as arm and leg cycling on cervicolumbar coupling after incomplete SCI and stroke have been outlined previously, little is known about severe cases when the injury to the spinal cord is clinically complete. Pairing tSCS with A&L cycling may allow for similar improvements in interlimb connections after complete SCI or multiple sclerosis. However, the impact of tSCS on propriospinal connectivity has yet to be investigated. Enhancing cervicolumbar connectivity by pairing A&L cycling with tSCS to improve mobility outcomes also remains a vital avenue for future research.

7. Trunk Stability Improvements with tSCS

While direct evidence of tSCS influencing remote segments of the spinal cord is limited, enhancing trunk stability—which is often an overlooked component—may provide indirect evidence of the influence of tSCS. Postural stability via regulation of trunk function is an integral part of locomotor control and a key element of the kinematic chain for reaching movements [102,103]. Modulation of lumbosacral networks via modulated tSCS has enabled individuals with various levels of SCI to stand without assistance from a therapist; more importantly, individuals showed improved postural control after repeated sessions of training, as demonstrated by an increased range of the center of pressure excursion during self-initiated body weight displacement [67]. As argued by the authors, biphasic pulses were perceived similarly to the sensation caused by monophasic pulses; however, biphasic stimulation could not enable unassisted standing, and was ineffective in producing motor output in the lower extremities, even at higher stimulation intensities [67]. Although critical, this observation is limited to one specific task of the lower extremity using only a modulated waveform; thus, future investigation is necessary in order to compare the effects and the underlying mechanisms of monophasic and biphasic tSCS paradigms, in an effort to uncover the best stimulation paradigm for improving functional outcomes. Moreover, modulated tSCS applied to the lumbar region increased the level of activity in the trunk muscles, adjusted the abnormal sitting posture, and extended the limits of multidirectional seated displacement, overall enhancing postural control [104]. The ability of lumbar tSCS to improve muscle activity in the trunk and postural control provides indirect evidence for tSCS inducing meaningful effects across multiple segments within the spinal cord. While further investigation is necessary in order to determine the specific pathways responsible for improved postural control, there is an additional incentive to pair tSCS with rehabilitation interventions in order to improve functions that are often overlooked in research and rehabilitation interventions.

8. Previously Developed Rehabilitative Approaches Are Enhanced through tSCS

Understanding the role of tSCS across multiple converging segments of the spinal cord is an important consideration when designing optimal rehabilitation interventions. The use of tSCS in conjunction with functional training appears imperative for optimizing functional recovery after SCI [17,27,32,105]. When tSCS (either modulated or unmodulated) and training are combined, functional changes emerge more rapidly and to a greater degree than with either method alone, making these strategies vital to one another's success [27,32]. Importantly, participants with SCI previously considered to be at maximal functional capacity following walking-based therapy were able to gain significant improvements in the 6 min and 10 m walking tests after incorporating unmodulated tSCS into a paired tSCS- and walking-based therapy intervention [32]. Likewise, unmodulated tSCS as an adjunct to locomotor training was shown to improve walking outcomes in individuals with subacute motor-incomplete SCI [106]. Furthermore, pairing modulated tSCS with walking using an exoskeleton can improve lower limb coordination [105]. Positive synergistic effects of tSCS neuromodulation and previously successful rehabilitation strategies are a promising avenue for increasing what is currently possible for recovery after neurotrauma. This may be enhanced by further understanding of the unique properties of tSCS, and how it may modulate spinal circuitry differentially based on stimulation parameters and waveforms, muscles of interest, and desired functional outcomes.

9. Conclusions and Future Directions

Collectively, this work demonstrates that the activation of spinal cord networks with tSCS alters excitability over multiple segments of the spinal cord, with differential properties depending on the site, waveform, and parameters of tSCS. Importantly, multiple sites of tSCS converge to enhance the modulation of spinal reflex and corticospinal pathways in neurologically intact individuals. Clinical data also indicate that multi-segmental functional improvements occur after SCI. This highlights the potential for engaging cervical and lumbar spinal cord networks through tSCS to enhance the effectiveness of rehabilitation interventions. An essential next step in the evolution of tSCS research is determining the unique contributions of cutaneous mechanoreceptors, islet cells, dorsal root afferents, interneuronal projections, and large afferents in the dorsal horn, all of which likely contribute to neuromodulation with tSCS. Understanding the mechanisms of action with tSCS and potential differences in recruitment between modulated and unmodulated tSCS will provide the foundation with which to establish optimal concomitant rehabilitation therapy to improve sensorimotor function after neural injury or disease [66,107]. In general, tSCS appears to be a safe approach for modulating the excitability of neural networks of the spinal cord. Several studies have reported that the stimulation is well tolerated with minimal skin irritation or adverse changes in blood pressure, heart rate, spasticity, and/or incontinence [2,27]. However, two studies have reported side effects after tSCS including unintentional voiding during standing, skin damage and redness, fluctuation of blood pressure and heart rate, and nausea [63,105], and significant work is still required in order to ensure the safety of tSCS, including its application to locations where autonomic nerves are located. Specific caution should also be taken to ensure that tSCS is not applied to areas containing a metal implant or medical device, because the potential interactions between tSCS and such devices have not been explored [106].

Author Contributions: Conceptualization, T.S.B. and V.K.M.; literature review, T.S.B., J.P. and B.P.; writing—original draft preparation, T.S.B.; writing—review and editing, T.S.B., J.P., B.P. and V.K.M.; visualization, T.S.B., J.P. and B.P.; supervision, V.K.M.; project administration, V.K.M.; funding acquisition, T.S.B. and V.K.M. All authors have read and agreed to the published version of the manuscript.

Funding: This research was funded by the Canadian Institutes of Health Research (CIHR; PS 166015), the Canada Foundation for Innovation (CFI; LEF 30852), and the Spinal Cord Injury Treatment Centre Society (SCITCS). B.P. holds a Faculty of Medicine and Dentistry Dean's Doctoral Scholarship. T.B. holds a CIHR Postdoctoral Fellowship, a Craig H. Neilsen Postdoctoral Fellowship Research Grant, and an Alberta Innovates Postgraduate Fellowship in Health Innovation. V.K.M. is a Canada Research Chair in Functional Restoration.

Institutional Review Board Statement: All studies from our lab discussed in this review were conducted according to the guidelines of the Declaration of Helsinki, and approved by the Human Research Ethics Board of the University of Alberta.

Informed Consent Statement: Informed consent was obtained from all study participants involved in published works completed within our laboratory and discussed in this review.

Data Availability Statement: Processed data from reference [60] can be accessed in the Open Data Commons for Spinal Cord Injury at http://dx.doi.org/10.34945/F5B59T that was published on 9 March 2021 [108], and last accessed on 17 January 2022.

Acknowledgments: The experiments described in this review were conducted in the Rehabilitation Innovations Core of the Sensory Motor Adaptive Rehabilitation Technology (SMART) Network at the University of Alberta.

Conflicts of Interest: The authors declare no conflict of interest.

References

1. Balykin, M.V.; Yakupov, R.N.; Mashin, V.V.; Kotova, E.Y.; Balykin, Y.M.; Gerasimenko, Y.P. The influence of non-invasive electrical stimulation of the spinal cord on the locomotor function of patients presenting with movement disorders of central genesis. *Vopr. Kurortol. Fizioter. Lech. Fiz. Kult.* **2017**, *94*, 4–9. [CrossRef] [PubMed]
2. Inanici, F.; Samejima, S.; Gad, P.; Edgerton, V.R.; Hofstetter, C.P.; Moritz, C.T. Transcutaneous electrical spinal stimulation promotes long-term recovery of upper extremity function in chronic tetraplegia. *IEEE Trans. Neural Syst. Rehabil. Eng.* **2018**, *26*, 1272–1278. [CrossRef] [PubMed]
3. Harkema, S.; Gerasimenko, Y.; Hodes, J.; Burdick, J.; Angeli, C.; Chen, Y.; Ferreira, C.; Willhite, A.; Rejc, E.; Grossman, R.; et al. Effect of Epidural stimulation of the lumbosacral spinal cord on voluntary movement, standing, and assisted stepping after motor complete paraplegia: A case study. *Lancet* **2011**, *377*, 1938–1947. [CrossRef]
4. Angeli, C.A.; Boakye, M.; Morton, R.A.; Vogt, J.; Benton, K.; Chen, Y.; Ferreira, C.K.; Harkema, S.J. Recovery of over-ground walking after chronic motor complete spinal cord injury. *N. Engl. J. Med.* **2018**, *379*, 1244–1250. [CrossRef] [PubMed]
5. Fregni, F.; Grecco, L.; Li, S.; Michel, S.; Castillo-Saavedra, L.; Mourdoukoutas, A.; Bikson, M. Transcutaneous spinal stimulation as a therapeutic strategy for spinal cord injury: State of the art. *J. Neurorestoratol.* **2015**, *3*, 73–82. [CrossRef]
6. Dimitrijević, M.R. Residual motor functions in spinal cord injury. *Adv. Neurol.* **1988**, *47*, 138–155.
7. Courtine, G.; Roy, R.R.; Raven, J.; Hodgson, J.; Mckay, H.; Yang, H.; Zhong, H.; Tuszynski, M.H.; Edgerton, V.R. Performance of locomotion and foot grasping following a unilateral thoracic corticospinal tract lesion in monkeys (Macaca mulatta). *Brain* **2005**, *128*, 2338–2358. [CrossRef]
8. Friedli, L.; Rosenzweig, E.S.; Barraud, Q.; Schubert, M.; Dominici, N.; Awai, L.; Nielson, J.L.; Musienko, P.; Nout-Lomas, Y.; Zhong, H.; et al. Pronounced species divergence in corticospinal tract reorganization and functional recovery after lateralized spinal cord injury favors primates. *Sci. Transl. Med.* **2015**, *7*, 302ra134. [CrossRef]
9. Rosenzweig, E.S.; Courtine, G.; Jindrich, D.L.; Brock, J.H.; Ferguson, A.R.; Strand, S.C.; Nout, Y.S.; Roy, R.R.; Miller, D.M.; Beattie, M.S.; et al. Extensive spontaneous plasticity of corticospinal projections after primate spinal cord injury. *Nat. Neurosci.* **2010**, *13*, 1505–1512. [CrossRef]
10. Courtine, G.; Song, B.; Roy, R.R.; Zhong, H.; Herrmann, J.E.; Ao, Y.; Qi, J.; Edgerton, V.R.; Sofroniew, M.V. Recovery of supraspinal control of stepping via indirect propriospinal relay connections after spinal cord injury. *Nat. Med.* **2008**, *14*, 69–74. [CrossRef]
11. Gerasimenko, Y.; Musienko, P.; Bogacheva, I.; Moshonkina, T.; Savochin, A.; Lavrov, I.; Roy, R.R.; Edgerton, V.R. Propriospinal bypass of the serotonergic system that can facilitate stepping. *J. Neurosci.* **2009**, *29*, 5681–5689. [CrossRef] [PubMed]
12. Asboth, L.; Friedli, L.; Beauparlant, J.; Martinez-Gonzalez, C.; Anil, S.; Rey, E.; Baud, L.; Pidpruzhnykova, G.; Anderson, M.A.; Shkorbatova, P.; et al. Cortico-reticulo-spinal circuit reorganization enables functional recovery after severe spinal cord contusion. *Nat. Neurosci.* **2018**, *21*, 576–588. [CrossRef] [PubMed]
13. Mayr, W.; Krenn, M.; Dimitrijevic, M.R. Epidural and transcutaneous spinal electrical stimulation for restoration of movement after incomplete and complete spinal cord injury. *Curr. Opin. Neurol.* **2016**, *29*, 721–726. [CrossRef] [PubMed]

14. Formento, E.; Minassian, K.; Wagner, F.; Mignardot, J.B.; Le Goff-Mignardot, C.G.; Rowald, A.; Bloch, J.; Micera, S.; Capogrosso, M.; Courtine, G. Electrical spinal cord stimulation must preserve proprioception to enable locomotion in humans with spinal cord injury. *Nat. Neurosci.* **2018**, *21*, 1728–1741. [CrossRef]

15. Herman, R.; He, J.; D'Luzansky, S.; Willis, W.; Dilli, S. Spinal cord stimulation facilitates functional walking in a chronic, incomplete spinal cord injured. *Spinal Cord* **2002**, *40*, 65–68. [CrossRef]

16. Wagner, F.B.; Mignardot, J.B.; Le Goff-Mignardot, C.G.; Demesmaeker, R.; Komi, S.; Capogrosso, M.; Rowald, A.; Seáñez, I.; Caban, M.; Pirondini, E.; et al. Targeted neurotechnology restores walking in humans with spinal cord injury. *Nature* **2018**, *563*, 65–93. [CrossRef]

17. Kou, J.; Cai, M.; Xie, F.; Wang, Y.; Wang, N.; Xu, M. Complex Electrical Stimulation Systems in Motor Function Rehabilitation after Spinal Cord Injury. *Hindawi* **2021**, *2021*, 2214762. [CrossRef]

18. Gill, M.L.; Grahn, P.J.; Calvert, J.S.; Linde, M.B.; Lavrov, I.A.; Strommen, J.A.; Beck, L.A.; Sayenko, D.G.; Van Straaten, M.G.; Drubach, D.I.; et al. Neuromodulation of lumbosacral spinal networks enables independent stepping after complete paraplegia. *Nat. Med.* **2018**, *24*, 1677–1682. [CrossRef]

19. Courtine, G.; Gerasimenko, Y.; Van Den Brand, R.; Yew, A.; Zhong, H.; Song, B.; Ao, Y.; Ichiyama, R.M.; Lavrov, I.; Roy, R.R.; et al. Transformation of nonfunctional spinal circuits into functional states after the loss of brain input. *Nat. Neurosci.* **2009**, *12*, 1333–1342. [CrossRef]

20. Angeli, C.A.; Edgerton, V.R.; Gerasimenko, Y.P.; Harkema, S.J. Altering spinal cord excitability enables voluntary movements after chronic complete paralysis in humans. *Brain* **2014**, *137*, 1394–1409. [CrossRef]

21. García, A.M.; Serrano-Muñoz, D.; Taylor, J.; Avendaño-Coy, J.; Gómez-Soriano, J. Transcutaneous Spinal Cord Stimulation and Motor Rehabilitation in Spinal Injury: A Systematic Review. *Neurorehabil. Neural Repair* **2020**, *34*, 3–12. [CrossRef] [PubMed]

22. Gerasimenko, Y.; Gorodnichev, R.; Moshonkina, T.; Sayenko, D.; Gad, P.; Edgerton, V.R. Transcutaneous electrical spinal-cord stimulation in humans HHS Public Access. *Ann. Phys. Rehabil. Med.* **2015**, *58*, 225–231. [CrossRef] [PubMed]

23. Martin, R. Utility and Feasibility of Transcutaneous Spinal Cord Stimulation for Patients with Incomplete SCI in Therapeutic Settings: A Review of Topic. *Front. Rehabil. Sci.* **2021**, *2*, 724003. [CrossRef]

24. Hofstoetter, U.S.; Freundl, B.; Binder, H.; Minassian, K. Common neural structures activated by epidural and transcutaneous lumbar spinal cord stimulation: Elicitation of posterior root-muscle reflexes. *PLoS ONE* **2018**, *13*, e0192013. [CrossRef] [PubMed]

25. Ladenbauer, J.; Minassian, K.; Hofstoetter, U.S.; Dimitrijevic, M.R.; Rattay, F. Stimulation of the human lumbar spinal cord with implanted and surface electrodes: A computer simulation study. *IEEE Trans. Neural Syst. Rehabil. Eng.* **2010**, *18*, 637–645. [CrossRef]

26. Danner, S.M.; Hofstoetter, U.S.; Ladenbauer, J.; Rattay, F.; Minassian, K. Can the Human Lumbar Posterior Columns Be Stimulated by Transcutaneous Spinal Cord Stimulation? A Modeling Study. *Artif. Organs* **2011**, *35*, 257–262. [CrossRef]

27. Gad, P.; Lee, S.; Terrafranca, N.; Zhong, H.; Turner, A.; Gerasimenko, Y.; Edgerton, V.R. Non-invasive activation of cervical spinal networks after severe paralysis. *J. Neurotrauma* **2018**, *35*, 2145–2158. [CrossRef]

28. Zheng, Y.; Hu, X. Elicited upper limb motions through transcutaneous cervical spinal cord stimulation. *J. Neural Eng.* **2020**, *17*, 036001. [CrossRef]

29. Inanici, F.; Brighton, L.N.; Samejima, S.; Hofstetter, C.P.; Moritz, C.T. Transcutaneous Spinal Cord Stimulation Restores Hand and Arm Function after Spinal Cord Injury. *IEEE Trans. Neural Syst. Rehabil. Eng.* **2021**, *29*, 310–319. [CrossRef]

30. Gerasimenko, Y.P.; Lu, D.C.; Modaber, M.; Zdunowski, S.; Gad, P.; Sayenko, D.G.; Morikawa, E.; Haakana, P.; Ferguson, A.R.; Roy, R.R.; et al. Noninvasive Reactivation of Motor Descending Control after Paralysis. *J. Neurotrauma* **2015**, *32*, 1968–1980. [CrossRef]

31. Solopova, I.A.; Sukhotina, I.A.; Zhvansky, D.S.; Ikoeva, G.A.; Vissarionov, S.V.; Baindurashvili, A.G.; Edgerton, V.R.; Gerasimenko, Y.P.; Moshonkina, T.R. Effects of spinal cord stimulation on motor functions in children with cerebral palsy. *Neurosci. Lett.* **2017**, *639*, 192–198. [CrossRef] [PubMed]

32. McHugh, L.V.; Miller, A.A.; Leech, K.A.; Salorio, C.; Martin, R.H. Feasibility and utility of transcutaneous spinal cord stimulation combined with walking-based therapy for people with motor incomplete spinal cord injury. *Spinal Cord Ser. Cases* **2020**, *6*, 104. [CrossRef] [PubMed]

33. Hofstoetter, U.S.; Freundl, B.; Danner, S.M.; Krenn, M.J.; Mayr, W. Transcutaneous Spinal Cord Stimulation Induces Temporary Attenuation of Spasticity in Individuals with Spinal Cord Injury. *J. Neurotrauma* **2020**, *493*, 481–493. [CrossRef] [PubMed]

34. Madhavan, S.; Hofstoetter, U.; Estes, S.P.; Iddings, J.A.; Field-Fote, E.C. Priming neural circuits to Modulate spinal reflex excitability. *Front. Neurol.* **2017**, *8*, 17. [CrossRef]

35. Cogiamanian, F.; Ardolino, G.; Vergari, M.; Ferrucci, R.; Ciocca, M.; Scelzo, E.; Barbieri, S.; Priori, A. Transcutaneous Spinal Direct Current Stimulation. *Front. Psychiatry* **2012**, *3*, 63. [CrossRef]

36. Powell, E.S.; Carrico, C.; Salyers, E.; Westgate, P.M.; Sawaki, L. The effect of transcutaneous spinal direct current stimulation on corticospinal excitability in chronic incomplete spinal cord injury. *NeuroRehabilitation* **2018**, *43*, 125–134. [CrossRef]

37. Taylor, C.; McHugh, C.; Mockler, D.; Minogue, C.; Reilly, R.B.; Fleming, N. Transcutaneous spinal cord stimulation and motor responses in individuals with spinal cord injury: A methodological review. *PLoS ONE* **2021**, *16*, e0260166. [CrossRef]

38. Mushahwar, V.K.; Horch, K.W. Proposed specifications for a lumbar spinal cord electrode array for control of lower extremities in paraplegia. *IEEE Trans. Rehabil. Eng.* **1997**, *5*, 237–243. [CrossRef]

39. Mushahwar, V.K.; Jacobs, P.L.; Normann, R.A.; Triolo, R.J.; Kleitman, N. New functional electrical stimulation approaches to standing and walking. *J. Neural Eng.* **2007**, *4*, S181–S197. [CrossRef]
40. Shealy, C.N.; Mortimer, J.T.; Reswick, J.B. Electrical Inhibition of Pain by Stimulation of the Dorsal Columns: Preliminary Clinical Report. Available online: https://pubmed.ncbi.nlm.nih.gov/4952225/ (accessed on 19 November 2020).
41. Cook, A.W.; Weinstein, S.P. Chronic dorsal column stimulation in multiple sclerosis. Preliminary report. *N. Y. State J. Med.* **1973**, *73*, 2868–2872.
42. Richardson, R.R.; Cerullo, L.J.; McLone, D.G.; Gutierrez, F.A.; Lewis, V. Percutaneous epidural neurostimulation in modulation of paraplegic spasticity—Six case reports. *Acta Neurochir.* **1979**, *49*, 235–243. [CrossRef] [PubMed]
43. Richardson, R.R.; McLone, D.G. Percutaneous epidural neurostimulation for paraplegic spasticity. *Surg. Neurol.* **1978**, *9*, 153–155. [PubMed]
44. Pinter, M.M.; Gerstenbrand, F.; Dimitrijevic, M.R. Epidural electrical stimulation of posterior structures of the human lumbosacral cord: 3. Control Of spasticity. *Spinal Cord* **2000**, *38*, 524–531. [CrossRef] [PubMed]
45. Andersson, O.; Forssberg, H.; Grillner, S.; Lindquist, M. Phasic gain control of the transmission in cutaneous reflex pathways to motoneurones during "fictive" locomotion. *Brain Res* **1978**, *149*, 503–507. [CrossRef]
46. Grillner, S.; Zangger, P. On the central generation of locomotion in the low spinal cat. *Exp. Brain Res.* **1979**, *34*, 241–261. [CrossRef]
47. Huang, H.; He, J.; Herman, R.; Carhart, M.R. Modulation effects of epidural spinal cord stimulation on muscle activities during walking. *IEEE Trans. Neural Syst. Rehabil. Eng.* **2006**, *14*, 14–23. [CrossRef]
48. Carhart, M.R.; He, J.; Herman, R.; D'Luzansky, S.; Willis, W.T. Epidural Spinal-Cord Stimulation Facilitates Recovery of Functional Walking Following Incomplete Spinal-Cord Injury. *IEEE Trans. Neural Syst. Rehabil. Eng.* **2004**, *12*, 32–42. [CrossRef]
49. Rattay, F.; Minassian, K.; Dimitrijevic, M.R. Epidural electrical stimulation of posterior structures of the human lumbosacral cord: 2. Quantitative analysis by computer modeling. *Spinal Cord* **2000**, *38*, 473–489. [CrossRef]
50. De Noordhout, A.M.; Rothwell, J.C.; Thompson, P.D.; Marsden, C.D. Percutaneous electrical stimulation of lumbosacral roots in man. *Neurosurg. Psychiatry* **1988**, *51*, 174–181. [CrossRef]
51. Shapkova, E. Spinal locomotor capability revealed by electrical stimulation of the lumbar enlargement in paraplegic patients. *Prog. Mot. Control* **2004**, *3*, 253–289.
52. Minassian, K.; Persy, I.; Rattay, F.; Dimitrijevic, M.; Hofer, C.; Kern, H. Posterior root–muscle reflexes elicited by transcutaneous stimulation of the human lumbosacral cord. *Muscle Nerve* **2007**, *35*, 327–336. [CrossRef] [PubMed]
53. Hofstoetter, U.; Hofer, C.; Kern, H.; Danner, S.; Mayr, W.; Dimitrijevic, M.; Minassian, K. Effects of transcutaneous spinal cord stimulation on voluntary locomotor activity in an incomplete spinal cord injured individual. *Biomed. Tech.* **2013**, *58*, 2–3. [CrossRef] [PubMed]
54. Hofstoetter, U.S.; McKay, W.B.; Tansey, K.E.; Mayr, W.; Kern, H.; Minassian, K. Modification of spasticity by transcutaneous spinal cord stimulation in individuals with incomplete spinal cord injury. *J. Spinal Cord Med.* **2014**, *37*, 202–211. [CrossRef] [PubMed]
55. Ward, A.R.; Robertson, V.J. Sensory, motor, and pain thresholds for stimulation with medium frequency alternating current. *Arch. Phys. Med. Rehabil.* **1998**, *79*, 273–278. [CrossRef]
56. Manson, G.A.; Calvert, J.S.; Ling, J.; Tychhon, B.; Ali, A.; Sayenko, D.G. The relationship between maximum tolerance and motor activation during transcutaneous spinal stimulation is unaffected by the carrier frequency or vibration. *Physiol. Rep.* **2020**, *8*, e14397. [CrossRef]
57. Jensen, M.P.; Brownstone, R.M. Mechanisms of spinal cord stimulation for the treatment of pain: Still in the dark after 50 years. *Eur. J. Pain (United Kingdom)* **2019**, *23*, 652–659. [CrossRef]
58. Burke, D. Clinical uses of H reflexes of upper and lower limb muscles. *Clin. Neurophysiol. Pract.* **2016**, *1*, 9. [CrossRef]
59. Lavrov, I.; Gerasimenko, Y.P.; Ichiyama, R.M.; Courtine, G.; Zhong, H.; Roy, R.R.; Edgerton, V.R. Plasticity of spinal cord reflexes after a complete transection in adult rats: Relationship to stepping ability. *J. Neurophysiol.* **2006**, *96*, 1699–1710. [CrossRef]
60. Wu, G.; Ringkamp, M.; Murinson, B.B.; Pogatzki, E.M.; Hartke, T.V.; Weerahandi, H.M.; Campbell, J.N.; Griffin, J.W.; Meyer, R.A. Degeneration of myelinated efferent fibers induces spontaneous activity in uninjured C-fiber afferents. *J. Neurosci.* **2002**, *22*, 7746–7753. [CrossRef]
61. Ahmed, S.; Yearwood, T.; De Ridder, D.; Vanneste, S. Burst and high frequency stimulation: Underlying mechanism of action. *Expert Rev. Med. Devices* **2018**, *15*, 61–70. [CrossRef]
62. Arle, J.E.; Mei, L.; Carlson, K.W.; Shils, J.L. High-Frequency Stimulation of Dorsal Column Axons: Potential Underlying Mechanism of Paresthesia-Free Neuropathic Pain Relief. *Neuromodulation* **2016**, *19*, 385–397. [CrossRef] [PubMed]
63. North, J.M.; Hong, K.S.J.; Cho, P.Y. Clinical Outcomes of 1 kHz Subperception Spinal Cord Stimulation in Implanted Patients with Failed Paresthesia-Based Stimulation: Results of a Prospective Randomized Controlled Trial. *Neuromodulation Technol. Neural Interface* **2016**, *19*, 731–737. [CrossRef] [PubMed]
64. Chakravarthy, K.; Richter, H.; Christo, P.J.; Williams, K.; Guan, Y. Spinal Cord Stimulation for Treating Chronic Pain: Reviewing Preclinical and Clinical Data on Paresthesia-Free High-Frequency Therapy. *Neuromodulation* **2018**, *21*, 10–18. [CrossRef] [PubMed]
65. Benavides, F.D.; Jo, H.J.; Lundell, H.; Edgerton, V.R.; Gerasimenko, Y.; Perez, M.A. Cortical and subcortical effects of transcutaneous spinal cord stimulation in humans with tetraplegia. *J. Neurosci.* **2020**, *40*, 2633–2643. [CrossRef]
66. Sayenko, D.G.; Rath, M.; Ferguson, A.R.; Burdick, J.W.; Havton, L.A.; Edgerton, V.R.; Gerasimenko, Y.P. Self-assisted standing enabled by non-invasive spinal stimulation after spinal cord injury. *J. Neurotrauma* **2019**, *36*, 1435–1450. [CrossRef]

67. Milosevic, M.; Masugi, Y.; Sasaki, A.; Sayenko, D.G.; Nakazawa, K. On the reflex mechanisms of cervical transcutaneous spinal cord stimulation in human subjects. *J. Neurophysiol.* **2019**, *121*, 1672–1679. [CrossRef]

68. Sayenko, D.G.; Angeli, C.; Harkema, S.J.; Reggie Edgerton, V.; Gerasimenko, Y.P. Neuromodulation of evoked muscle potentials induced by epidural spinal-cord stimulation in paralyzed individuals. *J. Neurophysiol.* **2014**, *111*, 1088–1099. [CrossRef]

69. Parhizi, B.; Barss, T.S.; Mushahwar, V.K. Simultaneous Cervical and Lumbar Spinal Cord Stimulation Induces Facilitation of both Spinal and Corticospinal Circuitry in Humans. *Front. Neurosci.* **2021**, *15*, 379. [CrossRef]

70. Al'joboori, Y.; Hannah, R.; Lenham, F.; Borgas, P.; Kremers, C.J.P.; Bunday, K.L.; Rothwell, J.; Duffell, L.D. The Immediate and Short-Term Effects of Transcutaneous Spinal Cord Stimulation and Peripheral Nerve Stimulation on Corticospinal Excitability. *Front. Neurosci.* **2021**, *15*, 749042. [CrossRef]

71. Sasaki, A.; de Freitas, R.M.; Sayenko, D.G.; Masugi, Y.; Nomura, T.; Nakazawa, K.; Milosevic, M. Low-intensity and short-duration continuous cervical transcutaneous spinal cord stimulation intervention does not prime the corticospinal and spinal reflex pathways in able-bodied subjects. *J. Clin. Med.* **2021**, *10*, 3633. [CrossRef]

72. Macefield, G.; Gandevia, S.C.; Burke, D. Conduction velocities of muscle and cutaneous afferents in the upper and lower limbs of human subjects. *Brain* **1989**, *112*, 1519–1532. [CrossRef] [PubMed]

73. Rijken, N.H.M.; Vonhögen, L.H.; Duysens, J.; Keijsers, N.L.W. The effect of spinal cord stimulation (SCS) on static balance and gait. *Neuromodulation* **2013**, *16*, 244–250. [CrossRef] [PubMed]

74. Dubuisson, D. Effect of dorsal-column stimulation on gelatinosa and marginal neurons of cat spinal cord. *J. Neurosurg.* **1989**, *70*, 257–265. [CrossRef] [PubMed]

75. Zehr, E.P.; Stein, R.B. What functions do reflexes serve during human locomotion? *Prog. Neurobiol.* **1999**, *58*, 185–205. [CrossRef]

76. Zehr, E.P.; Barss, T.S.; Dragert, K.; Frigon, A.; Vasudevan, E.V.; Haridas, C.; Hundza, S.; Kaupp, C.; Klarner, T.; Klimstra, M.; et al. Neuromechanical interactions between the limbs during human locomotion: An evolutionary perspective with translation to rehabilitation. *Exp. Brain Res.* **2016**, *234*, 3059–3081. [CrossRef]

77. Beekhuizen, K.S.; Field-Fote, E.C. Massed practice versus massed practice with stimulation: Effects on upper extremity function and cortical plasticity in individuals with incomplete cervical spinal cord injury. *Neurorehabil. Neural Repair* **2005**, *19*, 33–45. [CrossRef]

78. Duysens, J.; Pearson, K.G. The role of cutaneous afferents from the distal hindlimb in the regulation of the step cycle of thalamic cats. *Exp. Brain Res.* **1976**, *24*, 245–255. [CrossRef]

79. Vallbo, Å.B.; Hagbarth, K.-E. Activity from Skin Percutaneously Mechanoreceptors in Awake Human Recorded Subjects. *Exp. Neurol.* **1968**, *289*, 270–289. [CrossRef]

80. Guiho, T.; Baker, S.N.; Jackson, A. Epidural and transcutaneous spinal cord stimulation facilitates descending inputs to upper-limb motoneurons in monkeys. *J. Neural Eng.* **2021**, *18*, 046011. [CrossRef]

81. Kaneko, N.; Sasaki, A.; Masugi, Y.; Nakazawa, K. The Effects of Paired Associative Stimulation with Transcutaneous Spinal Cord Stimulation on Corticospinal Excitability in Multiple Lower-limb Muscles. *Neuroscience* **2021**, *476*, 45–59. [CrossRef]

82. Barss, T.S.; Parhizi, B.; Mushahwar, V.K. Transcutaneous spinal cord stimulation of the cervical cord modulates lumbar networks. *J. Neurophysiol.* **2020**, *123*, 158–166. [CrossRef] [PubMed]

83. Ferris, D.P.; Huang, H.J.; Kao, P.C. Moving the arms to activate the legs. *Exerc. Sport Sci. Rev.* **2006**, *34*, 113–120. [CrossRef] [PubMed]

84. Hundza, S.R.; Zehr, E.P. Suppression of soleus H-reflex amplitude is graded with frequency of rhythmic arm cycling. *Exp. Brain Res.* **2009**, *193*, 297–306. [CrossRef] [PubMed]

85. Zhou, R.; Parhizi, B.; Assh, J.; Alvarado, L.; Ogilvie, R.; Chong, S.L.; Mushahwar, V.K. Effect of cervicolumbar coupling on spinal reflexes during cycling after incomplete spinal cord injury. *J. Neurophysiol.* **2018**, *120*, 3172–3186. [CrossRef]

86. Dietz, V. Do human bipeds use quadrupedal coordination? *Trends Neurosci.* **2002**, *25*, 462–467. [CrossRef]

87. Zehr, E.P.; Hundza, S.R.; Vasudevan, E. V The quadrupedal nature of human bipedal locomotion. *Exerc. Sport Sci. Rev.* **2009**, *37*, 102–108. [CrossRef]

88. Frigon, A.; Collins, D.F.; Zehr, E.P. Effect of rhythmic arm movement on reflexes in the legs: Modulation of soleus H-reflexes and somatosensory conditioning. *J. Neurophysiol.* **2004**, *91*, 1516–1523. [CrossRef]

89. Frigon, A. The neural control of interlimb coordination during mammalian locomotion. *J. Neurophysiol.* **2017**, *117*, 2224–2241. [CrossRef]

90. de Ruiter, G.C.; Hundza, S.R.; Zehr, E.P. Phase-dependent modulation of soleus H-reflex amplitude induced by rhythmic arm cycling. *Neurosci. Lett.* **2010**, *475*, 7–11. [CrossRef]

91. Gorodnichev, R.M.; Pivovarova, E.A.; Pukhov, A.; Moiseev, S.A.; Savokhin, A.A.; Moshonkina, T.R.; Shcherbakova, N.A.; Kilimnik, V.A.; Selionov, V.A.; Kozlovskaia, I.B.; et al. Transcutaneous electrical stimulation of the spinal cord: Non-invasive tool for activation of locomotor circuitry in human. *Fiziol. Cheloveka* **2012**, *38*, 46–56.

92. Gerasimenko, Y.; Gorodnichev, R.; Puhov, A.; Moshonkina, T.; Savochin, A.; Selionov, V.; Roy, R.R.; Lu, D.C.; Edgerton, V.R. Initiation and modulation of locomotor circuitry output with multisite transcutaneous electrical stimulation of the spinal cord in noninjured humans. *J. Neurophysiol.* **2015**, *113*, 834–842. [CrossRef] [PubMed]

93. Sayenko, D.G.; Atkinson, D.A.; Floyd, T.C.; Gorodnichev, R.M.; Moshonkina, T.R.; Harkema, S.J.; Edgerton, V.R.; Gerasimenko, Y.P. Effects of paired transcutaneous electrical stimulation delivered at single and dual sites over lumbosacral spinal cord. *Neurosci. Lett.* **2015**, *609*, 229–234. [CrossRef] [PubMed]

94. Rothwell, J. *Control of Human Voluntary Movement*, 2nd ed.; Chapman & Hall: London, UK, 1994.

95. Yamaguchi, T. Descending pathways eliciting forelimb stepping in the lateral funiculus: Experimental studies with stimulation and lesion of the cervical cord in decerebrate cats. *Brain Res.* **1986**, *379*, 125–136. [CrossRef]

96. Juvin, L.; Simmers, J.; Morin, D. Propriospinal circuitry underlying interlimb coordination in mammalian quadrupedal locomotion. *J. Neurosci.* **2005**, *25*, 6025–6035. [CrossRef]

97. Juvin, L.; Le Gal, J.-P.; Simmers, J.; Morin, D. Cervicolumbar coordination in mammalian quadrupedal locomotion: Role of spinal thoracic circuitry and limb sensory inputs. *J. Neurosci.* **2012**, *32*, 953–965. [CrossRef]

98. Zhou, R.; Alvarado, L.; Ogilvie, R.; Chong, S.L.; Shaw, O.; Mushahwar, V.K. Non-gait-specific intervention for the rehabilitation of walking after SCI: Role of the arms. *J. Neurophysiol.* **2018**, *119*, 2194–2211. [CrossRef]

99. Klarner, T.; Barss, T.S.; Sun, Y.; Kaupp, C.; Loadman, P.M.; Zehr, E.P. Exploiting Interlimb Arm and Leg Connections for Walking Rehabilitation: A Training Intervention in Stroke. *Neural Plast.* **2016**, *2016*, 1517968. [CrossRef]

100. Klarner, T.; Barss, T.; Sun, Y.; Kaupp, C.; Loadman, P.; Zehr, E. Long-Term Plasticity in Reflex Excitability Induced by Five Weeks of Arm and Leg Cycling Training after Stroke. *Brain Sci.* **2016**, *6*, 54. [CrossRef]

101. Kaupp, C.; Pearcey, G.E.P.; Klarner, T.; Sun, Y.; Cullen, H.; Barss, T.S.; Zehr, E.P. Rhythmic arm cycling training improves walking and neurophysiological integrity in chronic stroke: The arms can give legs a helping hand in rehabilitation. *J. Neurophysiol.* **2018**, *119*, 1095–1112. [CrossRef]

102. Robertson, J.V.G.; Roby-Brami, A. The trunk as a part of the kinematic chain for reaching movements in healthy subjects and hemiparetic patients. *Brain Res.* **2011**, *1382*, 137–146. [CrossRef]

103. Cetisli Korkmaz, N.; Can Akman, T.; Kilavuz Oren, G.; Bir, L.S. Trunk control: The essence for upper limb functionality in patients with multiple sclerosis. *Mult. Scler. Relat. Disord.* **2018**, *24*, 101–106. [CrossRef] [PubMed]

104. Rath, M.; Vette, A.H.; Ramasubramaniam, S.; Li, K.; Burdick, J.; Edgerton, V.R.; Gerasimenko, Y.P.; Sayenko, D.G. Trunk Stability Enabled by Noninvasive Spinal Electrical Stimulation after Spinal Cord Injury. *J. Neurotrauma* **2018**, *35*, 2540–2553. [CrossRef] [PubMed]

105. Gad, P.; Gerasimenko, Y.; Zdunowski, S.; Turner, A.; Sayenko, D.; Lu, D.C.; Edgerton, V.R. Weight bearing over-ground stepping in an exoskeleton with non-invasive spinal cord neuromodulation after motor complete paraplegia. *Front. Neurosci.* **2017**, *11*, 333. [CrossRef]

106. Estes, S.; Zarkou, A.; Hope, J.M.; Suri, C.; Field-Fote, E.C. Combined transcutaneous spinal stimulation and locomotor training to improve walking function and reduce spasticity in subacute spinal cord injury: A randomized study of clinical feasibility and efficacy. *J. Clin. Med.* **2021**, *10*, 1167. [CrossRef]

107. Wu, Y.K.; Levine, J.M.; Wecht, J.R.; Maher, M.T.; LiMonta, J.M.; Saeed, S.; Santiago, T.M.; Bailey, E.; Kastuar, S.; Guber, K.S.; et al. Posteroanterior cervical transcutaneous spinal stimulation targets ventral and dorsal nerve roots. *Clin. Neurophysiol.* **2020**, *131*, 451–460. [CrossRef] [PubMed]

108. Parhizi, B.; Barss, T.S.; Mushahwar, V.K. Alteration of H-reflex and MEP amplitude in the flexor carpi radialis muscle with transcutaneous spinal cord stimulation during static and leg cycling tasks in neurologically intact male and female humans. *Front. Neurosci.* **2021**, *15*, 379. [CrossRef]

MDPI

St. Alban-Anlage 66

4052 Basel

Switzerland

Tel. +41 61 683 77 34

Fax +41 61 302 89 18

www.mdpi.com

Journal of Clinical Medicine Editorial Office

E-mail: jcm@mdpi.com

www.mdpi.com/journal/jcm